Praise for

DNA Nanoscience:
From Prebiotic Origins to Emerging Nanotechnology

The subjects of Ken Douglas' book may be nanometers in scale, but their stories have Mega-impact. *DNA Nanoscience* takes us on a journey into the future, where sub-microscopic gadgets built from DNA may be used to detect specific molecules one at a time or to deliver therapeutic drugs specifically to cancer cells. Looking in the other direction, the journey takes us back 4 billion years to a time when the self-organization of DNA into liquid crystals may have facilitated the reproduction of what would become our genetic material, arguably the key step in the origin of life.

While so many books on science avoid the personalities involved, Douglas embraces them. Ned Seeman (New York University) is a master at origami, but instead of folding paper into storks he folds DNA into arrays of shapes reminiscent of those in M.C. Escher prints. His perseverance in the face of technical difficulties is particularly remarkable considering that for many years, the DNA nanostructure field consisted of the Seeman lab alone. The other protagonists are Noel Clark (University of Colorado-Boulder) and Tommaso Bellini (University of Milan). Their chance meeting in Italy led to the discoveries of liquid crystal phases of nanometer-length DNA and of the ability of these phases to select molecules and facilitate their chemical reaction into longer DNA chains.

DNA Nanoscience is scholarly and full of technical figures. But the science is accompanied by clear explanations that make it accessible to college students and science-savvy citizens. It is a pleasure to find a book that is so true to the science while being so enjoyable to read.

—Thomas R. Cech
Distinguished Professor, University of Colorado–Boulder
Director, BioFrontiers Institute
Nobel Laureate (Chemistry, 1989)

Kenneth Douglas's *DNA Nanoscience* is something of a miracle. With intelligence, care, and even charm, he lays out a crisp account of an emerging field and the origin of life. The emerging field is DNA nanostructures from tiles and brick to nanorobots that can deliver chosen antibodies specifically to cancer cells. He reports new work on the origin of life based on liquid crystals that offers the hope of synthesis of reproducing DNA double-stranded molecules in an entirely unexpected way. If you wish insight into the novel technologies of DNA nanostructures and an expanded view of the origin of life, read this fine book.

—Stuart Kauffman
Emeritus Professor Biochemistry and Biophysics, University of Pennsylvania
Affiliate Professor, The Institute for Systems Biology, Seattle
Author of *At Home in the Universe*

This book changed my life. Every seven years, as my sabbatical approaches, I search about for a new direction to focus my research and Ken Douglas' book, *DNA Nanoscience,* appeared just in time. Through this book, I read of the work of Hendrik Dietz, whose lab in just a few short years has engineered molecules to pump ions across membranes and is working toward creating molecules that transform chemical energy to perform work inspired by proteins that nature took a billion years to evolve. I became so excited by the prospects of DNA nanotechnology compellingly described in this book, that I arranged to leave Brandeis University and spend 2017 conducting research with Dietz in Munich.

This book has three distinct parts, but I feel as if all three sections were written expressly for me, someone who is ignorant of the technical details of the cutting-edge science of DNA nanotechnology, but is curious of the history, desirous to be exposed to the outstanding questions, and motivated to learn enough of the requisite knowledge to begin research. On one hand, the book reads like a history of science, replete with entertaining anecdotes of the profound discoveries of the field. On the other hand, the book plays the role of first textbook, with excellent pedagogic written explanations and clear illustrations of the key concepts.

The first part of the book provides a reader with a general scientific background sufficient to understand the field of DNA nanotechnology. The second part delves into the details of DNA nanotechnology. It thankfully explains clearly all the background information needed to understand the breakthrough work that took decades of hard labor to elucidate. The third part moves from technology to one of the top three scientifically important questions: what is the origin of life? An astounding tale is told here. It recounts how two liquid crystal scientists, Noel Clark and Tomasso Bellini, turned the story of DNA on its head. Normally, one says that the structure of double-stranded DNA is linear because that is the best way that information can be stored, like a sequence of symbols on an instruction tape of a Turing machine. It turns out that such linear polymers form liquid crystals, which means that at high concentration all the DNA will line up next to each other. This is what allowed Rosalind Franklin to prepare the exquisite X-ray diffraction images of DNA that Watson and Crick deciphered to solve the structure of the double helix. Conventional wisdom holds that DNA forms a liquid crystal because DNA evolved to be a linear chain and that stiff linear chains serendipitously happen to make liquid crystals. Instead, disregarding all convention, Clark and Bellini posit that double-stranded DNA composed of complementary base pairs was created by purely inanimate, cyclic processes over time based on temperature- and concentration-driven liquid crystalline phase transitions. According to Clark and Bellini, it is the physics of liquid crystallinity that led to DNA, not DNA leading to liquid crystals. As a liquid crystal scientist, thanks to Clark and Bellini, now I can hold my head high with pride, buoyed by the knowledge that liquid crystal scientists are on the verge of explaining the origin of life! As in all great science, what is new and original is provocative. It is too early to tell if Clark and Bellini's bold assertion is true, but this book gives us the tools and motivation to ask the important questions that will lead us all forward.

—Seth Fraden
Professor of Physics
Director, The Bioinspired Soft Materials Center
Brandeis University

DNA is known for being the medium for storing genetic information. However, in Kenneth Douglas' book, *DNA Nanoscience,* two emerging topics are treated that put DNA in a very different light. One is the field of DNA nanotechnology that has revolutionized the art and science of fabricating structures on the nanoscale. Moreover, the author describes how these nanoscopic objects have been taught by scientists to act as molecular machines and even how to perform logic operations on cancer cells. On the other hand, the book details how short pieces of DNA and RNA form liquid crystals, a completely different type of nanostructure. Contrary to what one might imagine, they do not represent building blocks in future displays but are rather discussed as being early molecular building blocks during the origin of life. The author elegantly connects both topics. These subjects are narrated in a very lively fashion with many personal anecdotes of scientific heroes in the field of DNA nanoscience. This gives the reader an excellent impression of how scientific progress is achieved with all its frustration and successes.

Instructive like a textbook and exciting like a novel! For everybody interested in modern natural sciences, this book is a must-read.

—Andreas Herrmann
Professor of Polymer Chemistry and Bioengineering
Chair of the Board, The Zernike Institute for Advanced Materials
University of Groningen, The Netherlands

I have admired Noel Clark's science for more than forty years. I had heard Tommaso Bellini speak at the conference celebrating Noel's 70th birthday and found it very good. I knew of Nadrian Seeman's beautiful work and I was particularly thrilled by his collaboration with Paul Chaikin and David Pine in which they start to succeed to cook up self-replicating colloidal objects. So when I was asked to write a few words about the book *DNA Nanoscience: From Prebiotic Origins to Emerging Nanotechnology,* I accepted immediately, without even starting to read the manuscript. This was pretty unprofessional since, after all, the book could very well have been deceptive. I didn't know Kenneth Douglas!

I was lucky: I enjoyed his unique style. The many anecdotes ranging from almost hilarious to really moving give a sense of real life tightly intertwined with real science. Usually it is one or the other. I spent a summer in the subterranean 2B level of the Duane Physical Laboratories of the University of Colorado in Boulder, and Douglas' description is just perfect. You get an exact feel of what it's like and of the atmosphere floating there! I trust the other anecdotes are just as faithful, and I learned a lot in reading them.

But don't be mistaken: this is serious science. All major concepts and techniques are explained in the simplest possible terms keeping enough rigor to be meaningful. Extensive references and exercises are proposed at the end of each chapter.

DNA nanoscience and engineering are carefully introduced, with the necessary conceptual tools such as system free energy and topology, allowing us to understand base pairing, the role of sticky ends, the troublesome floppiness, the success of cubic catenanes, and the very clever invention of double-crossover molecules. Douglas brings us all the way to nanomachines, 3D crystals, and artificial ion channels, a beautiful tour in DNA nanotechnology.

Grasping Clark and Bellini's input in the field of prebiotic origins requires a good understanding of liquid crystal physics. Again concepts involving free energy and entropy are important. Douglas teaches us about the important liquid crystal phases, explaining in the most efficient and simple way the Onsager criterion, and how and why the ordering of short, paired DNA violates its prediction. He subsequently explains how mixtures of pairing and non-pairing nanoDNA can separate, respectively, into columnar and isotropic phases, and the key role of the subtle depletion interaction in this process. That phase separation between isotropic and liquid crystal phases has played a role in prebiotic conditions is very likely, since it still plays a role in current cell behavior.

Then comes the importance of autocatalytic self-ligation. Douglas describes the experiments and the reasoning in a careful and fully convincing way. The columnar structure increases significantly the ligation probability. This explains a missing link: people could understand the appearance of nucleic acid monomers or short oligomers but not longer ones. I buy easily these arguments since I know that radiation-resistant bacteria, which are found in nuclear reactor pools, have their genome organized in a columnar phase in contrast to standard bacteria. Radiation damage does occur, but healing is almost immediate using a process akin to the Clark/Bellini ligation process. Douglas is also very careful to describe alternative ways of thinking, which puts into perspective the originality and power of the Clark/Bellini picture.

To sum up, this is both a lively and profound book, the reading of which I strongly recommend.

—Jacques Prost

Director Emeritus, CNRS (Le centre national de la recherche scientifique)
and founder, Physical Chemistry Laboratory, Institut Curie, Paris
Distinguished Professor, National University of Singapore
Director General Emeritus, ESPCI (École Supérieure de Physique et de
Chimie Industrielles de la Ville de Paris)
Co-author with P. G. de Gennes of the book *The Physics of Liquid Crystals* (over 15,000 citations)

Ken Douglas' book, *DNA Nanoscience,* was a very pleasant surprise for me. The book tells a fascinating new story about DNA. Douglas weaves two tales, one tale about how DNA can and will be used in the near future to make nano- and micro-scale objects and devices, and a second tale about how the liquid-crystal physics of DNA was crucial for its chemical evolution billions of years ago. The book is written as a kind of novel for the general public that affectionately describes the personalities of key contributors, along with their struggles and triumphs. However, it also teaches the science with a rigor that will appeal to readers seeking to increase their depth of understanding about these topics. Thus, we not only learn about DNA and state-of-the-art features of DNA nanoscience, but we also learn about liquid crystals and soft materials, about seminal accomplishments in nanotechnology, and about techniques such as X-ray diffraction, electron, optical and atomic force microscopy that permit visualization of DNA constructs and elements. The subject matter also stretches as needed into biology to teach basic ideas about cell membranes and metabolism. I believe the book will prove interesting for a broad audience ranging from the layperson to students and practitioners in science, engineering, and medicine. It provides a wonderful taste of DNA nanoscience at the research frontier.

—Arjun G. Yodh
James M. Skinner Professor of Science, Endowed Chair
Director, PENN Laboratory for Research on the Structure of Matter
University of Pennsylvania

"DNA" is one of those things that is so familiar—millions of people have had their DNA analyzed to determine their origins, heard on the news that an innocent person was released from prison based on DNA evidence, or about the wondrous prospects of personalized medicine made possible by the human genome project and the reduction in cost of having your own genome sequenced. But what is DNA? Douglas answers this question in an accessible, yet technically detailed and accurate way in his new book, *DNA Nanoscience: From Prebiotic Origins to Emerging Nanotechnology,* by connecting the science to the scientists who made this new genetic revolution possible. The words, nomenclature, and fundamentals of this field need to become as routine to the scientifically curious public as how to design a web page or master a cell phone, as a basic understanding of DNA will start to influence our daily lives more and more.

As Douglas points out, DNA discovery was a remarkable sequence of triumphs and tragedies, mostly unknown even to modern-day scientists. Oswald Avery, Erwin Chargaff, James Watson and Francis Crick, and Frederick Sanger, relative unknowns compared to our modern celebrities, made seminal discoveries about how this remarkable DNA molecule was organized and its ability to convey the information necessary to make a complete bacterium, or with minimal rearrangement, a human being. There is also the tragedy of Rosalind Franklin, who obtained the first high-resolution X-ray diffraction patterns showing the characteristic double helix structure, but died before achieving the recognition she deserved. Douglas takes us on this historical journey to show how DNA performs its almost magical chemistry and physics and why DNA was a necessary ingredient to direct the origin of life. Part of the fascination is that our knowledge of DNA is completely within the lives of our own or our parents' generation! This is not old science or conventional wisdom. Our understanding of DNA came about almost in parallel with that other paradigm of modern life—microelectronics. Shockley, Bardeen, and Brattain were inventing the transistor about the same time as Avery and Chargaff were discovering the role of DNA in the gene. Crick and Watson were working at the same time as Kilby and Noyce were inventing the integrated circuit that led to microprocessors, the combination of which made sequencing the human genome possible. We are reminded that the people are what make the science so fascinating.

In the second part of the book, Douglas again highlights the personalities of the scientists to make sense of the science. Nadrian Seeman is generally recognized as the father of DNA nanotechnology. There is something very human about taking what nature gives us and trying to make it into something else—something that we have decided is "more useful" or more profitable. Seeman wants to build structures from the bottom up by having them organize themselves by programming a DNA sequence. The day-to-day equivalent would be to have lumber, concrete, wiring, shingles, etc., just show up at a job site by themselves, wait a few days, and have a complete house appear. Douglas takes us through the very basic steps required to change DNA from an information molecule into a combined information-structure molecule according to the rules Seeman has discovered over the past 30 years. While DNA can be quite rigid over short sequences, making corners, connections, and even motors requires an understanding of how the DNA chemistry influences the DNA structure, and how the topology of tangled bits of DNA can lead to complex, three-dimensional objects, a thousandth of the diameter of a strand of hair in size. To do so required the marriage of an ancient art form of origami with the rules imposed by sequencing and folding DNA. The jury is still out on whether or not this new science will find its "killer app." It could eventually show why nature decided that DNA and RNA were best suited to information storage and transmission, while structural issues were best left to the lipids and proteins.

In the final sections of the book, Douglas tells the newest "creation" story, starting with liquid crystals—phases in which non-spherical molecules and especially those that are rod or disc shaped, prefer to line up more or less parallel to each other while staying in a fluid state. Among those that form liquid crystals are long DNA double-helix polymers in solution, which soft matter scientists Noel Clark and Tommaso Bellini took to be a footprint of the origin of DNA in early life. They were surprised to find that even the shortest DNA duplexes could form liquid crystals—DNA's unique end-to-end molecular adhesion apparently can make short DNA act like long DNA and help segregate the molecules into their own special liquid crystalline phase. Concentrating and organizing in this way makes it easier for the DNA to "ligate" or grow in length and complexity, in turn stabilizing the liquid crystal order, an autocatalytic loop. This is the basic problem of how our original building blocks got together out of the primordial ooze, and it is a surprising, yet physically very sound explanation. This liquid crystal hypothesis explains rather nicely how DNA became DNA out of a sloppy mixture of molecularly rather simple building blocks. We still need to figure out how the DNA managed to start coding for proteins and lipids and all of the other things that make up living creatures. But this is a fun start.

In sum, Douglas does an excellent job of capturing the reader's interest in one of the fundamental chapters of science that the current and future generations need to understand. This book can shorten the distance between the scientist and non-scientist by making the science personal, fun, and somewhat intuitive, except where the findings are so counterintuitive. The only way that the general public will continue to trust the proclamations of the scientific establishment is through books like this one—where the foibles and fears and eccentricities of the scientists are shown to be the same as those of the artist, musician, and businessman. Scientists are just artists who want to work with mother nature, without the freedom to make up new worlds as we go along. The real world is magical enough for them.

—Joseph A. Zasadzinski
3M Harry Heltzer Chair in Multidisciplinary Science and Technology
Chemical Engineering and Materials Science
University of Minnesota

DNA Nanoscience

From Prebiotic Origins to Emerging Nanotechnology

Kenneth Douglas
Department of Physics
University of Colorado–Boulder, USA

CRC Press is an imprint of the
Taylor & Francis Group, an **informa** business

All original and adapted illustrations created by: Marjorie Leggitt, Leggitt Design; Lexie Foster, Foster Design; Nicolle R. Fuller, Sayo-Art LLC; Ikumi Kayama, MA, Studio Kayama LLC.

Cover design: Kenneth Douglas (© Kenneth Douglas) using

- DNA model – modified from Caroline Davis2010 at Flickr; Creative Commons (CCBY 2.0), https://www.flickr.com/photos/53416677@N08/4973532326/

- Janus – modified from 123rf stock image. Image ID 23896706

- Mount Rinjani volcano – modified from Shutterstock, stock image ID 108347660

- Respirocytes Flowing Through a Blood Vessel – modified from CG4TV,

stock image SKU respirocytes-3DI5m

Information Technology (IT) support: J. Bryan Kelley, Dreadnaught, LLC

Interior design: Rebecca Finkel, F + P Graphic Design, fpgd.com

CRC Press
Taylor & Francis Group
6000 Broken Sound Parkway NW, Suite 300
Boca Raton, FL 33487-2742

Printed in Canada on acid-free paper
Version Date: 20160714

International Standard Book Number-13: 978-1-4987-5012-7 (Paperback)

Visit the Taylor & Francis Web site at
http://www.taylorandfrancis.com

and the CRC Press Web site at
http://www.crcpress.com

To my mother, my father, and my sister

and

To Noel Clark, mentor and friend

Table of Contents

A Note to the Reader

This book looks at the first 35 years of DNA nanotechnology to better appreciate what lies ahead in this emerging field and looks back about 4 billion years to the possible origins of DNA which are shrouded in mystery. On the subject of DNA nanotechnology, it's necessary to comment about what is *not* included in this account.

There are a large and growing number of scientists who have made valuable contributions to the field including Milan Stojanovic (Columbia University), Darko Stefanovic (University of New Mexico), Thomas LaBean (North Carolina State University), Andrew Turberfield (Oxford University), Hiroshi Sugiyama and Masayuki Endo (Kyoto University), Bernard Yurke (Boise State University), Tim Liedl (Ludwig Maximilian University of Munich), Anne Condon (University of British Columbia), John Reif (Duke University), Rebecca Schulman (Johns Hopkins University), and Kurt Gothelf (Aarhus).

Several of the scientists listed above (and others whose names do not appear above) are referred to in this book but the specifics of their contributions are not elaborated upon. In this volume, five chapters are devoted to the exploration of DNA nanotechnology. Of those five chapters, three are focused on the contributions of the founder of the field, Nadrian Seeman (although the work of other DNA nanotechnologists—such as Paul Alivisatos and Erik Winfree—appear in these chapters). The other two chapters on DNA nanotechnology present work by Paul Rothemund (Caltech), Peng Yin (Harvard University), William Shih (Harvard University), Hao Yan (Arizona State University), Hendrik Dietz (Technical University of Munich), Friedrich Simmel (Technical University of Munich), and the collaboration of Shawn Douglas, Ido Bachelet, and George Church (Harvard University). Thus, approximately 60% of the material presented on DNA nanotechnology (three chapters out of five) is devoted to the work of Seeman and 40% to the work of others. In the roughly 35-year history of the field, Seeman's lab was virtually alone for the first 20 years—60% of the time period covered in this book—and the allocation of the discussion of DNA nanotechnology in these pages reflects that historical fact.

In addition to the scientists whose work has not been covered, there are many topics in DNA nanotechnology that have not been described. Among these are artificial enzymes, photonic molecular circuits, nanoscale organization of ligands and proteins, biosensors, DNA nanotubes and many others.

The author is grateful to the anonymous reviewers of the manuscript who pointed out the omissions in this book and hopes that what has been left out will find a home in another introductory text.

Preface

In 1959 the British physicist and novelist C.P. Snow gave the Rede Lecture at the University of Cambridge titled, "The Two Cultures and the Scientific Revolution." He lamented the cultural divide separating science and the humanities, specifically the ignorance of science and technology on the part of those who govern as well as the populace at large. Fifty years on, the average citizen is challenged to have informed opinions about varied scientific issues such as stem cells and climate change; challenged because they are hard-pressed to judge for themselves the merits of the arguments.

In a developed society, people should be able to make intelligent decisions on the public support of both basic and applied science. To do this they need to understand how the scientific process works. They must also be able to discern the merits of differing views, those that are supported by the quality and preponderance of evidence and those that are not. To become knowledgeable requires effort but need not be tedious. The rewards for making judicious funding choices can be handsome for the citizenry, their children, and their children's children.

In this book the reader will learn about DNA nanoscience. We will consider forward-looking technological applications and also scientific inquiries of a philosophical nature that speak to our common ancestry. Part I of this book will introduce these two lines of investigation. The first is the emerging use of DNA as a building material for nanoscale engineering of both structures and devices, rather than as a genetic material. Nadrian Seeman first conceived of the idea. His insight was to design stably *branched* (rather than linear) strands of DNA containing precise base sequences. Seeman's work created the field of structural DNA nanotechnology, an endeavor now pursued by some 60 or 70 labs worldwide. The second line of research is the materials science study of short lengths of DNA (called nanoDNA). Noel Clark and Tommaso Bellini found that nanoDNA exhibits liquid crystal phases. Their experiments suggested that liquid crystal formation acts as a structural gatekeeper by orchestrating a series of staged self-assembly processes using nanoDNA. This led to an explanation of the linear polymer structure of our bio-information carriers and of how life may have emerged from the prebiotic clutter.

Despite the subtitle of this book, we begin with emerging nanotechnology rather than prebiotic origins. The order of presentation could have accurately reflected the subtitle. However, DNA nanotechnology is real, while any origins proposal however well supported by laboratory experiments is, perforce, speculative; life's origins will remain a mystery. We choose to begin with a look at the concrete—the first 35 years of DNA nanotechnology that whets our appetite for what might develop in the coming decades. We choose to close with a look at the conjectural—an ingenious proposal rooted in experiment that might be the best explanation for what happened some 4 billion years ago.

So the reader can better appreciate the work of Seeman and of Bellini and Clark, Part I includes individual chapters devoted to the subjects underpinning their research. We have an historical and scientific look at DNA, at the broader field of nanoscience, liquid crystal basics, and the experimental tools common to both research groups. In addition, a sprinkling of Exercises throughout the book gives readers a chance to check their understanding of the material. In harmony with the publisher's policy, the answers to the Exercises are compiled in a *Solutions Manual* available from the publisher rather than placed at the end of the book.

Part II will be devoted to the work of Nadrian Seeman and other bio-inspired nanotechnologists. We will travel down culs-de-sac with him and witness his painstaking route to the founding of a new field of science. We also present the work of some of his intellectual disciples and their embryonic successes in connecting to real-world applications.

Part III will turn the spotlight on the (initially) serendipitous experiments of Bellini and Clark that may provide the explanation for how information-bearing nucleic acid chains arose from the primordial soup. They have illuminated the processes of the hierarchical self-assembly of nanoDNA as quintessential examples of the phenomenon of emergence wherein new forms of self-organization—such as life—spontaneously appear.

Knowledge of these topics and of their value to society is more than an intellectual exercise; this is research that the public ultimately pays for. As to the division of the two cultures, it may come as a pleasant surprise that the elevating power of art is present throughout the scientific enterprise that we examine, both as muse and as outcome.

Author Biography

Kenneth Douglas is a member of the research faculty in the Department of Physics at the University of Colorado–Boulder. He attended the University of Chicago where he earned a B.A. in mathematics and an M.S. in physics, and the University of Colorado–Boulder where he earned a Ph.D. in physics. His earliest publications were in experimental quantum magnetooptics. This was meat-and-potatoes solid-state physics. Experiments used laser light to stimulate quantum mechanical transitions of electrons and holes in semiconductors placed in high magnetic fields obtained by cooling a superconducting magnet to cryogenic temperatures. It was fun and a good way to gain hands-on experience with lots of useful tools and techniques. But a heretofore suppressed interest in interdisciplinary science led to an appointment as a guest scientist in a cell biology laboratory for three and one-half years (before Douglas started his own laboratory in the Department of Physics at the University of Colorado). He created a strategy of employing the surface layers (S-layers) of bacterial extremophiles—such as *Sulfolobus acidocaldarius* and *Sulfolobus solfataricus*—as masks to periodically pattern an inorganic substrate on a nanometer-length scale. Equipped with incubator shakers, ultracentrifuges, and the like, the Douglas lab grew its own extremophiles and isolated their S-layers. The lab laid the foundation for employing biomolecular masks as patterning elements to generate ordered nanostructured materials. As one of the early examples of biomimetic nanotechnology, the project put to good use the newly emerging assays of scanning tunneling microscopy (STM) and atomic force microscopy (AFM) along with traditional transmission electron microscopy (TEM). Douglas is one of three co-inventors of the first-ever U.S. patents for parallel fabrication of nanoscale multi-device structures. His work has

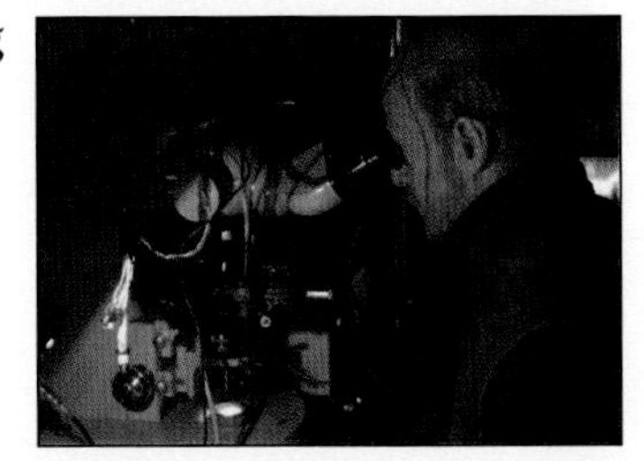

appeared in *Science, Nature, Biophysical Journal, Applied Physics Letters, Physical Review B, Surface Science, FEMS Microbiology Reviews, Journal of Applied Physics,* and *Popular Science* among other publications. He has authored multiple book chapters and seven U.S. patents.

Acknowledgments

In science you can do control experiments and address the "what if" question. Life is not like that. Once you've set out on a path you can't do a control experiment to see what would happen if you had set out on a different path. Would this book—regardless of its merit or lack thereof—have come into being if it were not for the people I met along the path and whom I now take great pleasure in thanking?

If you've read Douglas Adams' *The Hitchhiker's Guide to the Galaxy* you recall Zaphod Beeblebrox, the former President of the Galaxy—the sentient being who had two heads (and three arms). I feel like I need a multiplicity of heads to speak simultaneously of those I thank, but instead I have to make a choice and start with one. I'll start with Noel Clark, my post-doctoral mentor, friend of some 40 years, and the person without whom this book would not be necessary.

I thank Noel for his innumerable patient explanations of concepts that I struggled with. And I've drawn so much inspiration from him. I'll illustrate this with a conversation we had in his office one afternoon.

"You've had that poster up on your wall for years. How did it get there?" I asked.

"It goes back to when I was writing a lot of proposals to get the liquid crystals lab started. There are times you need to be reminded that rowing may be necessary."

The poster had a simple line drawing of a man in a boat with oars in his hands and it read, "When there's no wind, row." Noel does that with consummate skill and his resourcefulness helps those who work with him become more resourceful themselves.

Noel's work on nanoDNA was the raison d'être for the book. But on that subject I also thank others who are a part of the team. Tommaso Bellini has been a wonderful resource for conversations and ideas and has generously provided images that appear in the book. Others in the Milan portion of the team have been similarly magnanimous. I thank Giuliano Zanchetta and Tommaso Fraccia for answering many questions and allowing me to use their illustrations. On the Boulder side I thank the late Michi Nakata for her seminal contributions to the discovery and characterization of nanoDNA. I thank Dave Walba for his insights and for coming up with the fine stereoisomer images that appear in Brief Interlude II. I thank Greg Smith for discussions and for his seminar presentations of work in progress that were so well done and informative.

I've been fortunate to work with such talented people in assembling the many drafts of the manuscript. J. Bryan Kelley is a superb information technology specialist. For historical reasons the manuscript was written in MS Word for Mac. I'm now well aware that this software is not the best choice for page layout. Despite this, Bryan was masterful in combating all manner of text wrapping and formatting issues as well as an eclectic range of other technical challenges. His patience and reliability in doing this was invaluable. Remarkably, he was even adept at teaching me a modicum of skills to use when he was absent.

Lead artist, Marjorie Leggitt, is a gifted scientific illustrator who was personally responsible for many of the original figures. Moreover, she had the organizational ability to manage the project. She secured additional artists, compiled and distributed artwork assignments, created spreadsheets, managed edit and revision phases with artists, managed the budget and invoicing, and more. She was a gracious host, allowing meetings to be held in the dining room of her beautiful home that brought Marjorie, and our interior book designer, and myself together.

In fact, it was Marjorie who brought up the idea of an interior book designer in contrast to using a standard template from the publisher. And she found Rebecca Finkel to fill that position. Rebecca is a formidable book designer. She has great aesthetic sensibilities and came up with stylish touches that made the book more inviting to the reader.

And, apart from original illustrations, what would my interior book designer need to flow text around if it wasn't for the high-resolution images that I was given? In this category, Ned Seeman stands at the pinnacle, if only for the sheer volume of images he sent me. Ned could not have been more gracious and cooperative even though we were both fully aware that he would be using many of the images he gave me in his own book on structural DNA nanotechnology.

In addition to Ned, there were a slew of people who took time from their hectic schedules to respond to my requests for high-resolution images, permissions, and explanations of technical points that had me stumped. I thank all of them for making it possible to put the book together. I thank Mary Beth Aberlin, Cynthia Aguilera, Oleksii Aksimentiev, Fredrik Ali, Aseem Ansari, Kyoichi Araki, Tim Atherton, Hidemitsu Awai, Ido Bachelet, Jeff Barnes, Chris Beaudoin, Jeroen Beeckman, Karin Beesley, Petra Bele, Steven Block,

Amy Bradley, Raegan Carmona, Zora Catterick, Alice Chen, Jen Christiansen, George Church, Bill Clyne, Oliver Coles, Selena Coppock, Tina Cuccia, Richard Dabb, Ingo Dierking, Hendrik Dietz, Shawn Douglas, David Dunmur, Don Eigler, Ilene Ellenbogen, Sherri Elsworth, Ken Eward, Rachel Kramer Green, Hongzhou Gu, Mairi Haddow, Jim Hainfeld, Boualem Hammouda, Ali Jackson, Thomas Jorstad, Jerry Joyce, George Kelvin, Richard Kiehl, Hajin Kim, Chris King, Paavo Kinnunen, Laura Konyha, Daniel Kulp, Jörg Langowski, Karen Lee, David Leigh, Robert Leighton, Vineta Lewis, Françoise Livolant, Borin Van Loon, Chengde Mao, Paul Melnyk, Roger Miesfeld, Chad Mirkin, Althea Morin, Christine Morrow, Rod Nave, Elisa Neckar, Hilary Newman, Dennis Niedfeldt, Bernd Nilius, Lisa Pallatroni, Jane Park, John Pelesko, Sean Pidgeon, Isabelle Quinkal, Annie Reiser, Kevin Renn, Martín Martínez-Ripoll, Paul Rothemund, Mark Sansom, Sarah Saunders, Bohdan Senyuk, Oliver Sherman, Fritz Simmel, Lianne Smith, Sharon Smith, Andrzej Stasiak, Joe Stroscio, Lubert Stryer, Andrew Swift, Claire Taylor, Karin Tucker, Margareth Verbakel, Sylvia Wallace, Chris Watters, Erik Winfree, Hao Yan, and Peng Yin—and anyone I inadvertently neglected to call out.

I also thank Claire Smith and Melissa Rose of the Nature Publishing Group for helping me out of all proportion to their job description. They were exceptional.

Speaking of exceptional, Erika Bagley at the Copyright Clearance Center gave me service and support that made my task of acquiring copyright permissions immeasurably less challenging. And always did so with the most pleasant attitude.

Elizabeth Sandler in the permissions department of *Science* was outstanding in helping me to acquire permissions for papers published in *Science*. She couldn't have been more generous and kind.

The personnel at Taylor & Francis do their company proud. I thank my acquisitions editor, Hilary LaFoe, for giving me the opportunity to become a T&F author indulging in vicarious science and for her continued interest and involvement throughout the process. She was willing to work with me when thorny issues came up, and I'm so grateful for her cooperation and support. Suzanne Lassandro, production manager at T&F, was endlessly helpful and a delight to work with. When I learned that her sister holds a doctorate in Musical Arts in vocal performance and loves opera as I do, we formed a bond. Now, whenever I listen to Mozart's *Die Zauberflöte* and the "Queen of the Night" aria comes around, I marvel not only at Mozart's musical genius, but also that Suzanne's sister can sing this most challenging piece—and it's her favorite role. Jill Jurgensen, my production coordinator at T&F, was great fun and always on top of everything. If she didn't have the answer to a question you could be sure she would find it and get it to you in jig time.

Ah, we return to the question of time. Many colleagues kindly gave me time they didn't have to read drafts of the manuscript. In particular, Neil Ashby did an extraordinarily detailed job of what amounted to line editing. His candid and insightful comments on some of the stylistic elements I employed at the time he read the draft led to substantial changes for the better. I could not have hoped for a more valuable reading.

Years before I began to write the present book I had the privilege of studying writing with three gifted teachers (and authors): Paulette Bates Alden, Anna Maria Spagna, and Thomas Jude White. Each of them was an outstanding writing mentor. Paulette was the first and without her I would never have gotten any traction. She was accustomed to working with people who had an MFA in creative writing and I didn't. Moreover, I couldn't have been more ignorant. She is a candidate for literary canonization for taking me on as a student and for gently introducing someone as raw as me to the writer's basic toolbox.

I'm the beneficiary of endorsements of the book from a number of first-class scientists whose schedules were impossibly full but who managed to make the time to read and comment on the book. I thank Tom Cech, Seth Fraden, Andreas Herrmann, Stuart Kauffman, Jacques Prost, Arjun Yodh, and Joe Zasadzinski.

I hope the efforts of all those who were a part of this undertaking will result in the book finding its audience.

INTRODUCTION

Grandma Needs a Walker

"Grandma needs a walker now, and those poor teen-age mothers on 'Oprah' could use some job training." In 1995, Katherine Dowling was a family physician at the University of Southern California School of Medicine. Dr. Dowling wrote an opinion piece for the *Los Angeles Times* called "There's No Future Without Research." [1] She said that it's easier to comprehend and vote for people-related issues than for those that offer only the possibility of future benefits. Lawrence Livermore National Laboratory of the Department of Energy, renowned for its work on national security needs, weaponry, nuclear fusion, and the like, was *also* engaged in subcellular research and the spinoffs, though unpredictable, might help in understanding how malignancies develop. Dr. Dowling vented her opinion that with money becoming more scarce, she would rather see Lawrence Livermore get their lasers than Grandma get her walker. She reasoned that if Grandma got her walker, "her false teeth, her Winnebago; if Uncle Fred is declared totally disabled because his back hurts and he is too lazy to do the exercises to make it better—then. . . my grandchildren will sit on a polluted old planet bereft of energy. And unable to reach the stars."

Scientific research is essential to society's well-being. The knowledge we derive from such inquiry is the lifeblood of prosperity in the broadest sense of that term. And fundamental advances are frequently stimulated by curiosity-driven basic investigations. The interaction between basic research and applied research is said to be autocatalytic (it feeds or catalyzes itself).

This book is about DNA nanoscience—both applied and fundamental. The book came about because of my amazement at the studies being done collaboratively by Tommaso Bellini, a professor of physics at the University of Milan, and Noel Clark, a professor of physics at the University of Colorado. They have proposed a radical new thesis for one of the unsolved mysteries of science: how life emerged on the early Earth. Specifically, they have suggested a materials science solution for how molecules of DNA may have started out as small snippets—nanoDNA—and went on to become long-chain polymers capable of carrying genetic information. And they did laboratory experiments that backed up every step of their work.

I met with Bellini and Clark and said I'd like to write a book that laid out their line of argument to a wider audience than those who read scientific journals. At that meeting they made a terrific suggestion: expand the breadth of the book to include not only their look back in time but also a look forward to the emerging field of structural DNA nanotechnology. This discipline, founded by Nadrian Seeman, a professor of chemistry at New York University, is the remarkable use of synthetic DNA as a construction tool to build structures and devices on the nanometer-length scale (1 nanometer = 1 billionth of a meter). Seeman's insight was to synthesize stably branched (rather than linear) strands of DNA containing precise base sequences. He designed these strands to have *sticky ends* in which one strand of the double helix extends for several unpaired bases beyond the other strand. This enabled him to use the strong preferential binding properties of the bases that make up DNA (base complementarity) to join strands together and form nanoconstructions having a planned pattern. Complementarity refers to those base pairs that form bonds between two single strands and hold them together to form double-stranded DNA. The base adenine (A) is complementary to the base thymine (T), and guanine (G) is complementary to cytosine (C). Seeman's initial goal was to find a systematic way to form crystals, but the power of the technique he invented led to much more in the way of DNA nanostructures and nanodevices. The scope of this work is broad: the construction of DNA lattices that hold a periodic array of identical biological molecules for X-ray crystallography to determine their structure—a vital step in the rational design of drugs; DNA nanorobots that deliver instructions to cell surface signaling receptors of disease cells to arrest their growth or induce their self-destruction. These are among the numerous areas being actively investigated by Seeman and others who have joined the field. The reader will take away from this book a feel for the first 35 years of structural DNA nanotechnology.

The reader will also take away from this chronicle a vision of our Earth in the distant past, some four billion years ago—Bellini and Clark's liquid crystals hypothesis for the origin of DNA polymers. They surmise that on the prebiotic Earth, nanoDNA self-assembled into liquid crystal domains that served as microreactors. Within these vessels the nanoDNA segments became strongly chemically connected, forming elongated chains. Under conditions of cycled temperature or hydration—widely believed to be present on the primordial Earth—the formation and dissolution of liquid crystal domains was iterated many times, each cycle beginning anew with an elongated feedstock of nanoDNA. This liquid crystal autocatalysis was the engine that drove the polymerization of nanoDNA building blocks into the long information-bearing molecules we know, resulting in the origin of life.

Combining the stories of Seeman and of Bellini and Clark makes sense; their efforts are intimately connected. Seeman exploits hierarchical self-assembly in his creation of DNA nanoconstructions and nanodevices, and Bellini and Clark are examining how the self-assembly properties of the molecule DNA came into being. All of DNA nanotechnology is predicated on base complementarity; adenine (A) binds highly selectively with thymine (T), and guanine (G) binds highly selectively with cytosine (C). Seeman's branched DNA enables nanoconstructions and nanodevices to be built *because different strands of DNA can be designed with complementary base sequences that will bind to one another and form double helices,* and these double helices self-assemble into the desired nanostructures/nanodevices. Base complementarity is *also* at the heart of what Bellini and Clark are studying. They found that in a mixture of single strands of nanoDNA, a spontaneous series of self-assembly steps influenced only by temperature and nanoDNA concentration leads to the formation of liquid crystals. *Only those single strands of nanoDNA that have complementary base sequences will form double-stranded nanoDNA* (duplex nanoDNA) and *only the duplex nanoDNA will lead to liquid crystal formation.* Bellini and Clark's thesis is that liquid crystal formation is the gatekeeper that enabled the elongation of short lengths of nucleic acids into much longer ones on the primordial Earth. *So, as is true for Seeman's work, base complementarity is at the core of Bellini and Clark's study of how DNA came to be DNA.*

When diving into Seeman's extensive body of work I was in for a surprise. In particular, in the experiments of his first 15–20 pioneering years, his lab single-handedly faced scientific challenges that would have stopped someone with less grit. I found the mix of setbacks and success compelling. In the pages that follow I've chronicled hard-earned triumphs as well as failures and culs-de-sac. The challenges were no less daunting for Bellini and Clark during their odyssey to establish the connection between the formation of liquid crystals of duplex nanoDNA and the possible origins of life. DNA is the headliner in the tales that follow. But the struggles of the scientists and their answering moments of piercing insight provide the momentum that drives the narrative.

[1] http://articles.latimes.com/1995-02-16/local/me-32475_1_lawrence-livermore

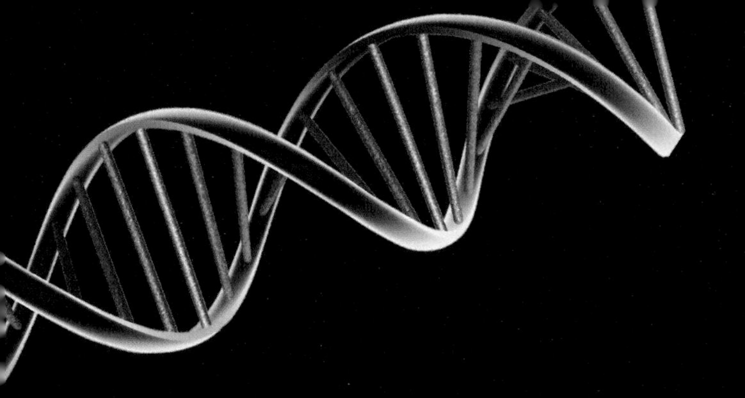

PART I

The Story Line and Its Underpinnings

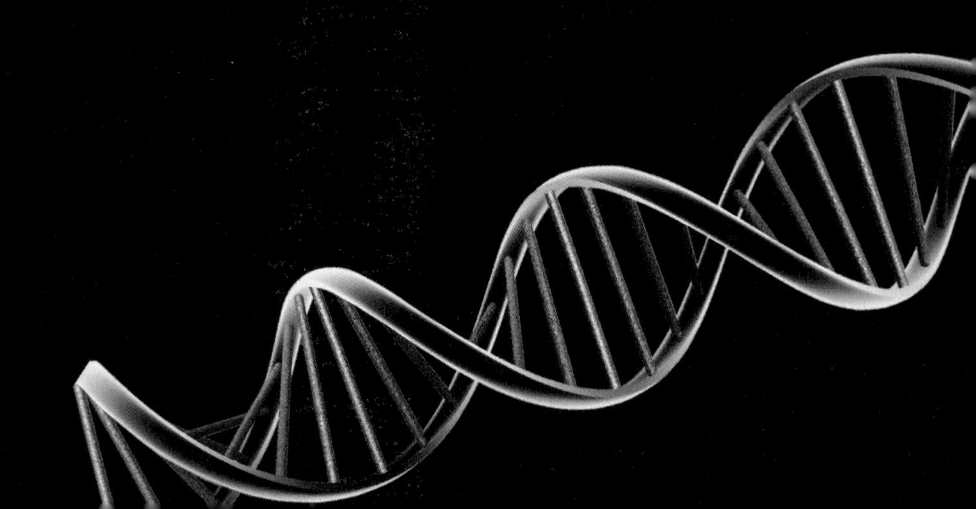

CHAPTER ONE

Down the Road and the Gemisch

So a guy walks into a bar and grill. He looks around and finds all the booths and tables full. He sits down at a barstool and the bartender asks him, "What'll it be?"

"Gimme a bowl of primordial soup."

"You want a large or a small?"

"Large."

"Plain or with all the works?"

"Load it up."

The bartender turns around and hollers into the pass-through window to the kitchen, "I need one *Gemisch*! Large!"

Gemisch is the German word for mixture. *Gemisch* is also the Yiddish word for a messy, complex mixture consisting of a myriad of different types of components.[1] Many scientists believe that RNA (**r**ibo**n**ucleic **a**cid) emerged from water enriched with organic molecules (basically, molecules containing both carbon and hydrogen) on the early Earth—the primordial soup, also called the *Gemisch*. That led, several hundreds of millions of years later, to DNA (**d**eoxyribo**n**ucleic **a**cid). RNA and DNA are biochemical information chains. Initially, RNA, or some precursor of modern RNA, played a critical role, acting as both a coding molecule and a catalyst for biochemical reactions. At a later stage in evolution DNA took over the key coding function in nature's exquisite scheme for transferring biological information.

In this book we're going to look at two lines of research using DNA. The first story is the manipulation of DNA as a design tool for future technologies, and the second story looks back four billion years at how DNA itself may have first formed. Surprisingly, perhaps, neither account focuses on DNA in its biological context. In the first case, the emphasis is on synthetic DNA and using the properties that make DNA such a great molecule for bio-inspired nanoscale control of the structure of matter. The second line of work uses the materials science study of short lengths of DNA in an attempt to understand one of life's most basic features, the linear polymer structure of its genetic material. Later on we'll see other similarities in the two tales. For now, we'll just say that the stories have common beginnings. The first started in a pub in Albany, New York, in 1980 and the other in a bar in Milan, Italy, in 1989.

Polymer A molecule consisting of many small units called monomers.

Nucleic acid A biopolymer, i.e., a long biological molecule, composed of monomers called nucleotides. The major types of nucleic acids are DNA and RNA.

DNA A nucleic acid in which the nucleotides each consist of the sugar deoxyribose, a phosphate group, and one of the four chemical bases adenine (A), cytosine (C), guanine (G), and thymine (T).

RNA A nucleic acid in which the nucleotides each consist of the sugar ribose, a phosphate group, and one of the four chemical bases adenine (A), cytosine (C), guanine (G), and uracil (U).

Dramatis Personae, Part I: Nadrian Seeman

To Albany by way of Chicago and Pittsburgh. Nadrian ("Ned") Seeman's scientific interests were kindled while in high school. He learned high school biology as a freshman from John E. Broming, a teacher who spent the first third of a semester talking about atoms and molecules because he saw life as a chemical phenomenon.[2] "I've been entranced with the edge of life ever since, and have spent my entire career on that cusp, sometimes venturing into biology, and sometimes into chemistry."[3]

Because of these dual interests Seeman decided to make biochemistry his college major at the University of Chicago. But there were problems. He didn't much like it. Biochemistry at Chicago in the mid-1960s had been developed by organic chemists and was rooted in the study of the metabolic transformations of small molecules. The field was focused on the biochemical cycles of nutrition and metabolism. There were huge charts that displayed cycle after cycle after cycle. One small molecule led to another small molecule that led elsewhere via intricate pathways. To do well required lots and lots of memorization. Seeman was bored silly. He tried his hand at physical chemistry but didn't do too well there either. In 1966 he graduated from Chicago with a B.S. in Biochemistry. But his grades were poor. He did well on the national Graduate Record Exam (GRE) tests, but his good scores weren't enough to overcome his poor grades. Every graduate chemistry department he applied to rejected him. Nevertheless, he managed to fast-talk his way into Chicago's graduate Biochemistry Department.[4]

Unfortunately, he didn't do much better in the basic biochemistry courses the second time around. But he did catch a break. His graduate student advisor, John H. Law, told him that his problem was not a lack of talent; he was just not interested in what was being offered. Law told Seeman that he should leave, and he helped him to get into a new crystallography/biochemistry training program at the University of Pittsburgh.[4] And Pittsburgh was a totally different experience. The crystallography research was a lot of fun and he did very well. He took his Ph.D. in crystallography/biochemistry in 1970. It was the structural beauty of crystallography along with the informational nature of the nucleic acids DNA and RNA that seduced Seeman into working at their interface. "Being able to make things, rather than just analyze them, vectored me into structural DNA nanotechnology."[5]

What do these last three words mean? Seeman's answer: "This term refers to the construction of molecules with robust topological or geometrical features from DNA components."[6] It's clear that Seeman wanted to make things. We've seen countless pictures of DNA as a double-stranded helix, a straight or nearly straight chain like a long piece of twine. We can imagine using it to make longer lines or circles, maybe tangled up or knotted in some way. It's not so easy to envision using it to make designed complex structures in multiple dimensions. However, the same interactions that enable DNA to excel as genetic material also enable its use as the fundamental component molecule for construction on the nanoscale. (We will discuss the term *topological* later in this book. For now, think of topological features as ones that don't change when an object is twisted or stretched. For example, a hole is a hole, whether in a donut or a coffee cup. In geometry, by contrast, numerically measured features such as lengths or angles can change when an object is continuously deformed.)

Topological When two objects can be continuously deformed from one to the other, they are said to be topologically equivalent. The classic illustration of two such objects is a donut and a coffee cup.

Years after Seeman switched fields and vectored into structural DNA nanotechnology he observed that even though it worked out well, it was not necessarily the smartest move. He was not at the beginning of his scientific career: he was 35 years old and an assistant professor—in fact, in the fourth year of his five-year probationary period as a faculty member.[7,8] It took pluck and courage to pioneer a new field of research and to do so at that point in his professional life. In subsequent chapters we'll explore the impact of the field that Seeman founded—an impact that might touch us all sometime down the road. But before continuing with Seeman's trajectory, we look at *starter kits* for DNA and nanotechnology. We'll give each of these topics a fuller treatment in Chapters Two and Three. Here we just have a peek.

DNA Starter Kit

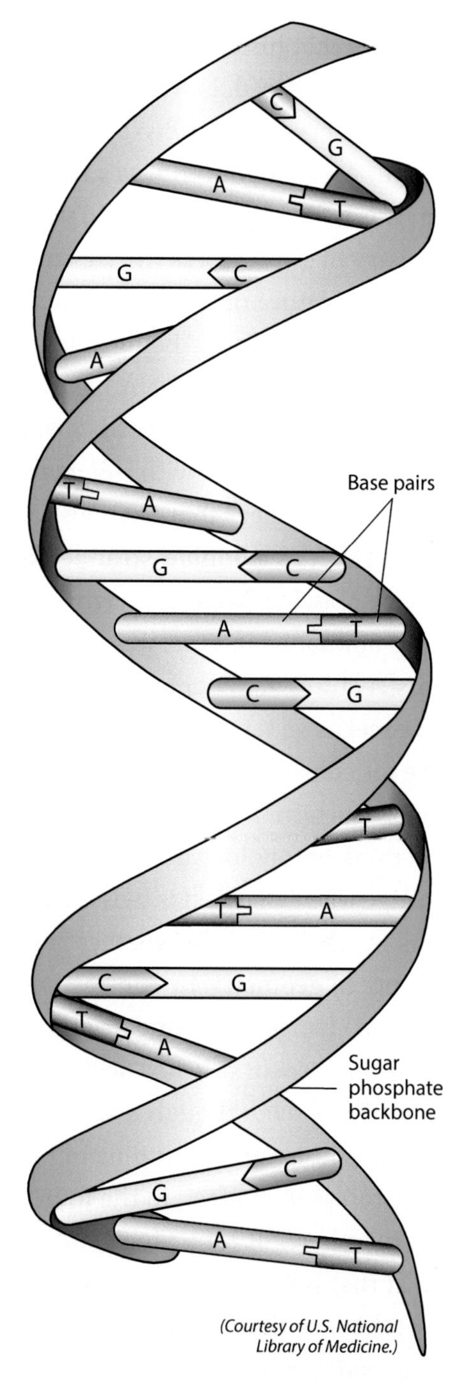

(Courtesy of U.S. National Library of Medicine.)

- DNA is a long, unbranched polymer. It's composed of four types of subunits called nucleotides. Each is made up of the sugar deoxyribose, a phosphate group, and one of the four chemical bases adenine (A), cytosine (C), guanine (G), or thymine (T).
- The DNA polymer assembles into two chains (the sugar-phosphate backbones) that twist into a double helix.
- The bases of the nucleotides connect the chains and match up in pairs, referred to as complementary bases. Each base pair is held together by hydrogen bonds (i.e., shared hydrogen atoms). The base shapes and their chemical makeup impart rules: A (orange) will only fit with T (red) and G (yellow) only fits with C (green).
- During DNA replication, the hydrogen bonds break apart. Free nucleotides found in the cell attach to each strand in accord with the base-pairing rules. This results in two identical molecules of DNA.
- The human genome is the collection of all the genetic material of *Homo sapiens*. It has over 3 billion DNA base pairs.
- In the first four decades of the twentieth century, many scientists felt that proteins carried the genetic code. But work done in the 1940s and 1950s proved that DNA carries the genetic code. In fact, a gene is a piece of DNA carrying instructions for making a protein.
- There are some organisms, bacteria are one example, in which virtually all DNA encodes proteins. But experiments in the 1960s and 1970s found many organisms in which sections of DNA do not encode proteins. Non-coding sequences are also found within genes, interrupting the protein-coding regions. It's estimated that only about 5% of human DNA encodes protein. Many non-coding regions have important biological functions, other non-coding sections have likely but undetermined functions, and some non-coding DNA may be without function.

As we see in the next starter kit offering an overview of nanotechnology, there are many ways to schematically represent the DNA double helix using a variety of color schemes and images.

Nanotechnology Starter Kit

- Nanotechnology is a very broad term encompassing the creation, exploration, and manipulation of materials measured in nanometers (billionths of a meter).
- As an umbrella term, nanotechnology embraces words such as nanodevices, nanoparticles, nanomaterials, nanomanufacturing, carbon nanotubes, molecular computation, nanophotonics (or nano-optics), nanofluidics, nanosensors, nano-biotechnology, nanomedicine, and nanoconductors.
- The term *nano-* derives from the Greek word *nanos*, meaning "dwarf."
- How small is a nanoscale object? The width of a human hair is about 80,000 nanometers. Here's another perspective: the smallest cellular life forms, the bacteria of the genus *Mycoplasma*, are around 200 nm in length.

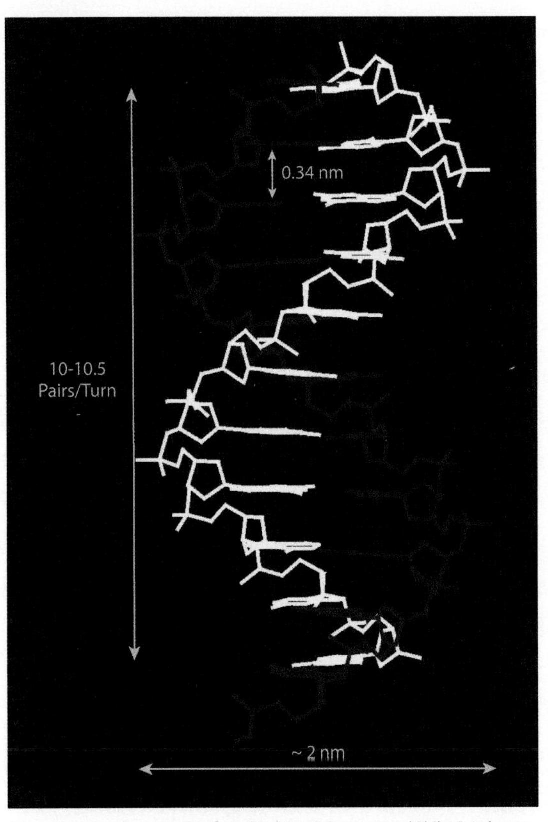

(Reproduced with permission from Nadrian C. Seeman and Philip S. Lukeman, "Nucleic Acid Nanostructures: Bottom-Up Control of Geometry on the Nanoscale," Rep. Prog. Phys. 68 (2005), 237–270.)

- The DNA double helix is inherently a nanoscale object: its diameter is about 2 nm, the separation of the bases is 0.34 nm, and the helical periodicity is 10–10.5 nucleotide pairs per turn, or about 3.5 nm per turn. Nadrian Seeman invented structural DNA nanotechnology, a subfield of nanotechnology that uses DNA as a nanoscale construction tool.
- One reason the nanoscale is of great interest is that "biology is nanotechnology that works."[1] That is, many of the inner mechanisms of cells occur at the nanoscale. Nanotechnology researchers often draw their inspiration from naturally occurring phenomena such as protein-based molecular motors, molecular ion pumps, and molecular walkers performing tasks such as muscle contraction and sensory transduction.

[1] Nadrian C. Seeman, "In the Nick of Space: Generalized Nucleic Acid Complementarity and DNA Nanotechnology," *Synlett No. 11* (2000), 1536–1548.

Equipped with the rudiments of DNA and nanotechnology, we return to Nadrian Seeman. Following postdoctoral work at Columbia University and MIT, Seeman took his first independent position at the State University of New York at Albany in 1977. A few years later he started to think about an original and unorthodox way to form a biomolecular crystal—that is, a material that has a large number of identical biological molecules arranged in a definite, repeating pattern that extends in all three dimensions. He wanted to make DNA boxes to serve as a host lattice for molecules of great interest. Figure 1.1 shows biological molecules (blue and pink) in boxes constructed from DNA (shown as brown); the branch point label will be explained shortly. Seeman proposed to use a technique called X-ray diffraction (as explained in Chapter Five) on the entire contents of the crystal to deduce the structure of the individual molecules that make up the crystal. X-ray diffraction reveals the detailed structure of the molecule, and this can open many doors, for example, the rational design of drugs that mesh exactly with their intended targets.

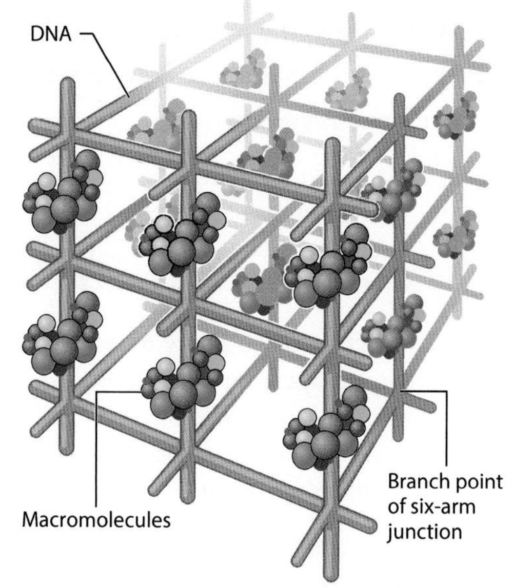

Figure 1.1 Cages made of DNA to enable crystallization of biological macromolecules. *(Reproduced with permission from Nadrian C. Seeman, "Nanotechnology and the Double Helix," Scientific American 290, No. 6 (2004), 64–75 and artist Jen Christiansen.)*

Molecular Crystals—Inspiration From Escher

Perhaps this all sounds like a lot of trouble just to obtain a molecular crystal. It isn't. Seeman has said that doing *conventional* crystallization experiments is arguably the dumbest experiment in modern science because it's so difficult to control. "Basically you're trying to get molecules to line up in a crystalline arrangement. What interactions are you trying to promote? Who knows? I don't have the patience for it, and I'm not quite enough of a jerk to force my students to do that kind of experiment for me."[9] He wanted to avoid doing what he called "piss-in-a-pot-and-pray" crystallization experiments.[10] But well into his first faculty appointment he had managed to crystallize nothing. His department's decision of whether to grant him academic tenure loomed. His job security was on the line. So, "to save my butt," Seeman invented a new way to make molecular crystals.[11] And his new way grew out of his fascination with a phenomenon in DNA called branching.

As we saw in the starter kits, naturally occurring DNA forms a linear chain (i.e., a chain that resembles a line) and appears as a spindly ladder that has been twisted. But during certain cellular processes, DNA passes through a stage in which it exists as a *branched* molecule. You might visualize it as the intersection where two highways cross at right angles. This happens when DNA's zipper unzips—when the comparatively weak hydrogen bonds break. These are the bonds that hold together the pairs of A (adenine) and T (thymine) and pairs of G (guanine) and C (cytosine). Seeman recalled that he was talking to an undergraduate student in his laboratory when a thought struck him. It was as if he'd been slapped between the eyes with a wet fish. "Wow, if I had synthetic branched molecules, I could study these things and test various hypotheses . . . I was really excited about that and I told everybody I knew."[12] Sometime afterwards, he went over to the campus pub, lost in a reverie about synthesizing DNA molecules having six-arm junctions.

Figure 1.2 (a) M.C. Escher's woodcut *Depth*. The center of each fish is just like the branch point of a six-arm junction. *(Reproduced with permission from The M.C. Escher Company.)* **(b)** Sticky ends are short *single*-strands of *unpaired* DNA extending from the ends of DNA double helices. They are called sticky because the unpaired bases can still form complementary base pairs and thus join the two double helices together. *(Reproduced with permission from Nadrian C. Seeman, "Nanotechnology and the Double Helix," Scientific American 290, No. 6 (2004), 64–75 and artist Alice Y. Chen.)*

That day in the Albany pub his thoughts flashed to an M.C. Escher wood engraving called *Depth* (panel (a) of Figure 1.2). And he made the leap from flying fish organized like the molecules in a molecular crystal to the six-arm junction molecules. He realized that the center of each fish in that picture was just like an idealized picture of the branch point of a six-arm junction. Six features extend from the center point on the fish: a head and a tail, a top fin and bottom fin, a left fin and a right fin. The fish are organized in the same way as the molecules in a molecular crystal. He mentally joined neighboring fish together head to tail, lateral fin to lateral fin, and vertical fin to vertical fin. They would produce a three-dimensional scaffolding of cages such as shown in Figure 1.1. "It struck me that if I held junctions together using sticky ends [panel (b) of Figure 1.2], I might be able to organize matter on the nanometer scale in the same way that Escher held his school of fish together using his imagination."[13] Seeman's motivation was to give him the technical capacity to make complex structures. He could not make intricate arrangements in two and three dimensions using unbranched DNA.

Perspiration and Reinvention

Immobile DNA branched junction A convergence of strands of synthetic DNA at a junction such that the intersection point does not change.

Sticky end DNA A fragment of double helix DNA in which a terminal portion of one strand has a stretch of unpaired bases (also called an overhang) that extend beyond the end of the other strand.

From that inspiration it took a decade of work to realize the first DNA construction. Seeman had assumed a faculty position in the Department of Chemistry at New York University. It was there in 1990 that he and his student Junghuei Chen put together a DNA molecule shaped like a cube (Figure 1.3).[14] This was the first non-trivial structure they fashioned, the first multiply connected object, which we will return to in Chapter Six. The edges of the cube consist of double-helical DNA and the vertices correspond to the branch points of three-arm branched junctions. Seeman's group had to devise many tricks to make the cube while coping with the daily failure of doing research. As he reflected later, "It was a nightmare!"[15]

Although Seeman did succeed in making a three-dimensional object constructed from DNA, he found that his hope of using single-stranded sticky ends led to structures that weren't rigid. He found this to be true even when trying to construct stick figures in two dimensions and the floppiness of the constructions bedeviled him. He pushed on with sheer determination to fashion the molecule that had cube-like features. But to progress toward his goal of three-dimensional DNA crystals, he had to abandon his initial strategy of simply adding sticky ends to branched DNA molecules. He was compelled to reinvent his tactics.

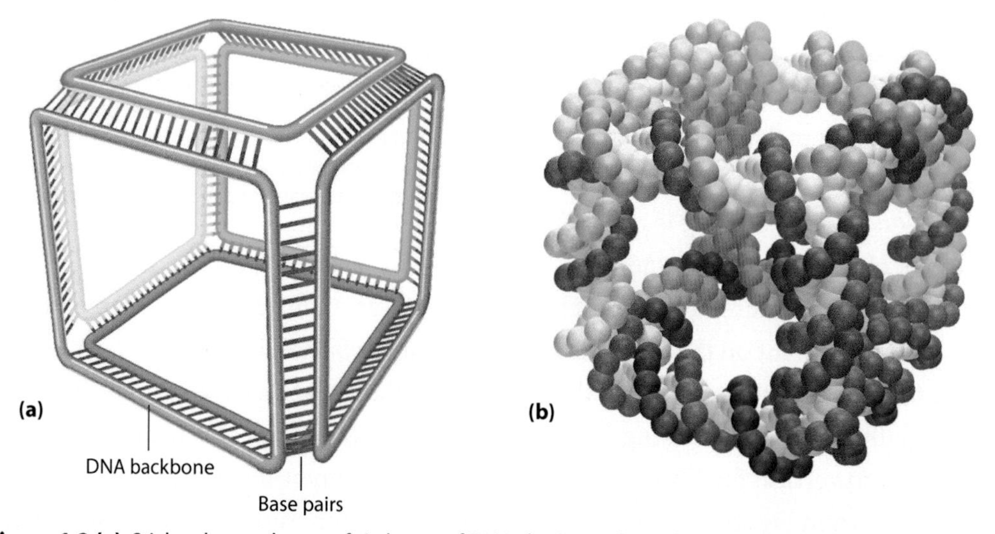

Figure 1.3 (a) Stick cube made out of six loops of DNA depicting how the strands are connected but omitting the helical twists. *(Reproduced with permission from Nadrian C. Seeman, "Nanotechnology and the Double Helix," Scientific American 290, No. 6 (2004), 64–75 and artist Alice Y. Chen.)* **(b)** The same cube with the backbone of each DNA strand shown as colored spheres (a different color for each strand) and the bases as white spheres. (*Note*: These are artistic renderings inferred from autoradiograms of electrophoretic gels. We will describe the results more thoroughly in Chapter Six.) *(Reproduced with permission from Nadrian C. Seeman, "Nanotechnology and the Double Helix," Scientific American 290, No. 6 (2004), 64–75 and artist Ken Eward.)*

He did so by using parallel DNA helices that are yoked together creating what he called a double-crossover molecule. With this building block he was able to achieve the stiffness he needed to create two-dimensional DNA crystals. With the advent of the robust double-crossover motif, he was also able to dream up ways of making nanomachines, a feat that had eluded him for years. In time he would use another rigid motif known as a tensegrity triangle to achieve the milestone of a three-dimensional DNA crystal. By dint of gumption, perspiration, and relentless reinvention, he was able to build nanoscale structures and devices of increasing complexity using successive stages of self-assembly. His enterprise was (and remains) on a roll.

Initially, structural DNA nanotechnology was mostly "a biokleptic enterprise."[16] But Seeman speculated the field would someday be less kleptomaniacal and more symbiotic with the field of biochemistry. He was right. DNA nanotechnology now includes the construction of a bewildering collection of nanostructures and nanomechanical devices. In fact, Seeman said that the most important thing that has happened to structural DNA nanotechnology is that it has not remained the preoccupation of a single laboratory.[17] And the dozens of fertile minds engaged in this surge of activity are taking the field far beyond what could have been done in only one laboratory, however inspired.

In later chapters we explore how DNA nanotechnologists design synthetic channels that might find a use in gene therapy (to inject material into a cell's interior). Others build molecular nanorobots to serve as drug-encapsulating devices to deliver payloads triggered by specific cell surface proteins. But the field is still young. It will take a long time to realize the grand dreams of some who see nanotechnology segued into our daily lives. As Seeman bluntly stated, "Our goal is to further in a realistic fashion the kinds of things that people have been talking about with nanotechnology for many years, most of which has been bullshit."[18]

This book takes a Janus-like approach to DNA, looking forward to the promise of DNA nanotechnology and backward to the possible origins of DNA on Earth. In ancient Roman mythology, Janus is the god of beginnings and transitions, often depicted as having two heads facing opposite directions. It's time now to introduce the human players who are looking to the past. They propose that the same self-assembly properties that make DNA nanotechnology possible may have enabled DNA to become DNA.

Dramatis Personae, Part II: Noel Clark, Tommaso Bellini

To Boulder by way of Cambridge, MA, and Milan, Italy. Noel Clark studied physics as a graduate student at MIT in the late 1960s, and his exploration of the light scattering spectroscopy of dilute gases earned him his Ph.D. After becoming a Research Fellow at Harvard University, he continued to collaborate with members of the MIT faculty and this led him to liquid crystals.

In 1969 Pierre-Gilles de Gennes had proposed a phenomenological theory for one specific phase transition (i.e., a transition from one state of matter to another) occurring in liquid crystals. De Gennes was already an eminent theorist and would later win the Nobel Prize. Experimental confirmation of de Gennes' 1969 theory was a plum that had eluded several talented scientists. But in 1972 Clark's skill at light scattering spectroscopy enabled him and his collaborators to succeed where others had not. The researchers not only found their measurements to be consistent with de Gennes' model but also obtained numerical values for the parameters in de Gennes' theory, giving it quantitative backing.[19] Clark's liquid crystals career was off to an auspicious start. His promise as a gifted experimentalist earned him a faculty appointment at Harvard. He fell in love with the beauty and mystery of liquid crystal phenomena, establishing his laboratory around that passion (Figure 1.4).

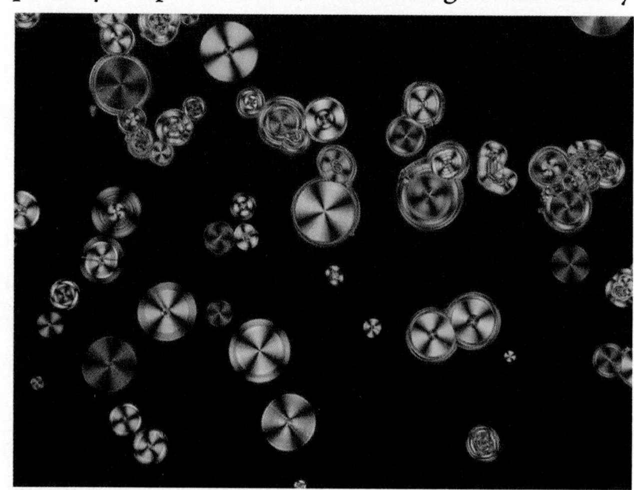

Figure 1.4 Liquid crystal domains nucleating from the compound P10PIMB. Photo by Michi Nakata. *(Courtesy of Noel A. Clark.)*

He also fell in love with Pauline and that passion led to marriage and three children. By 1977 Boulder beckoned. Clark packed his family into an orange and white VW minibus and trekked two-thirds of the way across the country to the University of Colorado. By the late '80s he was in demand as an invited speaker at international meetings.

In 1989 he gave several talks at a conference in Goteberg, Sweden and also chaired the Scientific Committee for the meeting. While on his trip, he accepted an invitation from Marzio Giglio, a colleague and friend from his graduate student days at MIT, to visit the University of Milan and present his latest research. His seminar in Milan led to a lucky encounter.

In the audience was a newly minted Ph.D. from the University of Pavia named Tommaso Bellini. He found Clark's talk stimulating. At a bar in Milan, Bellini and Clark sat down together to hoist a beer and chat about physics.[20] The following year, Bellini received a NATO Fellowship and joined Clark's laboratory in Boulder. After two years Bellini returned to Italy. He held appointments at the University of Pavia and the Polytechnic University of Milan. In due course he became a professor of Applied Physics at the University of Milan. Throughout this time his collaboration with Clark continued to flourish, and he often returned to Boulder to work in Clark's sprawling lab.

Before going further we look at our final starter kit, that of liquid crystals, a topic that we will treat more fully in Chapter Four.

Liquid Crystals Starter Kit

- Liquid crystals (LCs) are a state of matter whose order is intermediate between that of a liquid and that of a crystalline solid. LCs straddle the fence between fluidity and rigidity. The molecules are often rod shaped, and how they arrange themselves—their ordering—is a function of temperature. These different types of ordering are called *phases* and LCs will abruptly change phase (they will go through a *phase transition*) at certain temperatures.

- Orientational order characterizes the nematic phase; intermolecular forces cause the constituent molecules (shown as ellipsoids) to align themselves with respect to one another. The name *nematic* is derived from the Greek word for "thread." A nematic liquid crystal does not have any positional order. (The molecules are not constrained to occupy only specific positions.)

- The overall orientation of a LC material can be readily controlled with applied electric fields. For example, all the molecules may be parallel to the applied field or perpendicular to it, depending upon the choice of liquid crystal. This can produce profound effects in the material's optical properties. Nematics are the most commonly used phase in liquid crystal displays. Billions of displays are made each year for phones, computers, etc.

- Liquid crystals are pervasive in nature. The figure depicts in cross section a cell membrane with proteins in a lipid bilayer. The lipids are shown as spheres with two hydrocarbon tails, and proteins as embedded, bulbous objects. The lipids are in a liquid crystal phase (*the smectic phase*). The lipids are not perfectly ordered: the head groups are not arranged in an array, and the hydrocarbon chains are not rigid. The word *smectic* is derived from the Greek word for "soap"—the substance at the bottom of a soap dish is a smectic liquid crystal. In the smectic phase, molecules show a degree of translational order not present in the nematic. In the smectic state, the molecules maintain the general orientational order of nematics but may also have positional order, for example, aligning themselves in layers or planes. Motion is restricted to within these planes. Increased order means that the smectic state is more solid-like than the nematic.

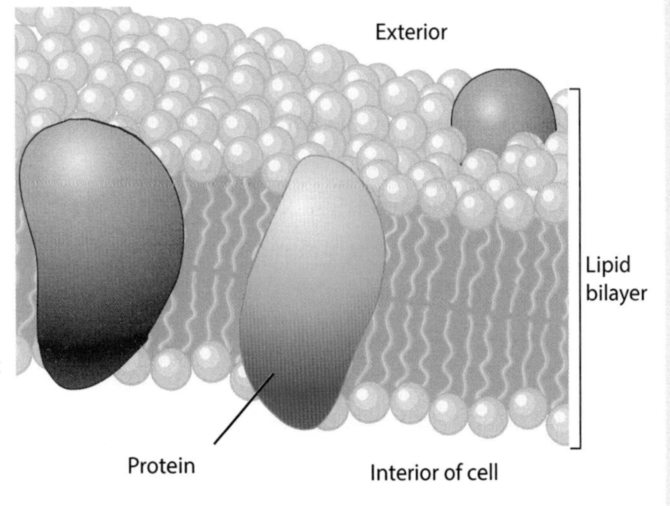

Liquid Crystals and Self-Assembly

Flash forward to 2006. One of Clark's postdoctoral students and one of Bellini's graduate students had been trying to use nucleic acids as a toolbox to design molecules with tunable length and rigidity as an approach to making new kinds of liquid crystals.[21] To their astonishment, very short lengths of DNA 6 to 20 base pairs in length, nanoDNA, produced liquid crystalline order. *This completely confounded the theoretical wisdom of the last half-century.* Prior to their experiments, the smallest DNA molecules to form liquid crystals were ~100 base pairs in length, giving them sufficient shape anisotropy (the ratio of their length to their diameter) to satisfy theoretical predictions. The first report of Clark and Bellini's findings appeared in 2007.[22] "We found that even tiny fragments of double helix DNA can spontaneously self-assemble into columns [a liquid crystal phase called *columnar*] that contain many molecules," Clark said (Figure 1.5).[23]

Figure 1.5 Columnar phase DNA liquid crystals 16 base pairs in length. Photo by Michi Nakata. *(Courtesy of Noel A. Clark.)*

Full disclosure: to look at Figure 1.5 and recognize it as columns of DNA molecules is not easy. It requires a great deal of practice and experience to interpret optical images of liquid crystals. Two key features are the *texture* and the colors. Texture, however, is not a precise term. It refers to the orientation of liquid crystal molecules in the vicinity of a surface. The texture can usually be understood in terms of different types of defects—points, lines, or regions in which the directional properties of a liquid crystal change, often abruptly. (In the case of Figure 1.5 the columns of DNA are quite convoluted and are oriented in many different directions.) Ultimately, examining the texture will provide details of the molecular order. For those who are interested, the Appendix of this book provides a discussion of the texture of liquid crystal images. Enough said about texture.

Texture A term used to describe the principal optical properties of a liquid crystal phase when viewed through crossed polarizers in a microscope. Texture is a qualitative term.

Now consider color. When looking at liquid crystals in a microscope one typically inserts the sample between crossed polarizers (Figure 1.6). A polarizer is an optical filter that only passes light of a specific polarization. (The polarization of an electromagnetic wave—such as light—is specified by the orientation of the wave's electric field.) Two polarizers are said to be crossed when no light passes through the system. The first polarizer assures that the incident light is polarized in one specific direction. Because the polarizers are crossed, no light would emerge through the second polarizer if the liquid crystals were absent. However, when the light passes through the liquid crystal sample its polarization state is altered—owing to the *birefringence* property of the liquid crystal—and thus some of the light does pass through the second polarizer. This is shown in panel (b) of Figure 1.6.

Birefringence is wavelength dependent. Thus the sample often appears in bright colors. That is, different colors of light have different wavelengths. Some wavelengths undergo a process of constructive interference, and some wavelengths go through destructive interference when passing through the birefringent liquid crystal. This enhances some colors and diminishes other colors.

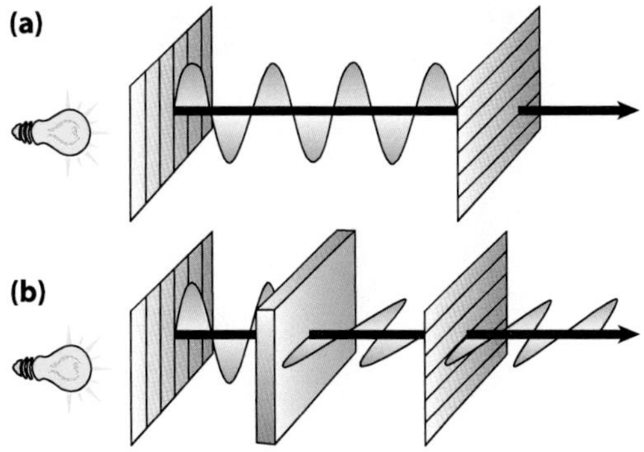

Figure 1.6 Crossed polarizers without and with a liquid crystal layer between them. *(Adapted from Chris Desimpel Ph.D. thesis Liquid Crystal Devices With In-Plane Director Rotation; used with permission from Jeroen Beeckman.)*

Yes, there is some optical physics that is necessary to explain the photographs of liquid crystals, and we have swept most of the explanation under the rug (for the present). We will expand on these subjects in Chapters Four and Five and in the Appendix, but for now we can all admire the beauty of the colors and textures in liquid crystal art.

Returning to the story of Bellini and Clark: the ability of long, double-stranded helical DNA to form liquid crystal phases had been known since the 1940s. But according to quantitative calculations, strands of DNA with fewer than 28 base pairs should be too short to form any liquid crystal phases, at least when the strands are modeled as hard rods.[24] Stimulated by their results, the Boulder/Milan team began a series of experiments to see how short the DNA segments could be and still show liquid crystal ordering. The team found that even a DNA segment as short as six base pairs could assemble itself into a liquid crystal (appearing in both nematic and columnar phases depending upon the concentration of the DNA in the solution). The cartoons in panel (a) of Figure 1.7, taken from the team's seminal paper, show how short segments of DNA pair up end-to-end to form liquid crystal phases. As we later explain, the researchers used X-ray diffraction combined with optical microscopy to perform structural analysis of the liquid crystals.

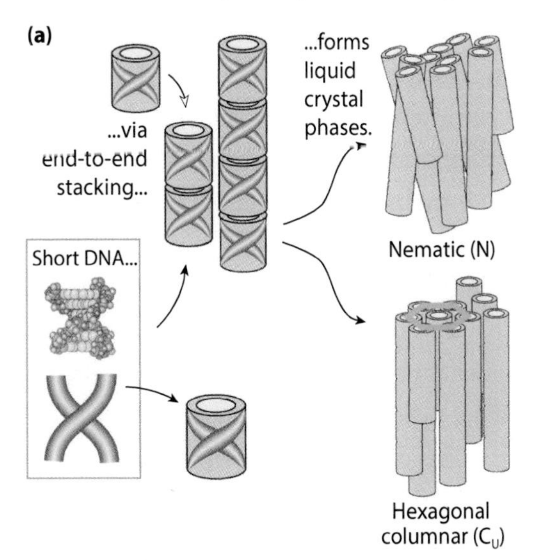

Figure 1.7 (a) Short segments of DNA (nanoDNA) pair up end-to-end to form liquid crystal phases. *(Adapted with permission from Michi Nakata, Giuliano Zanchetta, Brandon D. Chapman, Christopher D. Jones, Julie O. Cross, Ronald Pindak, Tommaso Bellini, and Noel Clark, "End-to-End Stacking and Liquid Crystal Condensation of 6– to 20–Base Pair DNA Duplexes," Science 318 (2007), 1276–1279.)*

Later experiments used a mixture of complementary and non-complementary DNA segments. (As we saw in the DNA Starter Kit, the term complementary

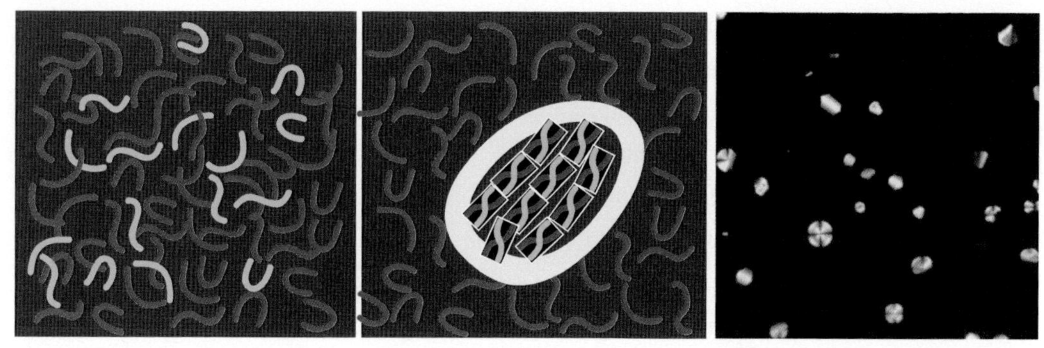

Figure 1.7 (b) Liquid crystal condensation of nanoDNA showing liquid crystal drops of complementary DNA in a background of non-complementary DNA (cartoons, left and center; micrograph, right). *(Adapted with permission from Michi Nakata, Giuliano Zanchetta, Brandon D. Chapman, Christopher D. Jones, Julie O. Cross, Ronald Pindak, Tommaso Bellini, and Noel Clark, "End-to-End Stacking and Liquid Crystal Condensation of 6– to 20–Base Pair DNA Duplexes," Science 318 (2007), 1276–1279.)*

specifies those base pairs that form hydrogen bonds and hold the two strands of helical DNA together.) The results of Bellini and Clark showed that the complementary DNA bits condensed out in the form of liquid crystal droplets, physically separating them from the non-complementary DNA segments (panel (b) of Figure 1.7). Clark said, "We found this to be a remarkable result. It means that small molecules with the ability to pair up the right way can seek each other out and collect together into drops that are internally self-organized to facilitate the growth of larger pairable molecules. In essence, the liquid crystal phase . . . selects the appropriate molecular components, and with the right chemistry would evolve larger molecules tuned to stabilize the liquid crystal phase."[25]

In fact, many years after Clark's speculation, the Boulder/Milan team showed that with correct chemistry, the liquid crystalline phase does indeed promote chemical bonding of stacked duplexes and their further elongation.[26] This result—known as ligation—is the keystone, and it will be the sole subject of Chapter Thirteen. The team had shown that liquid crystal formation is unquestionably a gatekeeper in laboratory experiments, but what about the role of liquid crystals on the prebiotic Earth?

This called for a leap of imagination. It doesn't invoke flying fish. It was evidence based but no less bold for that. In fact, it bordered on bizarre. The insight, as stated by Clark, was to pose the question: "Was there an era during the origin and evolution of early life during which the sole purpose of life was to make liquid crystals?"[27] This question addresses how linear nucleic acid polymers first appeared on Earth, one of the fundamental questions in science that remains unanswered. It is a

Hydrogen bonds Hydrogen atoms mediate an attractive force between two bases in double-helical DNA. The base *pair* G-C is stabilized by three hydrogen bonds while the base *pair* A-T is stabilized by two hydrogen bonds. Hydrogen bonds form links between an atom like oxygen or nitrogen present on one base to an oxygen or nitrogen present in another base.

Complementary Refers to those base pairs that form hydrogen bonds and hold the two helical strands of DNA together. The base adenine (A) is complementary to the base thymine (T), and guanine (G) is complementary to cytosine (C).

profound issue. As we discuss in Chapter Fourteen, many scientists study the origins question and often do so with experiments involving DNA because it is so easily obtained and manipulated. They believe that an understanding of how DNA came to be will reveal how its predecessor, RNA, came into being. That is the heart of the mystery: by what process did the first information-bearing molecule emerge out of the molecular clutter, the prebiotic soup? How did organic molecules in a prebiotic world assemble to produce "life"?

Origin of nucleic acid polymerization
Bellini and Clark contend that DNA and RNA became long, chain-like molecules because ultrashort lengths of both nucleic acids were able to self-assemble to form liquid crystals.

Bellini and Clark believe that on the primordial Earth, liquid crystal formation was the gatekeeper enabling the elongation of short lengths of nucleic acids into much longer ones. At first blush, it might appear that Bellini and Clark, in their capacity as liquid crystal physicists, were acting out psychologist Abraham Maslow's observation: "I suppose it is tempting, if the only tool you have is a hammer, to treat everything as if it were a nail."[28] But as we continue to examine their work in this book, we see that the question they posit cannot be so easily dismissed. They not only have a thesis, but they've backed it up with reproducible experiments that were performed under plausible prebiotic conditions.

On the other hand, they are new players in the field known as origins research, and caveats are necessary. Robert Hazen said it well: he chronicled a different approach to explaining the pre-RNA world and warned that it "presents an intriguing hypothesis but one that is highly speculative." But as he concluded, "Whatever the outcome, this story epitomizes the exhilarating process of scientific exploration."[29] The Bellini/Clark approach to explaining the pre-RNA world is an exhilarating process of scientific exploration and, most importantly, is supported by a decade of definitive experimental evidence.

Seeman, Bellini and Clark, and Base Complementarity

When Seeman took on the challenge of trying to invent an utterly new method for the crystallization of macromolecules, he was putting his academic career in jeopardy. It was a bold and risky decision. It took chutzpah, presumption plus arrogance, as Leo Rosten defined it.[30] There is a parallel to this in the work of Bellini and Clark. Though they were both well established when they serendipitously stumbled into the field of origins research, this discipline is rife with speculation and dispute. The intrigue of their early results unquestionably made their vibrissae tingle. But pursuing the answers to questions of such great scope is not for the faint of heart. It's been said that in the sixth century B.C., Thales of Miletus postulated the first explanation for the origin of life but that immediately after Thales' pronouncement, one of his disciples, Anaximander, disagreed with him.[31] "The bickering about beginnings has not ceased in ensuing millennia. . . . A definitive answer . . .

and how it could ultimately lead to a world populated by iPhones and reruns of *American Idol* still eludes today's natural philosophers."[31] Bellini and Clark took on a challenge of Brobdingnagian proportions, as did Seeman.

Now let's reflect on the science of the two tales. The efforts of Seeman and of Bellini and Clark are intimately connected. Seeman exploits hierarchical self-assembly in his DNA nanoconstructions, and Bellini and Clark are seeking to explain how the self-assembly properties of the molecule DNA came into being. Both are using the evolved properties of DNA as their starting point. Seeman uses the properties of DNA to design and build molecular architecture. But RNA and DNA have been making three-dimensional structures all along (RNA and DNA orchestrate protein synthesis). *Scientists can make three-dimensional structures using the molecules of life because the molecules of life can make three-dimensional structures.*[32] That may seem sophistic. But Seeman's work exists because of the very attributes of DNA that Bellini and Clark are investigating.

We'll expand on this because it's important to understand at the outset that our two tales have a common motif. The common element is complementarity, base complementarity to be explicit. DNA nanotechnology is dependent on the fact that adenine (A) binds highly selectively with thymine (T), and the same is true for cytosine (C) and guanine (G). If these specific affinities didn't exist then DNA nanotechnology as implemented by Nadrian Seeman and others would not work. (As something of a counterpoint to DNA constructions formed solely by means of base complementarity, recent work has demonstrated dynamic assembly and disassembly of three-dimensional structures on the basis of shape complementarity and without base pairing—though the three-dimensional structures themselves are built from a base-complementary technique called DNA origami.[33])

Of course in saying that DNA nanotechnology wouldn't work if base complementary didn't exist, we beg the question: if we didn't have base complementarity then double helix DNA would not exist in its present form. But let's not chase our tails. The point is that Seeman's branched DNA enables nanoconstructions and nanodevices to be built because different strands of DNA can be designed with complementary base sequences that will bind to one another and form a double helix. Base complementarity is also at the heart of what Bellini and Clark are studying. *Duplexing* is a concise way of saying that two single strands of DNA bind together to form the classic double helix (also called the DNA duplex). Duplexing is accomplished by means of base pairing, and it is the essential element in DNA liquid crystal formation (panel (b) of Figure 1.7). Bellini and Clark's thesis is that liquid crystal formation is the gatekeeper that enabled short lengths of nucleic acids to elongate into much longer ones on the primordial Earth. So it follows that their origins proposal, like Seeman's DNA nanotechnology, is grounded in base complementarity.

Seeman himself has called out as the central feature of his work the programmability of nucleic acid structure using the same Watson–Crick base pairs (A and T, C and G) that are responsible for the information content of all living systems.[34] (To say that nucleic acid strands are *programmable* means that by building strand sequences with appropriate combinations of complementary bases, the strands will self-assemble into a predesigned architecture.) Tying together applied and basic research, Seeman wrote that DNA nanotechnology is stimulating basic science by opening lines of inquiry that will lead to greater understanding of the molecules at the center of life.[35] Seeman has also invoked the "final blackboard" of the towering physicist Richard Feynman. Feynman's final blackboard contained the line, "What I cannot create, I do not understand."[36] Bellini and Clark are trying to create polymeric DNA from the minimal feedstock that is widely presumed to have been present on the prebiotic Earth. They are doing so to understand the self-assembly properties of DNA, those same properties that Seeman and all DNA nanotechnologists rely upon to ply their trade.

Programmable nucleic acid strands By building nucleic acid strand sequences with appropriate combinations of complementary bases, the strands will self-assemble into a predesigned architecture.

Conventional Wisdom and an Alternative View

Coming back to how DNA came to be DNA, in Chapter Fourteen we examine other scenarios for the origins of life, other attempts to fill the gap between the primordial soup and the existence of RNA. We then make the case that the liquid crystals explanation, while not decisively proven, has unique features that make it a compelling subject for examination. For now we note that while Bellini and Clark have focused on DNA because of its ready availability, they have also examined short segments of (double-stranded) RNA, a likely nucleic acid precursor to DNA. They found that RNA also self-organizes into liquid crystal phases.[37–39] Their overarching goal is to challenge the twentieth-century wisdom that our molecular carriers of genetic information, RNA and DNA, form liquid crystals because of their chain-like shape. They propose that it is the other way around: our carriers of genetic information are linear chain-like molecules because rod-shaped objects form liquid crystals. They believe that their findings can be taken as a paradigm of what could have happened on the prebiotic Earth based on the fundamental and simplifying assumption that "the origin of nucleic acids is written in their structure."[40]

Bellini and Clark are moving closer and closer to the edge of the prebiotic clutter that preceded the existence of nucleic acids. They've already pushed past the formation of nanoDNA 6 base pairs in length that marked the lower limit of their initial investigations. In their newest report they describe many liquid crystal phases of nanoDNA 4 base pairs long.[41] They also have preliminary evidence for nanoDNA duplexes that are only 2 base pairs in length.[42]

Their long-term ambition is to unveil the very emergence of complementarity, the heart of the matter concerning the prebiotic origin of nanonucleic acids when they arose from the realm of chemical noise. They propose to show that—when present in a variety of solutions that contain water, formamide, ammonia, oligosaccharides, oligopeptides, aliphatic and aromatic hydrocarbons, and other organic compounds—a single set of nucleotides that can duplex will also initiate self-assembly into liquid crystals. That is, they will conduct their experiments all the way back to Earth's original venue: the primordial soup. The *Gemisch* (Figure 1.8).

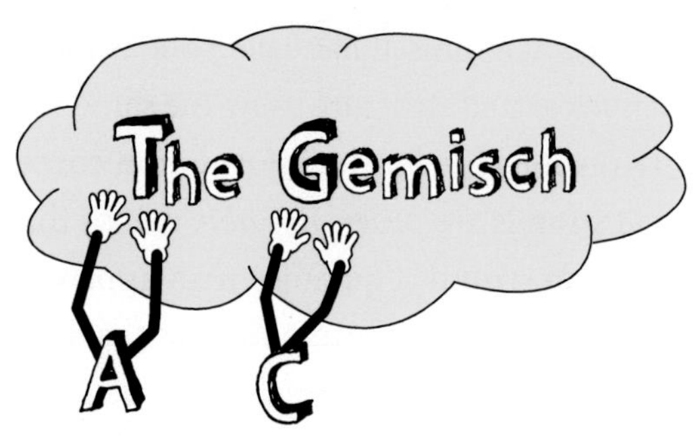

Figure 1.8 The emergence of complementarity. *(© Kenneth Douglas)*

ENDNOTES

[1] Christopher Wills and Jeffrey Bada, *The Spark of Life: Darwin and the Primeval Soup* (New York: Basic Books, 2001), 263.

[2] Ann Finkbeiner, "Crystal Method," *University of Chicago Magazine* (Sept–Oct/2011), http://mag.uchicago.edu/science-medicine/crystal-method

[3] Stuart Cantrill, "The Sceptical Chymist: Reactions—Ned Seeman," *The Nature Chemistry Blog* (May 11, 2007), http://blogs.nature.com/thescepticalchymist/2007/05/reactions_nadrian_seeman_1.html

[4] Nadrian C. Seeman, "At the Edge of Life: The Autobiography of Nadrian C. Seeman," http://www.kavliprize.org/sites/default/files/nadrian_seeman_autobiography.pdf

[5] Cantrill, "The Sceptical Chymist."

[6] Nadrian C. Seeman, "Biochemistry and Structural DNA Nanotechnology: An Evolving Symbiotic Relationship," *Biochemistry 42, No. 24* (2003), 7259–7269.

[7] Paul S. Weiss, "A Conversation With Prof. Ned Seeman: Founder of DNA Nanotechnology," *ACS Nano 2, No. 6* (2008), 1089–1096.

[8] Ned Seeman, "The Crystallographic Roots of DNA Nanotechnology," *ACA RefleXions, No. 2* (Summer, 2014), 22.

[9] Karen Hopkin, "3-D Seer," *The Scientist* (August 2011, Profile), 52–55.

[10] Ibid., 52.

[11] Finkbeiner, "Crystal Method."

[12] Weiss, "A Conversation With Prof. Ned Seeman," 1089.

[13] Nadrian C. Seeman, "Nanotechnology and the Double Helix," *Scientific American 290, No. 6* (2004), 64–75.

[14] Junghuei Chen and Nadrian C. Seeman, "The Synthesis From DNA of a Molecule With the Connectivity of a Cube," *Nature 350* (1991), 631–633.

[15] Weiss, "A Conversation With Prof. Ned Seeman," 1090.

[16] Seeman, "Biochemistry," 7268.

[17] Nadrian C. Seeman, "DNA: Not Merely the Secret of Life," http://www.kavlifoundation.org/sites/default/files/file/2010KavliLecture-NadrianSeeman.pdf

[18] Matthew Hutson, "The Godfather of *Really* Small Things," *NYU Alumni Magazine, No.15* (Fall 2010), https://www.nyu.edu/alumni.magazine/issue15/15_square_Chemistry.html

[19] T.W. Stinson, J.D. Litster, and N.A. Clark, "Static and Dynamic Behavior Near the Order Disorder Transition of Nematic Liquid Crystals," *Journal De Physique 33* (1972), C1–69 – C1–75.

[20] Personal communication; email from Tommaso Bellini to K.D. on 15 April 2011.

[21] Giuliano Zanchetta, "Spontaneous Self-Assembly of Nucleic Acids: Liquid Crystal Condensation of Complementary Sequences in Mixtures of DNA and RNA Oligomers," *Liquid Crystals Today 18, No. 2* (2009), 40–49.

[22] Michi Nakata, Giuliano Zanchetta, Brandon D. Chapman, Christopher D. Jones, Julie O. Cross, Ronald Pindak, Tommaso Bellini, and Noel Clark, "End-to-End Stacking and Liquid Crystal Condensation of 6– to 20–Base Pair DNA Duplexes," *Science 318* (2007), 1276–1279.

[23] University of Colorado Office of News Services, 22 November 2007.

[24] Peter Bolhuis and Daan Frenkel, "Tracing the Phase Boundaries of Hard Spherocylinders," *J. Chem. Phys. 106, No. 2* (8 January 1997), 666–686.

[25] University of Colorado, 22 November 2007.

[26] Tommaso P. Fraccia, Gregory P. Smith, Giuliano Zanchetta, Elvezia Paraboschi, Youngwoo Yi, David M. Walba, Giorgio Dieci, Noel A. Clark, and Tommaso Bellini, "Abiotic Ligation of DNA Oligomers Templated by Their Liquid Crystal Ordering," *Nature Communications 6, Article No. 6424* (10 March 2015), 1–7.

[27] Noel A. Clark, Christopher N. Bowman, Robert R. McLeod, and David M. Walba, Soft Materials Research Center, Renewal Proposal to the National Science Foundation, January 18, 2008, 4c1–2.

[28] Abraham H. Maslow, *The Psychology of Science: A Reconnaissance* (Chapel Hill, NC: Maurice Bassett Publishing, 2002), 15.

[29] Robert M. Hazen, *genesis: The Scientific Quest for Life's Origin* (Washington, DC: Joseph Henry Press, 2005), 222.

[30] Leo Rosten, *The New Joys of Yiddish* (New York: Harmony, 2001), 81.

[31] The Editors, "In the Beginning," *Scientific American 301* (2009), 35.

[32] Rephrasing of an observation made by Tommaso Bellini to K.D. and Noel Clark on 18 January 2012.

[33] Thomas Gerling, Klaus F. Wagenbauer, Andrea M. Neuner, and Hendrik Dietz, "Dynamic DNA Devices and Assemblies Formed by Shape-Complementary, Non-Base Pairing 3D Components," *Science 347* (27 March 2015), 1446–1452.

[34] Nadrian C. Seeman, Chengde Mao, and Hao Yan, "Guest Editorial: Nucleic Acid Nanotechnology," *Acc. Chem. Res., No. 47* (2014), 1643–1644.

[35] Ibid.

[36] http://caltech.discoverygarden.ca/islandora/object/ct1%3A483

[37] G. Zanchetta, M. Nakata, M. Buscaglia, N. A Clark and T. Bellini, "Liquid Crystal Ordering of DNA and RNA Oligomers With Partially Overlapping Sequences," *J. Phys.: Condens. Matter, No. 20* (2008), 1–6.

[38] Giuliano Zanchetta, Tommaso Bellini, Michi Nakata, and Noel A. Clark, "Physical Polymerization and Liquid Crystallization of RNA Oligomers," *J. Am. Chem. Soc., No. 130* (2008), 12864–12865.

[39] Zanchetta, "Spontaneous Self-assembly," 40–49.

[40] Fraccia et al., "Abiotic Ligation," 6.

[41] Tommaso P. Fraccia, Gregory P. Smith, Lucas Bethge, Giuliano Zanchetta, Sven Klussmann, Noel A. Clark, and Tommaso Bellini, "Liquid Crystal Ordering and Isotropic Gelation in Solutions of 4-Base-Long DNA Oligomers," submitted to *ACS Nano.*

[42] Unpublished: Noel Clark, "Liquid Crystals of Nanonucleic Acids: Hierarchical Self-Assembly as a Route to Prebiotic Selection, Templating, and Autocatalysis," Renewal Proposal for NSF Award Number DMR 1207606, D9.

CHAPTER TWO

DNA: The Molecule That Makes Life Work—And More

The early 1940s in New York City. Every morning a little before 9:00 A.M. a short, slight man of about 100 pounds, possessed of sparkling eyes and a bulky dome of a head and dressed impeccably in a suit and necktie, walks the few blocks from his 67th Street residence to the hospital at 66th Street and York Avenue. The hospital is one of four buildings comprising The Rockefeller Institute for Medical Research. Reaching his office on the sixth floor he sheds his jacket and dons a laboratory coat before entering the small bacteriology laboratory adjacent to his office. The fusty odor of bacterial cultures is everywhere, the breath of the microbial world. Both rooms are neat and clean but lack the photographs, pictures, unused books, mementos, and other items that often adorn (and clutter) such workspaces (Figure 2.1). The austerity of these rooms reflects how much doctor-turned-research-physician Oswald Avery has given up in all aspects

Figure 2.1 Oswald Avery at his desk. *(Courtesy of the Tennessee State Library and Archives.)*

of his life for the sake of complete concentration on his chosen professional goals. He exemplifies what he calls "the inwardness of research."[1]

Avery was born in Halifax, Canada. He graduated from Colgate University and earned his M.D. from the College of Physicians and Surgeons of Columbia University in New York City. In the early part of the twentieth century, pneumonia accounted for more than 50,000 deaths annually in the United States alone. Avery was hired by the Rockefeller Institute in 1913 and joined the hospital's pneumonia research program. While working on the diagnosis and treatment of pneumonia, Avery became intent on gathering detailed knowledge of the pneumococcal cell itself—its structure, chemical composition, physiological activities, immunological characteristics, genetic stability, and variability. This led to his focus on the phenomenon of pneumococcal *transformation*: a harmless strain of pneumococcus bacteria could become infectious when mixed with a virulent strain of the bacteria that had been killed. It appeared as though the dead bacteria provided some chemical that transformed the harmless bacteria and made them infectious. The changed bacteria were identical in virulence and type to the killed bacteria; the changes were permanent and inheritable. In 1944, Avery and his colleagues Maclyn McCarty and Colin MacLeod published an historic paper. They isolated the active transforming substance from samples of pneumococci. The substance was DNA.[2] This marked the onset of the DNA revolution. Avery's work transformed biology.[3]

Before continuing with our eclectic look at DNA, it's time to pause and ask a question that may have occurred to the reader: Why did DNA become the bearer of genetic information—why not RNA? Owing to numerous experiments, some of which we will touch on in Chapter Fourteen, there is widespread agreement that RNA (or something of the kind) was indeed the first molecule of heredity before DNA came onto the scene. However, in many of its biological roles RNA is a *single*-stranded helical molecule. The spaces between turns of the helix are larger in RNA than in (the double helix) DNA, and that makes RNA more vulnerable to attack by enzymes. Second, the backbone of RNA is a ribose-phosphate combination while the backbone of DNA consists of the sugar deoxyribose along with phosphate groups. The key difference between the sugars is that ribose contains hydroxyl groups (consisting of a hydrogen atom bonded to an oxygen atom) in places where DNA does not. These hydroxyl groups make RNA less stable than DNA, because, chemical details aside, they make RNA more prone to hydrolysis—that is, the rupture of chemical bonds by the addition of water. So (1) DNA doubles the existing RNA molecule—loosely speaking—making DNA less vulnerable to enzymatic degradation, and (2) DNA includes deoxyribose sugar instead of ribose making DNA less susceptible to hydrolysis. Thus, DNA evolved as a very stable molecule to pass genetic information with accuracy.

If you were to ask randomly chosen people what names come to mind when the words *DNA* or *gene* are mentioned, many would say Watson and Crick. However, they did not discover DNA nor did they identify it as the carrier of genetic information. The Swiss chemist Friedrich Miescher was the first to identify DNA in the late 1860s, and we have already seen the contribution of Oswald Avery and his team.[4] Avery's result transformed biology but heightened the mystery. How could DNA produce such varied biological outcomes—from amoebas to zebras? The goal now was to understand the structure of DNA, how it was put together and how it replicated itself. Before we add a few more words to the vast number that have been written about Watson and Crick and how they answered these questions, we'll have a look at several of the other players in the DNA sweepstakes. We begin with Erwin Chargaff.

Erwin Chargaff

Chargaff was a biochemist at Columbia University. He read Avery's 1944 paper and was thunderstruck. Chargaff wrote, "these are the words with which they concluded their paper: 'The evidence presented supports the belief that a nucleic acid of the desoxyribose type is the fundamental unit of the transforming principle of Pneumococcus Type III.'[5] It is difficult for me to describe the effect that this sentence, and the beautiful experimentation that had given rise to it, had on me."[6] Although Chargaff found it difficult to describe the effect the sentence had on him, he did exemplify Shakespeare's wisdom. In Act III of his tragedy *Coriolanus,* the Bard wrote, "Action is eloquence."[7] Chargaff acted with brisk decision. He abruptly concluded all the experiments he had been doing and plunged into a research program devoted to the chemistry of nucleic acids. This led to multiple papers in the *Journal of Biological Chemistry* and then a review paper that summarized his findings.[8] His results were intriguing and enigmatic. One was particularly puzzling. Chargaff wrote: "It is, however, noteworthy—whether this is more than accidental, cannot yet be said—that in all desoxypentose nucleic acids [DNA] examined thus far [extracted from calf thymus, beef spleen, yeast, avian tubercles bacillus, human spermatozoa, etc.] the molar ratios of . . . adenine to thymine and of guanine to cytosine, were not far from 1."[9] These Chargaff base ratios are sometimes summarized as A = T and G = C. That is, the nucleotides in DNA must be arranged so that there are about equal amounts of A and T, and about equal amounts of G and C. This is sometimes known as "Chargaff's rule." (To keep people on their toes, A, T, C, and G are referred to in the literature by different names: nucleobases, nucleotide bases, deoxyribonucleotide bases, nitrogenous bases, bases, and even nucleotides.) The discovery of these numerical relationships between bases implied that there was some sort of regularity in the composition of DNA. Chargaff's data would be vital to the solution of the structure of DNA. But the role played by the harmonious base proportions eluded him.

Rosalind Franklin

The next player we highlight is Rosalind Franklin (Figure 2.2). Franklin earned a Ph.D. in physical chemistry from the University of Cambridge. After taking her degree she mastered the technique of X-ray diffraction and made many improvements in the design of the equipment and in the mathematical techniques as well. We'll have another look at X-ray diffraction in Chapter Five because it's an important tool in the work of Seeman and of Bellini and Clark. Briefly, an X-ray beam bombards a sample (usually, a crystalline sample). Atoms in the crystal interact with the rays. Upon exiting the crystal, the rays reemerge as an elaborate pattern of beams that spread in many directions. The beam angles and intensities are recorded on an X-ray film. By mathematical analysis of the images on the X-ray film, one can work in reverse to determine the exact structure of the crystalline sample. Here's an analogy: Suppose you were to shine a powerful light at a chandelier and record where the scattered spots of light are found on a screen behind the chandelier. You might then be able to work out the arrangement of the chandelier's glass components by analyzing the pattern caused by its scattered light.[10]

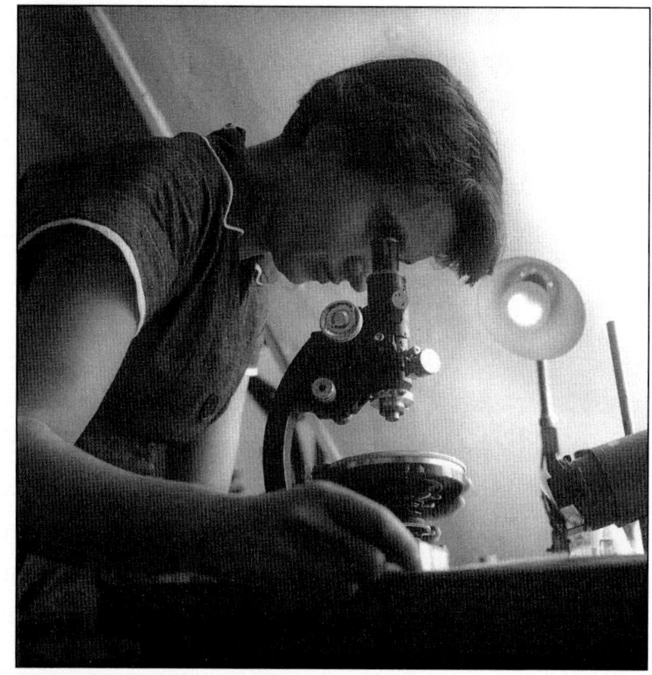

Figure 2.2 Rosalind Franklin.
(Used with permission, © Henry Grant / Museum of London.)

In January 1951 Franklin began a fellowship at King's College, London, using her skill at X-ray diffraction to study the structure of DNA. (The DNA was drawn out into thin, uniform fibers that were not truly crystalline—they were disordered crystals also referred to as *paracrystalline forms*. But the DNA fibers had sufficient order to be amenable to diffraction.) She was well aware of Erwin Chargaff's base ratios, and it is fascinating to read her laboratory notebooks as she wrestled with the need to reconcile Chargaff's analysis with her own X-ray diffraction information.[11] In March 1953, Franklin—along with her student Raymond Gosling—submitted a paper that included speculation that A (adenine) and G (guanine) were interchangeable as were C (cytosine) and T (thymine). They suggested that such interchangeability would make possible an infinite variety of nucleotide sequences and thus explain the biological specificity of DNA.[12] Alas, this was not the correct interpretation.

In the following month, however, the correct interpretation did appear in the journal *Nature*.[13] James Watson was a young postdoctoral biologist from the United States. Francis Crick was British, a lapsed physicist whose interests had turned to biology. The two shared Room 103 of the Austin Wing of the Cavendish Laboratory at the University of Cambridge where they had endless, energetic conversations on the mystery of DNA.[14] (The exchanges continued over lunch at The Eagle, the Cavendish's local pub—thus predating Seeman and Bellini/Clark in combining science with visits to ye olde watering hole.) They deduced the structure of DNA, and their model revealed the correct interpretation of Chargaff's analysis. The importance of Franklin's work on DNA was lost sight of for many years, partly because of her untimely death in 1958 from ovarian cancer. She was 37. Franklin has been spoken of as "the Sylvia Plath of molecular biology."[15]

What were Franklin's contributions to the solution of the DNA puzzle? Plenty. We won't mention all of them, but among the most dramatic was a Franklin X-ray diffraction photograph of exceptional quality. In his book, *The Double Helix*, Watson wrote, "The instant I saw the picture my mouth fell open and my pulse began to race."[16] The picture—with its telltale X shape—could only arise from a helical structure. Watson also learned of several key numerical parameters that were calculated from this image such as the separation of the bases of 0.34 nm and the helical periodicity of about 3.5 nm. This image was labeled by Franklin as "51" and has become famous as *Photo 51* (Figure 2.3).[17] In addition, because of the clarity of her X-ray diffraction picture, Franklin had determined the space group of the paracrystalline DNA she was examining. Crystals are classified by their symmetry, that is, by the shape of their unit cell—the simplest repeating unit that stacks to form the crystal—and there are a total of 230 possible space groups. Franklin had determined the exact dimensions of the unit cell and the crystal fell into the space group that was called monoclinic C2. When Watson and Crick learned of this, Crick immediately saw what Franklin had not yet tumbled to: since DNA belonged to the space group C2, DNA must look the same when turned upside down. (It is said to have a two-fold axis of symmetry.) This meant that there must be two helices and one helix must run up and the other down—they are anti-parallel, like up-and-down escalators.[18]

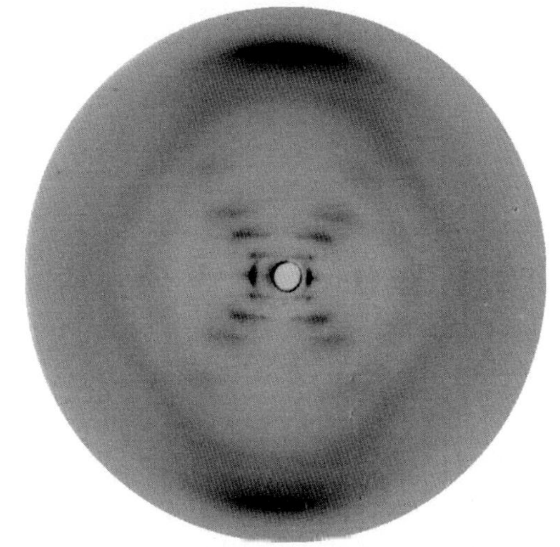

Figure 2.3 Franklin's X-ray diffraction image of DNA, *Photo 51*. *(Reproduced with permission from King's College London Archives.)*

The two helices form the backbone of DNA. Each helical chain is composed of many repetitions of a pentose sugar, deoxyribose, bound to a phosphate group—that is, phosphorus joined to four oxygens. These components are shown simply as blue bands running in opposite directions in Figure 2.4. Some six months before she took *Photo 51,* it was Franklin who corrected an erroneous model of Watson and Crick that had positioned the phosphates on the inside of the chains and the bases on the outside.[19] Though Franklin herself did not solve the mystery of the structure of DNA, the model building that Watson and Crick did was inspired by Franklin's diffraction evidence, her chemical reasoning that the phosphate groups had to be on the outside, and her determination that the space group was C2.

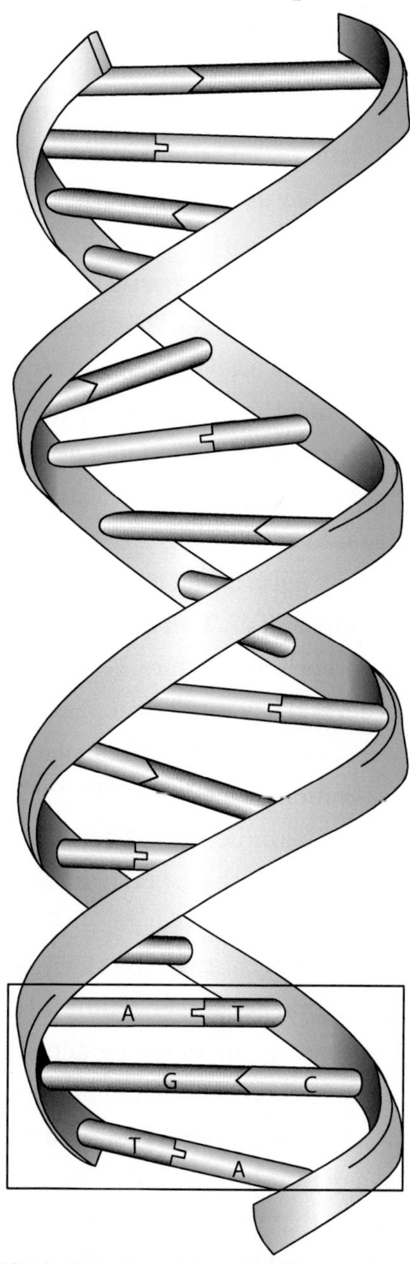

Figure 2.4. Complementary base pairing in DNA. The labels on the bottom three base pairs show A (adenine) pairing with T (thymine) while G (guanine) pairs with C (cytosine). These pairs resemble rungs on a ladder. The ladder twists around into a double-helical structure consistent with X-ray diffraction evidence. *(Courtesy of U.S. National Library of Medicine.)*

James Watson, Francis Crick, and Maurice Wilkins

Now we turn to Watson and Crick. They cracked the problem. They were able to take the available experimental evidence and make inductive leaps to determine its meaning. They were also adept at using physical models—cardboard cutouts in the shapes of the different bases (and precision-machined metal pieces for their final model in three-dimensions, Figure 2.5). Again and again Watson shuffled cardboard cutouts, concocting different combinations like pieces of a jigsaw puzzle. He tried many times without success. He was misled for a while because both the bases thymine (T) and guanine (G) can exist in more than one possible configuration of their atomic elements. (The carbon, nitrogen, hydrogen, and oxygen atoms can be connected differently.) Watson was using the wrong forms. By a lucky chance, a visiting American post-doc, physical chemist Jerry Donohue, suggested he try the alternative configurations of T and G.[20] Watson made new cardboard cutouts of the two bases. He suddenly found that now the combination of A joined by hydrogen bonds to T formed the same shape as the combination of C joined by hydrogen bonds to G: the two types of base pairs had the same overall size and shape. So when fitting these base pairs into a double-helical pattern, any order of A-T and C-G pairs could be accommodated while maintaining the regularity of the overall structure. The base pairs resemble horizontal rungs on a ladder (the sugar-phosphate backbones being the sidepieces of the ladder), and the ladder is twisted into a double-helical structure.

Base ratios Erwin Chargaff found that in DNA the amount of adenine (A) is equal to the amount of thymine (T), and the amount of guanine (G) is equal to the amount of cytosine (C). James Watson and Francis Crick later discovered that this indicated the specificity of base bonding in DNA. This became known as **base complementarity**: T bonds to A and G bonds to C in a pair of strands of DNA that form a double helix (a duplex).

In the Watson and Crick derivation of DNA's three-dimensional structure, the Chargaff base ratios came about naturally as a consequence of the model. At last, Chargaff's ratios had an explanation: complementary base pairing. Even more exciting was that complementarity suggested how DNA replicated itself. The base sequence of one chain automatically determined that of its partner. So if the double helix were split apart within a cell, each chain would serve as a template on which a new, complementary chain could be formed from free bases in the immediate proximity. Thus, base pairing—the real significance of Chargaff's analysis—held the key to how DNA passes information from generation to generation.

In their paper, Watson and Crick stated "Both chains follow right-handed helices. . . ."[21] If you imagine DNA as a screw, right-handed means that you need to twist your imaginary screwdriver clockwise to secure the DNA to an imaginary piece of wood.[22] How did they know DNA was right-handed? Their reasoning would lead us too far astray but is explained well elsewhere.[23] For now (to be revisited), let's assume that right-handedness is as accurate a descriptor as the base-pairing rules and the anti-parallel configuration of the two strands.

Figure 2.5 James Watson (left) and Francis Crick with their three-dimensional model of the DNA double helix. *(Reproduced with permission from Science Source Images.)*

Nine years after announcing their findings, Watson and Crick received the Nobel Prize in Physiology or Medicine. Since the prize is never awarded posthumously, Franklin was not eligible. But the prize can be shared by as many as three people. In this case the third person was Maurice Wilkins (whose autobiography is appropriately titled *The Third Man of the Double Helix*). Wilkins was a physicist who later turned to biology at King's College, London. He initiated the X-ray diffraction work on DNA at King's years before Franklin joined the group (Figure 2.6). For seven years after Franklin left DNA research and began to study the structure of viruses, Wilkins rigorously tested the Watson-Crick model. He built higher-resolution X-ray cameras and devised new analytic methods for refining the DNA model to fit the additional X-ray diffraction data.

Figure 2.6 Maurice Wilkins with X-ray crystallographic equipment. *(Reproduced with permission from King's College London Archives.)*

A small caveat. Because of its elegantly beautiful structure, the DNA double helix is idealized. However, the reality of DNA's physical existence is a different matter. It's rare that DNA looks this good. In fact, what you find instead in the cell nucleus appears as though it was a tangled mess.[24] If you took that mess—that compacted double helix—and stretched it out, the three billion or so base pairs (in the case of human DNA) would measure 1.8 meters. This strand, snipped into 46 chromosomes—a chromosome is a package of DNA combined with proteins—has to be packed into a cell nucleus that is only 6 microns across (a micron is one-millionth of a meter). Consequently, the DNA chains are far from the romanticized pictures of molecules with beautifully simple staircases of base pairs floating in an infinite solvent. They have a density comparable to that of a highly viscous gel. When touched by a glass rod, the gel can be pulled out into long strings. Maurice Wilkins remarked, "It's just like snot!" [25] On that note we turn to more decorous subjects.

The research described in Parts II and III of this book uses both DNA *sequencing* (determining the order of the nucleotides within a section of DNA) and DNA *synthesis* (constructing designer DNA segments). We will see a good example of the synergistic combination of the two in Chapter Eight when we examine DNA origami. In this nanofabrication technique, DNA folds into predesigned, intricate shapes. The researcher directs the folding using hundreds of small strands of synthesized DNA that are strategically combined with a single, long strand of naturally occurring DNA containing 7,000 bases. Thanks to base sequencing, the order of the 7,000 bases is known at every twist and turn. This sequence information dictates the synthesis of the small strands that are used to fold the long DNA into two-dimensional nanoconstructions of virtually any desired shape. Now it's time to learn how DNA sequencing is done, and then we'll turn to DNA synthesis.

DNA Sequencing

Early approaches to sequencing DNA were time consuming and labor intensive. In 1977 a sea change in sequencing was introduced by British chemist Frederick Sanger of the Medical Research Council Laboratory of Molecular Biology in Cambridge.[26] His technique is referred to by different names: chain-termination sequencing, dideoxy-sequencing, and Sanger sequencing. It is a splendidly simple method. Decades after its invention, Sanger sequencing was used—with technological improvements to achieve higher throughput—to sequence the human genome, our entire set of genetic material. For those of you keeping score, Sanger won his second Nobel Prize in Chemistry for this work, the only person to win two Nobel Prizes in Chemistry.

Primer A segment of single-stranded DNA consisting of a small number of nucleotides. A primer is designed so that its base sequence is complementary to the first portion of DNA that one wishes to sequence.

When Sanger created his approach to sequencing, molecular biology had already come a long way from the 1953 Watson/Crick/Wilkins/Franklin work. Sanger jump-started his method with knowledge and skills that had accumulated over the previous 25 years. One skill is to *amplify*—to make many copies of—the DNA segment to be sequenced. Once the DNA is amplified it is gently heated to break the hydrogen bonds so that the two strands separate. Then another skill from the molecular biology toolbox is employed. A short, synthetic *primer* strand is added to the solution of single-stranded DNA segments. The primer is a segment of single-stranded DNA consisting of a small number of nucleotides and is chosen so that its base sequence is complementary to the first portion of the target DNA, the DNA to be sequenced. Because of base complementarity between the primer and the first piece of target DNA, these two strands will bind to one another (panel (a) of Figure 2.7). The Sanger procedure is similar to the natural process of DNA replication, as we'll see in a moment.

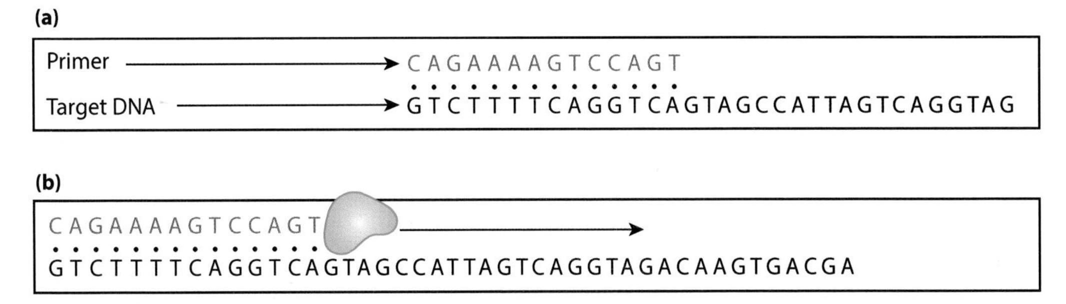

Figure 2.7(a) Primer DNA and target DNA. **(b)** Primer DNA and target DNA with polymerase ready to build a complementary strand.

The use of primers was an early example of the combined use of sequencing and synthesis of DNA. The short primer strands were synthesized DNA, and they initiated the sequencing of target DNA via the Sanger approach we are now discussing.[27] This is a case of second-strand synthesis. Later we'll look at synthesis that does not use an existing strand as a template.

Before proceeding with our primer-plus-target-DNA combination, we observe that Sanger sequencing is bio-inspired. That is, it's modeled after the natural process used by DNA to replicate itself. During DNA replication in nature, the two strands in the double helix separate, and this permits an enzyme called DNA polymerase to have access to each individual strand as shown in panel (c) of Figure 2.7. DNA polymerase plays the central role in the processes of life. This enzyme carries the responsibility of duplicating our genetic information. By the time the ingenious Dr. Sanger devised his sequencing method, DNA polymerase was a part of the molecular biology toolbox.

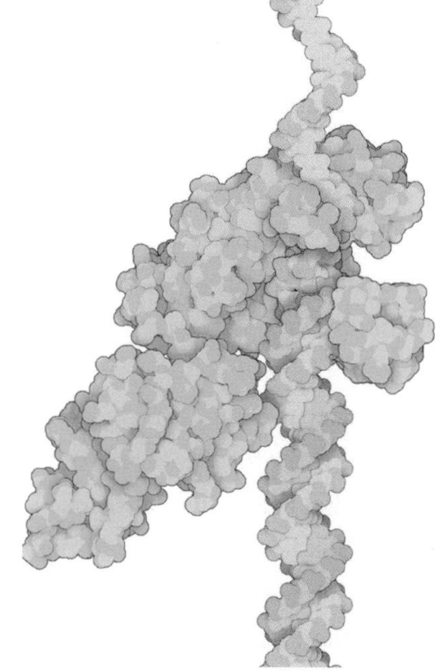

Figure 2.7(c) DNA polymerase moving along the length of a single strand of DNA. The polymerase (green) moves from bottom to top, forming a new double helix from a single strand of parent DNA. *(Part (c) adapted with permission from CC-BY-3.0 Dr. David S. Goodsell and the RCSB PDB.)*

Sanger exposed his primer-plus-target-DNA to a solution that contained DNA polymerase and all of the nucleotides—A, T, C, and G—that are required for synthesis of the strand that would be complementary to the target DNA. As illustrated in panel (b) of Figure 2.7, the polymerase was then ready, willing, and able to start adding free bases from the mixture to build a strand complementary to the target DNA.

Ah, but there was one very special ingredient that Sanger added as well. Nucleotides (in their unbound form) have the chemical name *deoxynucleotide triphosphates* and this is abbreviated as dNTP where N is a placeholder that can be any one of the familiar A, T, C, or G. But Sanger included special dummy nucleotides in his mixture as well as the genuine nucleotides. These special nucleotides are called ddNTPs where the *double-d* stands for *dide*oxynucleotide. They are chemically modified versions of the naturally occurring nucleotides and when incorporated into the growing strand they will inhibit further strand extension—they are *chain terminators.*

Take a specific example, say, the nucleotide T. Most of the time when DNA polymerase needs a T from the mixture to add to the growing DNA chain it will get a good one, that is, a dTTP (where we've specified that N = T in the generic dNTP). But sometimes the polymerase will instead grab a dideoxy-T (ddTTP), and when it does, the chain it's building can't be elongated any further and it simply breaks away from the polymerase. When the supply of ddTTPs in the mixture is exhausted, the result will be a large number of DNA strands of varying length (Figure 2.8). The reason for this is that the ddTTPs glom onto the strand at random. With a huge number of copies of the target (remember the amplification step), there will be strands that stop at every possible T along the way. These strands all have a terminal dTTP that indicates that an A (the base complementary to T) occurs in that position on the target strand. And every strand *starts* at exactly the same place because of the clever use of the primer segment.

Suppose that in addition to the mixture that contains ddTTP we make three other mixtures with each of the other ddNTPs (one with ddATP, one with ddCTP, and one with ddGTP). We will produce DNA chains of varying lengths in each test tube, and the chains will terminate with a T, A, C, or G depending on which dideoxynucleotide (ddNTP) we use. We are now on the doorstep of determining the sequence of the target DNA. What we need to do next is find out the sizes of all the terminated chains. To pull this off, we turn to the workhorse technique of polyacrylamide gel electrophoresis.

CAGAAAAGTCCAGTCCATCTGTTCACTGCT▲
CAGAAAAGTCCAGTCCATCTGTTCACT▲
CAGAAAAGTCCAGTCCATCTGTT▲
CAGAAAAGTCCAGTCCATCTGT▲
CAGAAAAGTCCAGTCCATCT▲
CAGAAAAGTCCAGTCCAT▲
CAGAAAAGTCCAGT▲

Figure 2.8 Oligonucleotide strands of varying length all terminated by dTTP.

Polyacrylamide Gel Electrophoresis (PAGE)

Polyacrylamide gel electrophoresis (PAGE) is a method used to analyze the size of DNA fragments (Figure 2.9). A polyacrylamide gel is made of a labyrinth of tunnels through a meshwork of polymer fibers. The DNA fragments are pipetted into small rectangles in the gel (called *wells*), and a vertically oriented electric field is placed across the gel. The DNA fragments (typically referred to as *oligonucleotides*) are negatively charged, and the smaller chains of DNA move down through the network of the gel faster than the larger chains. Multiple fragments of the same size will migrate in groups producing so-called bands in parallel lanes (the portion of gel beneath each well). After the bands have migrated they can be visualized by soaking the gel in a dye (ethidium

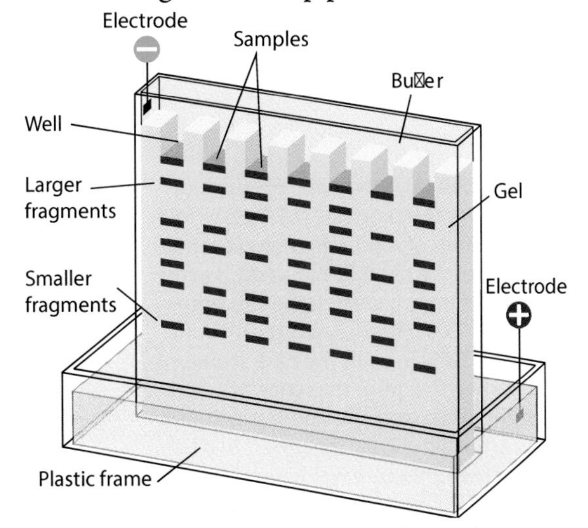

Figure 2.9 Polyacrylamide gel electrophoresis (PAGE) apparatus. *(Image courtesy of Bio-Rad Laboratories, Inc., © 2013.)*

bromide) that makes the DNA fluoresce under ultraviolet light.

Now to finish our Sanger sequencing experiment. We load the varying lengths of DNA fragments from each of the four mixtures into four different wells in the gel. We label these wells G, A, T, and C with the understanding that these are the positions of the bases in the target DNA. For example, the lane labeled as G will have all the fragments terminated by a ddCTP (because G and C are complementary bases), the lane labeled A will have all fragments that end with the chain terminator ddTTP (A and T being complementary), and so on. Figure 2.10 shows an actual image of the bands in four lanes of a PAGE experiment at the far right and a schematic of this on the left.[28] The target DNA sequence is shown on the far left as the vertical column of Ts, As, Cs, and Gs as read from the bands in the gel lanes. You can have the satisfaction of verifying the base assignments for yourself—use the TIPS in the caption to Figure 2.10.

PAGE (Polyacrylamide gel electrophoresis) A method to separate and/or identify biological macromolecules, typically proteins or nucleic acids. Molecules are separated on the basis of their electrophoretic mobility—that is, their rate of migration through a porous material under the influence of an applied electric field.

Oligonucleotides Short, single-stranded DNA or RNA molecules, i.e., short nucleic acid polymers.

We wrap up our conversation on sequencing by dipping into a review article by Sanger. This longish quote is too tempting to resist. "It is of course very exciting and gratifying to read of the method now being used . . . in the understanding of some of the fundamental problems of life. But I think I have derived even more pleasure from the development of the work—seeing the method gradually improving until we were able to read a sequence straight from an autoradiograph. Before this reading became automated, people complained that it was a tedious process, *but to me*

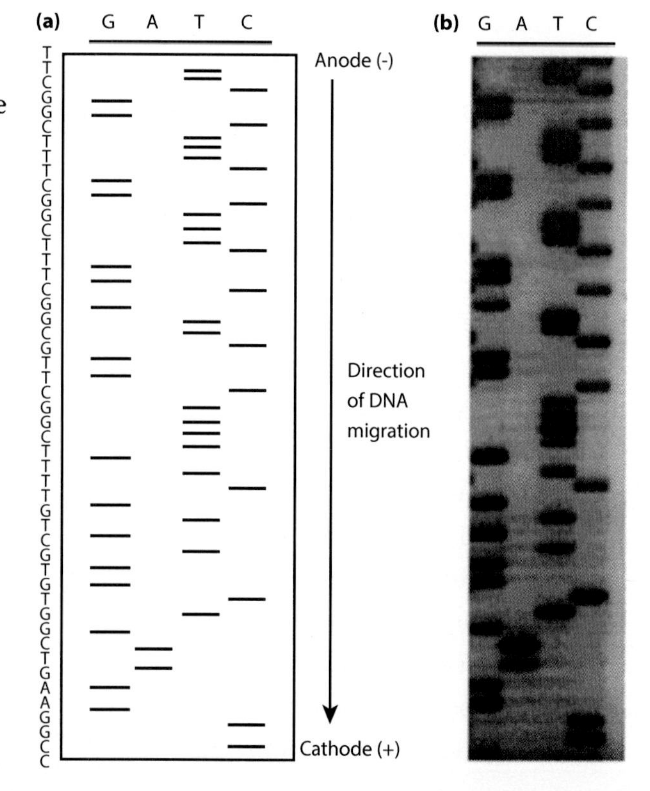

Figure 2.10 (a) Schematic of bands of Sanger reaction products. **(b)** PAGE of four vertical lanes of a gel with multiple bands for each of the bases G, A, T, and C. On the left side of the schematic is the sequence of the target DNA shown as a vertical column of bases. **TIPS**: To verify the base sequence, try reading the gel or the schematic from bottom to top. Note the lane in which each horizontal bar occurs: the lowest two bars are in lane C, next are two bars in lane G, then two in lane A, and so on. *(Adapted; permission from "DNA Sequencing" by Susan H. Hardin, Wiley Online Library, Copyright © 2001 John Wiley & Sons, Ltd. All rights reserved.)*

it was always a delight, having in the back of my mind the way we used to do sequences one residue [base] at a time by painstaking partial hydrolysis, fractionation, and analysis. *At one stage in the work I would take the autoradiographs home with me and look forward to the pleasure of reading them in the peace of the evening*"[29] [italics added]. That is someone whose job is a passion, not only a profession (Figure 2.11).

Figure 2.11 Frederick Sanger reading a DNA base sequence from an autoradiograph. *(Reproduced with permission from artist Borin Van Loon and from DNA: A Graphic Guide to the Molecule That Shook the World, by Israel Rosenfield, Edward Ziff, and Borin Van Loon. Copyright © 2011 Columbia University Press. Reprinted with permission of the publisher.)*

DNA Synthesis

Now we consider DNA synthesis. *Oligonucleotide* (i.e., DNA) *synthesis—oligo* is from the ancient Greek meaning "few"—requires lots of organic chemistry. We will pull the camera back to take a broader view of the *processes* involved and not the detailed chemical gymnastics. Initial chemical synthesis of nucleotides in the mid-1950s was done in a liquid state, solution-phase synthesis. Researchers later determined that binding the molecules to a solid support—solid-phase synthesis—made it easier to remove excess reactant or by-products. Issues of a technical nature limited the method and hampered its acceptance.

Then in the early 1980s, chemist Marvin H. Caruthers of the University of Colorado solved the problem with a technique called solid-phase phosphoramidite chemistry (Figure 2.12).[30] Phosphoramidites are chemically modified nucleosides and are used as the building blocks for the oligonucleotide synthesis. (A nucleo*side* is a nucleo*tide* without the phosphate group.)

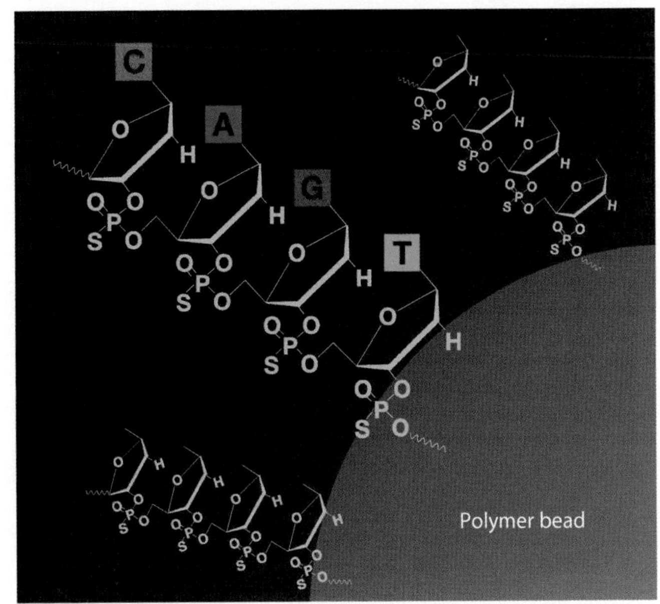

Figure 2.12 Oligonucleotide solid-phase synthesis on the surface of a polymer bead. *(Reproduced with permission from "Nucleic Acid Synthesis Model," Nitto Denko Corporation ("Nitto"), Japan.)*

The general synthetic strategy involves adding mononucleotides sequentially to nucleosides that are strongly attached via a covalent linkage to a polymer support. Caruthers collaborated with Leroy Hood of the California Institute of Technology to develop an *automated phosphoramidite DNA synthesizer*—popularly known as a gene machine—and reliable, automated oligonucleotide synthesis was born.

The synthesizer is a computer-controlled machine that delivers reagents and nucleotides and performs key chemical steps (follow these steps, top to bottom, in the central column of boxes in the flowchart, Figure 2.13).

To begin, the gene machine attaches the first nucleotide to the nucleoside that is attached to the solid support (topmost box, Figure 2.13). It's a fact of life that unmodified nucleotides contain reactive chemical groups (known as *trityl groups*, shorthand for triphenylmethyls) that can interfere with the growing nucleotide chain. Thus, the gene machine deprotects the nucleotide by removing that bothersome trityl group, a procedure

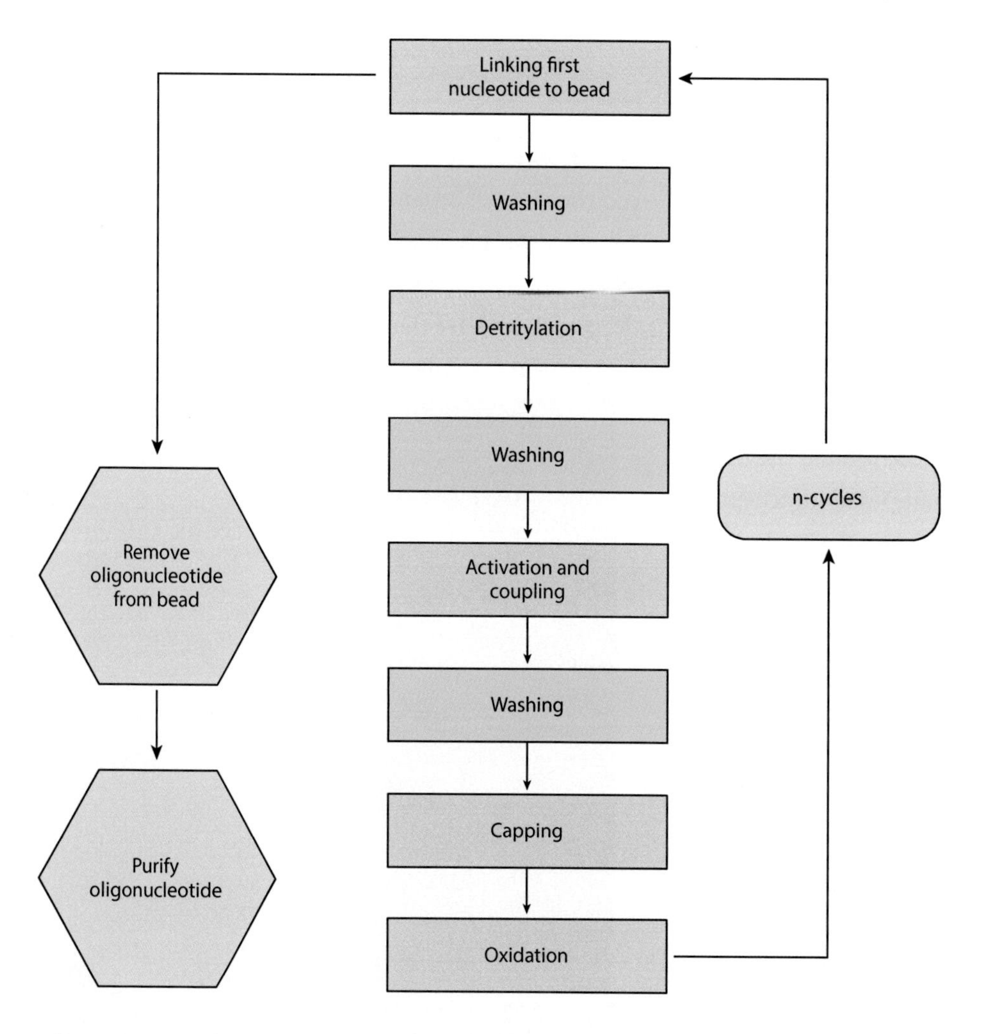

Figure 2.13 Synthetic cycle for preparation of oligonucleotides by phosphoramidite chemistry.

called *detritylation* (third box from the top in the flowchart). Then the machine snatches another nucleotide from solution and *activates and couples* it to the first one. Activation refers to the process of preparing a nucleotide for a subsequent reaction. (Conceptually, activation is the opposite of protection.) Some of the deprotected nucleotides evade activation and coupling. The machine prevents them from participating in further chain elongation by a trick called *capping* (seventh box from the top in the flowchart), rendering these renegade nucleotides inert to subsequent reactions. But after the activation and coupling operation, one of the desirable intermediate chemical groups is left in an unstable condition and prone to cleavage. So the machine performs an *oxidation* step to stabilize the bond between the first base and the second base. These maneuvers are repeated numerous times to form the desired nucleotide chain. Seeman, Bellini, and Clark routinely use the gene machine to create designer oligonucleotides.

In this chapter we've had a look at diverse subjects pertaining to DNA. The accomplishments of the scientists are compelling, and the properties of the DNA molecule are amazing. Since we've just finished sketching sequencing and synthesis as performed *in vitro*, we close with a few words in praise of nature and its astonishing abilities *in vivo*. Although the information we provide is from a few years back, the numbers cited are still pertinent and suggest that scientists have further to go before matching the marvels of the natural world. Reflect on the following. In living beings, the biological apparatus composed of enzymes such as polymerase can manufacture and repair DNA molecules at speeds of up to 500 bases a second, with error rates of about one base in a billion. That is a trillion-fold performance improvement in yield throughput (defined as output divided by error rate) over the best DNA synthesis machines, which add a base every 300 seconds.[31]

Commenting on the performance of DNA polymerase, molecular biologist David Goodsell put it this way: imagine reading a thousand novels and finding only one mistake.[32]

Exercises for Chapter Two: Exercise 2.1

We've considered a DNA polymerase and we've also mentioned RNA. Well, there is an *RNA polymerase* that's responsible for copying a DNA sequence into an RNA sequence. This process is known as *transcription*—the information stored in a molecule of DNA is copied (i.e., transcribed) into a new molecule called *messenger RNA*.

Suppose you fasten an RNA polymerase molecule to a glass slide and you've allowed it to initiate transcription on a template DNA strand *and* you've tethered the DNA to a magnetic bead as shown in Figure 2.14. The magnetization is in the vertical direction and does not prevent bead rotation. The bead is chemically decorated with smaller fluorescent beads that serve as markers of rotation. The rotations are observed with a technique called *epifluorescence optical microscopy*. (This neat experiment was actually done by Yoshie Harada et al. and reported in 2001.)[1]

If the DNA with its attached magnetic bead moves relative to the RNA polymerase as shown in Figure 2.14, in which direction will the magnetic bead rotate, clockwise or counterclockwise from the perspective of the magnet? Hint: Recall that DNA is a right-handed molecule.

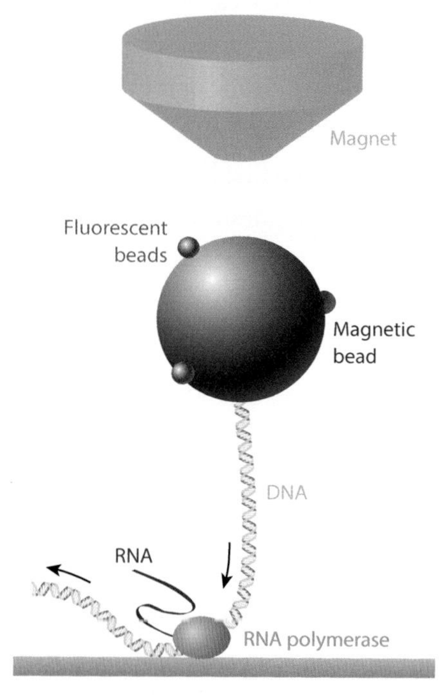

Figure 2.14 Observation of DNA rotation by RNA polymerase. *(Adapted with permission from Yoshie Harada, Osamu Ohara, Akira Takatsuki, Hiroyasu Itoh, Nobuo Shimamoto, and Kazuhiko Kinosita, Jr., "Direct Observation of DNA Rotation During Transcription by Escherichia coli RNA Polymerase," Nature 409 (4 January 2001), 113–115.)*

[1] Yoshie Harada, Osamu Ohara, Akira Takatsuki, Hiroyasu Itoh, Nobuo Shimamoto, and Kazuhiko Kinosita, Jr., "Direct Observation of DNA Rotation During Transcription by *Escherichia coli* RNA Polymerase," *Nature 409* (4 January 2001), 113–115.

Exercises for Chapter Two: Exercise 2.2

Forensic DNA profiling (also called DNA fingerprinting) is often in the news. It's used to identify perpetrators of crime and to exonerate those who may be wrongfully accused. Figure 2.15 shows an autoradiogram of various DNA samples in a rape investigation. The samples were loaded into the lanes as follows:

Lane 1—known blood sample from victim

Lane 2—known blood sample from defendant

Lane 3—DNA size markers, known as a DNA ladder (a set of DNA fragments with different sizes [in base pairs] that serve as standards)

Lane 4—female fraction from vaginal swab of victim

Lane 5—male fraction from vaginal swab of victim

Suppose you are the analyst. Your conclusion is the following:

A. The suspect is guilty.

B. The suspect might be guilty, but more probes should be used. (A probe is a sequence of single-stranded DNA that has been tagged with radioactivity so that the probe can be detected.)

C. The vaginal swab is from the wrong victim.

D. The suspect is excluded as a source of DNA in the evidence.

E. None of these.

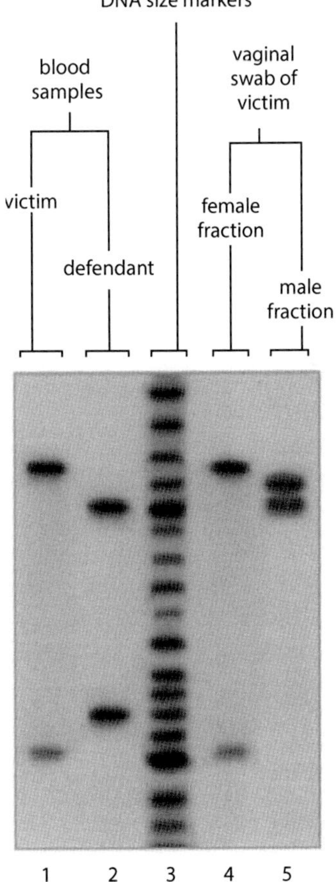

Figure 2.15 Forensic DNA profiling (DNA fingerprinting). *(Adapted with permission from The Biology Project, University of Arizona, Dr. Roger L. Miesfeld.)*

ENDNOTES

[1] René J. Dubos, *The Professor, The Institute, and DNA* (New York: The Rockefeller University Press, 1976).

[2] Oswald T. Avery, Colin M. MacLeod, and Maclyn McCarty, "Studies on the Chemical Nature of the Substance Inducing Transformation of Pneumococcal Types," *Journal of Experimental Medicine 79* (1944), 137–157.

[3] Ralph M. Steinman and Carol L. Moberg (For the Editors), "A Triple Tribute to the Experiment That Transformed Biology," *Journal of Experimental Medicine 179* (1994), 379–384.

[4] Dahm, R. "Discovering DNA: Friedrich Miescher and the Early Years of Nucleic Acid Research," *Human Genetics 122* (2008), 565–581.

[5] "Desoxy- became deoxy in the late 1950s or early 1960s. The decision to drop the (s) was made by an international nomenclature committee." Source: Dubos, *The Professor*, 142.

[6] Erwin Chargaff, *Heraclitean Fire: Sketches From a Life Before Nature* (New York: The Rockefeller University Press, 1978), 83.

[7] William Shakespeare, *Coriolanus*, Act III, Scene 2, Line 2253.

[8] Erwin Chargaff, "Chemical Specificity of Nucleic Acids and Mechanism of Their Enzymatic Degradation," *Experientia VI* (1950), 201–209.

[9] Ibid., 206.

[10] Charles Bond, "Layman's Guide," 2005, http://stein.bioch.dundee.ac.uk/~charlie/index.php?section=1

[11] The Rosalind Franklin Papers, "Laboratory Notebooks: Lab Notes on Possible DNA Structure (January–July 1953)," http://profiles.nlm.nih.gov/ps/retrieve/Collection/CID/KR

[12] Rosalind E. Franklin and R.G. Gosling, "The Structure of Sodium Thymonucleate Fibres. I. The Influence of Water Content," *Acta Crystallographica 6* (1953), 673–677.

[13] J.D. Watson and F.H.C. Crick, "Molecular Structure for Deoxyribose Nucleic Acid," *Nature 171, No. 4356* (25 April 1953), 737–738.

[14] Brenda Maddox, *Rosalind Franklin: The Dark Lady of DNA* (New York: HarperCollins Publishers, 2002), 158–159.

[15] Brenda Maddox, "The Double Helix and the 'Wronged Heroine'," *Nature 421* (23 January 2003), 407–408.

[16] James D. Watson, *The Double Helix: A Personal Account of the Discovery of the Structure of DNA*, (New York: W.W. Norton & Company, 1980), 98.

[17] Photo 51, NOVA, http://www.pbs.org/wgbh/nova/body/DNA-photograph.html

[18] J.E. Lydon, "The DNA Double Helix—The Untold Story," *Liquid Crystals Today 12, No. 2*, 1–9.

[19] Watson, "The Double Helix," 59–60.

[20] Ibid., 110–114.

[21] Watson and Crick, "Molecular Structure," 737.

[22] http://dwb4.unl.edu/Chem/CHEM869N/CHEM869NLinks/students.washington.edu/bafox/hubio514/handed.html

[23] Henry Rzepa, "The Handedness of DNA: An Unheralded Connection," http://www.ch.imperial.ac.uk/rzepa/blog/?p=3235

[24] Philip Ball, "Portrait of a Molecule," *Nature 421* (2003), 421–422.

[25] Maddox, *Rosalind Franklin*, 156.

[26] F. Sanger, S. Nicklen, and A.R. Coulson, "DNA Sequencing With Chain-Terminating Inhibitors," *Proc. Natl. Acad. Sci. 74* (1977), 5463–5467.

[27] Marvin H. Caruthers, "Gene Synthesis Machines: DNA Chemistry and Its Uses," *Science 230* (1985), 284.

[28] Hardin, Susan H., "DNA Sequencing," http://onlinelibrary.wiley.com/doi/10.1038/npg.els.003147/full

N.B. At the website referenced above, a base has been omitted in the sequence of bases to the immediate left of the schematic. As shown in the schematic of Figure 2.10, the sixth base down from the top is a "C." This "C" was omitted in the listing of bases in the online paper and the error has been corrected in Figure 2.10.

[29] Frederick Sanger, "Sequences, Sequences, and Sequences," *Ann. Rev. Biochem., 57* (1988), 1–28.

[30] Marvin H. Caruthers, "Gene Synthesis Machines: DNA Chemistry and Its Uses," *Science 230* (18 October 1985), 281–285.

[31] David Baker, George Church, Jim Collins, Drew Endy, Joseph Jacobson, Jay Keasling, Paul Modrich, Christina Smolke, and Ron Weiss, "Engineering Life: Building a FAB [Fabrication Technology] for Biology," *Scientific American 294, No. 6* (June 2006), 46.

[32] http://www.rcsb.org/pdb/101/motm.do?momID=3

CHAPTER THREE

Travels to the Nanoworld

As the decade of the 1950s draws to a close, over eight hundred scientists descend on the three-day *Winter Meeting in the West* of the American Physical Society at the California Institute of Technology (Caltech) in Pasadena.[1] Besides the 157 papers scheduled for presentation, researchers cluster in innumerable, intense discussions in the hallways of the meeting venues—Culbertson Hall, Bridge Laboratory, the Arms Laboratory, and Kerckhoff Laboratory—all on campus. The hotel situation is a mess due to the start of the Santa Anita Race Track season and the Rose Bowl Game. But now, on Tuesday evening, December 29, 1959, it's time for the much-anticipated banquet (and after-dinner speech) of the Society at the Huntington-Sheraton Hotel in Pasadena. The *Bulletin* of the American Physical Society had warned conference participants to get their tickets early: "We must remind members who plan to go to the banquet that they should purchase their tickets [$4.50 per person] as soon as possible, because recent dinners have been over-subscribed, and some members have been unable to attend who wanted to do so. This will be a problem all the more so, since our principal speaker will be Professor R. P. Feynman, of the California Institute of Technology, and the title of his talk will be 'There's Plenty of Room at the Bottom.' "[2] Indeed, the banquet room is packed. The dinner is over. Pungent pipe smoke mingles with that of cigarettes. An agreeable murmur floats over the scene like a benediction.[3] The ebullient, theatrical Richard Feynman, tall, slim, and dark haired, strides to the podium to elevate and entertain. The crowd is poised to be mesmerized. Showtime.

Attendee Paul Shlichta who was then at Caltech's Jet Propulsion Laboratory later said, "The general reaction was amusement. Most of the audience thought he was trying to be funny. . . . It simply took everybody completely by surprise."[4] Feynman presented his vision of the manipulation of individual atoms and molecules to achieve astonishing advances in microscopes, computers, mechanical and medical devices, and more. Caltech's magazine, *Engineering & Science*, printed the transcript of Feynman's talk and the general public read abbreviated accounts in *Saturday Review, Popular Science, Science News*, and *Life*.[5] Decades later, the 1959 talk has become a touchstone for everyone interested in nanotechnology. Thoughtful people wonder how influential it was in the genesis of the field. The answer appears to be: very little. Cultural anthropologist Chris Toumey reached this conclusion by a careful citation search of the scientific literature and personal interviews with people he called nanoluminaries—those who created formidable nano instruments and nano techniques.[6,7] Most likely, Feynman's talk has retroactive importance because it gives a magisterial founding myth to a young field and connects nanoresearch to the Feynman cachet. Toumey's observations aren't an evil anti-Feynman bias but rather an attempt to understand what catalyzed nanotechnology research.

So . . . where to begin our travels to the nanoworld? The nanoluminary Donald Eigler put it this way: "It was Binnig who blew life into nano by creating the machine that fired our imaginations. . . . When it comes to nano, start looking at Binning instead of Feynman."[8] The machine was the scanning tunneling microscope (STM) invented in 1981 by Gerd Binnig and Heinrich Rohrer and colleagues at the IBM Zurich Research Laboratory.[9] It set scientists buzzing like Rimsky-Korsakov's bumblebee on a spree. The STM will be our port of entry to the nanoworld.

The Scanning Tunneling Microscope (STM)

The STM is a tool that can visualize the surface of structures on an atomic level. The properties of the surface of a solid are often dramatically different from those of the interior of the solid. Neighboring atoms surround any particular atom within a solid, but an atom on the surface interacts only with other surface atoms, with atoms immediately beneath it, and with atoms above the surface. This is not a pedantic distinction. Surface atoms often arrange themselves differently from the other atoms in a solid, and the complexity of these reconstructions presents challenges to the microelectronics industry, for example.

A typical atom consisting of a nucleus and its surrounding electron cloud has a diameter of about 0.3 nm. The STM has a vertical resolution of about 0.001 nm and a lateral resolution comparable to the diameter of an atom. The STM is different from other microscopes in that it uses no free particles, so there is no need for lenses and special light or electron sources. It may seem perplexing at first, but the bound electrons existing in the sample under examination provide the only means of illumination.

This is because the STM puts quantum mechanics to use. The inventors of the instrument provided a nice analogy.[10] Imagine that the bound electrons are the water of a land-locked lake. Just as some of the lake water seeps into the surrounding land and forms groundwater, some of the electrons on the surface of the sample leak out and form an electron cloud—beyond the surface of the sample.

Classical physics tells us that there is no external electron cloud—the electrons are particles that are reflected by the boundary of the surface. But many of us have heard about the peculiar wave-particle duality phenomenon in quantum mechanics. All objects, both macroscopic and microscopic, have a wavelength as postulated by French physicist and Nobelist Louis de Broglie. However, the wave properties of matter are only observable for very small objects. If you calculate the de Broglie wavelength of a moving baseball you'll find that it's much too small to be observable. But this is not so for electrons. The electrons can behave both like particles and like waves. The electrons' wavelike behavior smears out their positions. This allows them to exist beyond the surface of the sample. Because the electrons seem to be digging tunnels beyond the surface boundary, we call this effect quantum-mechanical tunneling.

Wave-particle duality A principle in quantum mechanics that maintains that matter and light exhibit both wave-like and particle-like behavior.

de Broglie wavelength French physicist Louis de Broglie proposed that all matter displays wavelike behavior. The wavelength of an object is derived from an equation postulated by de Broglie.

To examine a surface with an STM, the experimentalist moves the tip of the instrument toward the sample. The electron cloud of the atom at the very end of the tip and the electron cloud of the atoms on the sample surface then overlap each other. The operator applies a voltage between the tip and the sample. This causes electrons to flow between the two clouds. The tip is swept across the sample in a series of horizontal lines (Figure 3.1). The variation in the tunneling current allows the STM to acquire a topographical likeness of the sample. After computer processing, a monitor displays data as atomic-scale, three-dimensional images. But peering into the nanoworld takes some getting used to. When Richard Feynman visited the IBM Thomas J. Watson Research Center in Yorktown Heights, New York, scientists showed him an STM and said they could see atoms. Feynman corrected them and said they were observing the tunneling of electrons.[11]

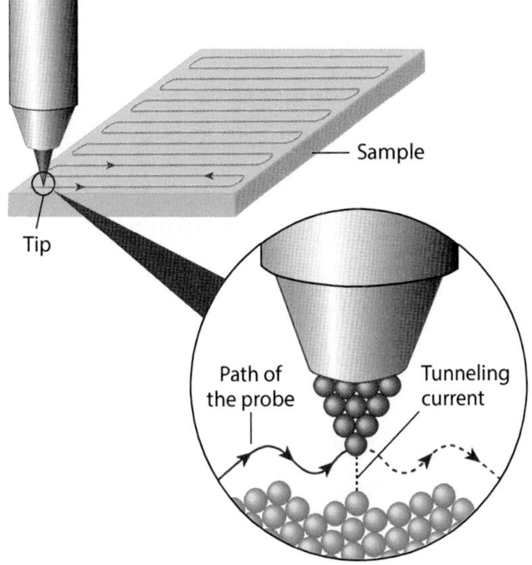

Figure 3.1 The principle of operation of a scanning tunneling microscope (STM). *(Adapted with permission from Nobelprize.org)*

Scanning Tunneling Microscope (STM)
An instrument for imaging surfaces at the atomic level based on the concept of quantum tunneling. A fine probe scans across the surface and the variation in the electron tunneling current between the sample and probe produces three-dimensional images of atomic topography and structure.

The STM is such an amazing instrument because its sensitivity to tunneling current changes exponentially with the gap distance between the tip and the sample. An estimate of this can be derived using quantum mechanics. The calculation actually gives the probability of an electron tunneling through the gap, *d*, and this is a dimensionless number. But it's not uncommon to refer to this dimensionless number as the tunneling current, *T*. The quantum mechanical approximation for the tunneling current is

$$T \sim \mathrm{e}^{-2\alpha d}.$$

In this expression, "e" is the fundamental mathematical constant that shows up in growth or decay rate situations. One unusual and creative way to explain the term α is: $\alpha = 1$ / (length of the exponential tail of the quantum mechanical wavefunction = how far the electron can tunnel).[12] The wavefunction referred to in this expression is the quantity that mathematically describes the wave characteristics of a particle.

The exquisite sensitivity of the STM is such that if the gap distance between the tip and the sample changes by just the diameter of a single atom, the tunneling current will change by almost a factor of 10. This is not as much of a conversation-starter as the butterfly effect in chaos theory, but notable in its own right.[13]

For those who are interested in the exercises at the end of this chapter, we'll go a little further. We observe that

$$\alpha = \frac{\sqrt{2m\Phi}}{\hbar},$$

where $\hbar$ is Planck's constant divided by 2π, and m is the mass of an electron. The term Φ is the tunneling barrier presented to an electron trying to tunnel through the gap, *d*, from the sample to the tip. Here the word *barrier* refers to energy flow. As a simple example, a pendulum at the very top of its arc is said to encounter a potential barrier. The pendulum must stop and reverse its direction of motion, thus turning some of its potential energy into kinetic energy, or energy of motion. A second detail, for those interested in the exercises, is the expression for the de Broglie wavelength, λ, of a particle. This is

$$\lambda = \frac{h}{p},$$

where h is Planck's constant (not divided by 2π), and p is the momentum of the particle.

STM tunneling current is also sensitive to the electronic structure of the surface and reveals atomic composition as well as topography. Each atomic element has its own unique electronic structure. The STM can be used to examine one element adsorbed onto the surface of another element, for example, cesium atoms on the surface of gallium arsenide. Figure 3.2 is a color-enhanced STM image of a 7 nm x 7 nm area of cesium atoms (red) on the surface of gallium arsenide (blue).[14] Gallium arsenide is a semiconductor compound and cesium is an elemental metal and that brings up an important limitation of the STM: it can only be used if both the sample and the tip are either conductors or semiconductors. STMs cannot image insulating materials. This shortcoming led to development of a family of scanning-probe microscopes. The atomic force microscope (AFM) is the most celebrated. It was invented in 1985 by Gerd Binnig, Calvin Quate, and Christoph Gerber.[15]

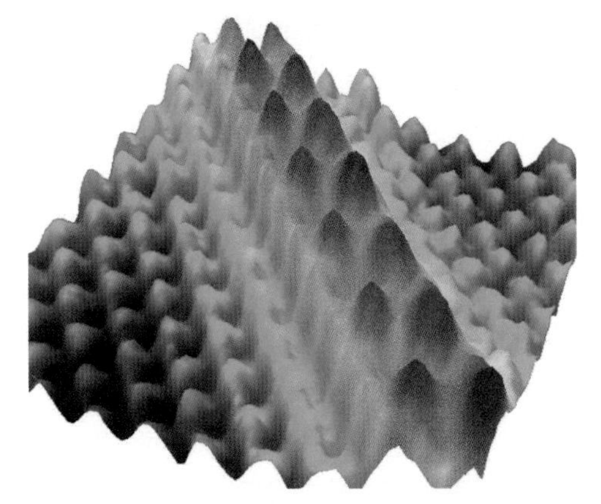

Figure 3.2 STM image of a zig-zag chain of cesium atoms (red) adsorbed onto the surface of gallium arsenide (blue). Total area shown is 7 nm x 7 nm. *(Reproduced with permission from "Geometric and Electronic Properties of Cs Structures on III-V (110) Surfaces: From 1-D and 2-D Insulators to 3-D Metals," L. J. Whitman, J. A. Stroscio, R. A. Dragoset, and R. J. Celotta, Physical Review Letters 66, (1991), 1338–1341.)*

The AFM is able to image insulators with sub-nanometer resolution and is a second major instrument that ushered in the age of nanoscience. Because the AFM is frequently used in the experiments of Seeman, Bellini, and Clark, we will postpone our look at AFMs until Chapter Five when we look at the whole toolbox these researchers employ.

Moving Atoms With an STM

But we're not yet done with the STM. Since we're looking back at the events that got the nanoworld into high gear, we must also turn the spotlight on the first manipulation of individual atoms using the STM. In 1989, nanoluminary Donald Eigler—whose name came up earlier in this chapter—used a tungsten STM tip to slide 35 xenon atoms (on a nickel surface) into precise positions to spell out the name of his employer, IBM (Figure 3.3).[16] The tip of an STM exerts a force on an adsorbate atom—called an adatom in surface science jargon. Eigler adjusted the position and voltage of the tip to fine-tune the magnitude and direction of this force. Then he pulled each adsorbed xenon atom across the surface while those atoms remained bound to the nickel underneath. Not a lot different than rearranging your living-room furniture but done on an atomic scale.

Adatom An atom adsorbed on a surface.

It's important to understand that Figures 3.2 and 3.3 are not computer-drawn art that is inferred from an experiment. We did see an example of the latter in the case of the DNA cubes in Figure 1.3 of Chapter One. However, Figures 3.2 and 3.3 of the present chapter are actual experimental data files. Just for fun and to achieve the most appealing presentation, the scientists who did this work used software to provide color enhancement

Figure 3.3 A patterned array of 35 xenon atoms on a nickel surface. Each individual atom was moved into place using an STM tip. *(Image originally created by IBM Corporation.)*

and an oblique perspective. (Discussions of this practice in science and technology occur in the literature.)[17]

In Figure 3.4 we see the raw images of xenon atoms on nickel without the added bells and whistles—and it still boggles the mind. A point to observe in panels (b)–(f) of Figure 3.4 is the exact periodicity of the xenon atom spacing. This is derived from the underlying nickel surface that has a rectangular unit cell and is oriented such that it measures 0.35 nm in the horizontal direction in the image and 0.25 nm in the vertical direction in the image. The xenon atoms are spaced on a rectangular grid that is four nickel unit cells long horizontally and five units cells long vertically, corresponding to 4 x 0.35 nm = 1.4 nm horizontally and 5 x 0.25 = 1.25 nm vertically. A letter is 4 x 1.25 nm = 5 nm from top to bottom. In Figure 3.4 take a look at the vertical portion of the letter "I" (just as an example) and you'll see four dark spaces between the five atoms. Each space measures 1.25 nm (but the crystalline structure of the underlying nickel surface is not resolved in Figure 3.4).

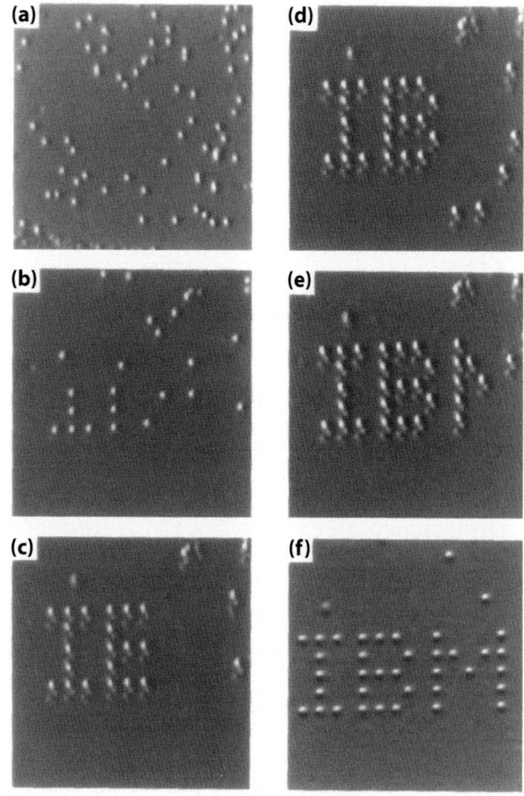

Figure 3.4 A sequence of STM images taken during the construction of a patterned array of xenon atoms on a nickel surface. The upper left panel **(a)** shows the surface after xenon dosing. Panels **(b)—(f)** show various stages during the construction. Each letter is 5 nm from top to bottom. *(Reproduced with permission from D.M. Eigler and E.K. Schweizer, "Positioning Single Atoms With a Scanning Tunneling Microscope," Nature 344 (5 April 1990), 524–526.)*

Standing Waves

Eigler produced another result in 1993 using his adatom sliding process that further fired the imagination of scientists throughout the world. Recall that when we learned about the principle of operation of the STM earlier in this chapter, we introduced the de Broglie wavelength. All objects have an associated de Broglie wavelength but wavelike behavior is usually not observable. However, when electrons are confined to lengthscales that approach their de Broglie wavelength, their behavior is dominated by quantum mechanical effects. Eigler and his colleagues used the tip of an STM to confine surface-state electrons to artificial structures at this lengthscale.[18]

First, let's look at an STM rendering of a single iron adatom on a copper surface, Figure 3.5. The concentric rings around the iron are *standing waves* resulting from the quantum-mechanical interference between incident and scattered *copper* surface-state electrons hitting the iron atom. When the incident and scattered waves combine they form a consistent pattern.

Figure 3.5 STM image of a single iron adatom on a copper surface. *(Reproduced with permission from M.F. Crommie, C.P. Lutz, and D.M. Eigler, "Confinement of Electrons to Quantum Corrals on a Metal Surface," Science 262 (8 October 1993), 218–220.)*

Standing waves are not unique to the quantum regime. They are widespread in classical physics and everyday life. An example of a standing wave in one dimension is a vibrating violin string. If you vibrate a violin string at the right frequencies and could look closely at the string you would see a standing wave pattern. Its name comes about because it doesn't appear to be traveling. The string simply oscillates up and down with a fixed pattern (Figure 3.6). The standing waves are produced at natural frequencies or resonant frequencies of the violin string. The lowest such frequency is called the *fundamental frequency* and for a simple string the higher resonant frequencies are called *harmonics.* (The fundamental frequency is the first harmonic.)

Standing waves Standing wave patterns are the result of the repeated interference of two waves of the same frequency that are moving in opposite directions. In our quantum-mechanical context, standing waves are patterns resulting from the interference between incident and reflected surface-state electrons hitting atoms adsorbed on the surface.

If you have an interest in the exercises at the end of the chapter, you'll need to know that the wavelengths of the standing waves have a simple relation to the length of the violin string

$$\lambda_n = \frac{2L}{n},$$

where L is the length of the violin string, λ_n are the wavelengths of the resonant frequencies, and $n = 1, 2, 3 \ldots$ If we want to find the frequency of each vibration, f_n, we use the relation

$$f_n = \frac{v}{\lambda_n},$$

where v is the wave velocity, the velocity of the standing wave. The wave velocity gives us the speed at which a fixed point on the wave propagates. Remember, the *pattern* formed by the incident and reflected waves—the standing wave—is what is stationary. It's an oscillating wave fixed in space. But the waves themselves have a velocity.

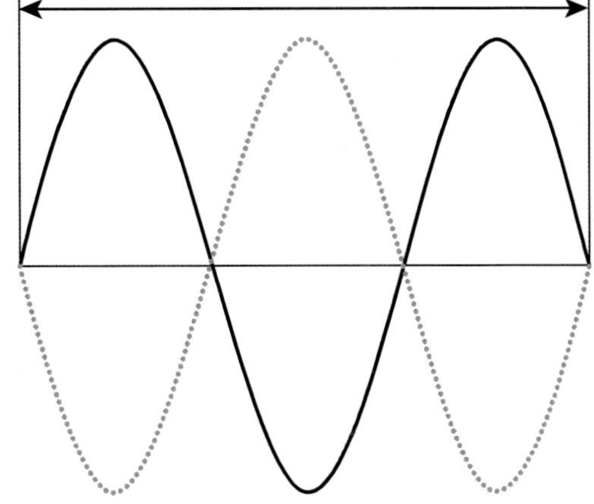

Figure 3.6 Example of a standing wave on a violin string.

Moreover, readers who are quite familiar with violins may know about standing waves in *two* dimensions. Physicist Ernst Chladni (famous, in part, for Chladni patterns) developed a technique in the late eighteenth century using two-dimensional standing waves to help in the design of violin bodies.[19] (More generally, the observation of standing waves probably goes back prior to recorded history. For instance, seiches are rhythmic oscillations of water in a lake or partially enclosed inlet.[20])

Quantum Corrals

Quantum Corrals An STM can be used to slide atoms across a surface and position them to form ringlike structures—quantum corrals—that enable the wavelike properties of electrons to be seen.

Returning to standing waves in the quantum realm, Eigler discovered a method for shaping the spatial distribution of surface-state electrons. He made quantum structures that possess a confinement property that is derived from the scattering of surface electron states that we saw in Figure 3.5. Eigler built what he called *quantum corrals* from iron adatoms that were individually slid into position on a copper surface. Figure 3.7 shows a 48-atom iron ring he constructed on the copper surface. The average diameter of the ring (from atom center to atom center) is 14.26 nm. The spacing between adjacent iron atoms varies from 0.88 nm to 1.02 nm. The striking feature in this image is the strong modulation of the distribution of surface electron states inside of the corral. Circular standing waves reveal the density distribution of electrons occupying quantum states of the corral. Let's elaborate on this last statement. de Broglie's wave nature of matter provided a base on which to build. Enter the Austrian physicist Erwin Schrödinger. He came up with an equation describing the motion of particles in terms of wave-like properties. (He also contributed the wave-function that we spoke about earlier in the context of tunneling between a surface and an STM tip.) In its full generality the equation is properly called the time-dependent *Schrödinger equation*. This is the quantum-mechanical analog of Newton's equation of motion in classical mechanics (Newton's Second Law) that describes the behavior of a particle in terms of its motion as a function of time.

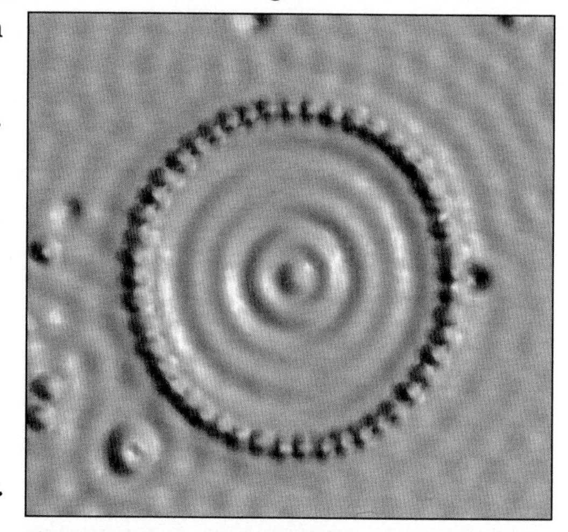

Figure 3.7 A quantum corral built of 48 iron adatoms on a copper surface. The adatoms were individually slid into position by an STM tip. Outside the corral, several defects in the copper surface are apparent in this image. *(Reproduced with permission from M.F. Crommie, C.P. Lutz, and D.M. Eigler, "Confinement of Electrons to Quantum Corrals on a Metal Surface," Science 262 (8 October 1993), 218–220.)*

For many situations in quantum mechanics, the system of interest has a fixed, unchanging total energy. In these cases the general Schrödinger equation reduces to a simpler equation called the time-independent Schrödinger equation. Standing waves are the solutions (the *eigenfunctions*) of the time-independent Schrödinger equation and the corresponding allowed energies are the energy *eigenvalues*. The term *eigenvalue* is from the German *Eigenwert* that means proper or characteristic value and eigenfunction is from *Eigenfunktion* meaning proper or characteristic function. One often finds the terms *eigenfunction* and *eigenstate* used interchangeably.

Schrödinger equation Austrian physicist Erwin Schrödinger formulated an equation that describes the motion of particles in terms of wavelike properties. This is the quantum-mechanical analog of Newton's equation of motion in classical mechanics (Newton's Second Law) that describes the behavior of a particle in terms of its motion as a function of time.

The corral in Figure 3.7 contains a spatial image of the eigenstates of a quantum corral, specifically, a round box. It is the physical realization of the mathematical solution of the Schrödinger equation for surface-state electrons confined to a round box. As Eigler explained it, "We build a box for our electrons . . . and we see how the electrons solve the Schrödinger equation in that particular environment."[21]

Corrals having different geometries give different distributions of eigenstates, Figure 3.8.[22] But now for a surprise: the bottom panel of the figure shows a *double-walled* quantum corral and yet the spatial distribution of eigenstates *within* the corral is virtually identical to that of the *single-walled* corral of Figure 3.7. The reason isn't instantly obvious. One might expect that if a single wall of iron atoms exhibits electron confinement, then a double wall should provide even more (by further narrowing of the spectral linewidths). The reason this is not the case can be found using multiple-scattering theory and is left to the very ambitious reader.[23]

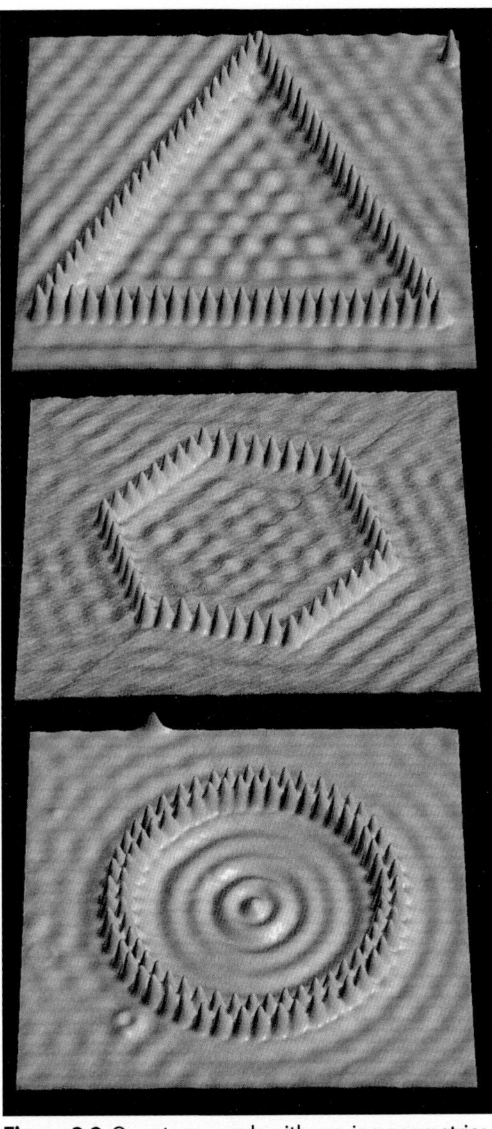

Figure 3.8 Quantum corrals with varying geometries. Observe the different distributions of the eigenstates of the surface electrons contained in the corrals. *(Image originally created by IBM Corporation.)*

Before closing the subject of quantum corrals, we'll present one last, celebrated image. The data of the STM image shown in Figure 3.7 can be rendered in false color and perspective to produce the stunning image shown in Figure 3.9.

Eigenvalues and eigenfunctions In quantum mechanics, if the system of interest has a fixed, unchanging total energy, then the general Schrödinger equation reduces to the *time-independent* Schrödinger equation. Standing waves are the solutions—the *eigenfunctions*—of the *time-independent* Schrödinger equation, and the corresponding allowed energies are the energy *eigenvalues.*

Figure 3.9 Quantum corral of 48 iron atoms on a copper surface, shown with false color and perspective. *(Image originally created by IBM Corporation.)*

Nanomethodology

Our travels to the nanoworld have featured the innovation of instrumentation (the STM) and investigations using this tool. Now we will turn to nanomethodology. Because we are considering the work of Seeman, Clark, and Bellini in this book, our focus will be on bottom-up fabrication techniques using the intrinsic properties of atoms or molecules to direct their self-organization into nanostructures. A superb chemist using such an approach is Chad Mirkin of Northwestern University.

Recall that in Chapter One we saw serendipity in the Clark laboratory when experiments had unexpected and spectacular results. Something along these lines happened in Mirkin's lab in 1996. Mirkin had asked two of his students to synthesize some DNA having an alkylthiol terminal group (a sulfur-containing group that sticks to gold) to form *thiolated* DNA. They wanted to make gold (chemical symbol Au) nanoparticles with DNA attached to them. The idea they were pursuing was to build materials using DNA as a construction tool (shades of Seeman, but actually a rather different approach). Now we'll draw on the knowledge gained about DNA in Chapter Two to dive right into Mirkin's work. His students synthesized the DNA capped with thiols and found the chemistry needed to immobilize it on gold nanoparticles. Then they prepared a second batch of gold particles with a non-complementary sequence of oligonucleotide bases. Last, they prepared a third oligonucleotide strand with a base sequence that would bring the two different gold particles together through hybridization—that is, complementary bases of the third strand pair with portions of each of the DNA sequences tethered to the two batches of gold particles. (This is the same *sticky ends* trick that we saw Seeman employ.) They put everything together—and were in for a huge surprise.

As Mirkin said, "a particle-assembly event took place. I'll never forget this. It was around 11 o'clock at night. Mucic [one of the students] came down to my office and said, 'You have to take a look at this. When I mix them together, the solution turns blue.

If I put it in the oven, it turns red.' This happened because the dispersed particles are red and the assembled particles are blue. What we were watching with the naked eye was, in effect, DNA raveling and forming the double helix and unraveling again when it was heated. At high temperature, the particles dispersed and were red again. Almost immediately, I remarked that this was a way of detecting DNA, a really simple way of doing so."[24] They had found a way of assembling gold nanoparticles into macroscopic materials using DNA. Chemical-specific Velcro was what Mirkin called this use of DNA as a construction material. The group had invented the enabling technology for a whole new class of medical diagnostics and possibly even therapeutics. Their initial results were published in 1996. In time, this led to the development of FDA-cleared molecular diagnostics systems for infectious diseases based on nanoparticle probes.[25] Their diagnostic tests (based on color change) are now used for diseases from cystic fibrosis to flu to detecting people who have a genetic predisposition to blood clotting. Better yet, all procedures are done at the point-of-care. Instead of sending samples to remote labs, the analysis is done right in the hospital and gets the information to the doctor in a hurry.

Spherical Nucleic Acids (SNAs)

Mirkin has had a huge influence in the nanoworld. We will focus on one of his most important contributions: structures called spherical nucleic acid (SNA) conjugates. An SNA is a three-dimensional conjugate (i.e., coupled object) consisting of functionalized and highly oriented nucleic acids covalently attached to the surfaces of nanostructures. When we use the verb functionalize, we mean the addition of functional groups—that is, an atom (or groups of atoms) that has similar chemical properties whenever it occurs in different compounds. And a covalent bond is a strong chemical bond in which one or more pairs of electrons are shared by two atoms.

Mirkin's lab has evolved and matured the unique nanostructures that had their roots in those dramatic late-night experiments. First we'll have a look at the anatomy of SNA nanostructures (Figure 3.10).[26] These materials often have inorganic (i.e., non-carbon-containing) cores such as gold, silver, iron oxide, etc. For simplicity, we'll assume that gold is the chosen nanoparticle (NP) core. However, the properties that make these structures so special stem in large part from the density and orientation of the oligonucleotides at the outer region of the nanostructure. SNAs can be prepared from both single- and double-stranded (duplex) nucleic acids and the nucleic acids can be either DNA or RNA.

In contrast to the work of Nadrian Seeman, SNA nanostructures can be synthesized independent of nucleic acid sequence and hybridization (base-pairing) interactions. SNAs are formed via chemical bonds and do not rely on the base-pairing of nucleotide

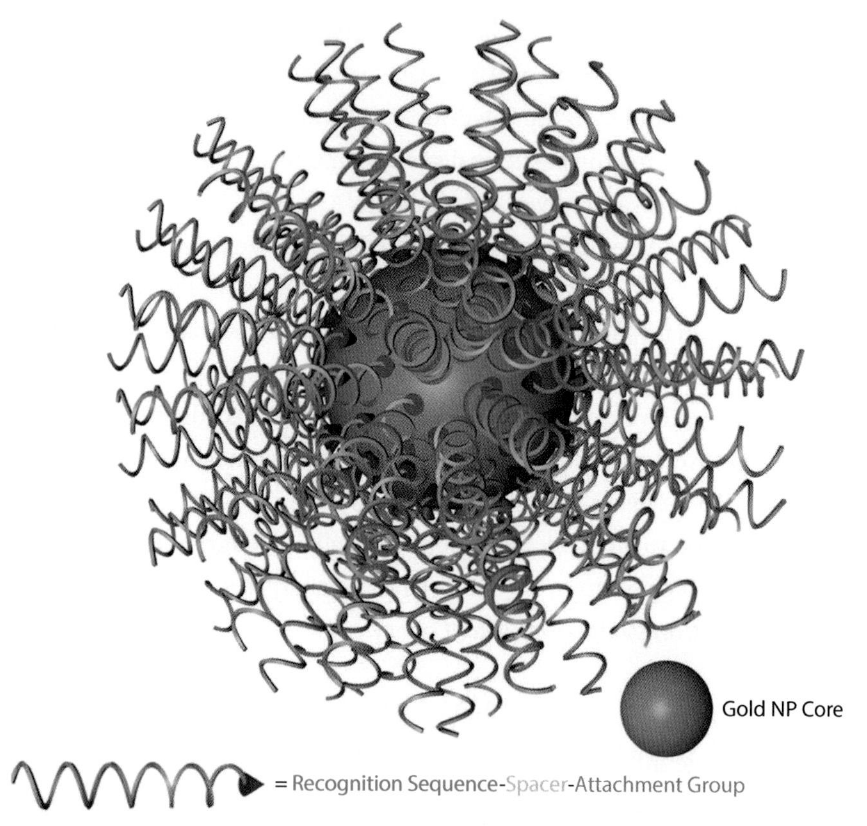

Figure 3.10 The anatomy of SNA nanostructures. An inorganic core is functionalized with oligonucleotides containing three segments: a recognition sequence (outermost), a spacer segment (middle), and a chemical-attachment group (innermost). Additionally, other functional groups such as drugs can be attached along any segment of the oligonucleotide. *(Reproduced with permission from Joshua I. Cutler, Evelyn Auyeung, and Chad A. Mirkin, "Spherical Nucleic Acids," Journal of the American Chemical Society 134 (2012), 1376–1391.)*

strands used in Seeman's constructions. However, a wise choice of base sequences is certainly relevant to SNA design. Specifically, the oligonucleotide portion of an SNA (shorthand for SNA-gold NP conjugate) consists of three main components: an attachment group, a spacer region, and a programmable recognition region. The attachment group is often some form of thiol, as mentioned earlier, a sulfur-containing group that sticks to gold. One chooses the attachment group to yield very high oligonucleotide densities on the surface of the gold NPs. The second segment of the oligonucleotide sequence, the spacer group, pushes the recognition region away from the gold NP surface. Last, the recognition portion of the oligonucleotide strand is tailored for each investigation or technological use and is generally the active segment that is available for additional base pairing with other strands of interest—for example, target strands in biodiagnostic detection assays. Think back to the phosphoramidite chemistry we looked at in Chapter Two when discussing DNA synthesis. The recognition portion of the SNA can be composed of any sequence that can be incorporated via phosphoramidite chemistry, and this is also true of the attachment group.

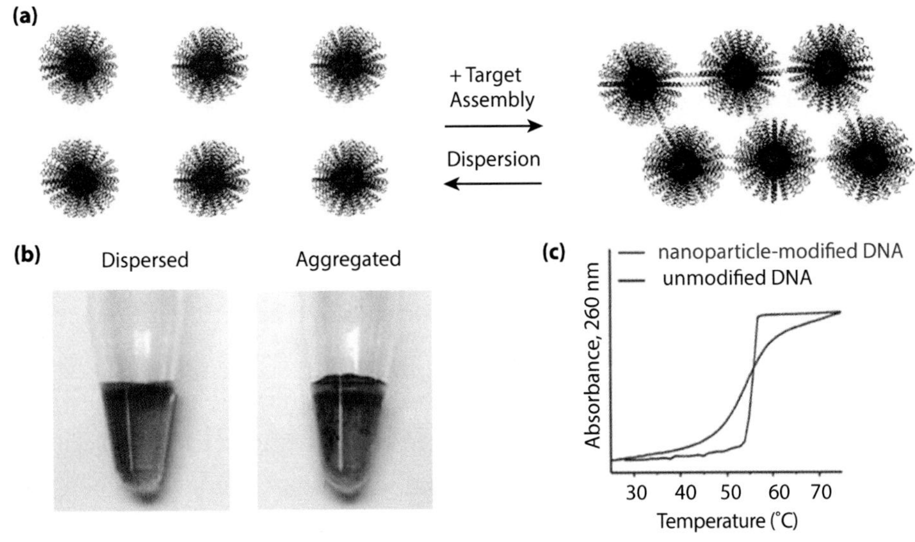

Figure 3.11 (a) The assembly and dispersion of SNA–Au NP conjugates and the corresponding shift in the wavelength (color) of light scattered from the Au cores. (Au is the chemical symbol for gold.) Dispersed particles are red, whereas aggregated particles are purple. Targets that induce assembly can be DNA, metal ions, or any molecule that the SNA shell has been programmed to recognize and bind. **(b)** Aggregation results in a visible red-to-purple color transition of the particles in solution. **(c)** Compared to duplexes of free-strand DNA that dissociate over a broad temperature range, the melting transitions of SNAs are sharp and occur over a very narrow temperature range. *(Reproduced with permission from Joshua I. Cutler, Evelyn Auyeung, and Chad A. Mirkin, "Spherical Nucleic Acids," Journal of the American Chemical Society 134 (2012), 1376–1391.)*

Biodiagnostic Detection Using SNAs

Figure 3.11 (a) and (b) show that SNAs can result in unusual optical properties that form the basis for a biodiagnostic detection scheme. When exposed to *target* DNA strands that are complementary to the DNA strands on the NPs, the NPs will assemble. They are held together by DNA linkage (base pairing). The color change from red to purple that accompanies aggregation is immensely important because it is an optical signature that can readily be detected spectroscopically, e.g., by measuring the absorbance of a particular wavelength of light—or even by eye. This color change provides a high sensitivity for the binding of complementary oligonucleotides.

The curious reader may ask: what controls the optical properties of SNA aggregates? The complete answer is rather long, but here are two salient points. The length of the oligonucleotides linking the NPs together is one determining factor. That is, the diagnostic method relies on the distance-dependent

Hybridization The process of joining two complementary strands of nucleic acids, e.g., two single strands of DNA, to form a double-stranded molecule (a duplex). In DNA the process utilizes the base complementarity of A with T and G with C (and in RNA, the complementarity of A with U and G with C).

Spherical Nucleic Acids (SNAs) Three-dimensional conjugates (i.e., coupled objects) consisting of densely functionalized and highly oriented nucleic acids covalently attached to the surfaces of nanoparticles (NPs).

Functionalize (verb) The addition of functional groups, that is, an atom (or groups of atoms) that has similar chemical properties whenever it occurs in different compounds.

Covalent bond A strong chemical bond in which one or more pairs of electrons are shared by two atoms.

optical properties of gold particles. But in addition to the interparticle distance, there is a second factor that came as a surprise to Mirkin and colleagues. The aggregate size—i.e., the number of NPs per aggregate structure—is also a governing factor, regardless of oligonucleotide linker length.[27] Speaking more generally, the electronic properties of the gold NPs are responsible for the color shift. When the NPs are separate, the electrons in a single particle can move more or less independently. But as the NPs come close together, electrons moving around one particle induce movements in the electrons of neighboring particles. It is this choreographed movement that influences which wavelengths of light the material absorbs.[28]

So far we have a wonderful colorimetric mechanism that lets us determine if a specific target strand is present. But there's more. The investigators made a critical and unanticipated observation: oligonucleotides pair much more suddenly when they are densely attached to NPs than when they are simply floating freely in solution. Pairing (hybridization) is a function of temperature. And so is *un*pairing. Because DNA linkages hold the NP aggregates together—the cores themselves do not interact or fuse—the aggregated SNAs can be reversibly dispersed through *de*hybridization (breaking of the complementary base pairing) by heating. The dehybridization is also referred to as melting. Take a look at Figure 3.11(c). An aggregate formed from a target nucleic acid made up of a base sequence that is perfectly complementary to the recognition region sequence of the SNA has a very narrow melting transition (shown compared to the broad temperature range for unmodified, duplex DNA). One can differentiate such a target sequence from other target strands containing just a single base-pair mismatch, insertion, or deletion (compared to the sequence of the SNA); that is all it takes to perturb the melting behavior of the aggregate.[29] This exquisite selectivity has consequences in the clinic: single nucleotide polymorphisms have a significant function in many diseases.

As was true of the optical properties of SNA aggregates, the reader might want to know: what controls the melting properties of SNA aggregates? And again, the full answer would require a lengthy explanation. However, we will once more present two key factors: The presence of multiple DNA linkers between each pair of NPs is vital; there is a cooperative melting effect that originates from short-range DNA duplex-to-DNA duplex interactions. The second factor may seem to come out of left field: the local salt concentration—yes, that's right, the amount of NaCl in the solution (more correctly, the salt ions Na^+ and Cl^-). Mirkin's lab studied the melting behavior of SNA aggregates as a function of NaCl concentration. They found that electrostatic interactions between NPs could be tailored by adjusting the salt concentration, and this impacts the hybridization and dehybridization of SNA aggregates. The influence of salt on the melting of SNA aggregates derives from the fact that DNA is a highly charged molecule and the local salt concentration near DNA changes after the dissociation of DNA double helices into single strands.[30] The conclusion: NP linkage and salt concentration result in sharp melting transitions.

We step back for a moment and consider Mirkin's work in a broader context, specifically the effort to use DNA in a bottom-up approach to the construction of well-defined nanoscale materials. As we later examine in Nadrian Seeman's work, early attempts to use DNA to make nanostructures were limited by the lack of rigidity of the structures that were created. Coincidentally, both Mirkin and Seeman overcame this problem simultaneously with milestone results circa 1996 but through chemically and conceptually different pathways. As we've seen, Mirkin used a rigid non-nucleic acid-based nanoparticle core (e.g., gold) to act as a template to immobilize and organize functionalized DNA strands perpendicular to the surface of the core.[25,31] In the other approach, Seeman achieved a rigid architecture using multiple DNA strand "crossovers" and the hybridization, i.e., base pairing, that stabilizes the crossover events (Chapter Seven).[31,32]

Back to the diagnostic use of SNAs, Mirkin's group increased the sensitivity of the NP-based colorimetric detection scheme and developed a chip-based method called the scanometric assay (Figure 3.12). Capture of a desired target results in the immobilization of the gold NP probes onto a glass slide. The different colors on the chip show the functionalization of the slide. The colors indicate different capture strands that will pair with different targets. After the chip is exposed to the target strands, the chip is then exposed to a solution of SNAs. These, in turn, pair (hybridize) to the targets if they are present on the chip. In Figure 3.12, the capture strands on the red square of the chip have hybridized with target strands and then the SNAs have hybridized to those targets. (It isn't obvious from Figure 3.12 but the target strand has two base sequences, one that pairs with the capture strand and the other that pairs with the SNAs. Thus the target strand links the capture strand to the SNA. Scheme 1 in the end-note shows this three-component sandwich assay explicitly.)[33]

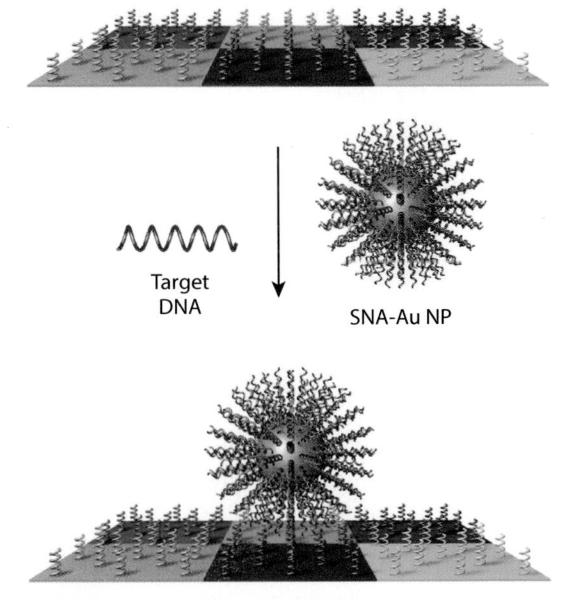

Figure 3.12 A scanometric detection assay. A chip is synthesized with capture strands for a number of different targets. The targets will hybridize to the appropriate spots if they are present. The chip is then exposed to a solution of SNA-gold NP probes, which will hybridize to the appropriate targets if they are on the chip. This assay enables detection of cancer markers in the target DNA. *(Reproduced with permission from Joshua I. Cutler, Evelyn Auyeung, and Chad A. Mirkin, "Spherical Nucleic Acids," Journal of the American Chemical Society 134 (2012), 1376–1391.)*

Figure 3.13 shows the scanometric detection of three protein cancer markers for eight different samples. The markers they used were human chorionic gonadotropin (hCG) that can test for testicular cancer, antibodies to prostate-specific antigen (PSA), and R-fetoprotein (AFP), a hepatic cancer marker. In Figure 3.13 the results for samples 1—8 are: (1) all targets present; (2) hCG and PSA; (3) hCG and AFP; (4) PSA and AFP; (5) AFP; (6) PSA; (7) hCG; (8) no targets present.[34]

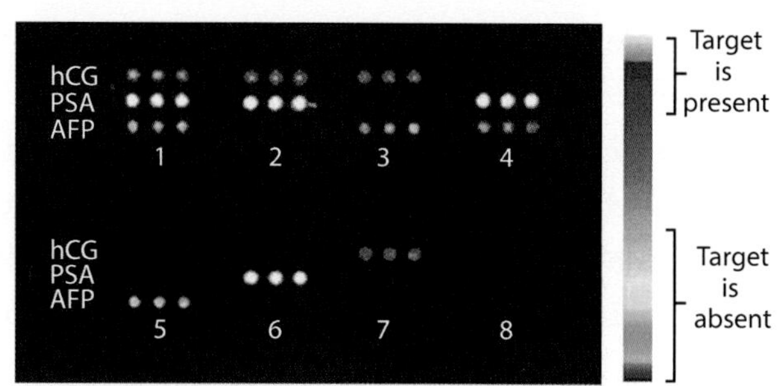

Figure 3.13 Read-out of a scanometric detection assay. If the target is present, the macroscopic structure can be detected via light scattering and a conventional optical flat-bed scanner. A bright spot indicates that the target is present, and the signal intensity permits quantification of target concentration. *(Reproduced with permission from Joshua I. Cutler, Evelyn Auyeung, and Chad A. Mirkin, "Spherical Nucleic Acids," Journal of the American Chemical Society 134 (2012), 1376–1391.)*

The diagnostic capabilities of SNAs are already in wide use. The therapeutic use of SNAs holds great promise. One reason is that they have an unusual ability to cross cell membranes. This is not a minor accomplishment. Nature has created a defense network for foreign nucleic acids. That has been a big problem for researchers who are trying to use synthetic oligonucleotides to treat genetic-based diseases such as many forms of cancer and neurological disorders. This strategy is known as regulation of gene expression or gene regulation. To try to get synthetic nucleic acids into cells, researchers have often had to resort to using transfection agents. (Transfection is the process of deliberately introducing nucleic acids into cells.) Unfortunately, transfection agents create problems: they can't be degraded naturally, they often provoke a severe immune response, and they're toxic at high concentrations.

SNAs, however, require no co-carriers to enter cells and yet are taken up in high concentrations. And Mirkin's group has gone a step further. They have made coreless SNAs. These particles—which are as close as one can get to "pure" SNAs—are composed only of cross-linked nucleic acids that are oriented in the same fashion as in the SNA-gold NPs (that is, perpendicular to the surface of the NP, as indicated in the "anatomical" Figure 3.10). They used the gold core as a scaffold, subsequently cross-linked the DNA strands at the base of the particle, and then dissolved the gold. (Hollow SNAs anticipate concerns about the potential toxicity of the inorganic gold core in therapeutic usage.)

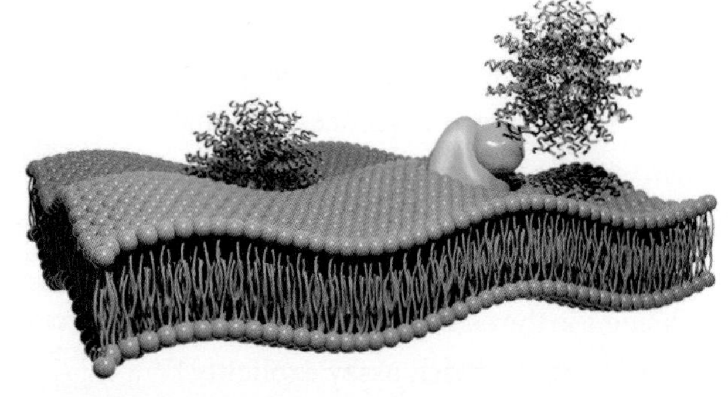

Figure 3.14 Hollow SNAs interacting with scavenger receptors in the cell membrane, which induces endocytosis of the particles. *(Reproduced with permission from Joshua I. Cutler, Evelyn Auyeung, and Chad A. Mirkin, "Sphercal Nucleic Acids," Journal of the American Chemical Society 134 (2012), 1376–1391.)*

A proposed mechanism for the ability of hollow SNAs to successfully outmaneuver nature is shown in Figure 3.14. A serum protein (green sphere) such as bovine serum albumin first adsorbs to the SNA's oligonucleotide shell. Next, the serum proteins are displaced by scavenger receptors (yellow) at the cell surface. (Scavenger receptors are cell surface proteins that can bind and then internalize many molecules; the latter function is called *endocytosis.*) The scavenger receptors, true to their name, initiate the endocytosis of the SNAs. Pulling off this feat of entering cells is impressive, and what's more, the SNAs are not digested after endocytosis but rather they retain their discrete character. Still better, they don't provoke an immune response and do not exhibit apparent toxicity. They just may be the perfect means of delivering nucleic acids into cells to perform gene therapy.[35]

We grab the opportunity to highlight a tie-in. Figure 3.14 shows the lipid bilayer that forms the membranes of our cells. We first saw this in Chapter One in the starter kit for liquid crystals. The lipid bilayer is a liquid crystal that separates the interior of the cell from the exterior.

Mirkin had this to say about the ability of SNAs to enter cells: "It's really remarkable—take a cell culture and feed it enormous concentrations of DNA, and the DNA will not go into the cells. If you take that same type of DNA and put it on the surface of a particle it goes in like gangbusters and is effective as a therapeutic even at one-hundredth the concentration that is typically applied! That is a truly important observation, not just in nanoscience, but in biology and chemistry in general. Essentially, it teaches us that, if you take one of the most important molecules ever synthesized and mass produced—DNA—and you simply arrange it on the nanoscale in a high-density, highly oriented, 3D format, it behaves completely differently from any other form of matter ever studied.... We have recently obtained extremely powerful results showing cellular entry and gene knockdown in animals, and now our sights are set on moving the SNA construct to clinical trials in humans. If such amazing results continue to be seen, SNAs have the potential to revolutionize the medical field, offering treatment options for almost any disease with a genetic basis, including many forms of cancer!"[36]

Consider brain cancer, for example. Malignant gliomas represent the most common and lethal brain tumors that involve the cerebral hemispheres of adults. In the United States, 17,500 new primary brain tumors are diagnosed in adults each year, and in children, primary brain tumors are second in number only to leukemia as causes of tumor mortality.[37] Mirkin, in collaboration with Northwestern University's Robert H. Lurie Comprehensive Cancer Center, has taken aim on this devastating disease. Cancer cells often over-express a gene called *survivin*. The survivin gene produces a set of proteins that stop the cell from dying; it has a role in making cancer cells immortal. Mirkin is pursuing the design of SNAs (in this case, using RNA rather than DNA as the nucleic acid) that can go into these types of cells and turn off each cell's ability to produce those kinds of proteins—to make it more like a normal cell. Mirkin said, "A surgeon takes out a lethal tumor. If he leaves even a few cells behind, that patient is likely going to die. In fact 95% of these patients do die within five years. So the question is, how do you make sure you clean up the last few cells? Well, here we're going to flood the area [with SNAs], and all the cells in the local area will pick up these particles. But we will target the cancer cells at the genetic level and therefore selectively kill them. We're targeting genes that are unique in cancers [*sic*] cells, so healthy cells will be unaffected."[38]

On December 2, 2011, Mirkin was interviewed on National Public Radio (NPR). Interviewer Ira Flatow asked what he called the $64 question: when will there be a product available? Mirkin replied that the more important question to ask would be: when will human clinical trials (with his approach) begin. He said he was hopeful that they would be into human trials in the very near future.[39]

We've certainly covered a good deal of nanoterrain—scanning tunneling microscopy, manipulating individual atoms, building quantum corrals, and learning about spherical nucleic acid nanoparticles. We began our travels to the nanoworld with a frank assessment of the place that Richard Feynman's 1959 after-dinner speech (and the published transcript) *didn't* have on the advent of the field of nanotechnology. To conclude, we reconnect with Feynman via one of his legion of admirers, nanoluminary Donald Eigler. He had read Feynman's paper: "I can not say for certain, but I believe I read, or came to be aware of 'There's Plenty of Room' in the late 1970s or early 1980s while I was a graduate student. I know for a fact that I had read it a long time before first manipulating atoms with the STM. The reason I say this is because, within weeks of manipulating atoms for the first time, I went back to dig up Feynman's paper. When I started reading the paper, I realized that I had read it a long time before."[40] However, Eigler also said "Feynman's work would be on a dusty shelf without Binnig." Although Eigler said that the technical aspects of his work had not been influenced by Feynman's paper, when he reread "There's Plenty of Room," he "found an extraordinary affinity between the written words of Feynman and my own thoughts . . . I was more than ever impressed with how prescient Feynman's thoughts were."

Eigler also clearly recalled a profound sense of sadness that Feynman had passed away before one of the provocative ideas from his after-dinner speech was achieved in the lab.[41] Eigler's research brought to fruition the idea in question. Although he well understood that "things do not simply scale down in proportion," Richard Feynman had also said: "But I am not afraid to consider the final question as to whether, ultimately—in the great future—we can arrange the atoms the way we want; the very *atoms*, all the way down!"[42]

Exercises for Chapter Three: Exercises 3.1–3.3

Exercise 3.1

While majoring in physics in college you had a part-time job running a flea circus called Monty Python's Flying Flea Circus. You became interested in wave-particle duality and considered the question: if a common flea has a mass of 0.008 gm and can jump vertically as high as 20 cm, what is the de Broglie wavelength of the flea immediately after takeoff? *Hint*: elementary classical mechanics gives expressions for velocity that assume constant acceleration; use such an equation.

(Adapted from Physics for Scientists and Engineers, *6E, by Paul A. Tipler and Gene Mosca, Copyright 2008 by W.H. Freeman and Company, New York. Used with permission of the publisher.)*

Exercise 3.2

You are using an STM to examine a surface with a gap, d, between the STM tip and the surface. Assume that Φ, the tunneling barrier presented to an electron trying to tunnel from the surface to the tip, is 4 eV (4 electron volts). Assume that the gap, d, is 0.3 nm. **(a)** How big is α, the inverse of the length of the exponential tail of the wavefunction = 1 / (how far the electron can tunnel)? **(b)** What is the value of the tunneling current? **(c)** Next, suppose the tip encounters a portion of the surface that is higher by approximately 0.1 nm than at the tip's previous location. What will the value of the tunneling current be now?

Hint: your solutions to parts (b) and (c) should reveal the remarkable sensitivity of the tunneling current to the gap distance.

Exercise 3.3

Combining art and science, you are a violinist with an interest in the science of music. You adjust a tuning peg so that a 0.32 m long violin string is tuned to 440 Hz (A above middle C), that is, you produce a 440 Hz vibration. Note that 1 Hertz (Hz) = 1 cycle per second. **(a)** What is the wavelength of the fundamental string vibration? **(b)** What are the frequency and wavelength of the sound produced? **(c)** Why is there a difference in the wavelengths in parts (a) and (b)?

ENDNOTES

[1] "Caltech Conference Ponders Bomb Tests," *The California Tech, Vol. LXI,* No. 12 (7 January 1960), 3.

[2] W. A. Nierenberg, *Bulletin of the American Physical Society, Series II,* Vol. 4, No. 8 (28 December 1959), 440–442 and 456.

[3] Inspired by P.G. Wodehouse, *Leave It to Psmith* (New York: Vintage Books, 1975), 97.

[4] Chris Toumey, "Apostolic Succession: Does Nanotechnology Descend From Richard Feynman's 1959 Talk?" *Engineering & Science, No. 1 / 2* (2005), 16–23.

[5] Richard P. Feynman, "There's Plenty of Room at the Bottom," *Engineering and Science* (February, 1960), 22–36; http://calteches.library.caltech.edu/47/2/1960Bottom.pdf

[6] Toumey, "Apostolic Succession," 18.

[7] Christopher Toumey, "Reading Feynman Into Nanotechnology: A Text for a New Science," *Techné 12:3* (Fall 2008), 133–168.

[8] Toumey, "Apostolic Succession," 21 and 23.

[9] G. Binning, H. Rohrer, Ch. Gerber, and E. Weibel, "Tunneling Through a Controllable Vacuum Gap," *Applied Physics Letters 40, No. 2* (15 January 1982), 178–180.

[10] Gerd Binnig and Heinrich Rohrer, "The Scanning Tunneling Microscope," *Scientific American 253, No.2* (August 1985), 50–56.

[11] Toumey, "Apostolic Succession," 23.

[12] http://www.colorado.edu/physics/phys2170/phys2170_fa08/Lecture_Notes/class37_STM_AlphaDecay.pdf

[13] Yaneer Bar-Yam, "Concepts: Butterfly Effect," New England Complex Systems Institute, http://necsi.edu/guide/concepts/butterflyeffect.html

[14] L.J. Whitman, J.A. Stroscio, R.A. Dragoset, and R.J. Celotta, "Geometric and Electronic Properties of Cs Structures on III-V (110) Surfaces: From 1-D and 2-D Insulators to 3-D Metals," *Physical Review Letters 66, No. 10* (1991), 1338–1341.

[15] G. Binnig, C.F. Quate, and Ch. Gerber, "Atomic Force Microscope," *Physical Review Letters 56, No. 9* (1986), 930–933.

[16] D.M. Eigler and E.K. Schweizer, "Positioning Single Atoms With a Scanning Tunneling Microscope," *Nature 344* (5 April 1990), 524–526.

[17] Thomas W. Staley, "The Coding of Technical Images of Nanospace: Analogy, Disanalogy, and the Asymmetry of Worlds," *Techné 12:1* (Winter 2008), 1–22.

[18] M.F. Crommie, C.P. Lutz, and D.M. Eigler, "Confinement of Electrons to Quantum Corrals on a Metal Surface," *Science 262* (8 October 1993), 218–220.

[19] Carleen Maley Hutchins, "The Acoustics of Violin Plates," *Scientific American 245* (October 1981), 170–186.

[20] J.R. Jackson, "On the Seiches of Lakes," *Journal of the Royal Geographical Society of London, 3* (1 January 1833), 271–275.

[21] Graham P. Collins, "STM Rounds Up Electron Waves at the QM Corral," *Physics Today 46* (11) (November 1993), 17–19.

[22] Corral Collage, STM Image Gallery, http://www.almaden.ibm.com/vis/stm/corral.html#stm17

[23] M.F. Crommie, C.P. Lutz, D.M. Eigler, and E.J. Heller, "Waves on a Metal Surface and Quantum Corrals," *Surface Review and Letters 2, No. 1* (1995), 127–137.

[24] "Northwestern's Chad Mirkin on Enabling Nanoparticles," scienceWATCH.com, January 2010, http://archive.sciencewatch.com/inter/aut/2010/10-jan/10janmirk/

[25] Chad A. Mirkin, Robert L. Letsinger, Robert C. Mucic, and James J. Storhoff, "A DNA-Based Method for Rationally Assembling Nanoparticles Into Macroscopic Materials," *Nature 382* (15 August 1996), 607–609.

[26] Joshua I. Cutler, Evelyn Auyeung, and Chad A. Mirkin, "Spherical Nucleic Acids," *Journal of the American Chemical Society 134* (2012), 1376–1391.

[27] James J. Storhoff, Anne A. Lazarides, Robert C. Mucic, Chad A. Mirkin, Robert L. Letsinger, and George C. Schatz, "What Controls the Optical Properties of DNA-Linked Gold Nanoparticle Assemblies?" *Journal of the American Chemical Society 122, No. 19* (2000), 4640–4650.

[28] Robert F. Service, "DNA Ventures Into the World of Designer Materials," *Science 277, No. 5329* (22 August 1997), 1036–1037.

[29] Cutler, Auyeung, and Mirkin, "Spherical Nucleic Acids," 1384.

[30] Rongchao Jin, Guosheng Wu, Zhi Li, Chad A. Mirkin, and George C. Schatz, "What Controls the Melting Properties of DNA-Linked Gold Nanoparticle Assemblies?" *Journal of the American Chemical Society 125, No. 6* (2003), 1643–1654.

[31] Matthew R. Jones, Nadrian C. Seeman, and Chad A. Mirkin, "Programmable Materials and the Nature of the DNA Bond," *Science 347, No. 6224* (20 February 2015), 1260901-1–1260901-11.

[32] X. Li, X. Yang, J. Qi, and N.C. Seeman, "Antiparallel DNA Double Crossover Molecules as Components for Nanoconstruction," *J. Am. Chem. Soc. 118* (1996), 6131–6140.

[33] T. Andrew Taton, Chad A. Mirkin, and Robert L. Letsinger, "Scanometric DNA Array Detection With Nanoparticle Probes," *Science 289* (8 September 2000), 1757–1760.

[34] Cutler, Auyeung, and Mirkin, "Spherical Nucleic Acids," 1385.

[35] Ibid., 1386–1387.

[36] Hannah Stanwix, "An Interview With Chad Mirkin: Nanomedicine Expert," *Nanomedicine 7* (5) (2012), 635–638.

[37] Alexander H. Stegh, Research Program, http://www.neurology.northwestern.edu/faculty/stegh /research/index.html

[38] "Northwestern's Chad Mirkin," scienceWATCH.com, 3.

[39] "Hitting the 'Off' Switch on Antibiotic Resistance," http://www.npr.org/2011/12/02/ 143055120 /hitting-the-off-switch-on-antibiotic-resistance

[40] Toumey, "Reading Feynman," 146.

[41] Toumey, "Reading Feynman," 145–146.

[42] Feynman, "There's Plenty of Room," 34.

CHAPTER FOUR

Liquid Crystals: Nature's Delicate Phase of Matter[1]

If you had a fondness for carrots and had chanced to be in Prague in the year 1888 you could not have done better than to visit the botanist Friedrich Reinitzer at the Institute of Plant Physiology. The man was obsessed with carrots.[1]

Reinitzer was busy extracting cholesterol from carrots and analyzing it on the (incorrect) assumption that cholesterol was chemically related to carotene and hence to chlorophyll. He formed the compound cholesteryl benzoate (cholesterol and benzoic acid) and found to his utter astonishment that the compound appeared to have not one but *two* melting points. Moreover, near both of these transition points there appeared dramatic flashes of color (recall the phase transitions from the liquid crystals starter kit in Chapter One).[2]

Reinitzer was understandably perplexed, his observations inexplicable. He realized he was out of his depth and wrote to Otto Lehmann, a physical chemist at the Polytechnical School of Aachen, and sent him some of his samples. For his doctoral research Lehmann had designed and built a special microscope. It was fitted with polarizers that could be readily slid in and out of the path of the light. One polarizer was below the sample and one was above and they were crossed (that is, they were at right angles to one another). And his microscope had a sample holder that could be heated or cooled. No commercially

[1] Expression taken from Peter J. Collings, *Liquid Crystals: Nature's Delicate State of Matter* (Princeton, New Jersey: Princeton University Press, Second edition, 2002).

available microscopes had this feature. After much examination of cholesteryl benzoate and other compounds that had the double-melting phenomenon, in 1889 Lehmann published an article titled "On Flowing Crystals." He advocated the coexistence of liquidity and crystallinity in the same material. Thus began the science of liquid crystals.[3] Lehmann published a great deal on the subject and drew scientific supporters for his novel work. He also drew critics. Vitriolic arguments raged for some fifteen years in print and at professional meetings. It was quite a kerfuffle.

Phase Transitions

Liquid crystal research is driven by scientific curiosity and technological applications. It has also become an important testing ground for theoretical physics. For example, the phenomenon of phase transitions (a change from one state of matter to another) that occurs in liquid crystals also occurs in superconductors, magnetic materials, superfluids, and other states of matter. Even though the physical properties involved and the underlying mechanisms causing the transitions are quite different, there are striking similarities. Since liquid crystals provide a wealth of different phase transitions, they have drawn intense interest from theorists seeking to understand the subject in greater generality. A discussion of the fundamental physics of liquid crystals would lead us too far afield from the stories in this book.

Phase transition A phase transition is a change from one state of matter to another without a change in chemical composition. Familiar examples include the melting of a solid to a liquid or the vaporization of a liquid to a gas. In liquid crystals, an example of a phase transition is the change from a state in which the properties of the liquid crystal molecules do not vary with direction (isotropic phase) to a state in which there is a preferred direction of orientation—a degree of orientational order—of the molecules (nematic phase).

However, there are palpable consequences of phase transitions that we cannot ignore. When Reinitzer wrote to Lehmann (his first letter was 16 pages long) he observed that as the temperature of his material changed, violet and blue colors appeared, rapidly vanished, and then reappeared only to vanish again. This light show was concomitant with the phase transitions that occurred at the melting points of the compound he was examining. Depending on the particular liquid crystal there can be multiple phase transitions. The associated optical effects are often beautiful and with modern techniques they can be recorded for study and contemplation. But there is a cautionary note: in any practical application of liquid crystals the phase transition temperatures are to be avoided. One must stay away from those temperatures and stay in the phase appropriate for the application—displays, sensors, spatial light modulators, and so on—in order to maintain predictable device operation. Liquid crystal synthesis allows one to find materials that have a wide temperature range within a given phase so that in operation the device is always far away from the transition temperature(s) and can function stably.

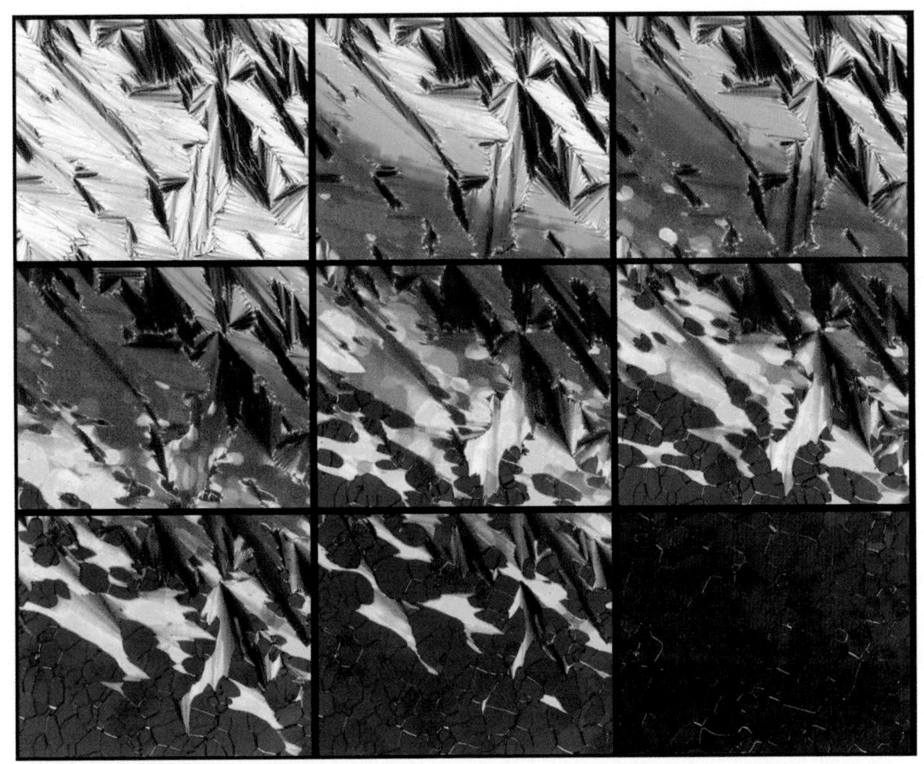

Figure 4.1 Photographs of the liquid crystal compound CE8 as it is slowly heated and undergoes a transition from the smectic A phase to the cholesteric phase. The phase transition temperature is approximately 145 °C. Beginning at the upper left, the temperature increases across each row until the lower right is reached. *(Courtesy of Oliver Coles.)*

The caveat to avoid phase transitions has a counterpoint in the celebration of their existence. After all, liquid crystal phase transitions are a natural phenomenon as remarkable as the *aurora borealis,* the celestial ballet also known as the northern lights. Several years ago, Oliver Coles was a graduate student in the physics laboratory of Professor Helen Gleeson at the University of Manchester. His work involved the use of liquid crystals within the capillaries of photonic crystal fibers (a type of optical fiber). Just for fun he was playing around with a liquid crystal known as CE8 when he came across a spectacular phase transition from the smectic A phase to the cholesteric phase. He placed the sample between crossed polarizers and slowly turned the temperature up. The still photographs of Figure 4.1, beginning from the upper left and reading across each row to the lower right, show the phase change that occurs at approximately 145 °C. The overall temperature change from the first picture to the final one is only about 0.5 °C.[4]

Classes of Liquid Crystals

We're going to have a look at liquid crystals in the natural world and their applications in display technology. In due course we'll return to their unusual optical properties—for example, that odd term *birefringence* that popped up in Chapter One. First, we explain what constitutes a liquid crystal. Liquid crystals possess many of the mechanical properties of a liquid—fluidity, inability to sustain shear forces (essentially, a lack of stiffness), formation and coalescence of droplets. Liquid crystals are also similar to (solid) crystals. Molecules show a degree of alignment and liquid crystals exhibit anisotropy (that is, unequal attributes in different directions) in their optical, mechanical, electrical, and magnetic properties.

Isotropic Having properties that do not vary with the direction of measurement.

Anisotropic Having properties that do vary depending on the direction of measurement. In liquid crystals, this is due to the alignment and the shape of the molecules.

The two major categories of liquid crystals are called *thermotropic* and *lyotropic*.[5] Thermotropic materials have liquid crystalline properties that change with temperature (thus, the prefix *thermo-*). Lyotropic materials have liquid crystal behavior in solutions and their properties change with the concentration of the molecules as well as with temperature. (The prefix *lyo-* derives from a Greek word and denotes *loosening* or *dissolving*.) Both thermotropic and lyotropic liquid crystals can form distinct phases of matter intermediate between the solid and liquid phases. The identifying feature of the phase is the nature of the internal molecular organization of the material. For example, in the nematic phase, molecules possess a degree of orientational order. (The molecules orient themselves with respect to one another.) In the smectic phase there is orientational order and a degree of translational order as well. (The molecules tend to align themselves in layers or planes.)

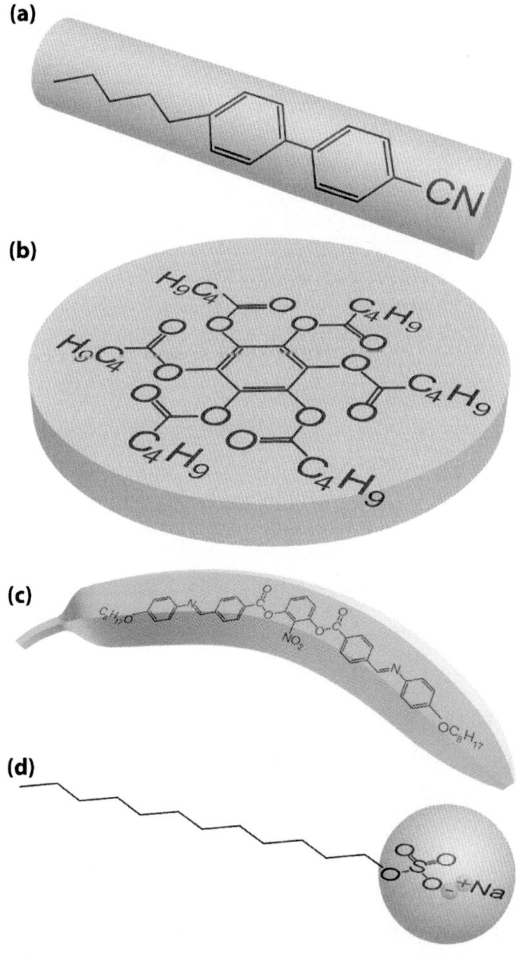

Figure 4.2 **(a)** Elongated liquid crystal building block called 5CB. **(b)** Disk-like building block, a derivative of benzene-hexa-n-alkanoate. **(c)** Banana-shaped (bent-core) building block called B7. **(d)** An amphiphilic molecule called sodium dodecylsulfate with both a hydrophilic (sphere) and hydrophobic (crooked line) portion. *(Courtesy of Bohdan Senyuk.)*

Liquid crystals that are thermotropic usually form from molecules with shape anisotropy—either elongated or disk-like, as well as molecules with bent-cores, popularly referred to as banana shaped (Figure 4.2). Lyotropic liquid crystals are generally two-component systems where an amphiphilic compound is dissolved in a solvent (typically water). Let's define amphiphilic.

An amphiphile is a molecule that contains discrete chemical groups that interact differently with water. The molecules that form lyotropic liquid crystals have two dissimilar parts: a hydrophilic (attracted to water) group and a hydrophobic (repelled by water) group. An example of this kind of molecule is sodium dodecylsulfate (panel (d) of Figure 4.2). In many cases the hydrophobic part of an amphiphile is a flexible hydrocarbon chain. In chemistry this is often depicted as a crooked line (as in panels (a) and (d) of Figure 4.2). It is understood, though not shown, that at each kink in the line there is a carbon atom (that can bind with two hydrogen atoms, hence *hydrocarbon* chain). A portion of the hydrocarbon tail is shown explicitly in all its glory in Figure 4.3; the black balls are carbon atoms and the blue ones are hydrogen. This is called a saturated hydrocarbon chain, meaning that all carbon-carbon bonds are single bonds. When carbon atoms form double—or even triple—bonds with each other, there are fewer bound hydrogen atoms and such chains are said to be unsaturated.

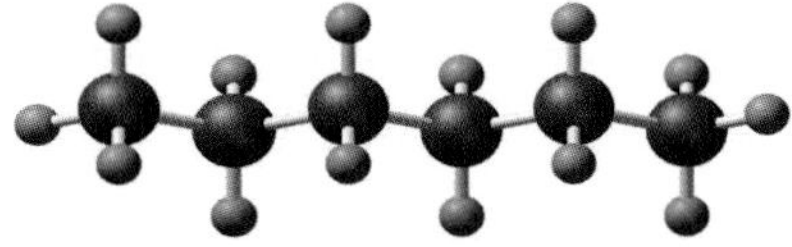

Figure 4.3 A hydrocarbon chain where black balls are carbon atoms and blue are hydrogen atoms.

Cell Membranes and the Langmuir Trough

That's a passel of vocabulary; now for some punch lines. Here's one: life. Cell membranes are liquid crystals. The cell membrane is essential in living beings and separates the inside of the cell from the outside of the cell. It maintains the integrity of the cell and yet is permeable to allow biomolecules in and out to perform their myriad functions. *The fact that the cell membrane is a liquid crystalline structure is arguably the most important occurrence of liquid crystals in the natural world.*[6] Consider the cell membrane of animals. It's composed of a phospholipid bilayer. Lipids are a type of amphiphilic molecule. Phospholipids have a phosphate group in their hydrophilic portion while their hydrophobic portion consists of not one but two hydrocarbon chains, one of them saturated and the other unsaturated. Not only is the cell membrane a lyotropic liquid crystal (in the smectic phase), but all the organelles within an animal cell also have membranes around them. That is, the cell nucleus, mitochondria, endoplasmic reticulum, Golgi apparatus—all those neat names you learned in middle school or high school biology—also possess lyotropic liquid crystalline membranes (although there are some variations in the type of phospholipids). The bilayer forms largely because of the phenomenon of

Thermotropic liquid crystals Liquid crystal molecules that exhibit temperature-dependent liquid crystalline behavior.

Lyotropic liquid crystals Materials in which liquid crystalline phases appear because of the presence of a solvent. The concentration of the solvent as well as the temperature will determine the type of phase (e.g., nematic).

Amphiphile A molecule with a hydrophilic head and a hydrophobic tail. Thus, it's a molecule with one end that attracts water and one end that repels water

Hydrophilic; Hydrophobic Water loving—describes a molecule that is attracted to water; Water fearing — describes a molecule that is repelled by water.

hydrophobic interactions (i.e., the aggregation of hydrophobic substances in aqueous solutions). The bilayer deters indiscriminant mixing of the water interior to the cell with the water outside it (Figure 4.4).

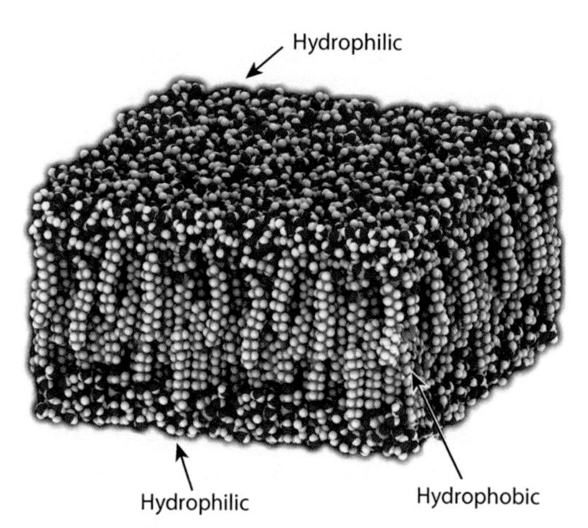

Figure 4.4 Phospholipid bilayer. *(123rf.com image, ID 18805971.)*

The discovery of the structure of the cell membrane is an interesting story. Moreover, there is a story within the story. Evert Gorter was a Dutch medical doctor in the early twentieth century. There are overtones in Gorter's life that are reminiscent of Oswald Avery (who appeared in Chapter Two). Gorter received his doctor's degree with a thesis on *tuberculosis bacillus*. He became a professor of pediatrics and had a strong desire to do basic research as well as to care for (young) patients. Partly because of his life-long battle with severe (familial) rheumatoid arthritis, he eschewed social events and recreation in order to lead a life focused on his two careers—one as a clinician and one as a researcher in physical chemistry.[7] Gorter was an avid reader of scientific journals and became fascinated with the usefulness of studying biological molecules at surfaces. He conceived of an experiment to make surface area measurements on cell membrane lipids. Before describing Gorter's experiment, we first mention that his method of choice was to use an apparatus—a type of trough—that is generally credited to the Nobel Laureate, chemist Irving Langmuir (though Gorter used a slightly modified version).[8]

Cell membrane A semi-permeable structure that surrounds the interior of a cell.

Lipid bilayer A two-layered structure composed of amphiphilic molecules in which the hydrophobic portions of each layer point toward each other.

But in fact, Miss Agnes Pockels designed the basic device. Pockels was a self-educated scientist (with *no* university education) who carried out her first scientific observations on surface films at the age of nineteen in the kitchen of her parents' home in Braunschweig, in north central Germany, in 1881.[9] She had invented the slide trough by the time she was twenty. In 1890 she learned that the eminent physicist Lord Rayleigh of Cambridge University was doing similar research and she wrote to him. After her letter was translated for him from German to English, he was so impressed that he wrote to the editors of *Nature* requesting that they publish her work. They did.[10] Rayleigh also adopted the Pockels technique in his own surface film measurements. Pockels continued to do independent research and publish her results for decades thereafter—despite the fact that her principal undertaking was the care of her invalid parents and the management of their household. Her scientific work had to take second place.[11] In 1939, Langmuir himself said that her homemade measuring device "laid the foundation for nearly all modern work with films on water" and it continues to be a standard tool for surface chemists and physicists in the twenty-first century.[12] Among its uses is the measurement of surface tension, which is the attractive force between molecules at the water surface that resists extension of the surface. Surface tension measurements are important in printing, coating, and adhesive applications, among others.

We return now to Gorter, his assistant François Grendel, and their challenging experiment using the apparatus that was generally referred to at that time as a Langmuir trough (panel (a) of Figure 4.5).[12] Gorter was interested in determining the molecular dimensions of biological molecules. He knew of earlier work suggesting that the material surrounding a cell contained lipid molecules. And Langmuir had shown that fats and fatty acids spread in a monomolecular layer film in a trough.[13] Gorter decided to find out the amount of lipid in the cell membranes of red blood cells. (Lipids are shown in orange in Figure 4.5.) This was a judicious choice of cell type because red blood cells have no nucleus or any other membrane-bound organelles within their interior. So any membrane lipids that are found must come from the outer membrane of the cell.

They used red blood cells taken from dogs, sheep, rabbits, guinea pigs, goats, and humans (to show that the result obtained was independent of the source and size of the cells) and extracted the lipids using acetone.[14] The scientists knew the number of cells from which they extracted the lipids and with the Langmuir trough they found how large an area the lipids could cover (panels (b)–(d) of Figure 4.5). They used a microscope equipped

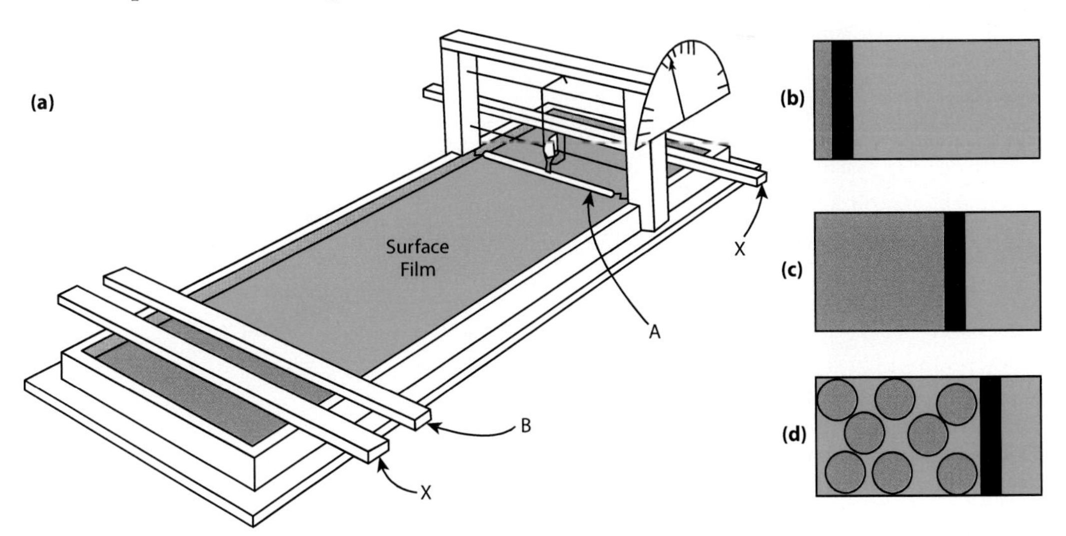

Figure 4.5 (a) A Langmuir trough, first designed by Pockels. "A" is a fixed barrier and "B" is a movable barrier that permits the area of the surface film to be changed. The barriers labeled "X" are used to sweep the surface to keep it immaculately clean. Blue denotes the water and orange denotes a lipid film placed on the water. **(b)** Water-filled trough with movable barrier and known amount of lipids (orange) to the left of the barrier. **(c)** Maximum coverage: the barrier is moved and increases the area covered by the lipids until the layer is only one molecule thick. **(d)** Overextended: the lipids can no longer cover the entire area and the continuity of the film is broken. *(Panel* ***(a)*** *adapted with permission from Ben Franklin Stilled the Waves, Charles Tanford (New York: Oxford University Press, 2004), Figure 16, page 147); panels* ***(b), (c), (d)*** *reproduced with permission from Scientific American, http://www.nature.com/scitable/topicpage/discovering-the-lipid-bilayer-14225438)*

with a drawing prism to draw the cells on graph paper and measure the size—i.e., the surface area—of a *single* red blood cell. Then they could calculate the *total* surface area of the blood cells that would be covered by membrane.[15] There was a difference of a factor of two in all cases: *the amount of lipid they had extracted could cover twice the surface area of the cells.* They concluded that the cell membrane is a *bilayer, two lipid molecules thick.*[16]

The Gorter and Grendel experiment is a classic. But there are multiple ironies. They made several experimental errors that by pure coincidence offset each other.[17] These errors were not discovered for about forty years.[18] And during those forty years their result was virtually ignored for subtle reasons.[19] Finally, in 1972 a consensus was reached that cell membrane lipids are organized as bilayers in the *fluid mosaic model* of cell membranes. (Membranes are viewed as two-dimensional fluids in which proteins are inserted into lipid bilayers.)[20,21] Figure 4.6 shows (in the words of biologist Michael Edidin) an exuberant picture of the fluid mosaic model.[22]

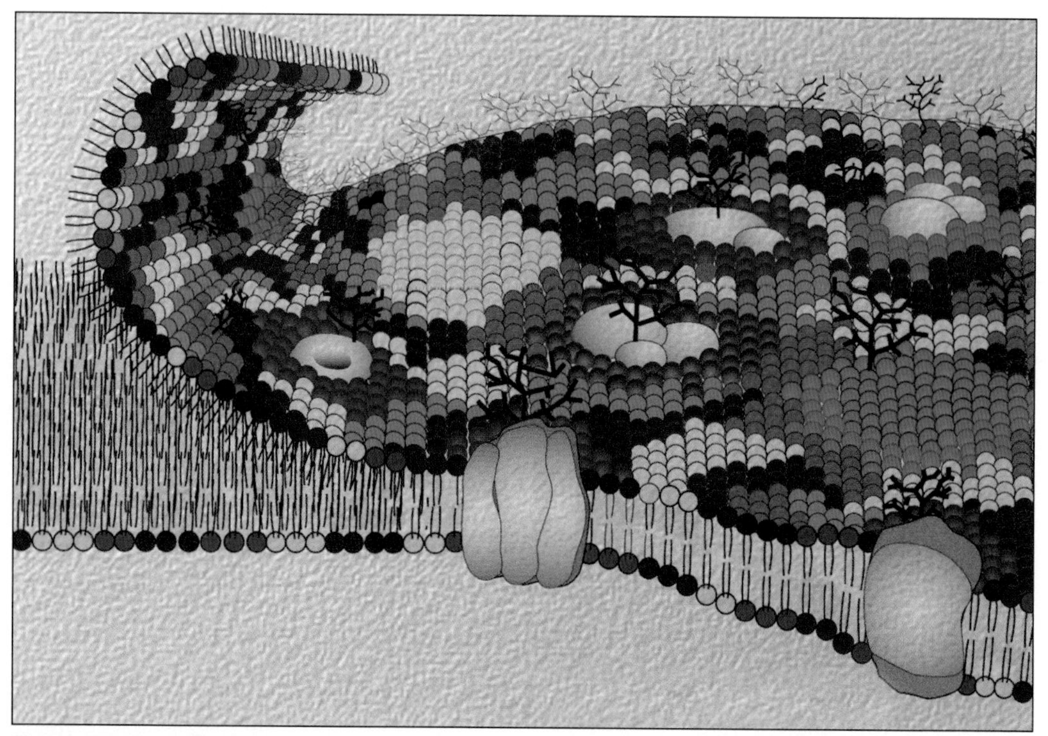

Figure 4.6 An exuberant picture of the fluid mosaic structure of the cell membrane showing the lipid bilayer with numerous attached proteins. *(Courtesy of Paavo Kinnunen.)*

As chemist Charles Tanford explained, the Gorter-Grendel paper was rediscovered only after everybody had already been convinced by more sophisticated methods of investigation (e.g., electron microscopy) that the phospholipid bilayer (along with a variety of membrane proteins) was the universal structural framework of biological membranes. In Chapter Three we presented Toumey's analysis of the role of Richard Feynman's talk on the conceptual foundation of nanotechnology. Tanford makes a similar case here.

"The [Gorter-Grendel] paper . . . is often cited as the 'foundation' of all membrane science. That, however, is a latter-day resurrection, intended as a teaching device and not as a historical statement. The Gorter-Grendel paper . . . is a superb *logical foundation*, even though it failed to be a foundation in the chronological sense" [italics added].[23]

Micelles

Micelle A spherical formation caused by an aggregation of amphiphilic molecules in a solution. The hydrophilic ends of the molecules tend to orient toward the outside of the sphere while the hydrophobic ends tend to orient toward the inside of the sphere.

We've discussed the phospholipid bilayer at length because it is the most pervasive instance of a liquid crystalline structure in the biological world. Better revise that to say, the current biological world. Before moving on to technological applications of liquid crystals, we consider the role liquid crystals may have played several billion years ago (though not quite as early as in the prebiotic investigations of Bellini and Clark).

Harvard biochemist Jack Szostak has made the case that an early version of RNA may have been enveloped in a *micelle of fatty acids*—another lyotropic liquid crystal (Figure 4.7). A fatty acid is an amphiphilic molecule with just one hydrocarbon chain (or tail) and a hydrophilic head called a carboxyl group made up of a carbon atom, two oxygen atoms, and a hydrogen atom. As we saw in the portion of a phospholipid bilayer (Figure 4.4), hydrophobic interactions determine that the hydrocarbon chains aggregate in the center of the micelle. A *micelle* (derived from the Latin word *mica,* meaning crumb or morsel) is simply a self-assembled collection of amphiphiles. Szostak has suggested that the earliest forms of life could have been fatty acid micelles that enveloped water as well as a rudimentary form of RNA.[24] If he's correct, then liquid crystals have been intimately associated with life right from the beginning until the present day. And, moreover, if Bellini and Clark are correct, liquid crystals may have been the driving force in enabling the polymeric, early version of RNA to come into being. We'll be examining Bellini and Clark's work in Part III of this book.

Figure 4.7 A fatty acid micelle. The yellow spheres are the hydrophilic head groups, and the kinked lines are the hydrophobic tails of the individual fatty acids. Because the molecules are amphiphilic they form lyotropic liquid crystals in water.

Before moving on, we will say a few words about invoking the hydrophobic interaction both in micelle formation and in lipid bilayer formation. The issue of polarity is inseparably linked to a description of these hydrophobic phenomena. The hydrophilic end of the fatty acid molecules and the solvent (water) that the fatty acids are in, are referred to as polar. The atoms in the head groups of the fatty acids are arranged such that one side of the head group has a positive electrical charge and the other side a negative charge. (The head group is said to be polar.) This is also true for the molecules of water. (The H_2O molecules are also polar.) In contrast, the hydrocarbon tails of the fatty acids don't have this charge separation—the tails are said to be non-polar.

Polarity Refers to a separation of electrical charge and can be used to describe an entire molecule (a polar molecule). This also applies to solvents. For example, a polar solvent like water is composed of molecules whose electric charges are unequally distributed, leaving one end of each molecule more positive than the other. If such charge separation is absent, the molecule or solvent is said to be non-polar.

Liquid Crystal Displays

Our look at the liquid crystalline state in the natural world has shown us that there is a lot more to liquid crystals than just present-day electronic displays. However, on the subject of displays, it's hard to ignore the fact that billions of liquid crystal displays are made each year for devices such as your mobile phone, your digital camera, your computer/tablet screen, you name it. So we now turn to the technological side of liquid crystals, specifically, the twisted nematic liquid crystal display that came into prominence in the early 1970s and marked the beginning of the modern liquid crystal display industry.[25] We'll first provide an overview by stitching together some observations and then go into more detail.

In the nematic phase of a liquid crystal there is a degree of orientational order of the molecules that doesn't exist in a pure fluid. The reason for this order is that in a nematic, intermolecular forces impart a tendency for the molecules to stay roughly parallel to one another. Moreover, scientists noticed that some materials, when placed in contact with a liquid crystal, forced the nearby molecules into a preferred direction. For example, a glass surface can be prepared so that the liquid crystal molecules near the glass are forced to lie parallel to the surface. In fact, it's surprisingly easy to make this happen: rubbing glass microscope slides on paper multiple times in the same direction is sufficient to define a preferred direction for the liquid crystal molecules to lie parallel to the surface. Suppose we do this with two glass slides. Then we rotate one slide by 90° with respect to the other and place a liquid crystal material in-between the slides—one that we know has a nematic phase. We'll call this a liquid crystal *cell.* (Cylinders representing the *average* alignment of the molecules indicate the liquid crystal orientation throughout the cell in Figure 4.8.) Now we place the cell between crossed polarizers on a microscope just like Professor Lehmann had, complete with a sample holder that can be heated or cooled. If we adjust the temperature of the cell so the liquid crystal is in the nematic phase, something neat happens. We get a twist in the orientation of the molecules between the plates. What's going on?

As you may have figured out, the intermolecular forces present in the nematic material produced the only solution they found possible when the molecules near the bottom glass slide are at an angle of 90° with respect to the molecules near the top slide. That is, they changed their orientation to accommodate both boundary conditions by forming a twist. The nematic phase behaved like an elastic medium. The molecules are guided into a twist because of the boundary conditions, the directionality imposed by the preparation of the two glass slides. A French organic chemist and crystallographer named Charles Mauguin performed some of the earliest experiments of this sort in 1911.[26] As a result, his name is associated with this effect. He couldn't actually see the twisted molecules because they have dimensions of nanometers, but he inferred that they twisted because of the way polarized light behaved when it was transmitted through the liquid crystal cell. Mauguin found that light coming out of the cell had its plane of polarization at a right angle to that of the incident light. From this observation he deduced that that the polarization direction of the light also twisted and did so in step with the twisted molecular orientation.

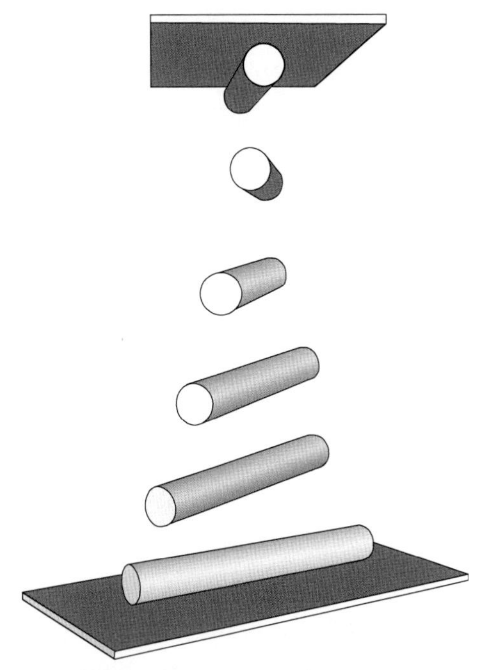

Figure 4.8 Mauguin's twisted structure. *(Reproduced with permission from David Dunmur and Tim Sluckin,* Soap, Science, and *Flat-Screen TVs: A History of Liquid Crystals (New York: Oxford University Press, 2011), 206.)*

He did a final experiment in which he applied a magnetic field along the direction of the light beam (that is, he reproduced the whole experiment in a magnet). When he did this, no light, regardless of its plane of polarization, was transmitted through the cell (for reasons we will shortly reveal). He concluded that the magnetic field had *un*twisted the molecules in the liquid crystal and reoriented them parallel to the direction of the light beam and the magnetic field. Mauguin had created an optical switch in which the molecules shift from a twisted state (light is transmitted) to an untwisted state (no light is transmitted) by application of a magnetic field.

This result languished for sixty years. Then scientists realized that the same experiment could be done with a modest electric voltage instead of a magnetic field. The resulting *twisted nematic display* flipped the switch to *on* for the liquid crystal display industry.

Let's back up now and have a tête-à-tête—Mauguin was French, after all. We'll talk about refraction, anisotropy, polarized light, and birefringence. We'll also get a handle on how the twisted nematic material can produce a twist in the polarization of light. After that we'll put it all together when we look at the insides of a twisted nematic display.

Twisted nematic effect The controllable, reversible alignment of liquid crystal molecules from one ordered molecular configuration to another in response to an applied electric field.

To begin, we call attention to three terms—refraction, anisotropy, and polarization—and then put them together. *Refraction*: The bending of light when it passes from one medium (e.g., air) to another medium (e.g., water) in which the velocity of the light is different. This is a familiar phenomenon that can produce some odd visual effects (Figure 4.9). In order to keep track of these changes in the velocity of light, we assign a *refractive index* to every medium. The refractive index is defined as the ratio of the speed of light in a vacuum divided by the speed of light in the medium (and thus is a dimensionless number).

Figure 4.9 Refraction of light at the boundary between air and water gives the effect of a bent pencil. *(Courtesy of Rod Nave, HyperPhysics Project.)*

Anisotropy: Earlier in this chapter we used the word *anisotropy*—unequal attributes in different directions. For example, consider a piece of wood. The wood has different strengths parallel to the grain and perpendicular to the grain.

Polarization: We spoke about polarization in Chapter One. The polarization of an electromagnetic wave—such as light—is specified by the orientation of the wave's electric field. As shown in Figure 4.10, we associate two directions with a light wave. They define a plane containing the direction of travel of the wave (left to right) and the direction of the light's electric field component that oscillates up and down. This is known as plane-polarized light (sometimes called linearly polarized light). In Figure 4.10 the light wave is propagating to the right and is plane-polarized in the plane of the paper.

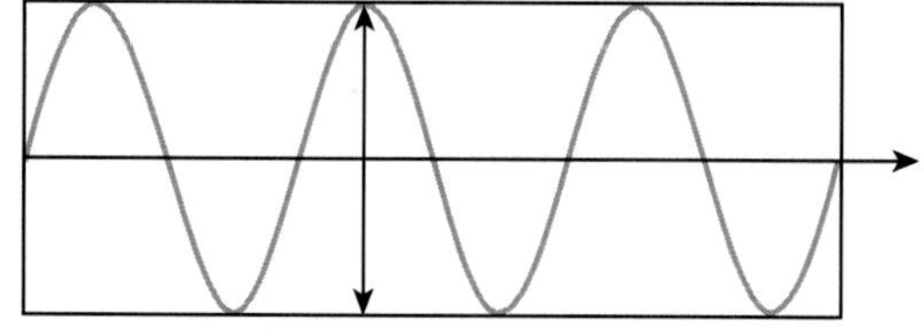

Figure 4.10 A light wave that is plane-polarized in the plane of the paper; i.e., the oscillating electric field is in the plane of the paper.

Armed with these three terms, we return to nematic liquid crystals. For nematics (actually, for *all* liquid crystal phases), the refractive index for light that is plane-polarized parallel to the molecular direction is different from the refractive index for light that is plane-polarized perpendicular to the molecular direction. The nematic has *two* refractive indices. This is known as *birefringence*, where *bi-* denotes the two refractive indices. Both refractive indices come into play at the same time only when the incident light is at an angle other than 0° or 90° with respect to the molecular direction. Because of birefringence, the nematic phase shows *optical* anisotropy.

Refraction The bending of light when it passes from one medium (e.g., air) to a second medium (e.g., water) in which the velocity of the light is different.

Polarization of light Describes the direction of the electric field vibrations in a light wave.

Polarizer (linear) A device that transmits electromagnetic radiation (e.g., light) and confines the vibration of the electric field to one plane.

Birefringent A material that has two indices of refraction is called birefringent. Liquid crystals are birefringent because of their anisotropic nature.

So the velocity of light polarized parallel to the molecular direction is different than the velocity of light polarized perpendicular to the molecular direction. In Mauguin's experiment, the twisted structure of the nematic caused the plane of polarization of light entering the liquid crystal cell to twist. There are rigorous mathematical explanations for why this happens but a qualitative argument will serve our purpose well (Figure 4.11).

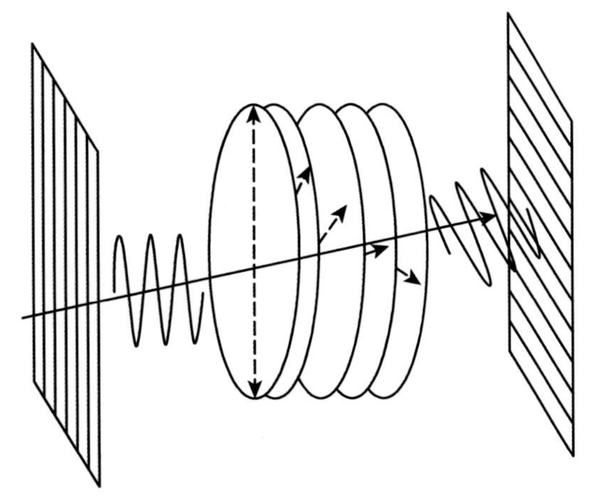

Figure 4.11 A qualitative schematic for Mauguin's twist experiment with plane-polarized light.
(Reproduced from David Dunmur and Tim Sluckin, Soap, Science, and Flat-Screen TVs: A History of Liquid Crystals *(New York: Oxford University Press, 2011), 74 with permission of Oxford University Press.)*

We can think of the twisted nematic structure as being composed of a series of very thin, parallel slabs. In each successive slab the average direction of the molecules is uniform but rotated by a small amount relative to the previous slab. Incident light that is plane-polarized along the direction of the molecules in the first slab will pass through without deviation. This is because the plane of polarization is parallel to the direction of orientation of the molecules in the first slab, so the light is sampling only one refractive index. But in the second slab, the molecules are at a slight angle to the plane-polarized light. So, in effect, part of the incident light is plane-polarized parallel to the molecular direction and part is plane-polarized perpendicular to the molecular direction. That is, the incident light can now be thought of as having two components.

Moreover, because of the optical anisotropy of the nematic, these two components will travel through the slab at different velocities—they are subject to the two different refractive indices of the nematic. When the light emerges from the second slab, the net result is a rotation of the plane of polarization. The amount of rotation is equal to the change in orientation of the molecules in the second slab compared to the first slab. So the light emerges from the second slab plane-polarized along the orientation of the molecules in the second slab. This argument can be applied to each of the slabs through the twisted nematic structure, as suggested schematically in Figure 4.11. Thus, after passing through the entire cell, the plane of polarization of the incident light is rotated by an angle of 90°.[27,28]

Some people say that the quintessential property of a liquid crystal is its anisotropy. We'll now see yet another manifestation of this attribute in explaining why no light—regardless of its plane of polarization—was transmitted through Mauguin's cell when a magnetic field was applied. We'll spell this out in the modern case where an electric field is applied but the answer is the same for a magnetic field. The new concept here is *dielectric* anisotropy. The dielectric properties of a material (and this applies to gases, liquids, and solids as well as liquid crystals) are a measure of how that material behaves when an electric field is applied. The molecules of a nematic liquid crystal are said to have a *positive dielectric anisotropy* if they tend to align parallel to the direction of an applied electric field (and negative dielectric anisotropy if they tend to align perpendicular to the field). In a twisted nematic display, a thin film of nematic liquid crystal with a positive dielectric anisotropy is sandwiched between two glass slides. The liquid crystal film is only a few micrometers thick; that is, only a few ten-thousandths of an inch thick. And the area of the glass slides enclosing the liquid crystal cell may be only 15 mm x 15 mm, that is, 0.6 inch x 0.6 inch. However, the individual liquid crystal molecules are only about 0.5 nm wide and about 2 nm long, so the volume between the glass slides contains more than a billion billion liquid crystal molecules (isn't the nanoworld amazing?).[29,30]

Dielectric property of a liquid crystal
The response of liquid crystal molecules to an applied electric field.

The glass slides each have a transparent conductive coating that acts as an electrode so an electric field can be established between the slides. The electrodes are connected to a voltage source that activates the liquid crystal film and thus changes the optical transmission of the cell. Or, to use some of our new vocabulary, the birefringence is controlled electrically.

There are two other vital points. The first is that each glass slide has been prepared à la Mauguin, so the top and bottom slides impart two preferred orientations—at an angle of 90° to one another—to nearby molecules. That is how the Mauguin twisted structure is created. An applied field does *not* create the molecular twist. The guiding is produced by the boundary conditions, the directionality imposed by the preparation of the two glass slides. The second point is that there are crossed polarizers placed on either side of the liquid crystal cell (recall that the cell consists of the glass slides with liquid crystal in between). A simple schematic is shown in Figure 4.12 where the alignment of the billion billion liquid crystal molecules appears as cylinders representing the *average* alignment of molecules throughout the cell.

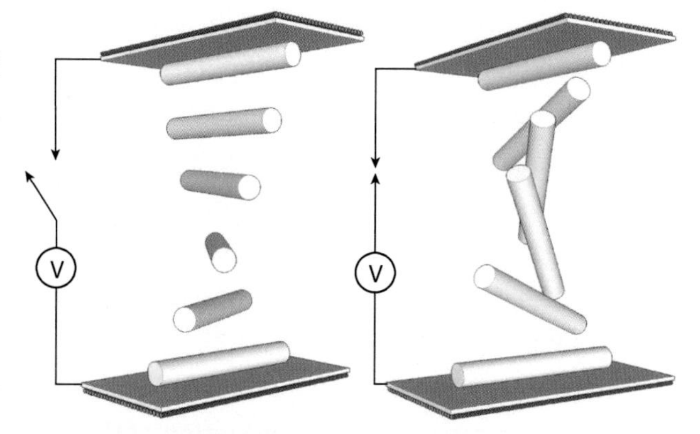

Figure 4.12 Schematic representation of the operation of a twisted nematic display cell. Light is transmitted when the electric field is off as shown on the left. No light is transmitted when the electric field is on, as shown on the right.
(Reproduced with permission from David Dunmur and Tim Sluckin, Soap, Science, and Flat-Screen TVs: A History of Liquid Crystals (New York: Oxford University Press, 2011), 214.)

In the left panel, light incident on the cell will be plane-polarized by the first polarizer. So the plane of polarization will be rotated by an angle of 90°. Thus, the light will pass through the exit polarizer. But when an electric field is applied (right panel) the liquid crystal twist is removed because of the positive dielectric anisotropy of the nematic liquid crystals. As a result, the molecules align along the electric field direction—a direction perpendicular to the glass plates. This destroys the guiding of the plane of polarization of the light. The cell no longer transmits the light through the exit polarizer (the process can be reversed by removing the electric field) because all the liquid crystal molecules are now oriented parallel to the plane of polarization. Thus, the plane-polarized light is not sampling both indices of refraction, only one. So the plane-polarized light retains its original plane of polarization (although its velocity changes). Since the exit polarizer is crossed with respect to the entrance polarizer, no light gets through the cell. A simple electric field of a few volts causes the liquid crystal cell to act as an optical switch. Numbers, letters, and complex pictures can all be formed from arrays of these optical shutters. Liquid crystal displays are much more compact than the cathode ray tubes that were the staple of the television industry, and they operate with very low voltage and consume less power than LEDs. Figure 4.13 summarizes our discussion.

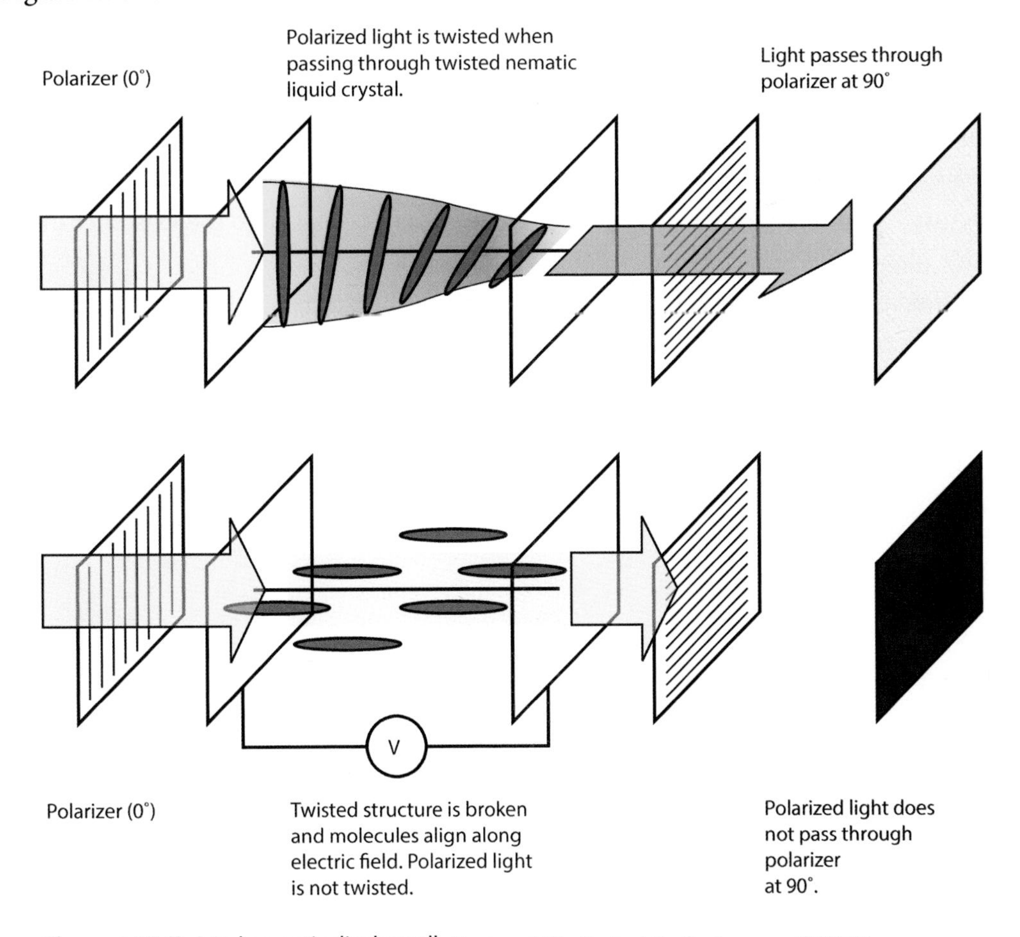

Figure 4.13 Twisted nematic display cell. *(Courtesy of Hidemitsu Awai, Creative Commons - CC BY 4.0.)*

Steady innovation has spawned super-twisted nematic displays, double super-twisted, film-compensated super-twisted, and on and on. There is also much more to making a practical display; specialized electronic components and processes come into play. But discussion of the original twisted nematic display is sufficient to give us an informed appreciation of the technology. Welcome to the club—we're now all fledgling high-tech cognoscenti.

Yes, the high-tech applications of liquid crystals are impressive and doubtless there will be more. But to return to the omnipresence of liquid crystals in the natural world, what can be more astonishing than a living organism? We've heard it said that biological organisms are 90% water. But water has no rigidity. How can organisms produce structures with some degree of rigidity—not just in the sense of stiffness, but the ability to confine the movement of material or provide a substrate for the ordering of molecules? And yet these same structures must be highly fluid so the molecules (e.g., enzymes and proteins in your body) can be produced in a given structure, move out of that entity, be transported to another part of the body, diffuse into another structure, and find the right place inside or near a particular cell. As Peter Collings wrote, "The ability of organisms to form fluid structures of some rigidity in an aqueous environment is central to the existence of life on this planet."[31]

Liquid crystals are the key to the delicate balance between rigidity and fluidity. And our realization of this started with Reinitzer's carrot juice. So hoist your blended vegetable smoothies and clink glasses in celebration of Reinitzer, Lehmann, and life. And liquid crystals.

Exercises for Chapter Four: Exercises 4.1–4.3

Exercise 4.1

When a sufficient concentration of amphiphilic molecules is in water (a polar solvent), they can form micelles (Figure 4.7). Suppose a sufficient concentration of the same amphiphilic molecules was placed in the non-polar solvent hexane. What simple structure could the molecules then form?

(Exercise reproduced with permission from University of Cambridge, Liquid Crystals Teaching and Learning Package, http://www.doitpoms.ac.uk)

Exercise 4.2

We spent quite some time on the Gorter-Grendel experiment in this chapter. One essential step was in extracting the lipid content of the red blood cells. Let's make the assumption that the lipid present in a red blood cell membrane has an average molecular weight of about 600 gm/mole and the protein in a red blood cell membrane has an average molecular weight of about 60,000 gm/mole.[32] (The molecular weight is the mass of one mole of a substance.) It's known that the mass of protein to lipid in red blood cells is approximately 1:1. For each protein molecule in a red blood cell membrane, how many lipid molecules would there be?

(Numerical information used with permission of Christopher Watters, http://cr.middlebury.edu/biology/labbook/membranes/frap/membranes/chap1.htm)

Note: For an illuminating and entertaining explanation of the numerology in this exercise and the underlying assumptions, be sure to look online at endnote 32, *The Lab Books on Membrane Structure and Fluidity* by Dr. Christopher Watters and Dr. Joseph Patlak.

Exercise 4.3

Far from being limited to liquid crystals, the phenomenon of birefringence is all around us. What properties do Scotch® Tape (adhesive cellophane tape), ice, airplane windows, some mineral crystals, and clear plastic food utensils have in common that enable them to exhibit birefringence? Please explain your answer.

ENDNOTES

[1] David Dunmur and Tim Sluckin, *Soap, Science, and Flat-Screen TVs: A History of Liquid Crystals* (New York: Oxford University Press, 2011), 18.

[2] Ibid., 19.

[3] O. Lehmann, "On Flowing Crystals," in, *Crystals That Flow: Classic Papers From the History of Liquid Crystals,* ed. Timothy J. Sluckin, David A. Dunmur, and Horst Stegemeyer (London: CRC Press, 2004), 42–53.

[4] Oliver Coles, ILCS (The International Liquid Crystal Society), featured liquid crystal artist, December 2011, http://www.ilcsoc.org/ILCS/page137/page135/page183/page201/page201.html and http://www.youtube.com/watch?v=0qsDahfulaQ

[5] B. Senyuk, "Liquid Crystals: A Simple View on a Complex Matter," 2005, http://www.personal.kent.edu/~bisenyuk/liquidcrystals/index.html

[6] Peter J. Collings, *Liquid Crystals: Nature's Delicate State of Matter* (Princeton, New Jersey: Princeton University Press, Second edition, 2002), 187–188.

[7] Charles Tanford, *Ben Franklin Stilled the Waves: An Informal History of Pouring Oil on Water With Reflections on the Ups and Downs of Scientific Life in General* (New York: Oxford University Press, 2004), 196–211.

[8] E. Gorter and F. Grendel, "On Biomolecular Layers of Lipoids on the Chromocytes of the Blood," *Journal of Experimental Medicine 41, No. 4* (April 1, 1925), 439–443.

[9] Tanford, *Ben Franklin,* 124.

[10] A. Pockels, "Surface Tension," *Nature 43* (March 12, 1891), 437–439.

[11] M. Elizabeth Derrick, "Profiles in Chemistry: Agnes Pockels," *Journal of Chemical Education 59, No. 12* (December 1982), 1030–1031.

[12] Tanford, *Ben Franklin,* 125. However, there is no source provided for this remark attributed to Irving Langmuir.

[13] Irving Langmuir, "The Constitution and Fundamental Properties of Solids and Liquids: II. Liquids," *Journal of the American Chemical Society 39* (1917), 1848–1906.

[14] Gorter and Grendel, "On Biomolecular Layers," 442.

[15] Ibid., 441.

[16] Mike Adams, "Discovering the Lipid Bilayer," *Nature Education 3, No. 9* (2010), 20.

[17] R.F.A. Zwaal, R.A. Demel, B. Roelofsen, and L.L.M. van Deenen, "The Lipid Bilayer Concept of Cell Membranes," *Trends in Biochemical Sciences 1, No. 2* (April/May, 1976), 112–114.

[18] Robert S. Bar, David W. Deamer, and David G. Cornwell, "Surface Area of Human Erythrocyte Lipids: Reinvestigation of Experiments on Plasma Membrane," *Science 153, No. 3739* (26 August 1966), 1010–1012, and references therein.

[19] Tanford, *Ben Franklin,* 204–210.

[20] Michael Edidin, "Lipids on the Frontier: A Century of Cell-Membrane Lipids," *Nature Reviews: Molecular Cell Biology 4, No. 5* (May 2003), 414–418.

[21] S.J. Singer and Garth L. Nicolson, "The Fluid Mosaic Model of the Structure of Cell Membranes," *Science 175, No. 4023* (18 February 1972), 720–731.

[22] Edidin, "Lipids on the Frontier," 417.

[23] Tanford, *Ben Franklin,* 211.

[24] Alfonso Ricardo and Jack W. Szostak, "Origin of Life on Earth," *Scientific American 301, No. 3* (September 2009), 54–61.

[25] M. Schadt and W. Helfrich, "Voltage-Dependent Optical Activity of a Twisted Nematic Liquid Crystal," *Applied Physics Letters 18, No. 4* (15 February 1971), 127–128.

[26] Ch. Mauguin, "On the Liquid Crystals of Lehmann," and "Orientation of Liquid Crystals by a Magnetic Field," in *Crystals That Flow: Classic Papers From the History of Liquid Crystals,* ed. Timothy J. Sluckin, David A. Dunmur, and Horst Stegemeyer (London: CRC Press, 2004), 100–127.

[27] Dunmur and Sluckin, "Soap, Science," 74.

[28] Sharon Ann Jewell, "Optical Waveguide Characterisation of Hybrid Aligned Nematic Liquid Crystal Cells," (Ph.D. diss., University of Exeter, 2002), 29–31. http://newton.ex.ac.uk/research/emag/pubs/saj_thesis.pdf

[29] Denis Andrienko, "Introduction to Liquid Crystals," http://www2.mpip-mainz.mpg.de/~andrienk/teaching/IMPRS/liquid_crystals.pdf

[30] Dunmur and Sluckin, "Soap, Science," 214.

[31] Collings, *Liquid Crystals,* 186–187.

[32] Christopher Watters and Joseph Patlak, *Lab Books: Membrane Structure and Function,* 1999, http://cr.middlebury.edu/biology/labbook/membranes/ and specifically, http://cr.middlebury.edu/biology/labbook/membranes/frap/membranes/chap1.htm

CHAPTER FIVE

Tools of the Trade

The late nineteenth century in Dartmoor, England. In a carriage bound for the small town of Tavistock, Inspector Gregory of Scotland Yard speaks to the London consultant:

"Is there any point to which you would wish to draw my attention?"

"To the curious incident of the dog in the night-time."

"The dog did nothing in the night-time."

"That was the curious incident," remarked Sherlock Holmes.

The famous scene from Sir Arthur Conan Doyle's short story "Silver Blaze" shows the acuity of Holmes' observational and reasoning ability. Throughout the canon of Sherlock Holmes we also see his use of a magnifying glass and even an optical microscope for trace evidence—e.g., tobacco ash, gunpowder residue, and fingerprints as well as hoof prints and bicycle tracks. He employs physical techniques to analyze ballistics, handwriting, and typewritten letters, analytical chemistry for blood residues, and knowledge of toxicology for the determination of poisons.

Seeman, Bellini, and Clark, in addition to their acute powers of observation and reasoning, share several technologies for detecting and recording the outcome of their experiments. Microscopy is a prominent tool and is employed in significant variations. Examples are depolarized light microscopy, transmission electron microscopy, and atomic force microscopy, and we will describe the fundamentals of each. Depolarized light microscopy, uses ordinary visible light while transmission electron microscopy uses a beam of electrons instead of light. Atomic force microscopy is not a radiation-based technique. An atomic force microscope *feels* what is happening at the nanoscale. It provides a three-dimensional profile of the object it identifies by measuring forces between a sharp probe and the sample surface. X-ray diffraction is another principal method in DNA nanoscience and is used to determine the *structure* of the very small molecules under investigation. In this chapter we'll get a feel for how the diffraction experiment is done and how the results are interpreted. We'll discuss both conventional X-ray diffraction and X-ray diffraction that uses synchrotron X-ray beams (produced by particle accelerators) available at facilities such as Brookhaven National Laboratory and Argonne National Laboratory. Polyacrylamide gel electrophoresis (PAGE) is frequently employed in Seeman's research as well as the Bellini/Clark research—particularly in their study of nanoDNA ligation (Chapter Thirteen). However, we introduced PAGE in Chapter Two so we won't discuss it in the present chapter. We will return to PAGE in Part II when we have a closer look at Seeman's work.

Polarized Light Microscopy

In fact, not only do we have a leg up on PAGE, but also on depolarized light microscopy. First a note on vocabulary: Bellini and Clark prefer the term *depolarized* light microscopy where most others use *polarized* light microscopy. We will stick to the Bellini and Clark convention. Regardless, the names refer to the same experimental setup. We described the technique implicitly in the previous chapter when we figured out how a twisted nematic display cell works. The particulars involving the Mauguin twisted structure are just particulars.

Depolarized light microscopy as used in studies of nanoDNA is all about looking at birefringent samples between crossed polarizers. A sample of liquid crystalline material is placed in a cell that is within a temperature-controlled stage. The stage is positioned between the crossed polarizers in the optical path of a microscope that leads to your eyes or to a camera as shown in Figure 5.1.

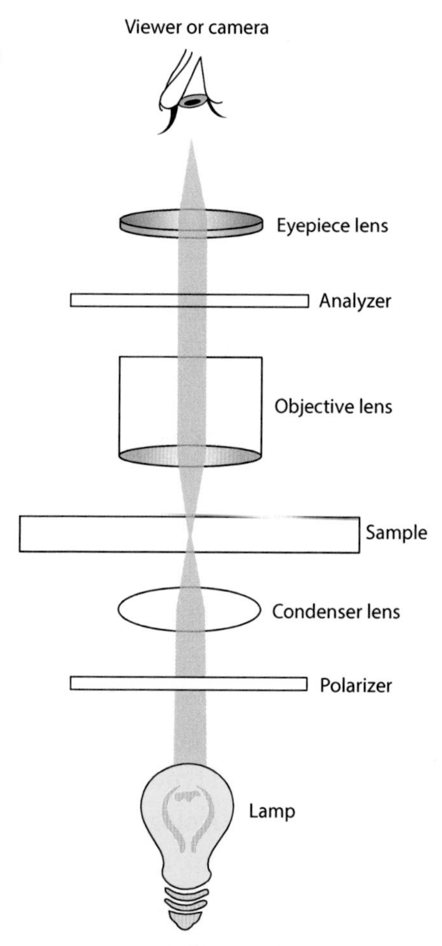

Figure 5.1 Depolarized light microscope (also called a polarized light microscope). A lamp illuminates the sample from below. Between the lamp and sample is a polarizer. The liquid crystal sample is in a temperature-controlled stage and oriented perpendicular to the light path. Following the sample and stage is another polarizer (also referred to as an analyzer) that can be rotated with respect to the first polarizer, but is generally positioned at 90° with respect to the first.

Polarized light interacts strongly with liquid crystal samples generating contrast that reveals birefringence. Birefringence is the classic signature of liquid crystals, resulting from the presence of two indices of refraction in the material. As we mentioned in Chapter One, the birefringence effect is wavelength dependent and thus the liquid crystal sample often appears in bright colors. To elaborate, different wavelengths of light correspond to different colors. Some colors will undergo destructive interference and some constructive interference in passing through the birefringent liquid crystal. There is also an intensity component accompanying the polarization and color information. One can see this by rotating the sample stage relative to the two polarizers. (The second polarizer is often labeled as the analyzer as in Figure 5.1.)

When the sample stage is rotated in a plane perpendicular to the direction of the light path, the light will encounter varying amounts of the two refractive indices. This produces variation in the extent of destructive and constructive interference and so the intensity of the polarization colors will vary. In addition to color and intensity, one can see distinctive textures appear in the liquid crystal sample. The contrasting areas in the textures correspond to regions where the liquid crystal molecules are oriented in different directions.

Depolarized light microscopy also called Polarized light microscopy A method of illumination employed to create contrast in birefringent materials such as liquid crystals. Samples are held in a temperature-controlled stage put between two polarizers that are most often at 90° with respect to one another (crossed polarizers). Polarized light then interacts strongly with the birefringent sample and enhances visualization.

Liquid Crystal Texture as Seen Through a Depolarized Light Microscope

In Chapter One we introduced the term *texture* when speaking of optical images of liquid crystals. The Appendix offers a fuller explanation. But we'll have a quick look at the interpretation of texture now because it is so relevant to depolarized light microscopy. Recall that Bellini and Clark were surprised to find liquid crystal ordering in very short segments of DNA. They found ultrashort DNA and RNA self-assembled to form duplexes with lengths (measured in base pairs) in the range of 6–20. These squat segments of DNA further self-assembled end-to-end, elongating to form rod shapes. This enabled liquid crystal ordering. Let's examine what the researchers saw when they looked at these stacked, ultrashort lengths of DNA. Figure 5.2 is an example, taken directly from one of their papers.[1]

First consider panel (a) of Figure 5.2. This is a schematic representation of the short, squat lengths of DNA that have stacked upon one another. The stacked nanoDNA is shown in a liquid crystal phase called *columnar* and Bellini

Constructive and destructive interference of light Consider a pair of light waves from the same source, with the same wavelength and the same amplitude. If the electric field vibrations associated with each wave occur in the same plane, then the light waves may *interfere*. Constructive interference will occur when the crests of the electric field vibrations of the two waves coincide. This interference doubles the amplitude of the electric field of the resultant wave. Destructive interference occurs when the crests of one wave coincide with the troughs of the other. In this situation the two waves will eliminate each other; the electric field vibration will have an amplitude of zero. Have a look at panel (a) of Figure 5.6.

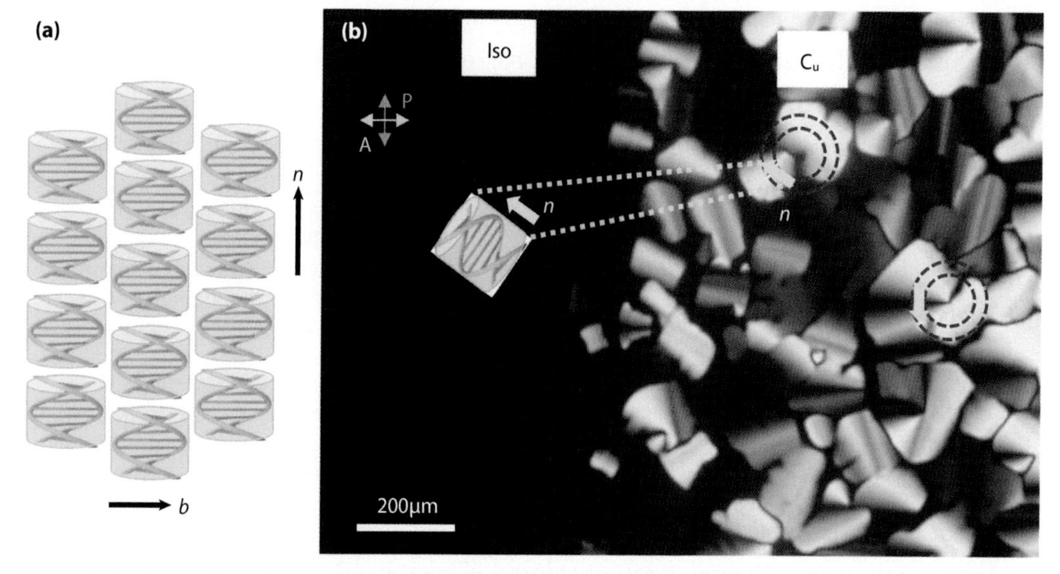

Figure 5.2 (a) Sketch of the nanoDNA columnar liquid crystal phase. **(b)** Optical texture of this phase obtained by depolarized light microscopy. *(Reproduced with permission from Dong Ki Yoon, Gregory P. Smith, Ethan Tsai, Mark Moran, David M. Walba, Tommaso Bellini, Ivan I. Smalyukh, and Noel A. Clark, "Alignment of the Columnar Liquid Crystal Phase of nano-DNA by Confinement in Channels," Liquid Crystals 39, No. 5 (May 2012), 571–577.)*

and Clark used synchrotron X-ray diffraction measurements (more on this later) along with depolarized light microscopy to determine this. The letter *b* has been chosen to denote the orientation of the base-pair planes. The base pairs are shown as multicolored horizontal bars joining the two strands of the double helices. The letter *n* denotes the direction of the axis of the cylinders that are formed by stacking of the nanoDNA duplexes. When we spoke about the twisted nematic cell we spoke of the "average, local alignment of the molecules." In liquid crystal jargon this is called *the molecular director* or simply *the director* and is denoted by *n* in Figure 5.2 (b).

The ringlike bands bounded by red dashes in panel (b) surround defects in the orientation of the director. These are line defects called *disclinations* and they run perpendicular to the plane of the photograph (i.e., straight out of the page). The orientational order in a liquid crystal has a discontinuity at a disclination. At these defect lines the director, *n*, changes abruptly or becomes ill defined. The stacked nanoDNA columns encircle these defects as indicated by the markings on the photograph.[1,2]

In this image the columnar phase is growing from the right to the left and so the left side is black and is labeled "Iso" which means isotropic. In the previous chapter we spoke often of anisotropic properties. The term *isotropic* describes properties that do not change with direction. If a liquid crystalline material is heated beyond a particular temperature called its clearing temperature, the orientation of the molecules becomes completely random. Thermal disorder rules and there is no orientational order and no positional order. The liquid crystal is then said to be in its isotropic phase. And in its isotropic phase it behaves like an ordinary liquid so incident plane-polarized light will not experience any rotation of its plane of polarization. If the isotropic material is placed between crossed polarizer and analyzer, the viewer will see blackness. That's what you see on the left side of panel (b). The experiment was done by first heating the entire sample above its clearing temperature and then cooling it down through a phase transition and into the temperature range in which the forces between stacks of DNA molecules are able to produce columnar order.

We'll elaborate on another language convention we mentioned in passing: the polarizer that is placed furthest from the light source is traditionally referred to as the analyzer. So when one speaks of a sample placed between crossed polarizer and analyzer it has the same meaning as saying that the sample is between crossed polarizers. That's why you see the pink and cyan double-headed arrows in panel (b) that are crossed perpendicular to each other and labeled as A and P.

Crossed polarizers also referred to as Crossed polarizer and analyzer
In depolarized light microscopy, one polarizer is placed before the sample in the optical path and one after the sample. These two polarizers are typically oriented at 90° with respect to one another. This is referred to as crossed polarizers. Often, the second polarizer is referred to as the analyzer.

Equating the microscope image in Figure 5.2 with its molecular interpretation may still feel elusive. To reiterate, the ambitious reader will find more offered in the Appendix on how to interpret such images. For example, what causes those beautiful curved fan shapes? However, if you've read this far and choose to skip the Appendix, you've still gotten a first impression of a challenging subject. Physicist Ingo Dierking of the University of Manchester is the author of a well-known book devoted to liquid crystal texture. Dierking wrote, "During years of teaching liquid crystal physics, in both lectures and the laboratory, I have gained the impression that, to many students, the textures of liquid crystals are very interesting and appealing on an aesthetic level, while at the same time they find it quite hard to relate the observed optic appearance to respective phases."[3] Most graduate students and younger researchers new to the field agree. If you also feel this way you have lots of company.

Transmission Electron Microscopy (TEM)

We now turn to transmission electron microscopy (TEM). In Chapter Three we spoke about the scanning tunneling microscope (STM) and how it jump-started nanotechnology. In 1986, the inventors of the STM, Gerd Binning and Heinrich Rohrer, were awarded one half of the Nobel Prize in physics. The other half was awarded to Ernst Ruska who designed and constructed the first electron microscope in 1931, fifty years prior to the first STM. The Nobel citation also acknowledged Ruska for his "fundamental work in electron optics," and with good reason. Electron optics is at the heart of the TEM. To approach this subject, think back to our discussion in Chapter Three concerning the wavelike behavior of electrons. One major difference between light optics and electron optics is that visible light has wavelengths of 400–700 nm (depending upon color) while electrons useful in microscopy have wavelengths of 0.001 and 0.01 nm (depending on electron velocity). The significantly smaller wavelength of electrons greatly improves spatial resolution and thus allows much higher magnification. As Harvard chemist George Whitesides wrote, "To tell the shape of something, one must use a probe that is smaller than the object. Fingers work well to tell the shape of a dog, but not the shape of its fleas."[4]

Spatial resolution The ability of a detection system (e.g., a microscope) to record details of the objects under study. It is often defined in terms of how close two features can be within an image and still be recorded as distinct.

The TEM can resolve two points that are approximately 0.2 nm apart, roughly the separation of atoms in a solid, whereas a *traditional* light microscope can only distinguish details that are 200 nm apart. (Recent advances have changed that, making *non-traditional* light microscopes into *nanoscopes*.[5]) Another distinction between light and electrons is that electrons carry a charge. So electromagnetic fields can be used as lenses to shape a beam of electrons and hence the importance of "fundamental work in electron optics" in Ruska's Nobel citation. An electromagnetic lens uses an electric current flowing through a coil of wire. The coil is placed within specially shaped magnetic material referred to as pole pieces. The current flow produces a magnetic field at right angles to the current direction. It is the force of this field that acts to direct and shape a beam of electrons traveling down the central axis of the coil. Electromagnetic lenses provide versatility because one can change the action of a lens for electrons by changing the current flow without having to physically change out or move the lens as one would with a light microscope.

A further difference between optical and electromagnetic imaging is that electrons are much more strongly scattered by air than is light. As a result, the column of an electron microscope must be under a vacuum with a pressure of no more than 10^{-8} that of atmospheric pressure at the surface of the Earth (about 10^{-5} Torr in a TEM column vs. about 10^{3} Torr at Earth's surface)—that is, a factor of one hundred million lower than atmospheric pressure.[6]

Transmission electron microscope (TEM) A TEM works on the same basic principles as a light microscope. However a TEM uses an electron beam instead of light. Electrons have a much lower wavelength compared to that of light. This gives the TEM a much higher spatial resolution. Objects less than a nanometer can be seen in a TEM. Electromagnetic lenses (an example of electron optics) are used instead of glass lenses.

You'll notice in Figure 5.3 that the TEM is oriented upside down compared to a conventional light microscope. The equivalent of a lamp in a light microscope is a source of electrons at the top of the evacuated column. This electron gun is often a tungsten wire that is heated (by passage of a current) to temperatures of several thousand degrees to produce so-called thermionic emission of electrons. The electrons that are released are accelerated to high velocities and electromagnetic lenses then shape and direct the electron beam. Typically, the samples are placed on a thin copper grid that is coated with a support film of Formvar (a polymer) and possibly carbon. The copper grids fit in a sample holder inserted into the column by means of an airlock. Most often the specimen features are imaged by means of a mass-thickness contrast mechanism. Regions of the sample that are thicker or of higher density scatter electrons more strongly—that is, more electrons will be deflected through a larger angle and will not make it through the aperture portion of the objective lens. These regions of the sample appear darker in the image.

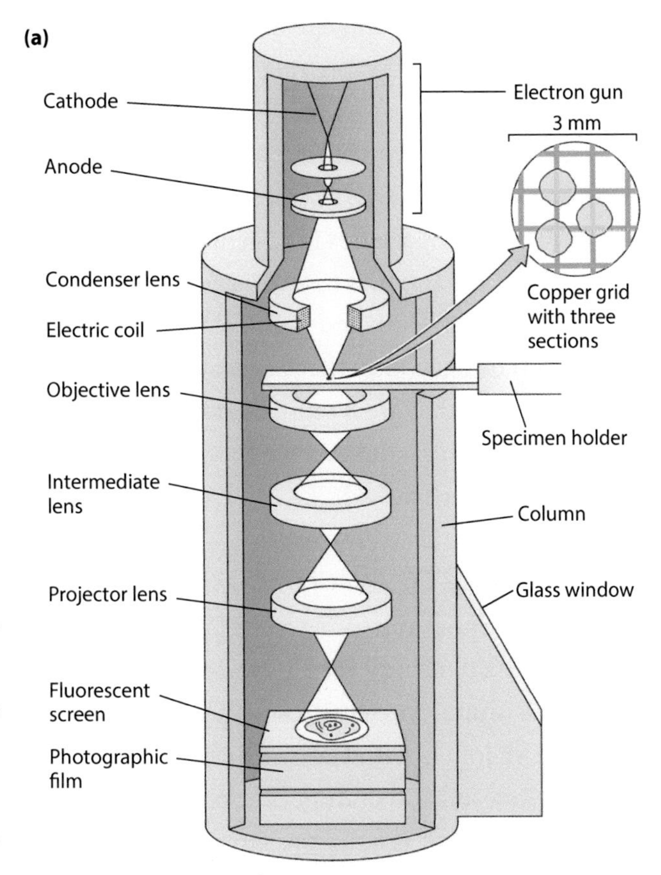

It's common practice to stain samples with heavy metal compounds such as uranyl acetate—a derivative of uranium known as depleted uranium because it has a lower content of the fissile isotope ^{235}U. This is done to enhance contrast provided by the density-dependent scattering of electrons. At the bottom of the microscope column the electrons that are not deflected or are only lightly scattered, hit a fluorescent screen. This creates a shadow image of the sample. Different details of the sample will appear in varying degrees of darkness depending on their density. The operator views these images on the fluorescent screen through a glass window and can record them by photography with a camera that is beneath the screen. Analytical TEMs may have electron detectors below the camera such as an electron energy-loss spectrometer (called an EELS)

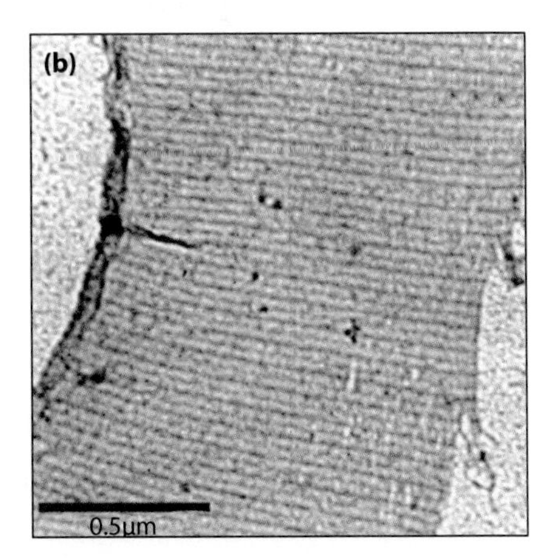

Figure 5.3 (a) Transmission electron microscope (TEM). **(b)** TEM image of two-dimensional DNA crystals stained with a 0.1% solution of uranyl acetate. The scale bar is 0.5 microns or 500 nm. *(**(a)** Reproduced with permission from* Junqueira's Basic Histology: Text and Atlas, 12/e *by Anthony Mescher, © 2010 McGraw-Hill Education;* ***(b)*** *Reproduced with permission from Shoujun Xiao, Furong Liu, Abbey E. Rosen, James F. Hainfeld, Nadrian C. Seeman, Karin Musier-Forsyth, and Richard A. Kiehl, "Self-assembly of Metallic Nanoparticle Arrays by DNA Scaffolding," Journal of Nanoparticle Research 4 (2002), 313–317.)*

that can measure the change in kinetic energy of the electrons after they've interacted with the specimen. This provides additional structural and chemical information about the sample.

The bottom portion of Figure 5.3 shows a two-dimensional DNA crystal that Nadrian Seeman and colleagues constructed to serve as a molecular scaffold (for the subsequent assembly of nanoparticles in specific geometrical arrangements).[7] The alternate dark-gray and light-gray bands in the image are the result of uranyl acetate staining. The major spacings (dark-to-dark and light-to-light) between gray bands are in agreement with the 64 nm design value that Seeman chose. This is also true for the minor spacings (dark-to-light) that are 32 nm, by design. Here is an example of the use of DNA as a programmable material that we spoke about in Chapter One—in this case a programmable molecular scaffold.

Atomic Force Microscopy (AFM)

We transition now from using some form of electromagnetic radiation in order to visualize small objects to using a tactile mode for that purpose. The imaging method we'll have a look at is atomic force microscopy, known as AFM. In 1986, Gerd Binnig, co-inventor of the STM, joined with Calvin Quate of Stanford University and Christoph Gerber of the IBM Research Laboratory (Zurich) to introduce a scanning-probe device that, unlike the STM, does not require the sample to be conducting.[8] Like the STM, the AFM uses a fine tip to scan back and forth over the surface of a sample. Rather than relying on an electrical signal (the tunneling current in the case of the STM), the AFM monitors the forces between the atoms in the tip and in the sample (Figure 5.4). And here the electron cloud that we spoke about in Chapter Three again comes into play. The AFM records contours of force, specifically, the repulsive force generated by the overlap of the electron cloud from the atom(s) at the very end of the tip with the electron clouds of surface atoms of the sample.

The probe of the AFM is a flexible cantilever that has been described as a diminutive diving board.[9] The sharp tip is attached to the bottom of the cantilever. The repulsive forces between tip and sample lead to a deflection of the cantilever. If the AFM is programmed to keep the force constant, referred to as the constant force mode of operation, an electrical feedback loop will keep the cantilever deflection constant.[10] The tip feels the surface, and the up and down vertical movement measures the atomic topography. Meanwhile, a laser is reflected off the cantilever and the path of the reflected laser beam changes. These changes are recorded in a position-sensitive detector that collects the reflected light. A computer then translates the data into an image.

There are a variety of different imaging modes available, for example, the tapping mode is gentle enough to image supported lipid bilayers such as we saw in the previous chapter. The AFM can also measure the elasticity or stiffness of a sample—*viscoelastic mapping*—as well as surface roughness and other benchmarks of specialized interest. There have been numerous spinoffs of the AFM, a world of different scanning-probe instruments that use specialized tips. One example is magnetic force microscopy where a magnetized tip scans a magnetic sample and the data are used to construct a map of the magnetic structure of the sample surface.

Seeman has made extensive use of the AFM. Figure 5.4 (b) shows the results of his experiments to optimize the annealing temperature for the formation of two-dimensional arrays of DNA origami tiles.[11] We'll speak at length about DNA origami in Chapter Eight. For now we'll say that it is an ingenious way of producing a large, precisely defined area built upon a two-dimensional DNA surface. As mentioned in Chapter Two, in DNA origami a long, single strand of DNA containing thousands of bases is folded into two-dimensional structures by using hundreds of smaller staple strands of synthetic DNA made up of about 30 to 50 bases. A favored scaffold strand has been derived from the virus named M13mp18 and contains over 7,000 bases whose exact sequence is known. The bases on the small DNA snippets bond to the bases on the scaffold to connect distant parts of the virus's genome and staple its folds together. Each staple strand occupies a specific position as a result of its unique sequence complementary to the long DNA scaffold strand. Basic structural units or building blocks that have been made using DNA origami are referred to as DNA origami tiles.

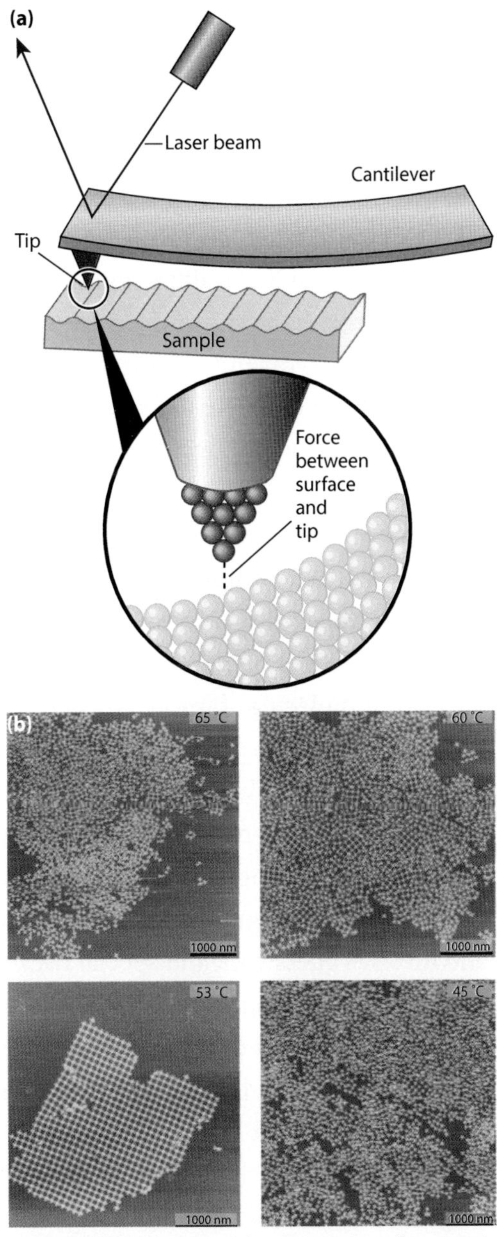

Figure 5.4 (a) The principle of operation of an atomic force microscope (AFM). **(b)** AFM images of experiments to optimize the annealing temperature for the formation of two-dimensional arrays of DNA origami tiles. *(Panel **(a)** adapted with permission from Sciencenter copyright 2012, illustrated by Emily Maletz for the NISE Network; Panel **(b)** reproduced with permission from Wenyan Liu, Hong Zhong, Risheng Wang, and Nadrian C. Seeman, "Crystalline Two-Dimensional DNA-Origami Arrays," Angew. Chem. Int. Ed. 50 (2011), 264–267.)*

Seeman employed origami tiles to get the AFM images in Figure 5.4. These experiments showed that the annealing temperature had to be carefully optimized to form ordered, periodic two-dimensional arrays of DNA origami tiles. Annealing at 45 °C, 60 °C, and 65 °C all failed; only annealing at 53 °C was successful, as shown in the lower left panel. (The temperatures may be a little hard to read; they are in the upper right of each of the four images.)

Atomic force microscope (AFM)
An instrument for imaging surfaces. A fine tip fixed to a cantilever moves back and forth over the surface of the sample. The AFM measures the forces between the atoms at the end of the tip and the surface atoms of the sample. The forces influence the cantilever's up-and-down motion and this information is converted to an image of the surface topography. Unlike the STM, the sample need not be an electrical conductor.

X-Ray Diffraction and Bragg's Law

It's time now to set off on a fresh track and take up the art and science of X-ray diffraction, a specialty that we gave a nod to in Chapter Two. We are familiar with the use of X-rays in medicine but perhaps not in the context of DNA nanoscience. X-ray diffraction is a principal tool of Seeman, Bellini, and Clark and is used to determine the structure of the very small molecules under investigation. Visible light won't do the job because the wavelength of visible light is very long compared with the dimensions of these objects—too coarse to reveal the details. As we said about transmission electron microscopy, it is not enough to image the dog—we must image its fleas. We need to use another form of electromagnetic radiation, X-rays, which have much shorter wavelengths. And we can't just examine a single molecule because it is so small that the reflections of the X-rays from it would be too weak to detect. We must use a large number of molecules for the X-rays to interact with so that these reflections reinforce one another and we can record them on a photographic plate.

If we happen to know the structure of a crystal, it is mathematically straightforward to calculate what X-ray pattern it would produce. But much of X-ray diffraction deals with the reverse problem: starting with the pattern of dots, diffuse bands, and streaks you find on a photographic plate and working back to the structure. This is not so easy and involves guesswork and special tricks of the trade. We'll try to get a feel for how the experiment is done and how the results are interpreted. The technique is quite remarkable: we can learn details of the structure of an individual molecule even though the X-ray images are generated by large numbers of identical molecules.

The basic idea of X-ray diffraction is shown in Figure 5.5. German physicist Max von Laue speculated on the possibility that electromagnetic radiation with a shorter wavelength than that of light (X-rays range in wavelength from approximately 10 to 0.01 nm) might cause some kind of interference phenomena when passing through a periodic arrangement of atoms. Walter Friedrich, Paul Knipping, and von Laue found that if they directed a beam of X-rays at a crystal that is turned in various directions, some of the X-rays do not travel in a straight line. When the transmitted rays hit a photographic film they produce a dark central spot and a pattern of fainter spots around it.

(The direct X-ray beam creates the center spot after its passage through the sample. It is the non-diffracted X-ray scattering.) The reason for this diffraction pattern is that X-rays are scattered or reflected by the electrons surrounding each atom in a crystal. Sometimes the reflected rays reinforce one another (constructive interference) and this produces one of the faint spots.

In 1912, von Laue published the mathematical formulation that explained this phenomenon.[12,13] Subsequently, British physicist and crystallographer William Lawrence Bragg along with his father William Henry Bragg provided an insightful interpretation of Laue's results and found a simpler way to analyze them.[14,15] Sir Lawrence Bragg was self-effacing about his contribution, "I first stated the diffraction condition . . . in 1912, and it has come to be known as *Bragg's law*. It is, I have always felt, a cheaply earned honor, because the principle had been well known for some time in the optics of visible light."[16]

To warm to our subject, we'll look at what the Braggs, father and son, had to say back in 1915.[17] "It is natural to suppose that the Laue pattern owes its origin to the interference of waves diffracted at a number of centers which are closely connected with the atoms or molecules of which the crystal is built, and are therefore arranged

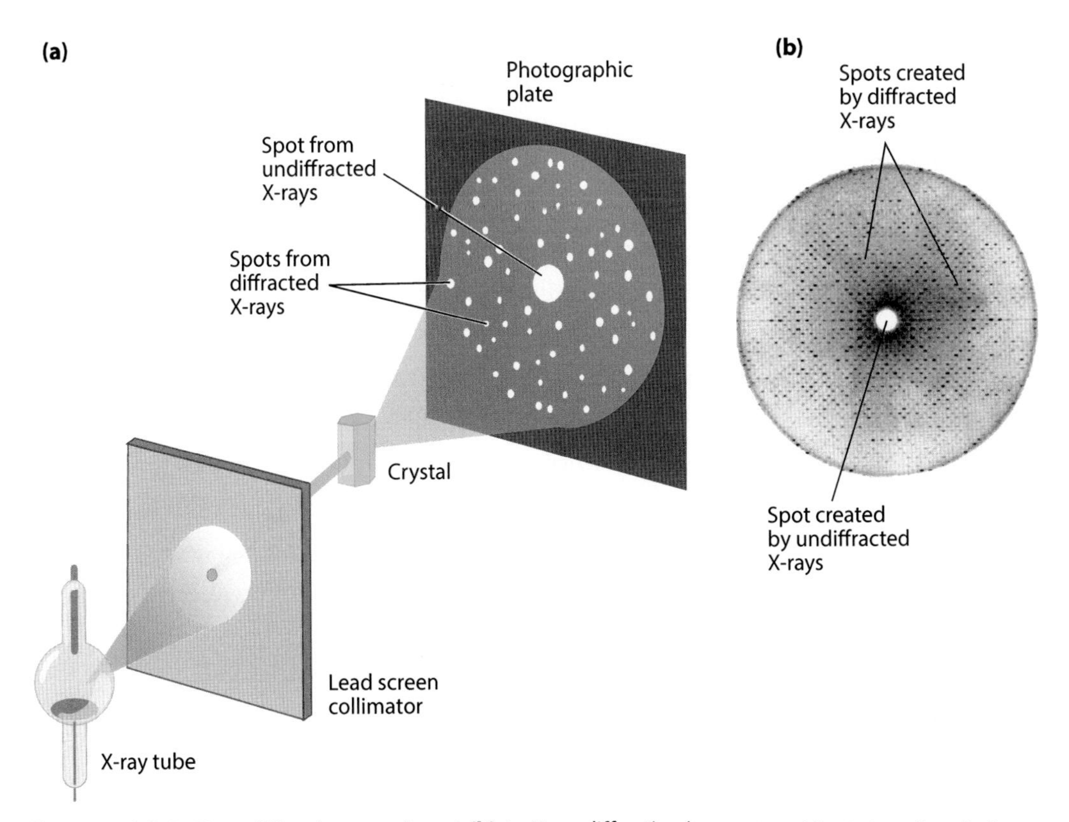

Figure 5.5 (a) An X-ray diffraction experiment. **(b)** An X-ray diffraction image. *(X-ray diffraction image from a bovine Cu,Zn superoxide dismutase crystal, SOD precession photo C2, by Dcrjsr (Own work), CC BY 3.0.)*

according to the same plan. The crystal is, in fact, acting as a diffraction grating." The optical case that Bragg referred to is shown in panel (a) of Figure 5.6. Here light is incident on a plate of glass or metal that is scored (i.e., ruled) with very closely spaced parallel grooves or slits (only two are shown). This is a diffraction grating. The explanation for what happens when light hits the grating goes way back to the seventeenth century Dutch physicist Christiaan Huygens and the nineteenth century French physicist Augustin-Jean Fresnel. Each slit becomes a source of a secondary cylindrical wave. You may find it helpful to think of this in terms of an ocean wave moving through a gap in a breakwater.[18]

X-ray diffraction A technique in which a beam of X-rays interacts with the atoms of a crystal. Recall the phenomenon of wave-particle duality. The uniform spacing of the atoms gives rise to an interference pattern as a result of the wave-like nature of the X-rays. The X-rays exit the crystal in the form of a complex group of beams. Then the intensities, angles of exit, and phases of these beams can be used to determine the detailed structure of the repeating unit of the crystal.

These waves interact to produce a complex pattern of varying intensity because of the superposition of the multiple waves. At certain places the wave crests or, alternatively, the wave troughs, will reinforce one another creating constructive or additive interference that results in intensity maxima (bright areas on the screen in panel (a) of Figure 5.6). At other places the wave crest from one secondary wave will overlap with the trough of another secondary wave and create destructive or subtractive interference (dark areas on the screen). Although the subject of interference came up earlier in this chapter, it is worth repeating because it occurs so frequently in different contexts.

We now use this approach to consider X-rays incident on a crystal. For convenience the incident X-rays are considered to be in phase. This means they are at the same point in their undulating cycle of movement at the same time. As Sir Lawrence Bragg recognized, the crystal acts as a complex diffraction grating—in place of slits are atoms lying in layers or planes (panel (b) of Figure 5.6). For layers of atoms that are at an angle θ to the X-ray beam, scattered X-rays will have constructive interference if the equation known as Bragg's law is satisfied. The equation requires that the difference between the path lengths of two scattered X-rays is equal to a whole number of X-ray wavelengths (panel (c) of Figure 5.6). When this condition is satisfied, a diffraction spot will appear just as we saw in Figure 5.5 that shows a schematic of an X-ray pattern and also the real deal.

During the experiment, the crystal is rotated. This brings different layers of atoms into play, layers that satisfy Bragg's condition and produce constructive interference. The resulting diffraction spot has an intensity related to the number and type of atoms in the layer. A typical diffraction experiment will measure thousands or even millions

Bragg's law The mathematical relationship connecting the spacing of atomic planes in crystals and the angles of incidence at which these planes give the most intense reflections of X-rays (i.e., constructive interference). The atomic planes of the crystal act on the X-rays just as a ruled grating acts on a beam of light.

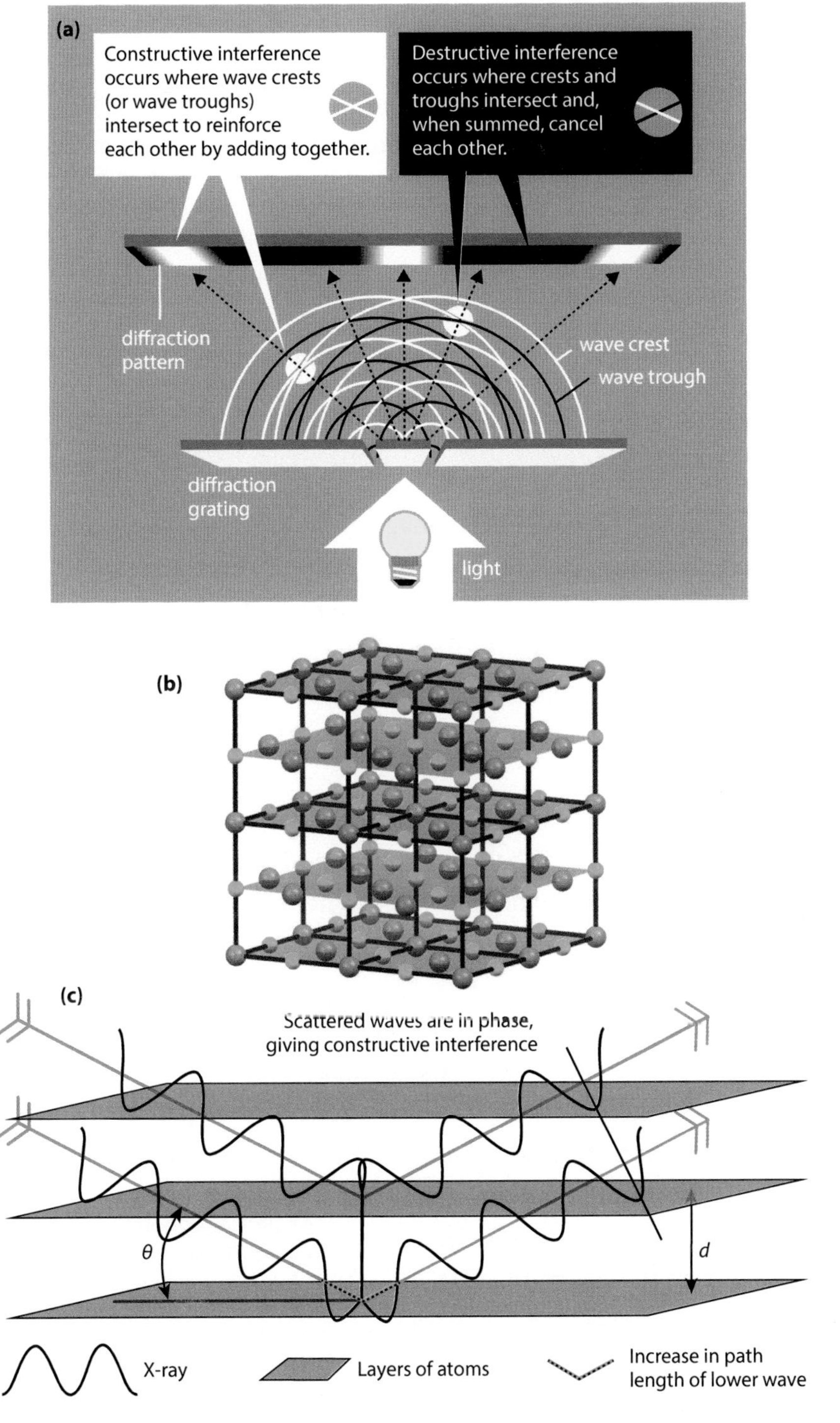

Figure 5.6 (a) Light passing through a diffraction grating. The resulting secondary waves interfere with each other producing a diffraction pattern consisting of intensity maxima (formed by constructive interference) and minima (formed by destructive interference). **(b)** Layers of atoms lying in planes act as slits in a diffraction grating. **(c)** The extra distance (green path) traveled by the lower X-ray must be a whole number of X-ray wavelengths to produce constructive interference and thus a diffraction spot. *(Panel **(a)** reproduced with permission from The Royal Swedish Academy of Sciences/Illustration: © Johan Jarnestad; Panel **(b)** courtesy of artist Mairi Haddow; Panel **(c)** courtesy of artist Mairi Haddow.)*

of reflections. If you choose to dip into the exercises at the end of this chapter, you'll need to know the expression for Bragg's law $n\lambda = 2d\sin\theta$, where $n = 1, 2, 3 \ldots$ The angle θ is measured with respect to the surface of the crystal; λ is the wavelength of the X-rays incident on the crystal; d is the distance between the layers of atoms.

The Phase Problem

Each type of crystal has its own characteristic arrangement of atoms (and hence electrons) and so will produce its own specific X-ray pattern. However, most often one takes a molecule of unknown structure and forms a crystal composed of many identical such molecules periodically arranged. After performing X-ray diffraction on this crystal, each spot on the diffraction pattern provides information *only* about the *intensity* of the X-ray waves that have been deflected to form that spot. But for large molecules this is insufficient to derive an accurate, high-resolution image of the molecule in question. The expermenter must also know the *phase* of the wave. Each X-ray wave executes an undulating cycle of movement from one wave crest to the next. The phase specifies where the wave is in that cycle when it hits the film and forms a spot in the diffraction pattern. Obtaining this information is known as *the phase problem.* (The use of the term *phase* in the context of X-ray diffraction has nothing to do with the term *phase* in the context of liquid crystals.)

Think of it this way. Suppose you're standing on the ground and you and your friend are playing catch with a Nerf® ball. But your friend is on a trampoline, bouncing up and down. You need to know where she'll be in her range of motion to give her a fair chance at catching the ball. You need to time your throw to match the phase of her motion.

Historically, the phase problem has been a formidable challenge. Approaches to solving the phase problem have evolved over many decades. In the mid-1930s, British physicist and crystallographer Arthur Lindo Patterson developed what quickly came to be known as the Patterson Function or just the Patterson. (In Chapter Two we referenced the notebooks of Rosalind Franklin—endnote 11—and their pages are rife with her use of Pattersons to find the structure of DNA.) The Patterson is a particular type of Fourier transform—and that takes us back some two hundred years to the French mathematician and physicist Jean Baptiste Joseph Fourier. He had a major impact on mathematical physics.

Fourier discovered that any waveform can be decomposed into a set of pure sine

The phase problem Relevant to an issue in X-ray diffraction. A diffraction pattern only shows the intensity of the diffracted X-rays. To work back to the exact molecular structure of the molecules that make up the crystal, it's also necessary to have the phase data—that is, the position of the wave crests and troughs of the diffracted X-rays relative to one another.

Patterson function In 1934, Arthur L. Patterson found a solution to the phase problem, and colleagues dubbed it the Patterson function.

and cosine waves of shorter and shorter wavelength and thus of higher and higher frequency that sum to the original waveform. The Fourier transform is the mathematical tool that decomposes the waveform into the simpler sine and cosine waves. (An example of a sine wave appears in Figure 4.10 of Chapter Four; it is a curve showing periodic oscillations of constant amplitude.)

Returning to Arthur Lindo Patterson, his Fourier-derived method is a workaround with respect to the missing phase information. (It doesn't actually require phase information—contrary to our earlier declaration about the necessity of phase.) It was used successfully to solve many simple crystal structures and is still important in the location of heavy atoms in a molecule of interest.[19] Although Patterson functions still have value, there are now a plethora of advanced techniques (that employ Fourier transforms too, but in a different way). They have names such as direct methods, isomorphous replacement, anomalous scattering, and molecular replacement. What is referred to as *phasing* is the most acronymic realm of crystallography with MIR (multiple isomorphous replacement), MAD (multi-wavelength anomalous diffraction), and SAD (single-wavelength anomalous diffraction) experimental methods.[20] Modern crystallographers use these tools to create an *electron density map* because X-rays scatter from the electron clouds of atoms in the crystal lattice. The electron density map is a three-dimensional description of the electron density within a crystal structure (actually, within the unit cell of the crystal). In practical terms, computers perform the analysis necessary to produce these maps of the crystallized molecule, and then fit the molecular structure into this calculated electron density map. This last step is called *interpreting* the map.

Figure 5.7 shows a region of the electron density map of a protein crystal before it is interpreted and the same map after its interpretation in terms of molecular structure.[21]

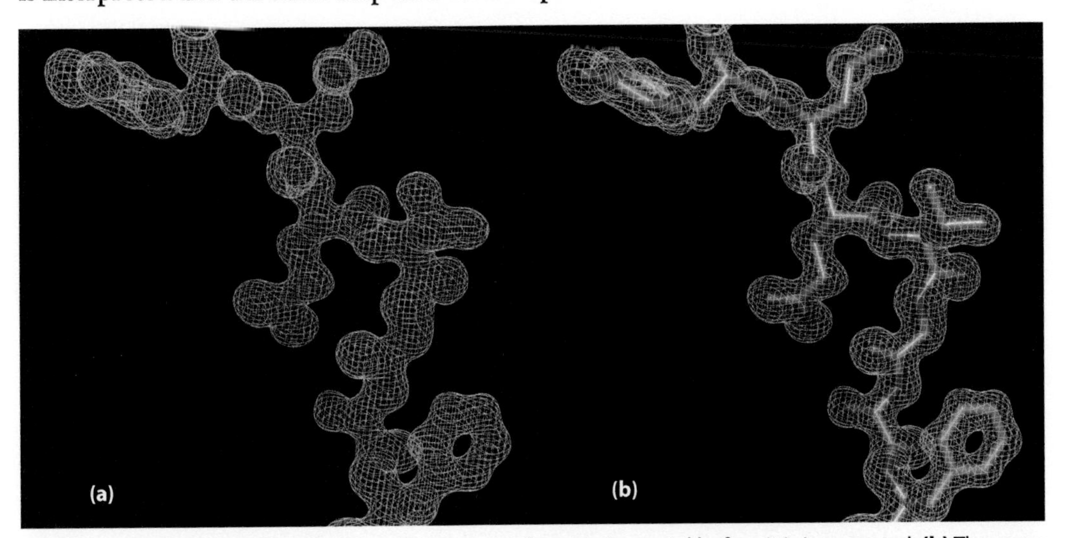

Figure 5.7 (a) Region of the electron density map of a protein crystal before it is interpreted. **(b)** The same electron density map after its interpretation in terms of molecular structure. *(Reproduced with permission from http://www.xtal.iqfr.csic.es/Cristalografia/index-en.html and Dr. Martin Martinez-Ripoll.)*

The map shows contours enclosing areas of higher electron density and the interpreted map uses sticks to indicate bonds between atoms. The map is interpreted by building a model that fits it using some prior knowledge of the chemical structure but not of the conformation—that is, the atomic spatial arrangement of the molecule under study. Thus, interpreting the image involves determining a chemically realistic conformation—bond lengths, bond angles, and so on—to fit the electron density map.

Electron density map A three-dimensional description of the electron density in a crystal structure determined from X-ray diffraction.

Synchrotron X-Ray Diffraction

Figure 5.5 presented the basic idea of X-ray diffraction. In a typical laboratory the source is an X-ray tube, an evacuated, sealed container in which electrons are produced by heating a metal filament and then accelerated and made to collide with a metal plate. These collisions produce both a continuous spectrum of X-rays as well as much more intense, sharply defined peaks at specific energies. The well-defined X-ray energies are used for the X-ray diffraction studies we've been discussing. But bear this in mind: the wavelength of these X-rays is determined by the choice of metal plate, copper for example. Thus, the X-ray wavelength from these sources cannot be tuned or changed. This is a snag for a zealous scientist wanting to mine as much information as possible. However, these X-ray sources can be used to solve crystal structures and, in recent decades, have been used to gather preliminary data and check crystal quality. But then the researcher—Seeman, Bellini, and Clark among them—takes the next big step. That step is to bring the sample to a *synchrotron* X-ray source at facilities such as Brookhaven National Laboratory and Argonne National Laboratory.

A synchrotron is a type of particle accelerator originally designed for high-energy physicists studying subatomic particles such as electrons. It is a circular, ringlike device in which charged particles travel in evacuated pipes. The particles are accelerated by alternating electric fields at cavities along the circumference of the ring. Guiding magnets are also positioned around the circumference to produce magnetic fields that bend the particles into a closed path. The magnetic fields are time dependent and must be *synchronized* with the particle acceleration to keep the particles on their path, hence the name synchrotron. Synchrotron radiation is the name given to light radiated by an electric charge following a curved trajectory. The emitted light is intense radiation composed of many wavelengths (it is polychromatic), but the radiation is tunable because one can select very narrow bands of wavelengths that are appropriate for particular experiments. The experimentalist can do this with monochromator crystals that filter the polychromatic X-rays and select only one of the wavelengths—making it monochromatic or nearly monochromatic—*by exploiting Bragg's law.*

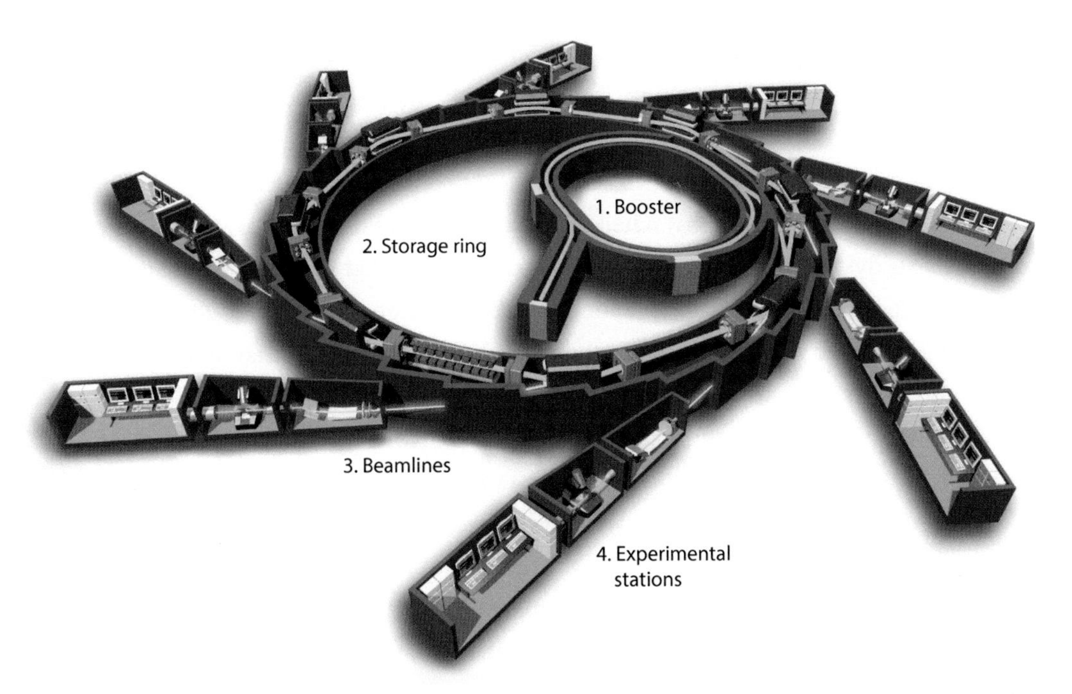

Figure 5.8 Synchrotron radiation light source. A diagram of *Synchrotron Soleil,* a synchrotron facility southwest of Paris, France. To get a feel for size, the storage ring is 354 meters (or 1,161 feet) in circumference. *(Adapted with permission from Synchrotron SOLEIL, © SOLEIL- EPSIM –J.F. Santarelli.)*

Synchrotron X-ray diffraction
The particle accelerators called synchrotrons provide tunable and high intensity X-rays. These features are lacking in standard laboratory X-ray apparatus.

Reflect on that. The same principle that *explained* X-ray diffraction patterns, Bragg's law, can also be used as a component in X-ray optics to *implement* X-ray diffraction. Scientists love that kind of serendipity. Who wouldn't?

There are many reasons that synchrotron radiation is such a valuable resource for X-ray diffraction, but the tunability of the wavelength and the high brightness of the X-ray beam are the most important characteristics.[22] The high brightness makes it possible to collect data in a much shorter time than with conventional X-ray sources. And when probing materials such as liquid crystal films, the extremely high-intensity radiation yields a reasonable number of scattered X-rays even if the effects of interest are intrinsically weak. (Regrettably, there is a lack of uniformity in the use of the terms *brightness* and *intensity.*[23])

In a synchrotron radiation light source, electrons are produced in an electron gun and go into a linear accelerator that connects to a booster section (which is actually a small synchrotron) where they are pre-accelerated before being injected into the largest component of a synchrotron, the storage ring (Figure 5.8).[24] Thus, a synchrotron light source is a combination of different electron accelerator devices. The ring is actually not a perfect circle, but a many-sided polygon. At each corner of the polygon, the steering magnets bend the electron stream and force it to stay in the ring (otherwise the particles would travel straight ahead and smash into the ring's wall).

Here's the key to synchrotron radiation: at each bend in the electrons' path, the electrons emit bursts of energy in the form of electromagnetic radiation that are directed along tangents to the beam. Thus, tangential to the storage ring are individual beamlines that lead to experimental stations where scientists conduct their research.

In the early years (ca. 1960s) when synchrotrons were designed and built for particle physicists studying subatomic particles, synchrotron radiation was regarded largely as a nuisance because it represented an energy leak that made it more difficult and expensive to push electrons to higher and higher energies.[25] Scientists other than particle physicists were parasitic users on those early machines.

How times have changed. Now many synchrotrons are specifically designed to optimize X-ray production for the use of thousands of biologists, chemists, environmental scientists, condensed matter physicists, and nanoscience researchers of every stripe. As an example, the National Synchrotron Light Source II (NSLS-II) at Brookhaven National Laboratory opened in 2015 with a storage ring about a half-mile long and, as construction continues, *60 to 80 beamlines.* The facility at Brookhaven will be busy as a Brueghel![26] Moreover, around the world there is a push to generate ever-smaller diameter beams and, yes, even nanobeams. Synchrotron X-ray beams with diameters less than 10 nanometers have already been demonstrated.[27]

We'll close this discussion of X-ray diffraction with a very revealing figure from Bellini and Clark.[28] Figure 5.9 shows both a depolarized light microscopy optical image (bottom) and synchrotron microbeam X-ray diffraction patterns (top) of selected regions of the same sample. The sample is a solution containing nanoDNA that is 16 base pairs long and held between glass plates and a polarizer and analyzer. Microbeams are X-rays that are focused down to beam diameters smaller than one micron (= 1000 nm).[29] This was an elegant and technically demanding experiment in which optical microscopy was performed simultaneously with synchrotron X-ray scattering (at the Advanced Photon Source at Argonne National Laboratory).

The top two panels of Figure 5.9 both show small domains of the sample that are 10 microns x 10 microns in area and 6 microns in thickness. The black shadow in each of these panels is caused by a piece of metal called a beamstop that blocks the direct X-ray beam from reaching the film (or X-ray detector) after its passage through the sample. This is often done so that the intensity of the direct X-ray beam doesn't obscure the nearby diffraction spots caused by low-angle scattering. The blocked beam is what showed up in the middle of Figure 5.5 (b) as the spot created by undiffracted X-rays.

The five panels of Figure 5.9 are intimately connected to one another. The depolarized light microscopy image in the bottom panel shows a columnar liquid crystal phase labeled as C_U as

Microbeam X-ray diffraction Synchrotron X-ray diffraction using X-ray beams focused down to beam diameters smaller than one micron (1 micron = 1000 nm).

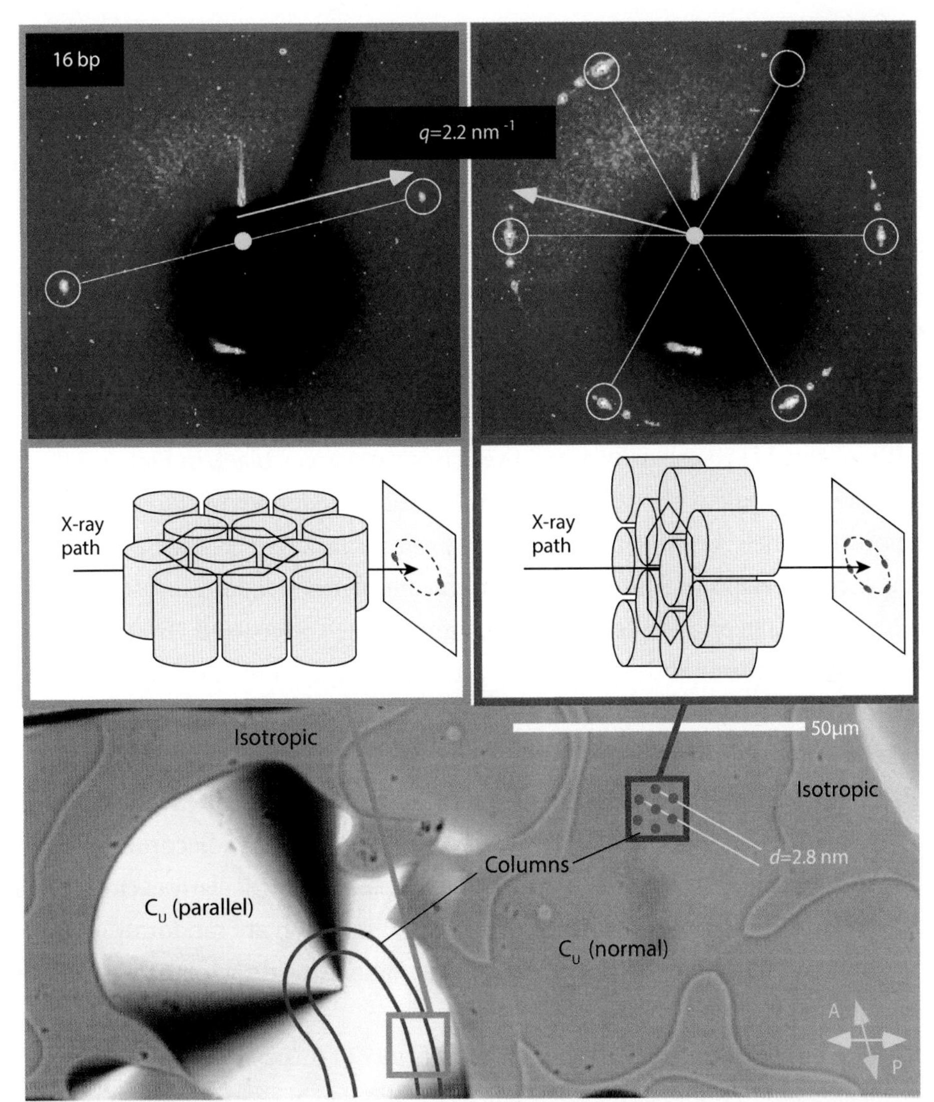

Figure 5.9 Depolarized light microscopy images of nanoDNA in a liquid crystalline columnar phase and synchrotron microbeam X-ray diffraction patterns of selected 10 micron x 10 micron domains. *(Adapted with permission from Michi Nakata, Giuliano Zanchetta, Brandon D. Chapman, Christopher D. Jones, Julie O. Cross, Ronald Pindak, Tommaso Bellini, and Noel A. Clark, Supporting Online Material for "End-to-End Stacking and Liquid Crystal Condensation of 6- to 20-Base Pair DNA Duplexes," www.sciencemag.org/cgi/content/full/318/5854/1276/DC1)*

well as an isotropic phase. First consider the columnar phase. (It may be helpful to have a look at the dodecamers [liquid crystals that are 12 base pairs long] in Figure 5.2 of this chapter when studying Figure 5.9.) The area labeled C_U (parallel) is a region where the columns of nanoDNA are oriented in the plane of the page (similar to Figure 5.2), and schematically shown by the curved blue lines drawn on the image. This region is sampled in the small purple square that connects to the drawing of the nanoDNA columns in the middle left of Figure 5.9. The diffraction pattern at the upper left of the figure is the result of microbeam X-ray scattering in this region of the sample where telltale diffraction spots have been enclosed in white circles.

Returning to the bottom panel, the area labeled C_U (normal) is a region of the sample where the columns of nanoDNA are oriented perpendicular to the page. They are schematically indicated by the blue dots that have been drawn on the image. This region is sampled in the small red square that connects to the drawing of the nanoDNA columns in the middle right of the figure. Microbeam X-ray diffraction scattering from this area of the sample results in the diffraction image in the upper right of the figure. Again, the locations of the diffraction spots are enclosed in white circles.

Now consider the areas of the figure that are not columnar—that is, they are designated as neither C_U (parallel) nor C_U (normal). They are isotropic. The label isotropic might seem perplexing. In our discussion of Figure 5.2 we said that isotropic areas should appear black. It's reasonable to ask: why aren't these isotropic areas black in Figure 5.9? To see why, take a close look at the pink and cyan double-headed arrows in the lower right of the bottom panel. At the same time, recall our remarks on crossed polarizers and birefringence in Chapters One and Four. In the depolarized light microscopy image of Figure 5.9, the analyzer and the polarizer (A and P) are *not* crossed. That is the answer. Here's what the authors of the paper say: "The apparently isotropic domains have the column axes normal to the glass plane: they become birefringent upon tilting the sample. . . ."[30] Thus, these regions of the sample do *not* appear black in the figure (as they would if the polarizers were crossed). Instead, by tilting the sample we are now viewing the liquid crystals through *un*crossed polarizers. The area in question then becomes birefringent and so it displays color. The texture in the isotropic areas isn't particularly exciting precisely because the area is isotropic. The liquid crystals in this region are randomly oriented and there are no sudden changes in the molecular orientation (the director) as we spoke about when we evaluated Figure 5.2.

It may be hard to appreciate the degree of difficulty of this experiment, just as many of us were challenged to understand the nuances of judging ice-skating compulsory figures before they were eliminated from Olympic singles competition in 1990. It's fair to say that these simultaneous optical and X-ray experiments and the interpretation of the results were of Olympic caliber.

It's time to size up where we're at in our narrative now that we've concluded Part I. We have met the main players—Seeman, Bellini, and Clark—and we have a familiarity with the subjects that engage their interests. Therefore, as Sherlock Holmes said—borrowing from Shakespeare—"the game is afoot." In Part II we confine our attention to Seeman and the colleagues who joined him in his quest to use DNA to build structures and devices on the nanoscale. Bellini and Clark will wait patiently until we turn the spotlight on them in Part III.

Exercises for Chapter Five: Exercises 5.1–5.2

Exercise 5.1

Suppose you direct a beam of X-rays with wavelength 0.154 nm at the planes of a crystal of silicon. As you increase the angle of incidence (starting at 0°) you find the first strong interference maximum from these planes when the X-ray beam makes an angle of 34.5° with the crystal planes. **(a)** How far apart are the crystal planes? **(b)** Will you find other interference maxima from these planes at larger angles of incidence? Why or why not?

(Exercise reproduced with permission from Hugh D. Young, Roger A. Freedman, and A. Lewis Ford, University Physics With Modern Physics Technology Update, 13th Ed., © 2014, p. 1208, Pearson Education Inc., New York, New York.)

Exercise 5.2

A rod cell in the eye has the structure shown in Figure 5.10. **(a)** Using a depolarizing microscope, at what angle with respect to the analyzer would the cell be most visible? **(b)** At this angle where are the brightest regions of the cell? **(c)** What simple modification could you make to the microscope to determine the orientation of the parallel discs? That is, how could you show that the discs are perpendicular and not parallel to the long axis of the cell?

(Exercise and Figure 5.10 reproduced with permission from Physical Biochemistry, *2/e by David Freifelder, Copyright 1982, p.69, by W.H. Freeman and Company, New York. Used with permission of the publisher.)*

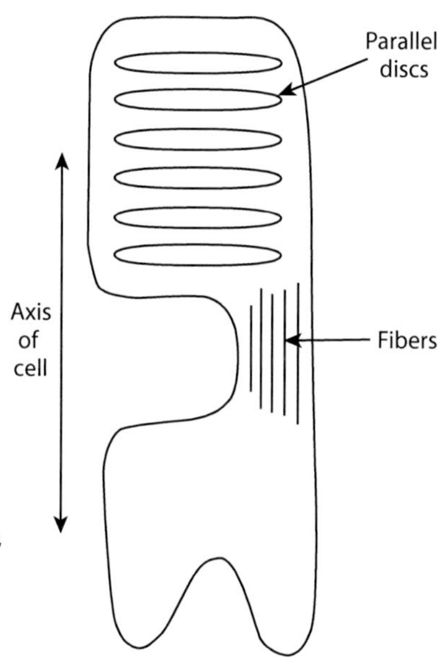

Figure 5.10

ENDNOTES

[1] Dong Ki Yoon, Gregory P. Smith, Ethan Tsai, Mark Moran, David M. Walba, Tommaso Bellini, Ivan I. Smalyukh, and Noel A. Clark, "Alignment of the Columnar Liquid Crystal Phase of nano-DNA by Confinement in Channels," *Liquid Crystals 39, No. 5* (May 2012), 571–577.

[2] F. Livolant, A.M. Levelut, J. Doucet, and J.P. Benoit, "The Highly Concentrated Liquid-Crystalline Phase of DNA Is Columnar Hexagonal," *Nature 339* (29 June 1989), 724–726.

[3] Ingo Dierking, *Textures of Liquid Crystals* (Weinheim, Germany: WILEY-VCH Verlag GmbH & Co., 2003), vi.

[4] Felice Frankel and George M. Whitesides, *On the Surface of Things: Images of the Extraordinary in Science* (San Francisco: Chronicle Books, 1997), 136.

[5] Ann Fernholm, "How the Optical Microscope Became a Nanoscope," The Nobel Prize in Chemistry 2014, The Royal Swedish Academy of Sciences (2014), http://www.nobelprize.org/nobel_prizes /chemistry/laureates/2014/popular-chemistryprize2014.pdf

[6] Michael J. Dykstra and Laura E. Reuss, *Biological Electron Microscopy: Theory, Techniques, and Troubleshooting* (New York: Springer Science + Business Media LLC, 2003), 323.

[7] Shoujun Xiao, Furong Liu, Abbey E. Rosen, James F. Hainfeld, Nadrian C. Seeman, Karin Musier-Forsyth, and Richard A. Kiehl, "Self-assembly of Metallic Nanoparticle Arrays by DNA Scaffolding," *Journal of Nanoparticle Research 4* (2002), 313–317.

[8] G. Binnig, C.F. Quate, and Ch. Gerber, "Atomic Force Microscope," *Physical Review Letters 56, No. 9* (3 March 1986), 930–933.

[9] "Atomic Force Microscope," Nanoscale Informal Science Education (NISE Network), http://www.nisenet.org/node/3449

[10] "Atomic Force Microscope," ETH Zurich, http://www.ferroic.mat.ethz.ch/research/labs_general/afm

[11] Wenyan Liu, Hong Zhong, Risheng Wang, and Nadrian C. Seeman, "Crystalline Two-Dimensional DNA-Origami Arrays," *Angew. Chem. Int. Ed. 50* (2011), 264–267.

[12] W. Friedrich, P. Knipping, and M. Laue in *Sitzungsberichte der Math. Phys. Klasse (Kgl.) Bayerische Akademie der Wissenschaften* (1912) 303–322.

[13] Max von Laue, "Concerning the Detection of X-ray Interferences," Nobel Lecture (12 November 1915), 347–355; note p. 352 for paper with Friedrich and Knipping submitted to the Munich Academy, http://www.nobelprize.org/nobel_prizes/physics/laureates/1914/laue-lecture.pdf

[14] W.L. Bragg, "The Diffraction of Short Electromagnetic Waves by a Crystal," *Proceedings of the Cambridge Philosophical Society XVII* (October 28, 1912–May 18, 1914), 43–57.

[15] W.L. Bragg, "The Specular Reflection of X-rays," *Nature 90, 2250* (12 December 1912), 410.

[16] Sir Lawrence Bragg, "X-ray Crystallography," *Scientific American 219* (July 1968), 58–70.

[17] W.H. Bragg and W.L. Bragg, *X Rays and Crystal Structure* (London: G. Bell and Sons, Ltd., 1915), 8–10, http://archive.org/details/xrayscrystalstru00braguoft

[18] http://proj.ncku.edu.tw/research/commentary/e/20111021/1.html

[19] "Patterson Methods," *The International Union of Crystallography Online Dictionary,* http://reference.iucr.org/dictionary/Patterson_methods

[20] Garry Taylor, "Introduction to Phasing," *Acta Cryst. D66* (2010), 325–338.

[21] CSIC Crystallography, Section 7. Structural Resolution, http://www.xtal.iqfr.csic.es/Cristalografia /parte_07-en.html

[22] Zbigniew Dauter website (Center for Cancer Research at Argonne National Laboratory), http://ccr.cancer.gov/staff/staff.asp?profileid=5859

[23] Richard Talman, *Accelerator X-ray Sources* (Weinheim, Germany: WILEY-VCH Verlag GmbH & Co., 2006), 71–74.

[24] http://www.synchrotron-soleil.fr/portal/page/portal/RessourcesPedagogiques/Soleil3Questions#CommentFonctionneSOLEIL

[25] Herman Winick, "Synchrotron Radiation," *Scientific American 257* (November 1987), 88–99.

[26] http://www.bnl.gov/ps/nsls2/about-NSLS-II.asp

[27] Gene E. Ice, John D. Budai, and Judy W.L. Pang, "The Race to X-ray Microbeam and Nanobeam Science," *Science 334* (2 December 2011), 1234–1239.

[28] Michi Nakata, Giuliano Zanchetta, Brandon D. Chapman, Christopher D. Jones, Julie O. Cross, Ronald Pindak, Tommaso Bellini, and Noel A. Clark, Supporting Online Material for "End-to-End Stacking and Liquid Crystal Condensation of 6- to 20-Base Pair DNA Duplexes," http://www.sciencemag.org/content/supl/2007/11/20/ 318.5854.1276.DC1

[29] http://www.aps.anl.gov/Science/Highlights/2000/microbeam.htm

[30] Nakata et al., "End-to-End," 5–6.

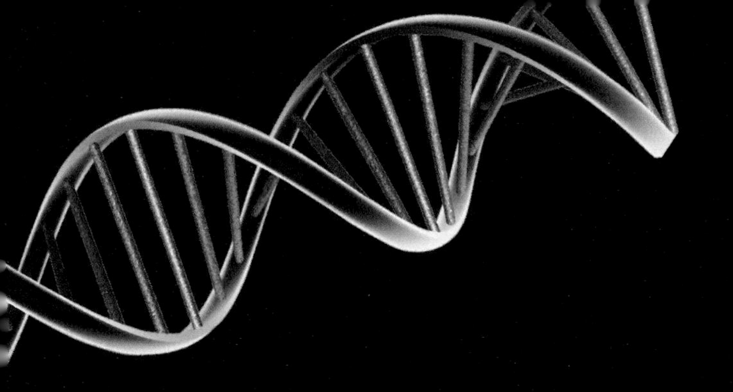

PART II

The Emerging Technology: Nanomaterials Constructed From DNA

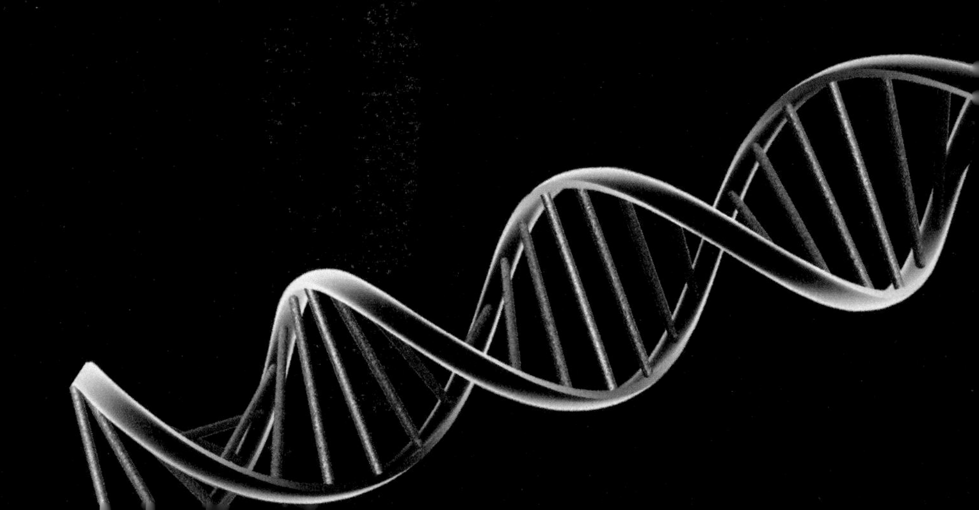

CHAPTER SIX

The Three Pillars of Structural DNA Nanotechnology

A bearded, bespectacled man stood by Exit 24 off the New York State Thruway waiting for the bus to take him to campus. His car had broken down and he was surrounded by snow three or four feet deep that stretched as far as the eye could see. In the late 1970s he'd taken the only job offer he had received—Assistant Professor of Biology at the State University of New York at Albany. He was reminded for the umpteenth time that he hated the place. When he and his friends got together he likened the conclave to something you'd see in World War II POW movies. All they talked about was "how the hell are we going to get out of this dump?"[1]

But Nadrian Seeman had grit. Following his appointment as assistant professor it took him seven years to find a qualified graduate student. "The graduate-student complement in that department [biology] was largely math-phobic, failed premeds."[2] He continued grinding away to bring his speculations into the light of day. His motivation was the desire to use DNA as a new and powerful way to conquer the challenge of crystallizing macromolecules. As we learned in Chapter One, he had a low opinion of doing conventional crystallization experiments. In such experiments you fill a container with a huge number of identical molecules—the molecule you want to crystallize. Then you concentrate the solution and hope that all the molecules will arrange themselves in a periodic pattern. As Seeman put it, "If you get a glop of crap, you have no idea what you did wrong."[3]

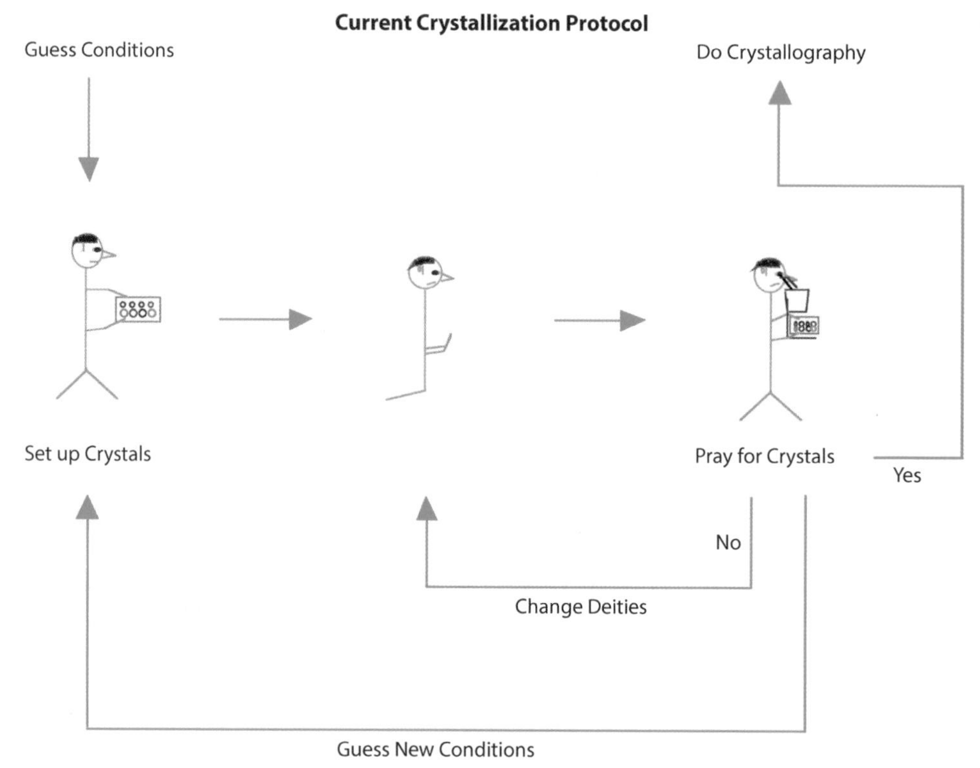

Figure 6.1 Seeman's view of conventional crystallization experiments. *(Conceived and drawn by Nadrian C. Seeman, used with permission.)*

Figure 6.1 is taken from a slide that Seeman has used to make his point to an audience.[4] He was desperate to find a new way to make crystals.

Branched DNA and DNA Junctions

Seeman's ambition sprouted in 1979 when he was considering a phenomenon called recombination in which two pieces of double helix DNA break and partially unravel.[5] In this configuration, each of the four strands switches from one partner to another. This conformation, which we'll shortly return to, is called a branch point or a junction. In recombining DNA, the branch point moves around; that is, it can occur with different combinations of complementary bases. The branch point is said to migrate. Seeman was working out a way to describe the nature of this branch point migration.[6] It was then that it struck him: it should be possible to construct synthetic DNA to make branched molecules whose branch points do not move. And these might be used to build lattice structures that could give him a new way to make molecular crystals.

To better understand his premise we introduce, in an *ad hoc* fashion, the term *free energy*. In thermodynamics, free energy refers to the energy available to do work rather than just be dispersed as heat.[7] This quantity is the driving force determining which product is favored in a chemical reaction that can have multiple possible outcomes.

Free energy Energy available to do work rather than just be dispersed as heat.

Work The energy transfer that occurs when an object is moved over a distance by a force at least part of which is applied in the direction of the displacement.

A chemical system tends to change toward the state that has the lowest free energy. Free energy also governs conformation—that is, the folds and joins of large molecules such as DNA and RNA. For two strands of nucleotides in which the bases are complementary, it happens that the free energy is minimized when they pair up to form a double helix. One might say that DNA's favorite structure is the conventional double helix. Seeman realized that the four strands of his hypothetical immobile junction could come together and form the maximum amount of conventional DNA double helices only by forming a branched molecule.

In nature an immobile, stable branch point is not favored because it increases the free energy of the molecule. But if one tinkered with synthetic DNA in just the right way, the increase in free energy by forming a junction could be outweighed by the greater energy saving in forming four arms of double helix DNA. In 1979 the challenge was a bit much for Seeman: creating stable, branched DNA molecules was state-of-the-art chemistry. He was a crystallographer, not an organic chemist. So he mostly just mused on the idea, stashing it in a small drawer in his mind.[8]

And then he had his epiphanic moment. This was his inspiration drawn from the Escher woodcut of flying fish described in Chapter One, the woodcut that Seeman saw as an idealized picture of branch points of a six-arm junction. In 1982 he published a paper in which he explained how it would be possible to form immobile junction structures of nucleic acids.[9] Maybe, just maybe, he "might be able to organize matter on the nanometer scale in the same way that Escher held his school of fish together using his imagination."[10] He had mused to great advantage and his idea was brilliantly thought out and elaborated. But at that time it was an entirely theoretical suggestion and so the paper appeared, appropriately enough, in the *Journal of Theoretical Biology.* In the subsequent thirty years the field of DNA nanotechnology has gone from a hypothesis to a burgeoning enterprise. As Robert F. Service said in 2011, "DNA nanotechnology has left its childhood behind and entered adolescence."[11]

The Seeman paper of 1982—and his work over the next decade to implement his vision—was the fountainhead for the rise of DNA nanotechnology. There have been remarkable and divergent advances since that time (for example, Paul Rothemund's DNA origami and Peng Yin's DNA bricks, both discussed later in this book). But Seeman's seminal concepts are still at the heart of the matter. So in this chapter and the next we examine Seeman's foundational ideas and his painstaking efforts to bring them into being. Here are his three pillars of structural DNA nanotechnology: (1) hybridization; (2) stably branched DNA, i.e., immobile DNA branched junctions; and (3) convenient synthesis of designed sequences.[12]

Immobile DNA branched junction A convergence of strands of synthetic DNA at a junction such that the intersection point does not change.

Hybridization The process of joining two complementary strands of nucleic acids, e.g., two single strands of DNA, to form a double-stranded molecule (a duplex).

Oligonucleotides Short, single-stranded DNA or RNA molecules, i.e., short nucleic acid polymers.

Hybridization is the process of joining two complementary strands of DNA by base pairing to form a double-stranded molecule. Stably branched DNA refers to synthesis of DNA sequences that will preferentially join to form immobile branches rather than the familiar intertwined linear chains of DNA that constitute the genetic material of living organisms (and which do not exhibit branch points). Last, the third pillar—convenient synthesis of designed sequences—refers to the oligonucleotide synthesis machines that are capable of creating DNA molecules made up of arbitrary base sequences. In previous chapters we've already had an introduction to hybridization and the synthesis of designer oligonucleotides. We will interweave these topics—the first and third pillars—in the context of junction structures of DNA. But we begin with a focus on the second pillar and examine junctions also called branch points. These are at the very heart of Seeman's endeavor.

Seeman knew of the replicational junction, often called the replication fork, which was implicit in the Watson-Crick proposal for the mechanism of DNA replication (Figure 6.2). We tacitly referred to this in Chapter One when we spoke about the unzipping of DNA's zipper—the breakage of the hydrogen bonds holding the double helix together—and we spoke further about it in Chapter Two. However, replication is an ephemeral process and so the triply branched replication junction is short-lived.

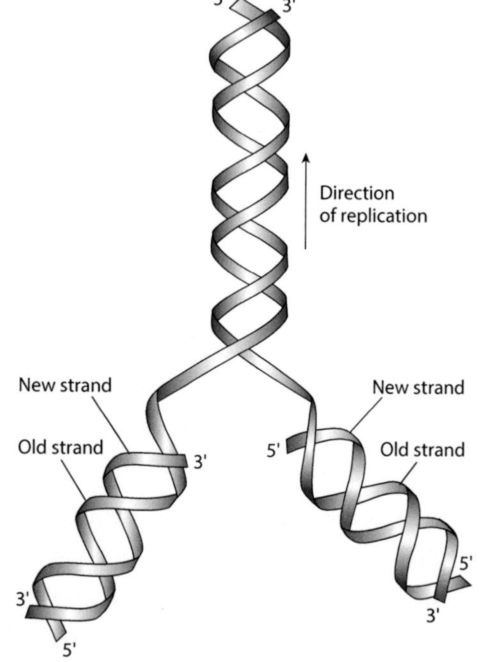

Figure 6.2 DNA replicational junction (replication fork). *(Reprinted with permission from Encyclopædia Britannica, © 2014 by Encyclopædia Britannica, Inc.)*

Another example of a naturally occurring branched DNA structure is the four-stranded DNA junction first proposed by British geneticist Robin Holliday in 1964 (panel (a) of Figure 6.3). It was hypothesized as a structural intermediate in a model to explain how yeast exchange genetic information.[13] In fact, this is a prevalent DNA structure and has been found to form whenever the breakage and rejoining of chromosomal DNA occurs.[14] While the details are not relevant to our present interest, the fact to observe is that we have a branched structure. The four-stranded intermediate junction is generally known as the Holliday junction. Again, note that this is a structural intermediate—that is, the existence of the junction is transient. One of the problems in learning more about such intermediate structures was that (ca. 1980) detailed structural data didn't exist. Seeman wrote a paper in which he and colleague Bruce Robinson did a computer simulation of DNA double-stranded branch point migration.[6] As they wrote, "The difficulty in understanding a complete molecular model of the junction has led us to represent it both schematically and by comparison with a familiar staircase."

DNA branch point migration As an example, consider four oligonucleotide strands with complementary base sequences that can form a branched DNA structure. The place where the four strands converge is known as a branch point or a junction. Branch point migration occurs if this junction can move (by the stepwise breakage and reformation of hydrogen bonds) and hence the DNA structure can change its shape.

Seeman's staircase (panel (c) of Figure 6.3) was a skillful way of visualizing his model. In the schematic representation of the junction (panel (b) of Figure 6.3), the four double helices are indicated by their helix axes shown as rod-like elements emanating from the central rhombus. This rhombus represents the junction itself. In the staircase analogy to the junction structure, the plateau where the four sets of stairs meet is suggestive of the rhombus. The banisters correspond to the backbone of the four strands which comprise the junction, while the steps are analogous to the base pairs (the four helix axes go up or down the middle of each flight of stairs). Look carefully to see the red and blue ribbons wrapped around the bannisters to suggest the alternately antiparallel nature of the backbones.

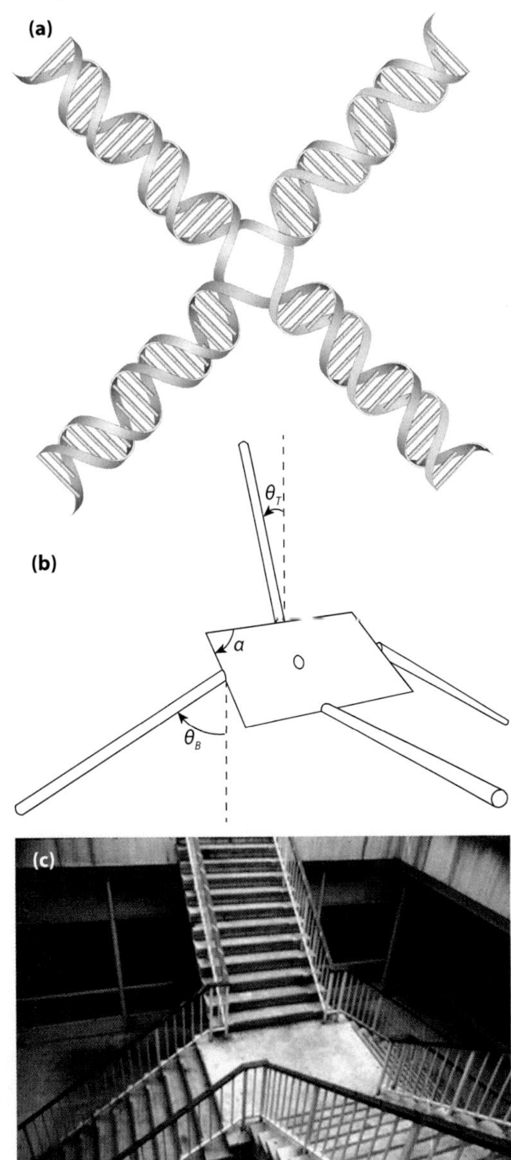

Figure 6.3 (a) Artistic rendering of a Holliday junction. **(b)** A schematic model of the junction. **(c)** Seeman's staircase is an inspired way of visualizing a four-arm branched junction of DNA by connecting it with a familiar object. *((**a**) reproduced with permission from Jon Reynolds, "Resolving a Holliday Romance," Nature Cell Biology 6, No. 3 (March 2004), 184: (**b**) (**c**) portion reproduced with permission from Nadrian C. Seeman and Bruce H. Robinson, "Simulation of Double Stranded Branch Point Migration," Proceedings of the Second SUNYA Conversation in the Discipline Biomolecular Stereodynamics 1, ed. Ramaswamy H. Sarma (New York: Adenine Press, 1981), 279–300; Photograph courtesy of Nadrian C. Seeman, used with permission.)*

Computer simulations of branch migration were necessary. It was not possible to study the structural and dynamic properties of these junctions because they were not stable. But then Seeman announced in his classic paper that it was possible to "generate sequences of oligomeric nucleic acids which will preferentially associate to form migrationally *immobile* junctions, rather than linear duplexes, as they usually do."[15] In order to accomplish this, he found that one needs to maximize Watson-Crick base pairing, that is, the pairing up of complementary bases A with T and C with G. It's also necessary to strive for a minimum of the sequence symmetry customarily present in naturally occurring DNA. When Seeman wrote his paper it was a decade after molecular biologists began to exploit enzymes that modify DNA to produce new arrangements of DNA base sequences. They used restriction enzymes that cut DNA at specific sites and enzymes known as ligases that fuse nicks, such as might be caused by radiation damage, in the DNA backbone—the blue bands shown in the DNA Starter Kit of Chapter One. (A nick is a place in a double helix where a discontinuity has occurred—that is, a lack of connection between adjacent nucleotides of one of the strands.) Molecular biologists using these enzymes were able to cut and splice DNA molecules in a manner that has been likened to a film editor using tools such as a cut away or a dissolve to reorder sequences of film to create a new overall effect.[16] This came to be known as genetic engineering and ushered in the age of biotechnology.

Sticky Ends

From Seeman's perspective the most important procedure of the molecular biologists was the use of single-stranded DNA overhangs—sticky ends—to specify the order in which molecules were to be assembled. We first encountered sticky ends in Figure 1.2 of Chapter One. A sticky end occurs when one strand of the helix extends for several unpaired bases beyond the other. Stickiness is the natural inclination of the overhanging piece to bond with a matching strand of another helix that has the complementary bases in the appropriate order: the base adenine (A) on one strand binds with thymine (T) on the opposite strand and cytosine (C) binds with guanine (G). Seeman showed that sticky-ended cohesion can be used to direct the assembly of stably branched structures, i.e., immobile junctions.

As depicted in Figure 6.4 (a) and (b), naturally occurring DNA junctions, such as the Holliday four-stranded branched structure, can undergo branch migration. The branch point can move around as indicated by the green and pale orange-colored trimers of base pairs. If we think of the strands of DNA as the lanes of a two-lane highway, a junction point would correspond to the lanes going through an intersection. Traffic could flow through four different lanes in a traditional intersection. In the United States, northbound traffic would turn east, westbound traffic would turn north, southbound traffic would turn west, and eastbound traffic would turn south. We can see the analog of this flow in the

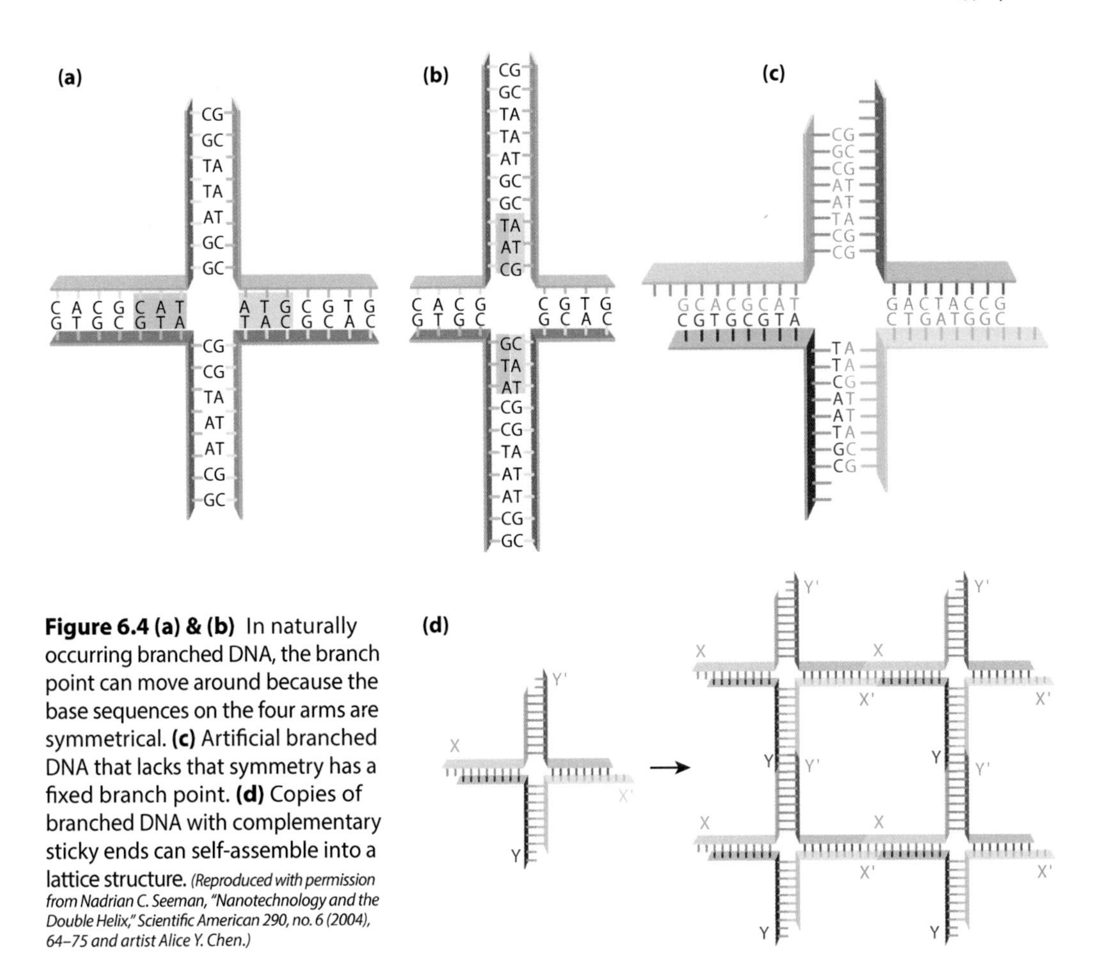

Figure 6.4 (a) & (b) In naturally occurring branched DNA, the branch point can move around because the base sequences on the four arms are symmetrical. **(c)** Artificial branched DNA that lacks that symmetry has a fixed branch point. **(d)** Copies of branched DNA with complementary sticky ends can self-assemble into a lattice structure. *(Reproduced with permission from Nadrian C. Seeman, "Nanotechnology and the Double Helix," Scientific American 290, no. 6 (2004), 64–75 and artist Alice Y. Chen.)*

branch point migration of Figures 6.4 (a) and (b).[17] Seeman worked out *sequence selection rules* that allowed him to design stable synthetic branched junctions such as shown in the four arm junction of Figure 6.4 (c).

To capture the idea of DNA as a highway more explicitly, Seeman used the illustrations shown in Figure 6.5.[18] In panel (a), the two antiparallel strands of the double helix are shown as analogous to the lanes of a road with two different directions of traffic flow—and a thin divider between the lanes. The divider corresponds to the helix axis and is clearly linear. Regardless of whether the road is straight (as shown) or curved, the divider remains linear—that is, it is unbranched. As Seeman remarked, thinking about DNA as a road makes it easy to imagine an intersection in the road as seen in Figure 6.5 (b). In this case, all of the traffic is shown to turn right when it reaches the intersection. (Figure 6.5 assumes that one is in a right-driving country such as the United States. It would look a bit different if one was in the United Kingdom, Japan, Australia and other left-driving countries!)

Note that for convenience, DNA is often drawn as a ladder-like molecule with parallel backbones (such as in Figure 6.4) rather than as a double helix. Figure 6.4 (d) shows how copies of the stably branched arrangement can self-assemble into a lattice structure.

However, this ladder-like representation of DNA obscures a vital property of branched DNA constructs: the relative orientations of two junctions joined by an edge of double-helical DNA are a function of their separation. (The separation of the junctions determines the torsion angle between them because of the double-helical nature of DNA.) As a consequence, the lattice described here is not limited to a two-dimensional system but can be realized as a three-dimensional system.[19] We have more to say about Seeman's sequence selection rules in the next chapter.

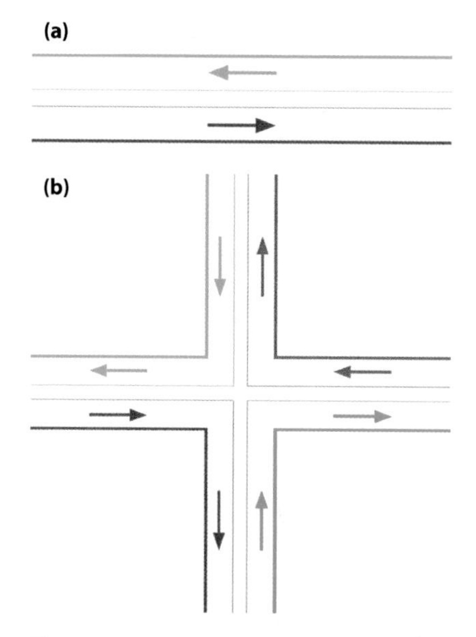

Figure 6.5 DNA as a highway. **(a)** An unwound double helix of DNA. The two lanes represent the two antiparallel strands of the double helix. The direction of traffic flow is indicated. **(b)** A four-arm branched junction as an intersection in a highway. The directions of traffic flow indicate the ways that the strands in a four-arm branched junction would extend. *(Reproduced with permission from Nadrian C. Seeman, "Another Important 60th Anniversary," Adv. Polym. Sci. 261 (2013), 217–228.)*

Immobile Four-Arm DNA Junction

In the year following his theoretical paper on nucleic acid junctions and lattices, Seeman published the first major experimental evidence in support of his vision (Figure 6.6).[20] He presented electrophoretic experiments (as well as ultraviolet optical absorbance measurements) that indicated the formation of a stable four-arm junction in solution. We introduced polyacrylamide gel electrophoresis (PAGE) in Chapter Two when discussing the work of Frederick Sanger. There we spoke extensively of the band patterns in an electrophoretic gel. There is more information that can be gleaned from gel results than we had reason to speak about earlier. Seeman's work makes use of electrophoresis-derived *Ferguson plots*. These come up again in our look at Seeman's research.

The method was developed by Australian veterinary scientist Kenneth Ferguson in 1964 and a colleague dubbed it the Ferguson plot.[21] Formally, it is the semi-logarithmic plot of mobility versus gel concentration. This mathematical technique measures the electrophoretic mobility, essentially the rate of migration of a molecule at several different gel concentrations. Loosely speaking, it measures how rapidly the bands move down the lanes. From this analysis one can learn about the molecule's size, its net charge density, and perhaps unexpectedly, its shape.

Ferguson plot A mathematical technique applicable to gel electrophoresis. This graphical analysis measures the electrophoretic mobility (essentially, the rate of migration) of a molecule at several different gel concentrations. From this one can learn about the molecule's size, its net charge density, and its shape.

A detailed look at Ferguson plots would lead us too far astray. Instead we will give an example of how mobility in a gel can give information about

DNA shape. As we noted in Chapter Two, when double-helical DNA moves through a gel it has to wiggle through small passages in the meshwork of the polymer comprising the gel. If there are any base mismatches in the DNA duplex they can create shape anomalies. For example, if one strand contains an extra base not found in the complementary strand, the extra base can remain within the helix or be rotated out of the helix. When the extra base remains in the helix it creates a bend that can slow electrophoretic mobility by making it more time consuming for the DNA to progress through the meshwork of the gel. This is an illustration of how electrophoretic mobility can shed light on DNA shape.[22]

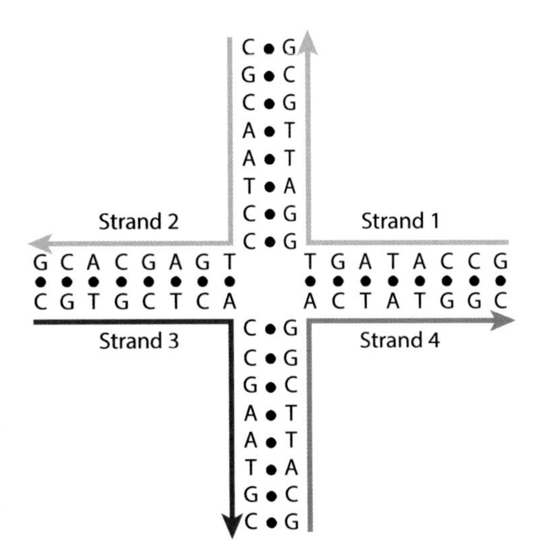

Figure 6.6 First demonstration of an immobile, four-arm nucleic acid junction. *(Adapted with permission from Neville R. Kallenbach, Rong-Ine Ma, and Nadrian C. Seeman, "An Immobile Nucleic Acid Junction Constructed From Oligonucleotides," Nature 305 (1983), 829–831.)*

Returning now to Seeman's stable four-arm junction, he used his sequence selection rules to engineer four hexadecanucleotides—i.e., each nucleotide is 16 (hexadeca-) bases in length. The sequences were commercially synthesized by the phosphoramidite chemistry we spoke of in Chapter Two. His paper demonstrated the formation of an immobile four-arm nucleic acid as shown in Figure 6.6. It is similar to Figure 6.4 (c). This is a stable analog of the Holliday junction that we discussed (Figure 6.3).

Two-Dimensional Ligation of DNA Junctions

Although Seeman's larger ambition was to form three-dimensional structures connected by immobile junctions, he approached this goal by working in two dimensions. His aim was to demonstrate that he could use many copies of an immobile nucleic acid junction to build larger complexes by linking the immobile junctions together. Thus, part (d) of Figure 6.4 would be the hoped-for result when working with four-arm junctions. He did not restrict himself to four-arm junctions, however, and also worked to use three-arm junctions to build up geometrical stick figures on the 1–1000 nm length scale. The vertices of such figures are formed by the junctions while their edges consist of double helical nucleic acids. An early test was to use junctions containing three arms, two of which terminated in complementary sticky ends (Figure 6.7).[23]

Then daybreak came to dissolve his utopian dreams. As shown in Figure 6.7, he did succeed in creating a series of linked junctions. But he discovered that while junction dimers formed that had free termini only (open structures), junction trimers and larger species were present in both open and cyclized (closed) forms.

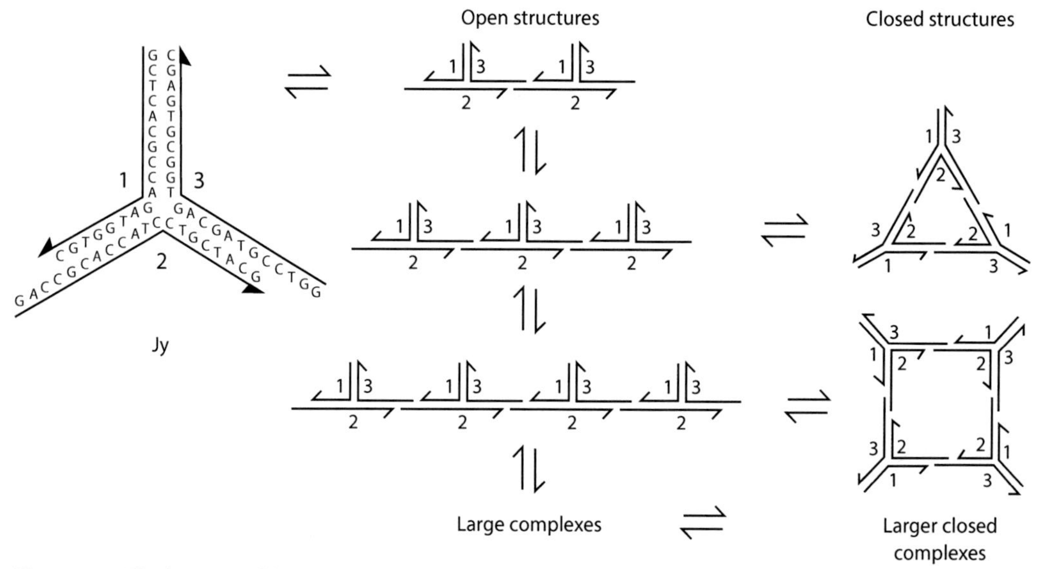

Figure 6.7 The ligation of three-arm nucleic acid junctions. *(Reproduced with permission from Rong-Ine Ma, Neville R. Kallenbach, Richard D. Sheardy, Mary L. Petrillo, and Nadrian C. Seeman, "Three-Arm Nucleic Acid Junctions Are Flexible," Nucleic Acids Research 14, No. 24 (1986), 9745–9753.)*

Time out. Before we can delve into the issues that are raised here, we need to explain the half-arrows that are all over the place in Figure 6.7. These indicate the 5′→ 3′ direction of the nucleotide chain. In Chapter Two we remarked that in double-stranded DNA, the two helical strands run anti-parallel to one another, like up-and-down escalators. Within each of the two strands of a DNA duplex, the nucleotide subunits all line up in the same direction giving each strand a chemical *polarity,* essentially a strand direction. One specifies the strand directionality of DNA by referring to the 5′ (pronounced five prime) and 3′ (three prime) carbon atoms on the deoxyribose sugar molecules. Deoxyribose is a *pentose* (five-carbon) sugar (Figure 6.8). The ends of DNA strands are called *the 5′ end* that has a terminal phosphate group—the P in the yellow circle bound to four oxygen atoms—and *the 3′ end* that has a terminal hydroxyl group—the OH. (The primes distinguish the pentose sugar carbons from the carbon atoms that are present in the bases; the sugar carbons are both numbered and primed.) Thus, the two ends of a DNA chain are chemically distinguishable.

A final item (call-out in Figure 6.8): carbons are at four of the vertices of the pentose sugar, but the 4′ carbon is said to "complete the ring via an oxygen" that bridges to the 1′ carbon, while the 5′ carbon hangs away from the ring.

Now we return to the two-dimensional constructs of Figure 6.7. Seeman created both the open and closed structures by using a process called *ligation* to join junctions together. In this

DNA strand polarity Describes how the nucleotide subunits of a single DNA strand are lined up in the same direction giving the strand a chemical polarity. Also, the two ends of a given strand can be chemically distinguished. The end having a phosphate group at its terminus is called the 5′ end, and the end having a hydroxyl group at its terminus is called the 3′ end. One speaks of the direction of the chain as running from the 5′ end to the 3′ end, or simply 5′ → 3′.

context, ligation is the covalent bonding of two ends of nucleic acids by use of an enzyme—in the present case, the enzyme designated T4 DNA ligase. Covalent bonding is the formation of a strong chemical bond in which electrons are shared by the atoms of each molecule. Note: In the present case, an enzyme is used to achieve the covalent bond. This will not be true with Bellini and Clark's non-enzymatic ligation of nanoDNA in Chapter Thirteen.

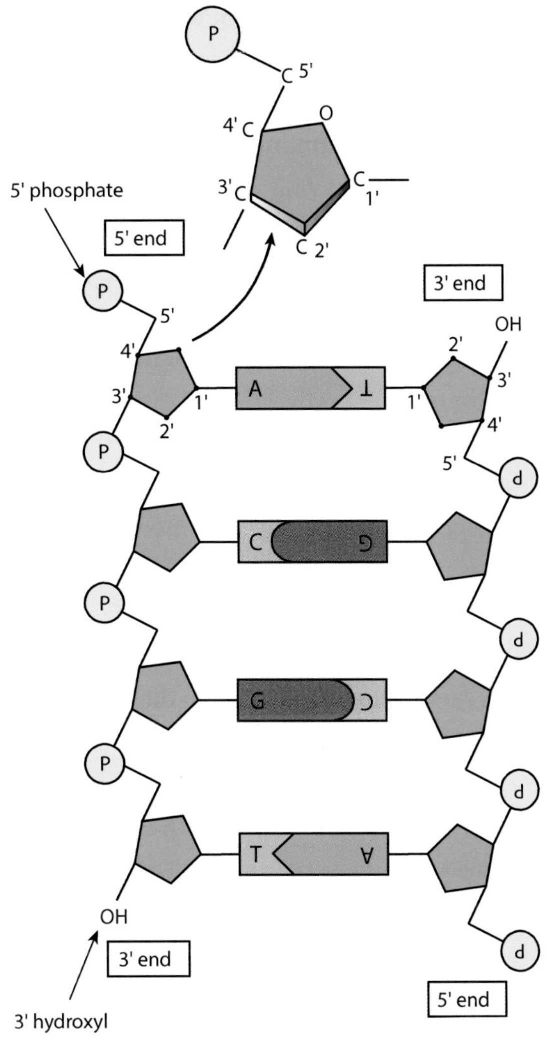

Figure 6.8 DNA strand polarity. Strands of DNA are said to run from the 5′ end to the 3′ end. *(Adapted with permission from Neil A. Campbell and Jane B. Reece, Biology, 6th Ed., ©2002, p. 296, Pearson Education, Inc., New York, New York.)*

Focus your attention on the closed structures in Figure 6.7. Seeman initially hoped that the formation of closed structures in his ligation experiments held promise that he could build large arrays by the self-assembly of multiple arm immobile junctions. Indeed, in 1985 he and his colleagues submitted an abstract to the *Biophysical Journal* in advance of the Biophysical Society's Twenty-Ninth Annual Meeting in Baltimore, Maryland. The abstract was titled "Nucleic Acid Junctions Are Not Highly Flexible." On the basis of a series of electrophoretic gel runs, Seeman and his colleagues had reason to be optimistic, and they concluded their abstract by observing, "These results suggest that the flexibilities of 3-arm and 4-arm nucleic acid junctions are not grossly different from that of linear DNA."[24]

Later that same year he published a paper in which his optimism seemed to be slightly tempered. He wrote, "Recent results on the concatenation of nucleic acid junctions show that these molecules can act as fairly rigid macromolecular valence clusters on the nanometer scale."[25] Translating this sentence, we observe that concatenation means linking together in a series. Seeman first used the term *valence cluster* in his foundational paper in the *Journal of Theoretical Biology,* drawing an analogy between the modes of self-assembly of nucleic acids and of atoms.[9] Valence is a term describing the ability of atoms to form a specific number of highly directional, covalent bonds to other atoms.

For example, a carbon atom will readily form bonds to four neighboring atoms located at the corners of a tetrahedron. When these neighbors are also carbon this produces a diamond lattice. Seeman envisioned clusters of DNA having valence, designed to organize into immobile multi-arm junctions. These junctions have directional selectivity (in addition to sequence selectivity) to link to neighboring junctions directly or with segments of linear DNA in between.

Enzymatic ligation of nucleic acids The covalent bonding of two DNA fragments (or two RNA fragments) catalyzed by the use of enzymes.

DNA Ligase An enzyme that catalyzes the formation of a covalent bond between two strands of DNA.

Covalent bonding The formation of a strong chemical bond in which electrons are shared by the atoms of each molecule.

But moving beyond the definition of terms, note the language he uses: "fairly rigid." In that same paper, he wrote that "the junction cluster system presents a system of defined, apparently stiff, building blocks. . . ."[26] Again, the language "apparently stiff" was more reserved than that in his Biophysical Society abstract had been, perhaps reflecting a growing awareness of flaws in his approach.

In 1986, just a year later, Seeman and colleagues reported additional studies done with both three- and four-arm junctions, in which the news was surprising and not so good. As we said earlier, junction dimers formed that had free termini only (open structures), while junction trimers and larger species were present in both open and cyclized (closed) forms (Figure 6.7). Seeman wrote, "The formation of a series of macrocyclic products which, surprisingly, begins with trimers and tetramers indicates that this junction is flexible around a bending axis, and perhaps twist-wise as well."[27]

As if to bookend his tales of woe, Seeman and his colleagues submitted an abstract for the Biophysical Society's Thirty-First Annual Meeting in New Orleans, Louisiana, just as they had done for the Twenty-Ninth annual meeting two years earlier. This time, however, the abstract was titled, "Ligation and Flexibility of 3-Arm and 4-Arm Nucleic Acid Branched Junctions."[28] This summary of the presentation they were to give said that all ligations involving adjacent arms resulted in a macrocyclic series that began with trimers. They found that over the time of the reactions, the arms of both types of junctions were able to form angles as low as 60° (in the case of the trimers). "This indicates a larger amount of flexural, and perhaps torsional, flexibility than previously believed for these structures."[28] New Orleans is a city known for its music—jazz, rhythm and blues. Seeman was singing the blues.

Deconstruction of Concatenated Nucleic Acid Junctions

We've seen that Seeman used ligation procedures initiated by the enzyme T4 DNA ligase to form both open and closed complexes of nucleic acid junctions. Over a period of a year these complexes went from a delight to a disappointment, a flip-flop determined by deconstruction.

Restriction enzymes Enzymes that recognize and bind to specific DNA sequences. Once they bind to their *recognition site/sequence,* restriction enzymes cut the sugar-phosphate backbones of the DNA strands.

Restriction site A sequence of DNA that is recognized by a restriction enzyme.

Exonuclease An enzyme that cuts nucleotides from the end of a DNA chain.

Endonuclease An enzyme that cuts the sugar-phosphate backbone of DNA within the length of the chain (rather than at the end of the chain).

Deconstructing the complexes, Seeman could examine their pieces separately. He took the complexes apart in a way that each strand could be excised in a tractable form. To do this he included restriction sites, also called restriction sequences, as part of the strands (more on this shortly). And he ran a *denaturing* gel—a gel including chemicals to unfold the DNA double helix, separating it into single strands and thus eliminating the influence of the helical shape on gel mobility. Next he took a sample of the products of the gel run and treated them with the restriction enzyme called exonuclease III. This process is called exonuclease digestion. (Exonuclease III is known to act at the 3′ terminus of DNA, so we have [tacitly] put to use our discussion of the 5′→ 3′ direction of the nucleotide chain.) Exonuclease III digests linear—that is, open—molecular structures but not closed molecular structures. Thus, after employing his enzyme and then running another electrophoretic gel, the only bands he observed corresponded to closed structures. He then excised these bands, nicked them using restriction enzymes called *endo*nucleases, and sized them against molecular weight standards on (yet another) gel to verify their length as measured in base pairs. (Note: endonucleases cleave within a nucleic acid chain, while exonucleases cleave at the end of a chain.) After this careful deconstruction analysis, he had to face the fact that the closed complexes formed a macrocyclic ladder, a progression of different size molecules having cyclic/ring structures. The results sounded the death knell for Seeman's original idea.

Finding solace in humor, Seeman wrote, "The closure reactions reported here indicate a surprising amount of flexibility for the 3-arm immobile nucleic acid junction. Thus, one appropriate paradigm for this 'soft' junction appears to be closer to a marshmallow impaled by three sticks, rather than a trigonal cluster such as the fork of a slingshot. (Other models are left to the reader's imagination.)"[29]

Macrocycles

Here's the quandary: his experiments created *more than one type* of closed structure. For example, Figure 6.7 shows both a trimer of three-arm junctions and a tetramer of three-arm junctions. *Seeman interpreted this to mean that the three-arm junction—the building block in the case of Figure 6.7—is flexible and capable of assuming a variety of conformations.* He found this same undesirable flexibility when ligating four-arm junctions as well as in junctions with even more arms. Flexible connections can result in heterogeneous structures because different angles can be formed between different repeating units.

For example, in Figure 6.7, the angle is 60° for trimers and 90° for tetramers—easily seen by looking at strand 2 in each case. This is anathema if you're trying to construct periodic matter, Seeman's primary goal. The formation of periodic matter relies on the same contacts from unit cell to unit cell. Flexibility can lead to networks of variable content and thus destroy the periodicity of the material. The multiplicity of different closed constructions that Seeman encountered is referred to as a series of macrocyclic products.[30] Essentially, this means a series of molecules having ring structures, macrocycles, of different sizes.

Macrocyclic products A macrocycle is a cyclic (ringlike) macromolecule. Macrocyclic products refer to a multiplicity of different size molecules having cyclic/ring structures.

Ligation-closure experiments DNA strands are ligated and their ligation products are assayed (typically via denaturing gel electrophoresis) to determine if cyclization has resulted in a unique outcome or in many different products. The presence of many cyclic products suggests that the angles at the junction points of the tested molecules are not rigid.

Seeman found that strand 2 annealed to itself during the ligation step and formed cyclic structures as is evident in Figure 6.7. (The individual strand components are maintained in the diagrams for clarity, but ligation would seal the gaps between successive strands.) Contrast the *exocyclic* segments (those *external* to the closed structure) on the right side of Figure 6.7 with the exocyclic segments on the right side of Figure 6.4 (d). There's little doubt that the sticky ends of X, X′, Y, and Y′ in Figure 6.4 (d) offer ample opportunity for continued periodic growth, but this is not so in Figure 6.7. As Seeman would remark years later, "A flexible system can cyclize on itself, thereby poisoning growth."[31] So it's vital to discover rigid DNA components in order to build periodic matter.

It's important to shout from the rooftops that we've presented the problems in his work—and we return to them in Chapter Seven—the better to appreciate Seeman's remarkable creativity and steadfast resolve in inventing rigid DNA components to overcome his challenges and galvanize a generation of DNA nanotechnologists.

At this time in our story, however, P.G. Wodehouse's one-liner is apt: if Seeman was not actually disgruntled, he was far from being gruntled.[32]

Three-Dimensional Constructions and Catenanes

Seeman was well aware that his approach was flawed because of the flexibility of the junctions he employed. But he still wanted to press on to build in three dimensions; he was the very antithesis of what Lady Macbeth called "infirm of purpose."[33] He wanted to show that it could be done before tackling the challenge of finding rigid branched junctions to use as construction tools. So we now move on to *three*-dimensional constructions driven by Seeman's hunch about the utility of stable branched junctions. To understand how Seeman achieved three-dimensional structures and how he demonstrated his accomplishment—his *proof of synthesis*—we introduce some new vocabulary.

The reader may have heard the word *topology*. It is the study of the properties of objects that are preserved in spite of the deformation of the objects by twisting or stretching. To take an example, a circle is said to be topologically equivalent to an ellipse; a circle can be deformed into an ellipse by stretching. To move to three dimensions, we present the stock mathematical joke on the subject. Question: What is a topologist? Answer: Someone who cannot tell the difference between a coffee cup and a donut because the objects are topologically equivalent.[34] *Groan*. In three dimensions, one can make different representations of molecular structures built from DNA that are topologically equivalent.

Catenane A group of organic compounds in which two or more ring structures are interlocked with each other (like a chain).

Topological equivalence When two objects can be continuously deformed from one to the other, they are said to be topologically equivalent. The classic illustration of two such objects is a donut and a coffee cup.

Molecular *architectures* are a second topological concept we need to become familiar with. They consist of mechanically interlocked molecules. These molecules are not connected through traditional bonding such as electron sharing, but rather as a consequence of their topology. The architecture we are most concerned with to appreciate Seeman's three-dimensional constructions is called a *catenane*. A catenane is a set of linked molecular rings (from the Latin word *catena* which means "chain"). Catenanes have been discovered in naturally occurring DNA and have also been generated in the lab. Chemist Edel Wasserman was the first to synthesize a (non-DNA) catenane—and prove that he had done so—in 1960.[35] A simple catenane cartoon is shown in Figure 6.9.

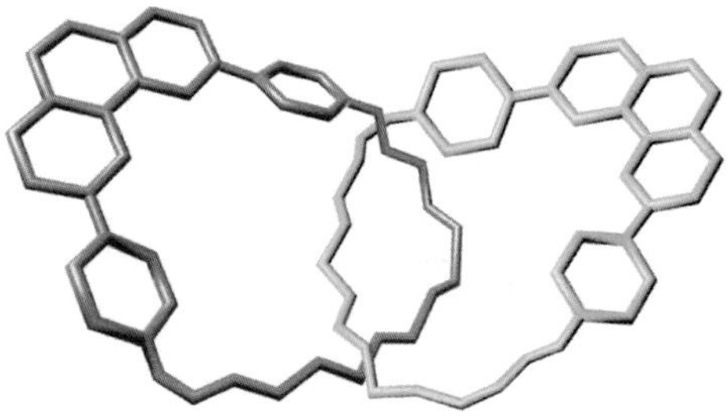

Figure 6.9 A simple catenane cartoon showing molecules linked as in a chain.

An example of DNA catenanes containing multiple links is shown in Figure 6.10. These are transmission electron micrographs (each accompanied by an instructive drawing)

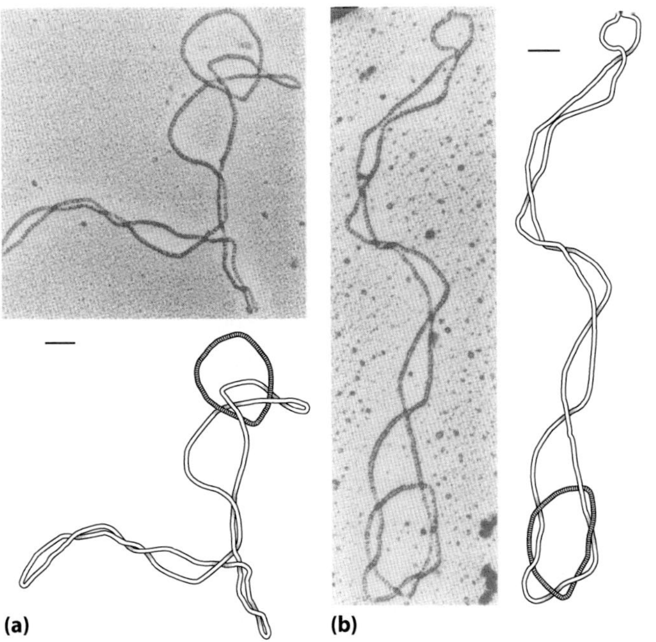

Figure 6.10 Transmission electron micrographs (and drawings) of multiply catenated DNA strands. Scale bars are 100 nm. *(Reproduced with permission from Sylvia J. Spengler, Andrzej Stasiak, and Nicholas R. Cozzarelli, "The Stereostructure of Knots and Catenanes Produced by Phage λ Integrative Recombination: Implications for Mechanism and DNA Structure," Cell 42 (August 1985), 325–334.)*

of DNA forming doubly and triply interlocked catenanes. The samples were stained with ethidium bromide for contrast; the scale bars show 0.1 microns (100 nm).[36]

Now we are prepared to examine Seeman's first three-dimensional DNA construction, a cube made from DNA—the result of his first dozen years of thinking and tinkering with DNA branched junctions.

The DNA Cube

Ironically, when Seeman succeeded in making his first three-dimensional figure he "almost threw in the towel at that point because, in order to make the cube, we had to throw out all the beautiful logic of DNA sticky ends to form the various edges of the polyhedron. . . . We thought of it as a cube. . . . It could have been rhombohedral-looking or whatever."[37] Perhaps the reader is surprised. Why bother with all the discussion of sticky ends if their value is compromised, you may ask. The use of sticky ends is one compelling deployment of hybridization or complementary base pairing. We will see that sticky ends and the importance of hybridization remained one of the three pillars of Seeman's approach to DNA nanotechnology.

And what to make of the uncertainty of the form he had constructed—cube, rhombohedron or whatever? A rhombohedron is a three-dimensional figure that is like a cube but its faces are not squares, they are parallelograms which means that they have two pairs of parallel sides. If all the angles between the sides are right angles (90° angles) then the faces are squares and the rhombohedron is a cube. As we now begin to describe, Seeman made three-dimensional objects that were *topologically* specified, rather than *geometrically* specified. When Seeman and his student Junghuei Chen published their paper on the floppy cube (more on the term floppy in Chapter Seven) it was greatly admired in the popular-science press. It also drew the attention of many scientists who saw that by treating DNA not as a biological object but as an architectural tool, *he had created a new field: structural DNA nanotechnology.*[38] Seeman's work was pioneering but its progress was not predictable. As our narrative zigzags we are describing how real science is done: with fits and starts and serendipity.

It may be helpful to have another look at Figure 1.3 (b) from Chapter One. In this representation the backbone of each DNA strand is depicted as colored spheres (a different color for each strand) and the bases are shown as white spheres. First we give a two sentence description of what Seeman did and then we provide some of the details of the synthesis. He constructed a covalently closed cube-like molecular complex containing twelve equal-length double helical edges arranged about eight vertices (though the depiction of twisting between strands in Figure 6.11 is confined to the central portion of each edge for clarity). Each of the six faces of the object is a single-stranded cyclic molecule that is doubly catenated to four neigboring strands, and each vertex is connected

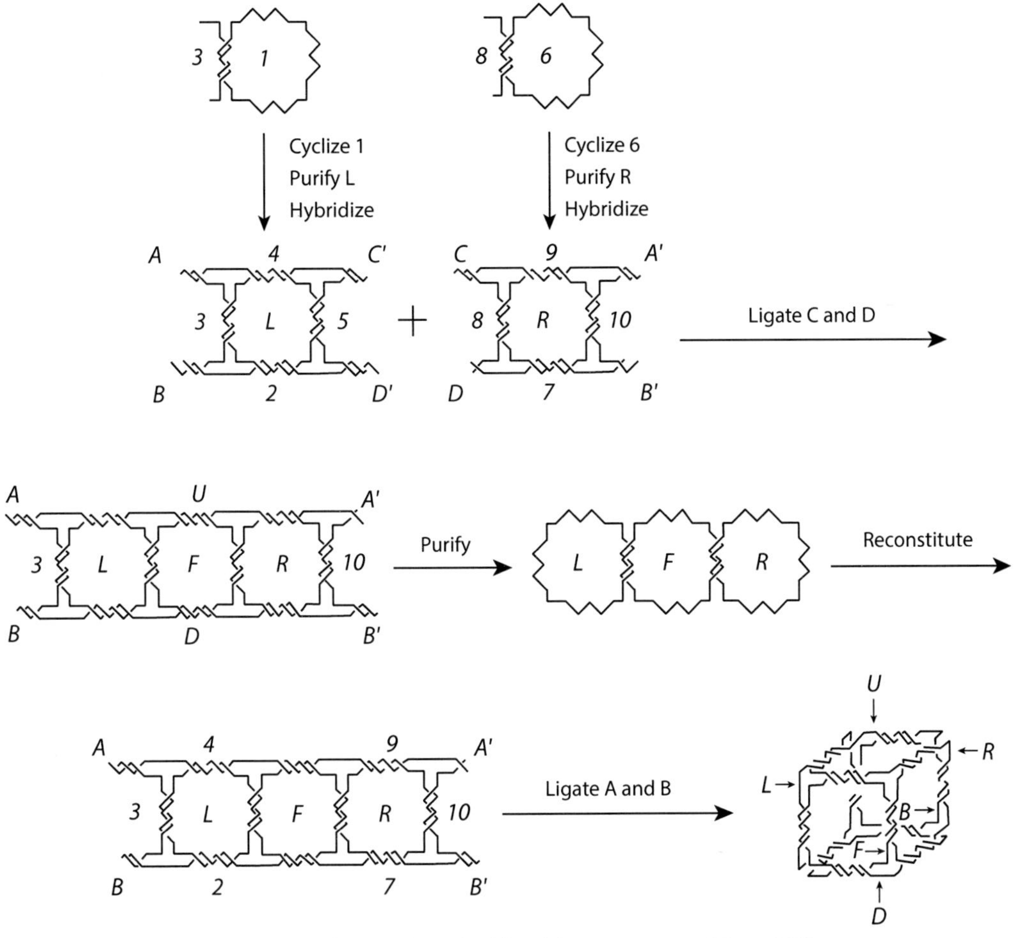

Figure 6.11 Scheme used to synthesize the cube-like object constructed from DNA. *(Reproduced with permission from Junghuei Chen and Nadrian C. Seeman, "Synthesis From DNA of a Molecule With the Connectivity of a Cube," Nature 350 (18 April 1991), 631–633.)*

by an edge to three others.[39] To follow the *synthesis scheme* (Figure 6.11) for the three-dimensional cube, we note that the term *cyclize* means the formation of a ring compound from a chain by formation of a new bond. Cyclization is accomplished by ligation. For example, ligation of square L to square R forms the tricyclic belt shown at the third stage of the figure.

In Figure 6.11 the numbers refer to strands but there is a convention used: as a new strand is formed by ligation, its identification *changes* from one or more numbers to a letter corresponding to its position in the final object. As an example, strand 1 is synthesized initially as a linear molecule, but is referred to as L once it is cyclized. The six final strands in the object are referred to as L (left), R (right), U (up), D (down), F (front), and B (back).

Notice that we said strand 1 was synthesized. This is the third pillar of DNA nanotechnology we spoke about. And the strand design incorporated the second pillar, namely, stably branched junctions: in the final object each vertex corresponds to the branch point of a three-arm junction. The first pillar, hybridization, is rife throughout the process and is specifically called out in the initial steps of the synthesis scheme. *So we truly have incorporated all three pillars, the bedrock of structural DNA nanotechnology.*

Seeman wrote a computer program (called SEQUIN) that encoded his sequence selection rules, and the program determined the base sequences of all the strands used in the fabrication.[40] By the way, he also did the chemical synthesis himself using grant money to purchase the recently invented gene machine.[41]

Figure 6.11 begins after all the individual strands have been synthesized and purified. There are five steps shown in the figure. These steps separate six stages in the synthesis of the three-dimensional DNA object. The first step has two parts: the cyclization of the full-length strands 1 and 6 to form L and R, respectively. These two cycles (rings) are then hybridized with strands 2–5 and 7–10, respectively, to form the squares that are shown in the second stage. These squares are ligated together at the complementary sticky ends C′ and C as well as D′ and D. This reaction forms strands U, F, and D. But U and D are discarded in a purification step (on a denaturing gel) that isolates the L-F-R triple complex. (This complex is a triple catenane.) The L-F-R-2-3-4-7-9-10 complex is then reconstituted by adding the missing six strands, and the final ligations are performed to close sites A and A′ as well as B and B′. Ligation also seals the nicks in U (4–9) and D (2–7).

The foregoing discussion presented a great many details. *The important point to note* is that once closed, the cube consists of six linked circles, corresponding to the six faces of the object, and labeled left (L), right (R), up (U), down (D), front (F), and back (B).

Another representation of the final cube-like object shows two triple catenanes that link to form the cube (Figure 6.12).[42] In addition, this illustration calls attention to the proof of synthesis. The cube was built in solution, and Seeman evaluated the results shown in the synthesis scheme (Figure 6.11) by breaking down both the partial and final products to a variety of constituent catenanes. He made extensive use of the electrophoretic-based Ferguson plot calculations that we referred to before.[39, 43] In order to break down

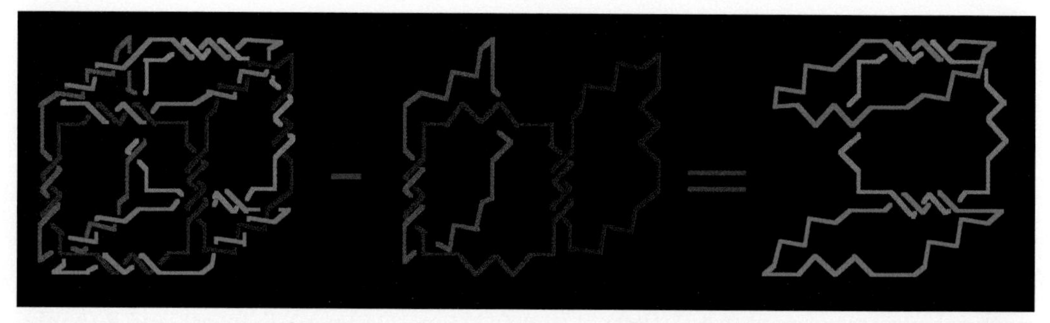

Figure 6.12 Two triple catenanes link to form the cube. The cube is shown at the left. The triple catenane shown at the center of the drawing corresponds to the left, front, and right faces of the cube. The schematic suggests that with this triple catenane removed we visualize the second triple catenane that corresponds to the up (i.e., top), back, and down (i.e., bottom) faces of the cube. *(Conceived and drawn by Nadrian C. Seeman, used with permission; Nadrian C. Seeman, "Single Stranded DNA," http://www.ams.org/meetings/lectures/seeman-lect.pdf, Nadrian C. Seeman, et al., "New Motifs in DNA Nanotechnology," Nanotechnology 9 (1998), 257–273.)*

the cube for analytical purposes, he incorporated restriction sites on each edge, also mentioned earlier. To reiterate: a restriction site is a sequence of DNA that is recognized by an enzyme such as the restriction enzyme exonuclease III. This is a protein that cuts DNA. The restriction site tells the enzyme where to cut the DNA. For example, the belt present at the last step of the synthesis scheme is destroyed by restriction of the L-F and F-R edges and results in the U-B-D triple catenane (shown in gel autoradiograms we have not presented). In Seeman's publication he calls this out as the most robust proof of the formation of the cube.[44] This step is visualized in Figure 6.12.

In the course of assembling the cube Seeman also made substructures to use as standards in order to monitor the progress of the synthesis. These included a five-cycle structure lacking one strand, two topologically equivalent versions (called topoisomers) of four-cycle structures, and two versions of three-cycle structures. One four-cycle structure was a cyclic belt around the cube but lacking a top and a bottom, while the other four-cycle version lacked two catenated strands such as the top and the front.[45] Yes, it's true that the final three-dimensional object was only topologically characterized and not geometrically characterized. But Seeman knew he was onto something big. He went all out to provide a completely convincing analysis of what he had created. And his thoroughness paid off.

This was the first ever construction of a closed polyhedral object from DNA. As Seeman said, "This was the key founding experiment of structural DNA nanotechnology. This *is* DNA nanotechnology, where the DNA sequence is used to program the structure assumed by the DNA molecules, and is not just the use of DNA whose complementarity allows it to be used as smart glue"[46] [italics added].

By his own account, Nadrian Seeman spent 3,983 days in the "uncomfortable small-town milieu of Albany."[47] But in 1988 he took a position in the Department of Chemistry at New York University. It was a more appropriate fit (and it had a lot of graduate students). In addition, he was in The Big Apple where he could go to the theater and to art museums—he got a lot of his ideas from art.[48] The bearded, bespectacled man with the broken down car that we met in Albany at the beginning of this chapter had relocated a distance of only 150 miles. But scientifically he had come a long way in a decade and a half. Still later—in 2011—some twenty years after the publication of his DNA cube paper, at least 60 or 70 other labs had joined him in the practice of DNA nanotechnology. At that time, Seeman was able to say that finally, "We don't have to make all the discoveries. We don't have to make all the mistakes."[49] *However, in our narrative we are still at ca. 1991*. Others had not harked to the piping of Pan: Seeman's work to date had not lured colleagues into the fold. The field had yet to blossom. There was no diffusion of responsibility. As we discuss early in the next chapter, Seeman's group continued to do every bit of it. They made all the discoveries and all the mistakes.

Exercises for Chapter Six: Exercises 6.1–6.2

Exercise 6.1

We live in a man-made digital age; moreover, a digital age that employs a binary code in which there are only two discrete values, 0 and 1. What if nature worked that way? Suppose that instead of the four-letter alphabet of DNA and RNA, the molecules of life, we consider a hypothetical DNA molecule that only has a two-letter alphabet, an alphabet consisting of two bases. We'll call these bases 0 and 1. Let's assume that 0 – 1 bonds can occur but not 0 – 0 or 1 – 1 bonds. Pretend that you live in this imagined world. You are asked to: **(a)** Design base sequences that allow branch migration in a four-arm junction of DNA. **(b)** Design base sequences that result in an immobile (stable) four-arm branch junction of DNA.

Note: This exercise appears in an expanded form as Exercise 8.2 in John A. Pelesko's wonderful book *Self-Assembly: The Science of Things That Put Themselves Together* (Florida: Chapman & Hall/CRC, 2007), 205, and is used with his kind permission. Written by a mathematician, *Self-Assembly* is a charming, eclectic book—highly recommended.

Exercise 6.2

In looking at Nadrian Seeman's two- and three-dimensional nanoconstructions we discussed the restriction enzymes/nucleases that cleave DNA. Restriction enzymes are so important that in 1978 the Nobel Prize in Physiology or Medicine was awarded to three scientists who discovered them and pioneered their use.[50] Restriction enzymes are divided up into several different types or classes based on technical considerations. Overall, there are several thousand known restriction enzymes. We consider three (of the total of twelve) restriction enzymes that Seeman used in his experiments resulting in the DNA cube publication. They are *Taq*I with recognition sequence (also called recognition site) 5′-TCGA, *Hae*III with recognition sequence 5′- GGCC, and *Alu*I with recognition sequence 5′- AGCT.

Suppose you are given the section of DNA shown in Figure 6.13. Your task is to decide whether this can be cut by the restriction enzymes we named. For those restriction enzymes that do cut the DNA, how many fragments will result?

```
5'—TAGAGGTCACCTTCGAACTGGAGGGACCCGCCGAAGGCCTCGACG—3
3'—ATCTCCAGTGGAAGCTTGACCTCCCTGGGCGGCTTCCGGAGCTGC—5
```

Figure 6.13

Exercises for Chapter Six: Exercise 6.3

Exercise 6.3

Figure 6.14

You have isolated what you think is a unique DNA molecule from mitochondria—that is, a catenane or two nicked circles linked as in a chain. In the TEM you see molecules that look like the drawing in Figure 6.14. You want to know what fraction of the molecules are catenanes. The simplest way is to count them. To get statistical accuracy, you do many spreadings and count the molecules in many different grid holes. The data obtained are shown in Table 6.1. **(a)** Why do you think that the fraction scored as catenanes is not the same in the three spreadings? *Hint:* Think about what situations could lead to Figure 6.14. **(b)** What fraction of the molecules are catenanes?

	Grid hole no.	Total molecules	No. of molecules that look like catenanes
	1	125	16
	2	123	18
Spreading 1	3	95	10
	4	90	8
	5	85	8
	1	62	5
	2	58	4
	3	57	5
Spreading 2	4	55	5
	5	40	3
	6	39	3
	7	38	3
	1	38	3
	2	38	3
	3	36	2
	4	35	4
Spreading 3	5	29	2
	6	28	3
	7	28	4
	8	27	3
	9	20	2

Table 6.1

ENDNOTES

[1] Karen Hopkin, "3-D Seer," *The Scientist* (August 2011, Profile), 52–55.

[2] Ibid., 53.

[3] Matthew Hutson, "The Godfather of *Really* Small Things," *NYU Alumni Magazine, No. 15* (Fall, 2010), https://www.nyu.edu/alumni.magazine/issue15/15_square_Chemistry.html

[4] Nadrian C. Seeman, "Single Stranded DNA Topology and DNA Nanotechnology," *Lecture Notes, AMS Short Course, San Diego* (5 January 2008), http://www.ams.org/meetings/lectures/seeman-lect.pdf

[5] Nadrian C. Seeman, "Nanotechnology and the Double Helix," *Scientific American 290, No. 6* (2004), 64–75.

[6] Nadrian C. Seeman and Bruce H. Robinson, "Simulation of Double Stranded Branch Point Migration," *Proceedings of the Second SUNYA Conversation in the Discipline Biomolecular Stereodynamics 1*, ed. Ramaswamy H. Sarma (New York: Adenine Press, 1981), 279–300.

[7] Peter Atkins, *The Laws of Thermodynamics: A Very Short Introduction* (New York: Oxford University Press, 2010), 63.

[8] Seeman, "Nanotechnology," 66.

[9] Seeman, "Nucleic Acid Junctions and Lattices," *J. Theor. Biol. 99* (1982), 237–247.

[10] Seeman, "Nanotechnology," 66.

[11] Robert F. Service, "DNA Nanotechnology Grows Up," *Science 332* (2011), 1140–1143.

[12] Nadrian C. Seeman, "Nanomaterials Based on DNA," *Annu. Rev. Biochem. 79* (2010), 65–87.

[13] Robin Holliday, "A Mechanism for Gene Conversion in Fungi," *Genet. Res. 5* (1964), 282–304.

[14] Franklin A. Hays, Jeffrey Watson, and P. Shing Ho, "Caution! DNA Crossing: Crystal Structures of Holliday Junctions," *Journal of Biological Chemistry 278, No. 50* (2003), 49663–49666.

[15] Seeman, "Nucleic Acid," 237.

[16] Stanley N. Cohen, Annie C.Y. Chang, Herbert W. Boyer, and Robert B. Helling, "Construction of Biologically Functional Bacterial Plasmids *In Vitro*," *Proceedings of the National Academy of Sciences 70, No. 11* (1973), 3240–3244.

[17] Nadrian C. Seeman and Philip S. Lukeman, "Nucleic Acid Nanostructures: Bottom-Up Control of Geometry on the Nanoscale," *Rep. Prog. Phys. 68* (2005), 237–270.

[18] Nadrian C. Seeman, "Another Important 60th Anniversary," *Adv. Polym. Sci. 261* (2013), 217–228.

[19] Seeman and Lukeman, "Nucleic Acid Nanostructures," 242.

[20] Neville R. Kallenbach, Rong-Ine Ma, and Nadrian C. Seeman, "An Immobile Nucleic Acid Junction Constructed From Oligonucleotides," *Nature 305* (1983), 829–831.

[21] Kenneth A. Ferguson, "The Origin of the Ferguson Plot," *Electrophoresis 28* (2007), 499–500.

[22] Arupa Ganguly, Matthew J. Rock, and Darwin J. Prockop, "Conformation-Sensitive Gel Electrophoresis for Rapid Detection of Single-Base Differences in Double-Stranded PCR Products and DNA Fragments: Evidence for Solvent-Induced Bends in DNA Heteroduplexes," *PNAS 90* (November, 1993), 10325–10329—and references 25 and 26 therein.

[23] Rong-Ine Ma, Neville R. Kallenbach, Richard D. Sheardy, Mary L. Petrillo, and Nadrian C. Seeman, "Three-Arm Nucleic Acid Junctions Are Flexible," *Nucleic Acids Research 14, No. 24* (1986), 9745–9753.

[24] Rong-Ine Ma, Neville R. Kallenbach, and Nadrian C. Seeman, "Nucleic Acid Junctions Are Not Highly Flexible," *Biophysical Journal 47* (1985), A14.

[25] Nadrian C. Seeman, "Macromolecular Design, Nucleic Acid Junctions, and Crystal Formation," *Journal of Biomolecular Structure and Dynamics 3, No. 1* (1985), 11–34.

[26] Seeman, "Macromolecular Design," 13.

[27] Ma et al., "Three-Arm Nucleic," 9745.

[28] Mary L. Petrillo, Colin J. Newton, Richard D. Sheardy, Nadrian C. Seeman, Rong-Ine Ma, and Neville R. Kallenbach, "Ligation and Flexibility of 3-Arm and 4-Arm Nucleic Acid Branched Junctions," *Biophysical Journal* 51 (1987), A569.

[29] Ma et al., "Three-Arm Nucleic," 9752.

[30] Ibid., 9745.

[31] Nadrian C. Seeman, Hui Wang, Xiaoping Yang, Furong Liu, Chengde Mao, Weiqiong Sun, Lisa Wenzler, Zhiyong Shen, Ruojie Sha, Hao Yan, Man Hoi Wong, Phiset Sa-Ardyen, Bing Liu, Hangxia Qiu, Xiaojun Li, Jing Qi, Shou Ming Du, Yuwen Zhang, John E. Mueller, Tsu-Ju Fu, Yinli Wang, and Junghuei Chen, "New Motifs in DNA Nanotechnology," *Nanotechnology 9* (1998), 257–273.

[32] P.G. Wodehouse, *The Code of the Woosters* (New York: W.W. Norton & Company, 2011), 3.

[33] William Shakespeare, *Macbeth*, Act II, Scene II, Line 714.

[34] Paul Renteln and Alan Dundes, "Foolproof: A Sampling of Mathematical Folk Humor," *Notice of the AMS 52, No. 1* (January 2005), 24–34.

[35] Edel Wasserman, "The Preparation of Interlocking Rings: A Catenane," *J. Am. Chem. Soc. 82, No. 16* (August 1960), 4433–4434.

[36] Sylvia J. Spengler, Andrzej Stasiak, and Nicholas R. Cozzarelli, "The Stereostructure of Knots and Catenanes Produced by Phage λ Integrative Recombination: Implications for Mechanism and DNA Structure," *Cell 42* (August 1985), 325–334.

[37] Paul S. Weiss, "A Conversation With Prof. Ned Seeman: Founder of DNA Nanotechnology," *ACS Nano 2, No. 6* (2008), 1090.

[38] Ann Finkbeiner, "Crystal Method," *University of Chicago Magazine* (Sept-Oct/2011), http://mag.uchicago.edu/science-medicine/crystal-method

[39] Junghuei Chen and Nadrian C. Seeman, "Synthesis From DNA of a Molecule With the Connectivity of a Cube," *Nature 350* (18 April 1991), 631–633.

[40] Nadrian C. Seeman, "*De Novo* Design of Sequences for Nucleic Acid Structural Engineering," *Journal of Biomolecular Structure and Dynamics 8, No. 3* (1990), 573–581.

[41] Ann Finkbeiner, "Crystal Method," 6.

[42] Nadrian C. Seeman, "Single Stranded DNA," http://www.ams.org/meetings/lectures/seeman-lect.pdf

[43] Junghuei Chen and Nadrian C. Seeman, "The Electrophoretic Properties of a DNA Cube and Its Substructure Catenanes," *Electrophoresis 12* (1991), 607–611.

[44] Chen and Seeman, "Synthesis from DNA," 633.

[45] Chen and Seeman, "The Electrophoretic Properties," 607–608.

[46] Ned Seeman, "The Crystallographic Roots of DNA Nanotechnology," *ACA RefleXions, No. 2* (Summer, 2014), 23.

[47] Ibid., 22.

[48] Eric Smalley, "NYU's Nadrian Seeman," *TRN's View From the High Ground* (May 4–11 2005), http://www.trnmag.com/Stories/2005/050405/View_Nadrian_Seeman_050405.html

[49] Finkbeiner, "Crystal Method," 8.

[50] http://www.nobelprize.org/nobel_prizes/medicine/laureates/1978/press.html

CHAPTER SEVEN

Motif Generation, Sequence Design, and Nanomechanical Devices

Nadrian Seeman is seated at his computer. His long hair is tousled, his salt and pepper beard profuse but not riotous. He wears sneakers, jeans, and a brightly colored short sleeve shirt. Coils of plastic tubing suspended by fishing wire hang from the ceiling around him. They appear to be DNA molecules embarked on a labyrinthine journey. Near his work desk is a table teeming with intricate assemblies of colored sticks and balls. Stacks of books and papers are piled high everywhere, like cairns on the path to knowledge. A successful scientist's feng shui.

Seeman is in an email conversation with an interviewer. Many questions are familiar from previous interviews: What got you interested in science and technology? What makes DNA particularly useful in nanotech? Can you speak about the trends in the applications of DNA nanotechnology? Then he is asked what books connected to science have impressed him in some way, and why. He replies that he has read biologist Sir Peter Medawar's *The Art of the Soluble*. Seeman says that the book has a clear message for scientists: "It is your job to solve problems, not merely to grapple with them."[1] Seeman seized this imperative at a defining moment in his career.

After a dozen years of work he had accomplished quite a lot. Publication of the first three-dimensional, closed DNA structure caused a sensation. One reviewer proclaimed, "This is founding a new field."[2] Nadrian Seeman proclaimed it "a floppy piece of crap, and you can't make a crystal out of something that's floppy."[3] As we touched on in the previous chapter, Seeman realized that the polyhedron that he'd built was akin to an assembly of toothpicks stuck into blobs of marshmallow at the corners. The edges had sufficient structural rigidity, but the angles at each corner were quite variable.

Flexible Junctions Redux

Take another look at the *two*-dimensional ligations of three-arm junctions in Figure 6.7 of Chapter Six. We showed closed trimers and tetramers as reaction products (and Seeman also found higher order hexamers, and so on). We consider here the trimer, the simplest of the closed cycles (panel (a) of Figure 7.1). To gain insight, Seeman synthesized three-arm junctions with different base pairs flanking the junction regions. He found that base-pair substitutions had no effect on the electrophoretic analysis. The autoradiograms seemed to leer up at him in mockery. The data shouted "Floppy!"[4, 5] To Seeman's well-trained mind the conclusion was inescapable. The bases at the branch sites were loosely paired or even unpaired. This introduced extreme flexibility. Then he did some molecular model-building of the trimer, panel (b) Figure 7.1. He found that if the bases that flank the three-arm junction are not completely paired, these junctions are not necessarily immobile regardless of the careful sequence selection he had employed.[4]

To underscore the point, in the previous chapter we wrote of the first demonstration of an immobile, four-arm nucleic acid junction. In fact, extensive studies showed this junction adopts an X-shape under physiological conditions (panel (c)

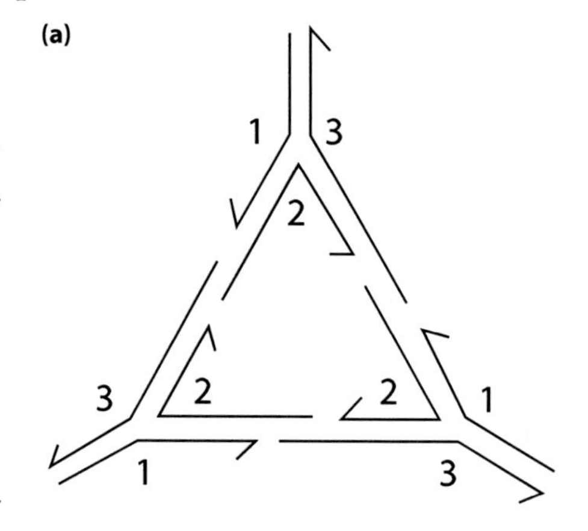

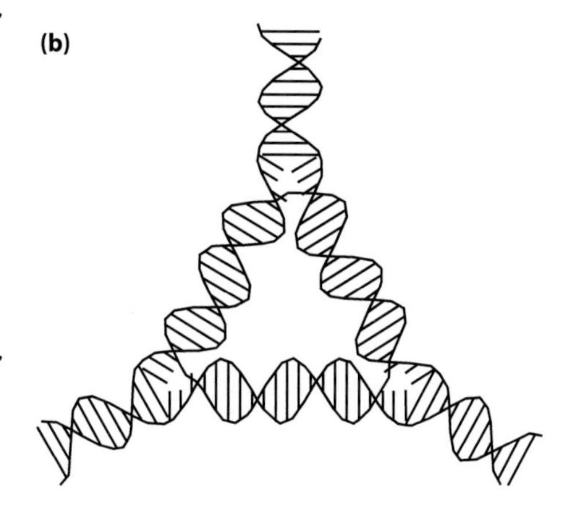

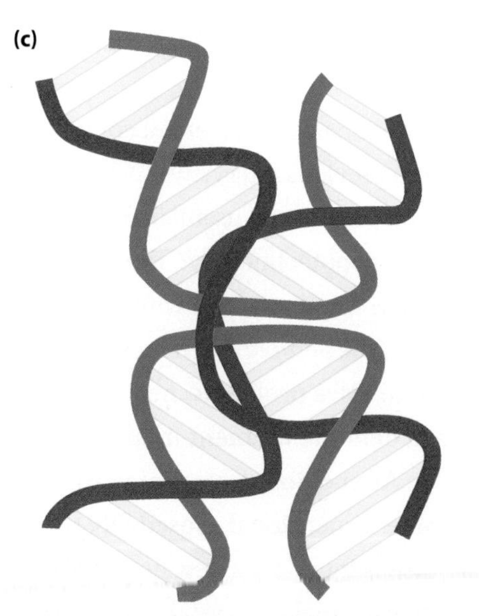

Figure 7.1 (a) Stick figure of a trimer of three-arm junctions. **(b)** Model of a trimer of three-arm junctions. Note the presence of unpaired bases. These might account for the flexibility of the junctions. **(c)** Three-dimensional structure of a four-arm nucleic acid junction as determined from experiments.

((a), (b) portions adapted/reproduced with permission from Rong-Ine Ma, Neville R. Kallenbach, Richard D. Sheardy, Mary L. Petrillo, and Nadrian C. Seeman, "Three-Arm Nucleic Acid Junctions Are Flexible," Nucleic Acids Research 14, No. 24 (1986), 9745–9753; (c) reproduced with permission from Mao C, "The Emergence of Complexity: Lessons from DNA," PLoS Biol 2, No. 12 (2004), e431.)

of Figure 7.1), and the angle between its two helical domains (the interhelical angle) can vary widely.[6,7] The unhappy implication is that the angles between the arms of the junction vary on the time scale (ca. 10 hours) of the ligation experiments. The production of a unique cyclic product—such as shown in Figure 6.4 (d) of Chapter Six—would suggest a fixed angle between the arms of the junction. But this is not obtainable with the simple branched junctions that Seeman employed. As we described previously, despite the flexibility he found in the junction ligation products, he decided to forge ahead anyway to prove that he could build in three dimensions.[8] Seeman achieved that goal with the cube-like figure that we examined. But as he later said, "Such structures might have uses, but building a regular lattice is not one of them."[9] Seeman spent *several years* trying to find a new building block, a new motif that was rigid. To say Seeman was doughty would be like saying that Rembrandt van Rijn dabbled in painting.

His epiphany of organizing DNA branched junctions was inspired by Escher's flying fish woodcut. Now he had a burning desire for another Escher-caliber construction design (Figure 7.2).[10]

"Escher! Get your ass up here."

Figure 7.2 Seeman needed another Escher-caliber construction. *(Reproduced with permission, © Robert Leighton/The New Yorker Collection 2013.)*

The Double-Crossover (DX) Molecule

After many failed attempts to find a motif that would assemble easily and be stiff, his breakthrough finally came in 1993. And as we will see, with one glorious burst of imagination he had created a superfluity of possible forms. Seeman designed the DNA motif known as the *double-crossover (DX) molecule*. This consists of two DNA double helices in which the strands are interwoven in a transverse fashion at what are called *crossover points*.[11] These crossovers yoke them together (Figure 7.3). The two crossovers prevent a duplex from twisting against its neighbor duplex. Thus, the interhelical angles between the two duplexes are fixed at 0° and the structure is rigid.[12] Here's another take: DX structures contain two Holliday junctions joined by two double-helical arms. The DX structure is a roughly rectangular-shaped molecule with four or sometimes five single strands of DNA providing sticky ends that allow the DX molecules to bind to each other in a predictable way: two DX molecules bind only if they have complementary sticky ends. *Seeman showed that sticky ends and immobile, branched junctions remain invaluable DNA construction tools when deployed in an appropriate motif.*

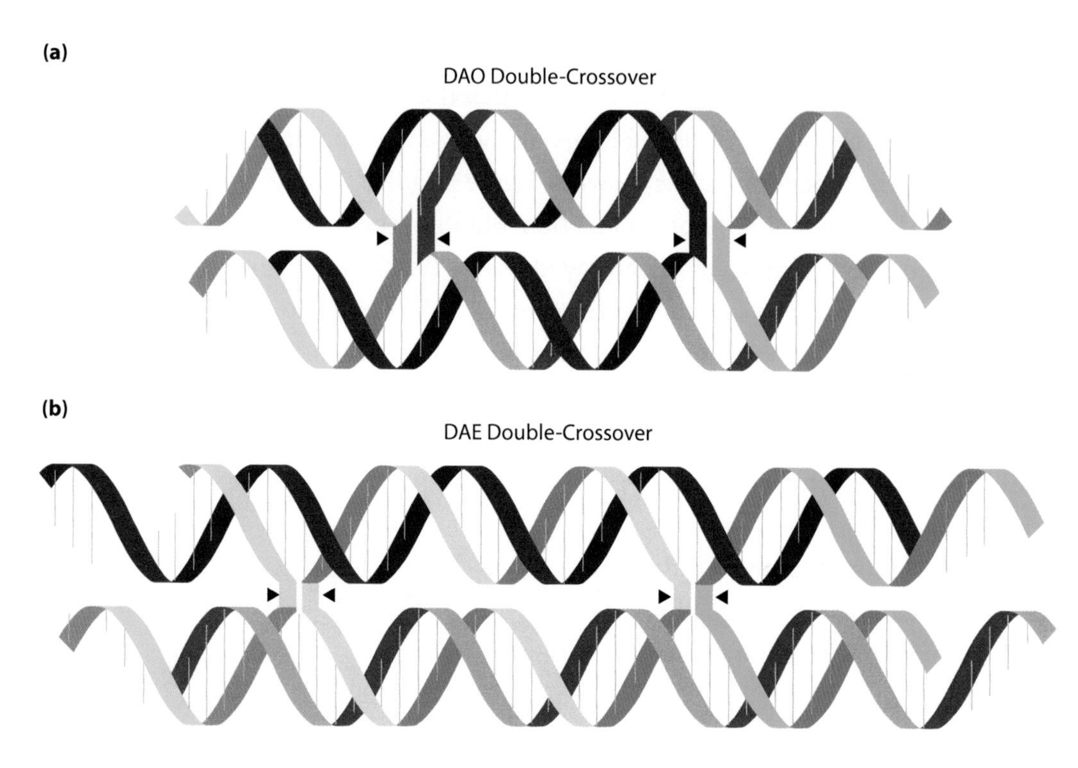

Figure 7.3 Two types of double-crossover (D) molecules, both in an antiparallel relative orientation (A), with an odd (O) or even (E) number of helical half-turns between crossover points. The *crossover points* are indicated with arrowheads. **(a)** The DAO molecule shown has three helical half-turns between crossover points and **(b)** the DAE molecule shown has four helical half-turns between crossover points. *(Adapted from Paul W.K. Rothemund, Nick Papadakis, Erik Winfree, "Algorithmic Self-Assembly of DNA Sierpinski Triangles," PLoS Biology, Vol. 2, No. 12, December 2004, e424, 2046.)*

Here's an additional feature: there are two different ways in which two helices can be combined in a transverse manner. This is because a double helix consists of two anti-parallel strands of DNA as we saw in Chapter Two. As we made explicit in Chapter Six, the 5′→ 3′ direction, the polarity of a strand, is opposite in the two strands of a double helix. To couple two helices together we can join strands of the same polarity or strands of the opposite polarity.

Motif In structural DNA nanotechnology, a motif is a distinctive design that specifies how DNA strands will link together. This pattern serves as the elemental building block for desired structures or devices.

Seeman designed base sequences that formed immobile junctions and that also determined the site of the crossover points. He found that there were a total of five different structural arrangements of DX molecules, three of which were in a parallel (P) orientation and two of which were antiparallel (A), where the parallel and antiparallel descriptors refer to the relative orientations of their two double-helical domains.[13] There are additional descriptors of these molecules, and he had to carefully evaluate each subtype for its suitability, most importantly, its rigidity. For example, he had to specifically build-in the distance between crossovers as multiples of helical half-turns to avoid torsional stress on the configuration. This led to further subdivisions: those double-crossover molecules separated by odd (O) and even (E) numbers of half-turns. So in Figure 7.3 (a), we have a DAO molecule, which is a double-crossover (D), in an antiparallel orientation (A), with an odd number (O) of helical half-turns between the two crossover points. Crossover points are indicated with arrowheads. DAO molecules consist of four strands of DNA, each of which participates in both helices. In Figure 7.3 (b) the DAE molecules consist of three strands that participate in both helices (yellow, blue, and green), and two strands that do not cross over (red and purple). In Figure 7.3, the 5′→ 3′ directions aren't shown, but the different colored strands that do crossover reverse their directions at the crossover points, a signature of antiparallel orientation of the duplexes.

Double-crossover (DX) molecule

The DX molecule consists of two double helices aligned side by side with strands crossing between the helices yoking them together. There are several types of DX molecules differentiated by the relative orientation of their helix axes, parallel or antiparallel, and by the number of double-helical half-turns (even or odd) between the two crossovers. The antiparallel versions are used extensively in structural DNA nanotechnology. They provide excellent rigidity for fabrication of extended nanostructures and nanomechanical devices.

Although Seeman made careful studies of all five possibilities, none of the three parallel orientations proved suitable and so we won't discuss them further. Figure 7.3 shows only antiparallel orientations. For those who are curious, the reason that there are more parallel possibilities than there are antiparallel is that there are two different types of parallel associations with an odd number of helical half-turns between crossover points (a detail related to the groove structure of DNA).[14]

The double-crossover (DX) motif was the first rigid DNA building block and has been aptly called one of the most fascinating creations in DNA nanotechnology.[15] How did Seeman come up with the double-crossover idea? As is true of many of his insights, it was bio-inspired.[16] The crossover process is a phenomenon that occurs naturally in genetic recombination although the crossovers are ephemeral, a means to an end (and the mechanisms in nature are more intricate than the technique that Seeman adopted). There are many details in the realm of double-crossover molecules and we won't cover them all. There are, however, several observations in order before moving on to a landmark self-assembly result enabled by DX molecules. The first observation is that in this book we are at home in the nanoworld. So, for the record, the DAO molecule has dimensions of ca. 2 x 4 x 13 nm while the DAE molecule has dimensions of ca. 2 x 4 x 16.0 nm.

Next, we'll make explicit the connection between the multi-arm immobile junctions we've spoken so much about and the double-crossover molecule. The antiparallel DX motif consists of two juxtaposed immobile four-arm junctions arranged so that at each junction the non-crossover strands are antiparallel to each other.[17] If you look at each crossover point in Figure 7.3 you'll find four DNA strands (even though strand colors may be the same, e.g., two yellow strands). That is, each crossover point is the site of an immobile four-arm junction.

Another point is that in order to characterize the suitability of double-crossover molecules for constructing periodic matter, Seeman used the same assay that we saw in Chapter Six, the ligation-closure experiment. His criterion for suitability was a lack of cyclization. As Seeman remarked, "This is a practical assay, because cyclization poisons the growth of periodic systems. In addition, cyclization at a series of different short lengths suggests high molecular flexibility." [18]

A final observation: in the present chapter we'll focus on DX molecules, but other rigid DNA motifs have been invented and used to advantage. These include triple-crossover molecules, a cross motif, a rhombus/parallelogram motif, and several triangle motifs. Now it's time to show some of the power of a stable, rigid DNA motif.

Design and Self-Assembly of Two-Dimensional DNA Crystals

Seeman's development of the DX motif stimulated a theoretical proposal by Erik Winfree of Caltech to use two-dimensional lattices of DX molecules for DNA-based computation.[19,20] The idea was to implement computation on the molecular level by means of what are known as Wang tiles.[21] These are rectangular tiles with programmable interactions and are a tool in the field of mathematical logic. DX molecules acting as molecular Wang tiles might self-assemble to perform desired computations.[19,22,23] Thus, the ability to create two-dimensional lattices assumed additional interest as a step toward DNA-based computation. This in turn drove the style of two-dimensional crystal design. Seeman and his colleagues used double-crossover molecules that would self-assemble into simple structures. For example, type A and type B units make a striped lattice (left side of panel (a), Figure 7.4).[17] They also made a set of four units that produced a striped lattice with a greater period—i.e., repeat distance (right side of panel (a), Figure 7.4).

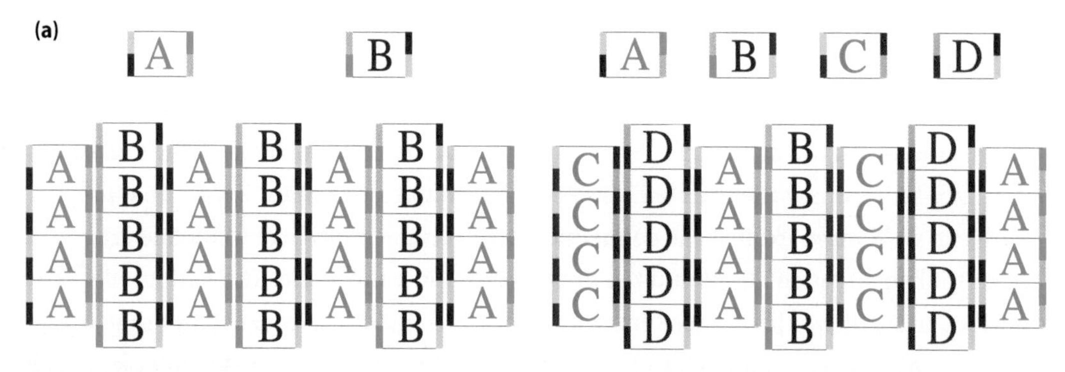

Figure 7.4 (a) The logical structure for two-dimensional lattices consisting of two units and four units. In the two-unit design, type A units have four colored edge regions each of which match exactly one colored region of the adjacent type B unit. Similarly, in the four-unit design, the edge colors are chosen uniquely to define the relations between neighboring units. *(Reproduced with permission from Erik Winfree, Furong Liu, Lisa A. Wenzler, and Nadrian C. Seeman, "Design and Self-Assembly of Two Dimensional DNA Crystals," Nature 394 (6 August 1998), 539–544.)*

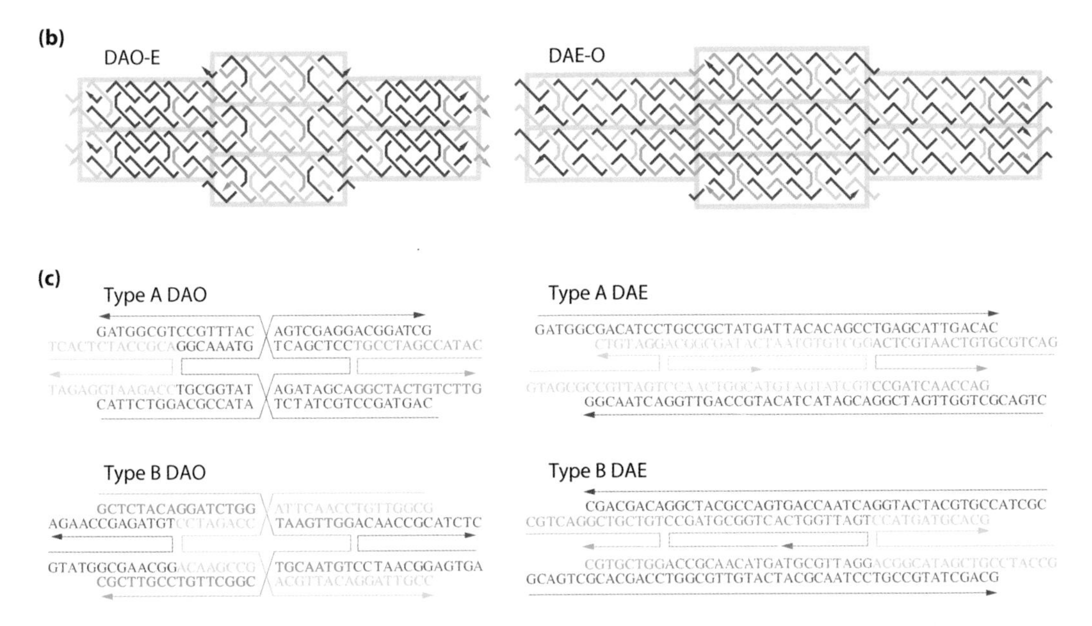

Figure 7.4 (b) The lattice topologies produced by the DAO-based system and the DAE-based system. Each individual DX (DAO or DAE) unit is highlighted by a gray rectangle. Seeman chose a unique color for each strand type that would be formed after covalent ligation of adjacent DX units. Arrowheads indicate 3′ ends of strands. **(c)** The sequences used to design type A DAO and DAE molecules and type B DAO and DAE molecules. The schematics accurately report oligonucleotide sequence and paired bases but are not geometrically *or* topologically faithful because they do not show the double-helical twist. *(Adapted with permission from Erik Winfree, Furong Liu, Lisa A. Wenzler, and Nadrian C. Seeman, "Design and Self-Assembly of Two-Dimensional DNA Crystals," Nature 394 (6 August 1998), 539–544.)*

When translated into molecular terms, they devised DX systems that self-assembled in solution into two-dimensional crystals. Each corner of each DX unit has a single-stranded sticky end with a unique sequence. Choosing sticky ends according to the familiar base complementarity rules controls the bonding of DX units.

We'll examine the comparatively simpler two-unit lattice. However, to make life interesting, Seeman used two separate systems to implement the two-unit lattice (the AB lattice). One system consisted of two distinct DAOs and the other consisted of two distinct DAEs (that is, distinct in their base sequences). These systems are called DAO-E and DAE-O, respectively, to indicate the number of half-turns between crossover points on adjacent units. He studied both the DAO-based system and the DAE-based system in parallel to show the versatility of his method: the same principles of self-assembly apply. The lattice topologies produced by these systems are, at first glance, a bit tough to wrap your head around (panel (b) of Figure 7.4).

Don't let the convolutions of panel (b) drive you crazy. Here's the upshot: in a two-dimensional crystal, each DX molecule contains four sticky ends distributed on its two component double helices (look at Figure 7.3). The complementarity of the sticky ends is designed in such a way that a DX molecule will interact with another four DX molecules through its four sticky ends. Any two DX molecules can interact with each other through only one pair of sticky ends. Any pair of sticky end interactions will position the two DX molecules in a conformation such that no other sticky ends from these two molecules are in sufficient proximity to interact.[24] This design results in the formation of regularly ordered two-dimensional arrays.

The important point is that the DX molecules are building blocks—as highlighted by the gray rectangles in panel (b) of Figure 7.4—and have programmed sticky end associations that control their self-assembly into a two-dimensional crystalline lattice.

Self-assembly of two-dimensional crystals
Two-dimensional crystals are designed by the self-assembly of double-crossover (DX) molecules. Each corner of each DX unit has a single-stranded sticky end with a unique sequence. Choosing sticky ends according to base complementarity rules controls the bonding of DX units.

Panel (c) of Figure 7.4 shows the base sequences used in the experiments. Even a quick look can be confirming of one's take on the lattice topologies shown in panel (b). For example, notice that in the type A and type B DAEs the red strand sequence and the purple sequence do not cross over (they go straight through). And these non-crossover strands are antiparallel to each other. You can also see this in the zigzag representation of the strands in the DAE-based system in the lattice topology representation where the arrowheads indicate the 3′ ends of strands. The schematics accurately report oligonucleotide sequence and paired bases but are not geometrically *or* topologically faithful because they do not show the double-helical twist. Note that for each type—A and B—there is a DAO version and a DAE version. To ensure that the component strands form the desired complexes, Seeman had to carefully design the strand sequences so as to avoid alternative conformations. In this context, remember our discussion in Chapter Six of the bane of multiple closed structures and also our remarks early in Chapter Six about free energy. Seeman sought the base sequences that maximized the free energy *difference* between the desired conformation and all other possible conformations. We'll return to the subject of sequence selection shortly.

The use of DX molecules provides great versatility in the construction of periodic two-dimensional lattices (Figure 7.5). Seeman and his colleagues made single two-dimensional crystals with areas up to 2 x 8 microns (that is, 2,000 x 8,000 nm) containing over 500,000 DX

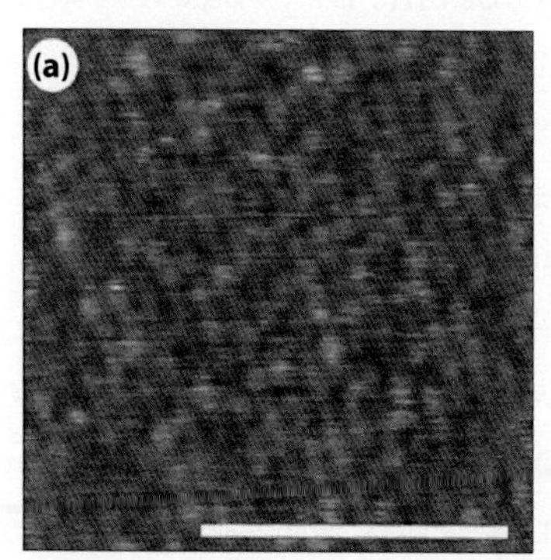

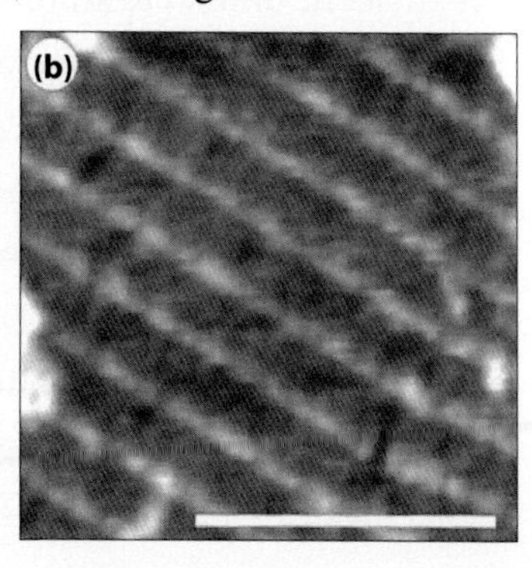

Figure 7.5 AFM images of **(a)** DAO molecules forming an AB style lattice where the stripes have a periodicity of 25 ± 2 nm. **(b)** DAE molecules forming an ABCD style lattice where the stripes have a periodicity of 66 ± 5 nm. Scale bars are 300 nm in both images. *(Reproduced with permission from Erik Winfree, Furong Liu, Lisa A. Wenzler, and Nadrian C. Seeman, "Design and Self-Assembly of Two-Dimensional DNA Crystals," Nature 394 (6 August 1998), 539–544.)*

molecules in which they modified the lattice periodicity. This demonstrated their ability to control construction at the nanoscale. They made direct physical observation of lattice assembly using atomic force microscopy (AFM). Figure 7.5 shows close-ups of two large crystals where (a) is composed of DAO molecules forming an AB style lattice and (b) is composed of DAE molecules forming an ABCD style lattice, both of the sort symbolized in the logical structure in panel (a) of Figure 7.4. The two-component lattice has a stripe every other unit and the four-component lattice was designed with a stripe every fourth unit.[25] To use the topographic sensitivity of the AFM, Seeman incorporated decorative labels such as *DNA hairpins* into appropriate DX molecules (more on hairpins in Chapter Eight). The DNA hairpins are unpaired loops of single-strand DNA that protrude above the plane of the lattice. The gray scale indicates that the decorated white stripes are 1–2 nm above the surface.

Seeman memorably said, "Biology is nanotechnology that works."[26] Invention of the double-crossover molecule launched biomimetic DNA nanotechnology that works—in two dimensions. At this point, the challenge of three dimensions still loomed.

Two-Dimensional Nanoparticle Arrays

Seeman put the DX motif to good use in other ways. We've had some exposure to gold nanoparticles via Chad Mirkin's work in Chapter Three. More generally, metallic and semiconductor nanoparticles exhibit optical and electronic properties that might be exploited in the design of future nanoelectronic devices. Proposed applications often entail deliberate organization of nanoparticles into specific structural arrangements.

Seeman, Paul Alivisatos, and colleagues used DX molecules to organize nanoparticles in two dimensions in a DX-based triangular motif (panels (a) and (b) of Figure 7.6). By using two of the directions of the motif to produce a two-dimensional crystalline array, one direction was free to bind gold nanoparticles. Identical motifs that were tailed in different sticky ends enabled the periodic ordering of 5 and 10 nm diameter gold nanoparticles.[27] Part (c) of the figure shows how a two-triangle array assembles in two dimensions. We see the connection of four triangles of two species. Only two domains (cyan bonding to magenta and brown bonding to red) are involved in array formation while the end of the third domain (dark blue or green) is free to act as a scaffold for gold nanoparticles. Part (d) shows the attachment of gold nanoparticles. Single-stranded DNA was bound to the gold nanoparticles using techniques similar to those of Mirkin in making thiolated DNA.

Now, like a molecular-based Julia Child, Seeman had all the ingredients prepared. He followed (invented, to be accurate) the recipe

> **Two-dimensional nanoparticle arrays**
> Metallic and semiconductor nanoparticles have optical and electronic properties that might be used in future nanoelectronic devices. Double-crossover molecules have been used to precisely organize nanoparticles into designed structural arrangements in two dimensions.

by adding DNA/gold conjugates to the solution containing the other strands needed to form the DX triangular motif. Only then did he mix the two sets of DX triangle/gold conjugates to form the two-triangle array.

In addition to the DNA/gold conjugate strands, Seeman designed and synthesized 21 other DNA strands that self-assembled into the desired motif. To get a feel for the intricacy of this concoction, Figure 7.7 shows the attachment site (for one of the molecules) of a gold nanoparticle and the complex sequence architecture of the strands comprising the triangular DX motif. Triangle edges contain 84 base pairs in each helix. Two of the directions terminate in a pair of 5′ sticky ends that are four bases long.[28]

Returning to Figure 7.6, the inset shows a transmission electron microscope image (TEM) of organized 5 and 10 nm gold nanoparticles. Because gold is so electron dense the nanoparticles appear dark in the electron micrograph without the need for the staining we mentioned in Chapter Five. However, the DNA is not electron dense and can't be seen in the micrograph without a stain to enhance contrast. The alternation of 5 nm particles and 10 nm particles is evident and the pattern mimics the parallelogram-like structure of the designed arrangement (panel (d), lower right).

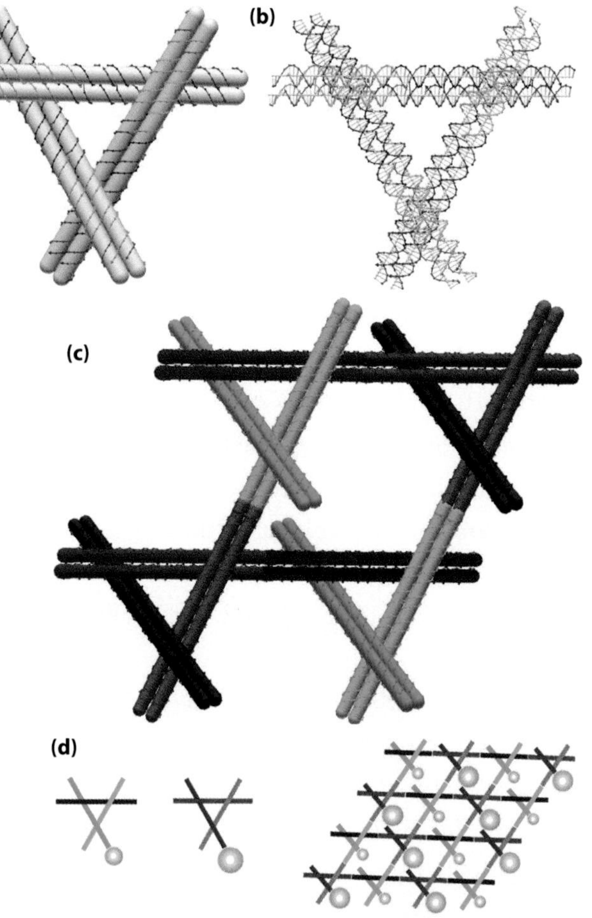

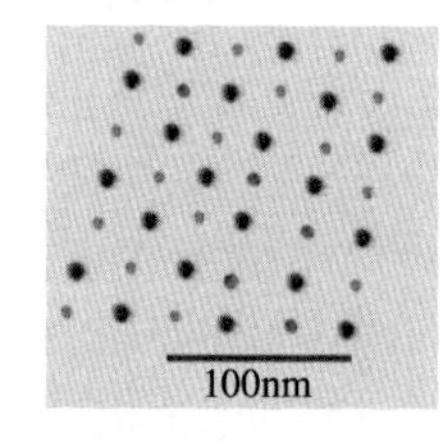

Figure 7.6 Motif to build gold nanoparticle arrays. **(a)** Design of three DX domains. Double helices shown as opaque rods with individual strands wrapped around. **(b)** Molecular structure. **(c)** Schematic of the formation of a two-component array (four triangles of two species). **(d)** Schematics of the attachment of 5 and 10 nm gold nanoparticles. Inset: Transmission electron micrograph (TEM) of two-dimensional arrays of gold nanoparticles. *(Reproduced with permission from Jiwen Zheng,Pamela E. Constantinou, Christine Micheel, A. Paul Alivisatos, Richard A. Kiehl, and Nadrian C. Seeman, "Two-Dimensional Nanoparticle Arrays Show the Organizational Power of Robust DNA Motifs," Nano Letters 6, No. 7 (2006), 1502–1504.)*

Sequence Design

Up to now we've had a good look at robust motifs but mostly just gawked at sequences. It's noteworthy that both motifs and sequence designs do not originate in the laboratory.

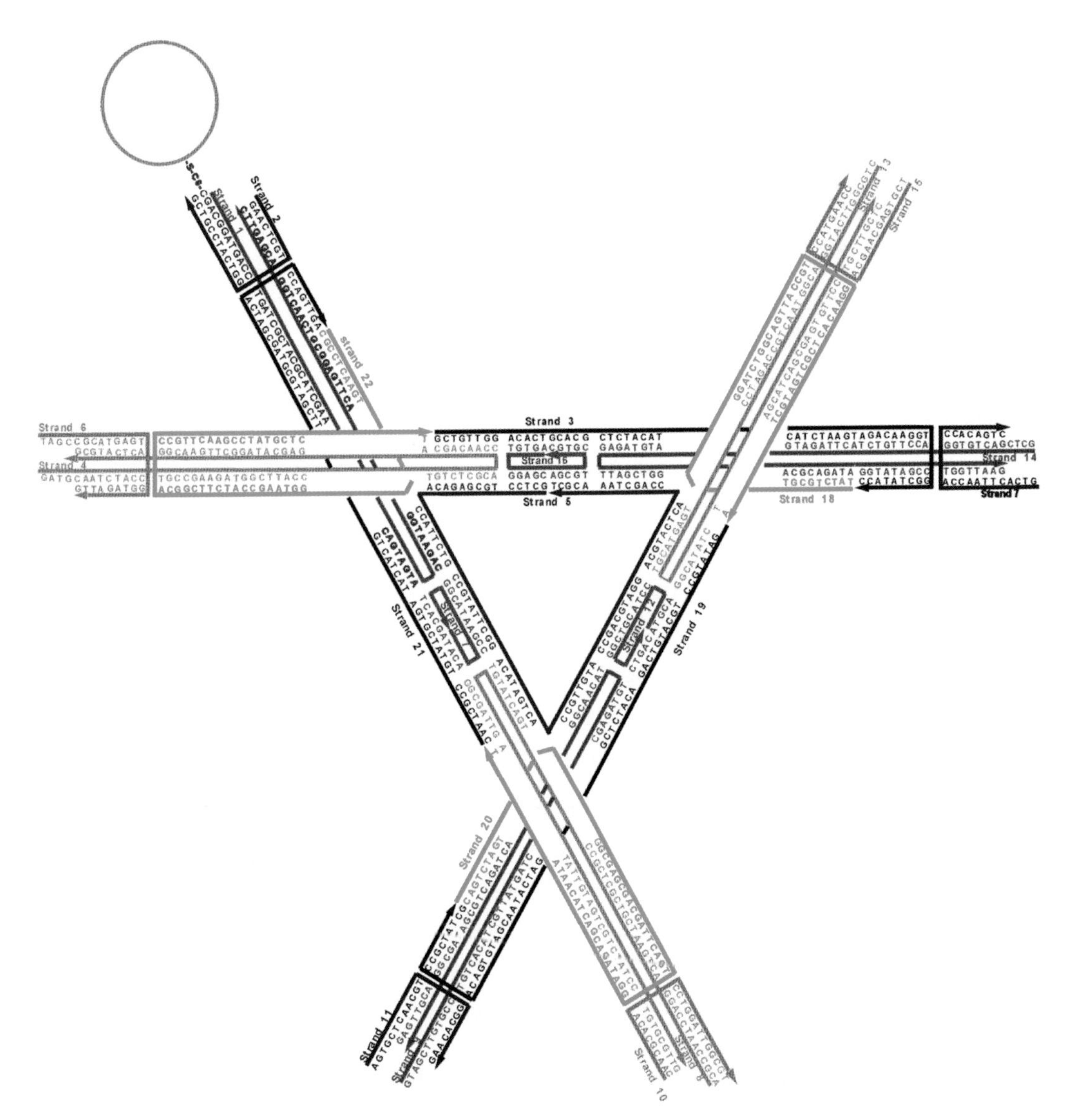

Figure 7.7 The sequence of the DX triangular motif. The helices shown in Figure 7.6 **(b)** have been unwound. The attached gold nanoparticle is shown as a gold-colored circle. *(Reproduced with permission from Online Supporting Information for Jiwen Zheng, Pamela E. Constantinou, Christine Micheel, A. Paul Alivisatos, Richard A. Kiehl, and Nadrian C. Seeman, "Two-Dimensional Nanoparticle Arrays Show the Organizational Power of Robust DNA Motifs," Nano Letters 6, No .7 (2006), 1502–1504.)*

Rather, they are determined on paper or on a computer and then the required DNA strands are synthesized. For nanotechnological purposes, once the sequence assignments are made the strands will self-assemble into the motif, thus highlighting the interplay between motif generation and sequence design.

Let's learn a bit about sequence design. Seeman's approach is rooted in what is called *sequence symmetry minimization*.[29,30] The basic idea is that DNA strands will maximize the double-helical structures that they form in order to have the largest number of paired bases (recall our remarks about free energy). The method can be illustrated using the example of a four-arm branched junction built from four strands (Figure 7.8).

Every strand contains 16 bases occurring as eight pairs denoted C • G, A • T, etc. Each of strands I-IV has been broken up into a series of 13 overlapping *tetramers*, so there are 52 tetramers in the entire molecule. To see this, note that in strand I the first two overlapping tetramers are shown as boxed: they are CGCA and GCAA. Sequence symmetry minimization insists that each tetramer be unique. The approach also insists that each tetramer spanning a branch point, such as the boxed CTGA, *not* have its linear complement (i.e., TCAG) present. The merit in being a hard taskmaster is that sequence symmetry minimization forbids the tetramers to form linear double helices that would destabilize the branched junction.

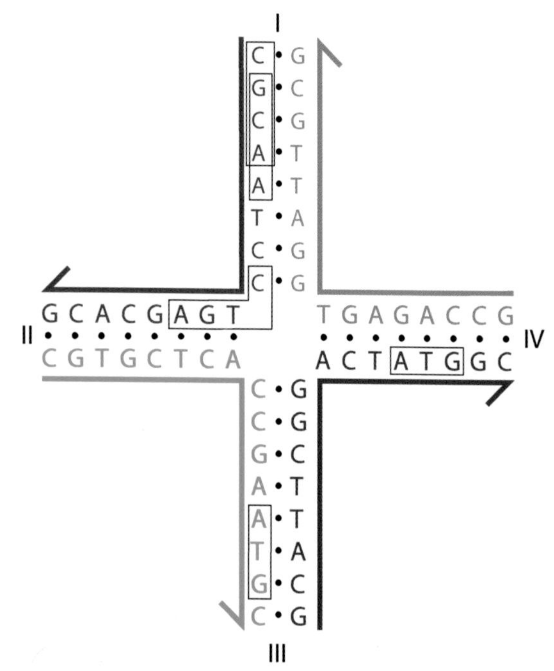

Figure 7.8 Sequence design illustrated with a four-arm branched junction. *(Reproduced with permission from Nadrian C. Seeman, "Biochemistry and Structural DNA Nanotechnology: An Evolving Symbiotic Relationship," Biochemistry, 42, No. 24 (2003), 7259–7269.)*

As a result, chemical competition during self-assembly with the four octamer double-helical targets (that is, the four sets of eight base pairs, one set in each junction arm) can occur only from trimers, such as the boxed ATG sequences. And the free energy difference between octamers (more paired bases, hence, lower free energy) and trimers (fewer paired bases, hence, higher free energy) carry the day: octomers win. (This is a bit simplistic because for larger constructs, units larger than tetramers and trimers would be required.)

DNA sequence symmetry minimization
This technique for the design of synthetic nucleic acid aims to minimize sequence similarities between segments of DNA strands. The goal is to decrease the chances of forming undesired structures instead of the desired target nanostructure.

Seeman's method of sequence symmetry minimization jump-started the creation of stable branched junctions. It measures up to the chore of designing sequences that work. However, in recent years there have been some issues involving symmetry minimization.[31] One of the most dramatic examples of *ignoring* sequence symmetry and just designing structures is Paul W.K. Rothemund's invention of DNA origami. We will discuss DNA origami at length in Chapter Eight.

Nanomechanical Devices

To close this chapter we'll return to the double-crossover motif and see how it enabled another benchmark achievement: the first robust *nanomechanical device*. A *robust* nanomechanical device is one that behaves like a macroscopic device. A robust device has well-defined endpoints and does not undergo component-changing reactions such as dissociation.[32] Before we hunker down with the first robust device, we step up to higher ground for an overview of the nanoterrain before us.

Seeman's initial endeavors in structural DNA nanotechnology were directed at making DNA objects such as the polyhedron (the cube) that we saw in the previous chapter. Once he generated the rigid DX motif it was possible to build two-dimensional arrays with programmable features. The original impetus for building arrays came from a desire to improve the practice of macromolecular crystallization; this was spelled out in Seeman's 1982 theoretical paper (Chapter Six). But goals also included organizing nanoelectronics and DNA-based computation.[19,33] We've seen how the DX motif contributed to these ambitions.

DNA polyhedra and arrays are admirable static objects and are a route to controlling nanoscale structure. But a key aspect of controlling the structure of matter is the ability to make it change its shape. Objects that change their shapes in response to an external stimulus are, in principle, capable of functional utility—they can operate as machines. Thus, once one has learned to control the structures of DNA objects, it makes sense to see if it's possible to get them to do some work.[21] The ability of a nanomechanical DNA *structure* to perform work is the rationale for Seeman's use of the term nanomechanical *device.*

DNA nanomechanical device A nucleic acid-based nanoconstruction that is capable of controlled mechanical movement.
Robust DNA nanomechanical device A nanomechanical device that behaves like a macroscopic device in that it has well-defined endpoints and does not undergo component-changing reactions, e.g., dissociation.

There has been an explosion of activity in the area of nucleic acid-based nanomachines. They work on several different principles and researchers have used a variety of physical techniques to demonstrate the nanomachines' movements. Seeman used his first device (a forerunner of the first *robust* device) to control DNA structure by changing the position of a branch point.[34] Seeman later referred to his effort as "successful, but cumbersome."[35] Moreover, he had hoped to couple the control of the position of a branch point to a more dramatic result: the conversion of a specific group of nucleotides from B-DNA to Z-DNA (terminology to be explained shortly).[36] He soon succeeded in making a nanomechanical device based on the B-Z *structural transition* of DNA. Even so, he acknowledged that the true power of using DNA is its programmability and that such single-trigger structural transitions can only lead to a few variants. So, *sequence-dependent* devices—taking advantage of the sequence specificity associated with DNA hybridization—are preferred over devices involving DNA structural transitions in achieving a multiplicity of possible responses.[37]

The first sequence-dependent nanomechanical device was constructed by Bernard Yurke (Bell Laboratories), Andrew Turberfield (University of Oxford), and colleagues.[38] Seeman knew a good idea when he saw it and was quick to adopt this strategy.[39]

This was an example of the changing landscape of DNA nanotechnology research that we alluded to in Chapter Six: the field had begun to blossom and others were contributing to the discovery process.

Since it's a big deal to crawl or walk compared to not moving at all under your own steam (just ask any parent), we've tried to provide some background for the advent of nanomechanical devices that followed the hegemony of motionless nanoconstructions. Now we'll zoom in and consider the specifics of the first robust DNA nanomechanical device (Figure 7.9).[40] It is a supramolecular (more than one molecule) structure consisting of two rigid DNA double-crossover (DX) molecules connected by 4.5 double-helical turns (i.e., 4.5 full turns of DNA) between the nearest crossover points. This is achieved with a long central helix that is flanked on either end by a DAO molecule. The DAO molecules are similar to what we saw in Figure 7.3 (a)—each is a double-crossover molecule whose helices are antiparallel to each other and whose crossover points are separated by an odd number (three) of double-helical half-turns. That's a lot to digest. So we now try to go from pemmican to purée.

The top illustration in Figure 7.9 is a molecular model constructed entirely from right-handed B-DNA (definition to follow). The bottom illustration shows the same supramolecular structure after it has undergone a transition in the form of a twist. The twist consists largely of a rotary motion around the central, long axis: in the top illustration, the two unconnected domains of the DX molecules lie on the same side of the central helix, while in the bottom illustration, these domains switch to opposite sides of the central helix. Note that this twist also increases the separation of the pink and green

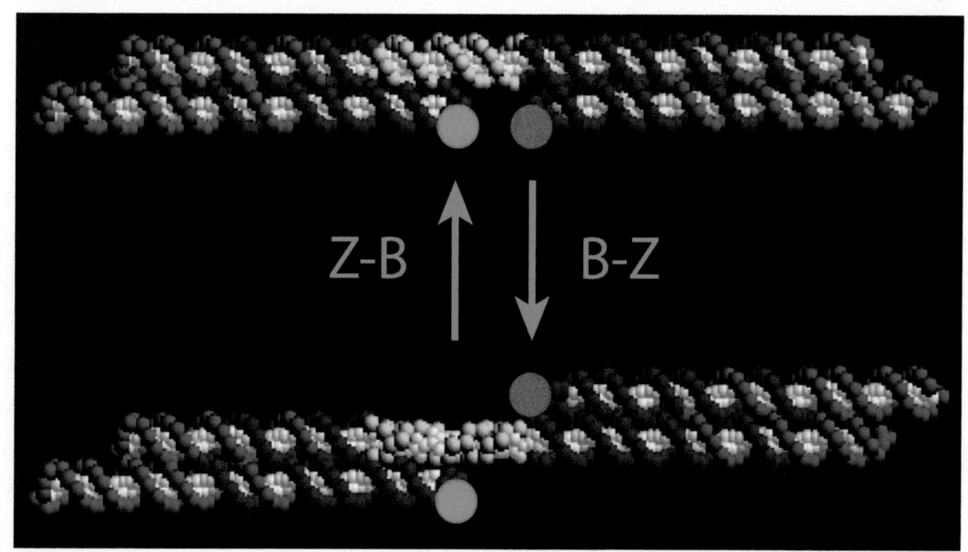

Figure 7.9 Design of the first robust nanomechanical device. Switchable motion in this assembly is accomplished using the (reversible) transition from the B form of DNA to the Z form of DNA. The key element that enabled the construction of the device was the discovery of the rigid DNA motif, the DX molecule. *(Reproduced with permission from Chengde Mao, Weiqiong Sun, Zhiyong Shen, and Nadrian C. Seeman, "A Nanomechanical Device Based on the B-Z Transition of DNA," Nature 397 (14 January 1999), 144–146.)*

circles. These are dye molecules that are used to detect which of the two states the supramolecular structure is in. We'll expand on this terse depiction, but it's time for some terminology.

> **B-form DNA** This is the most common form of DNA. It occurs as a right-handed double helix. The double helix winds to the right (clockwise when viewed end on).
>
> **Z-form DNA** A naturally occurring left-handed double-helical form of DNA in which the double helix winds to the left (counterclockwise when viewed end on).

B-DNA is the standard double-helical structure for DNA in aqueous solution. We have referred to it from the outset of this book simply as DNA. It's the most common form of DNA. The "B" is historical nomenclature that dates back to the studies of Rosalind Franklin whose work we saw in Chapter Two. What about the designation right-handed? Here's one way to understand this descriptor. The two DNA backbone polymers that we're familiar with wrap around each other in a helical spiral around an imaginary helix axis. If you were to look at the DNA from either end you would see a tight circle of atoms and the helix axis would be in the center of this circle. When viewed end on, both DNA strands spiral away from you around the helix axis in a clockwise direction. This signifies that it is right-handed. Z-DNA is a naturally occurring, three-dimensional, double-helical form of DNA in which the double helix winds to the left (counterclockwise when viewed end on). The "Z" came from the fact that there is a zigzag arrangement of the backbone of the molecule.[41]

Let's return to the top illustration in Figure 7.9. Each nucleotide is shown as two spheres, a colored one for the backbone portion of the nucleotide and a white one for the base portion. We see three cyclic strands: one in the center drawn as a red strand with a central yellow segment, and two blue strands on the ends that are each *triply catenated* to the red strand. Fluorescent dyes are drawn schematically as green (for the dye fluorescein) and magenta (for the dye known as Cy3) circles that are attached near the middle of the molecule. At the center of the connecting helix is a 20-nucleotide region of proto-Z DNA in the B-DNA conformation, shown in yellow. The term *proto* means that this nucleotide sequence is one from which a segment of Z-DNA can be formed. In fact, when the B-Z transition occurs, this same yellow portion becomes left-handed Z-DNA, as shown in the bottom illustration of Figure 7.9. The transition from B-form DNA to Z-form DNA is achieved by the addition of a particular salt (hexaamminecobalt(III) chloride) to the solution.[42] Removal of this salt causes the change back to the B-form. Accompanying the change in twist between the B and the Z forms, the two DX molecules change their relative positions; they switch to opposite sides of the helix (bottom illustration). This increases the separation of the dyes, and the change in the proximity of the two dye

> **B-Z transition** DNA of an appropriate nucleotide sequence is capable of forming both the right-handed double-helical form (B-form) and the left-handed double-helical form (Z-form). The two forms can be converted from one to the other by several means including changing the solution conditions.

molecules (a change of only a few nanometers) is detected by means of fluorescence resonance energy transfer (FRET).

Terminology time again. In our context, the FRET technique is used to measure the distance between fluorescent dye molecules (applicable only when they are separated by 10 nm or less). The experimenter uses a laser to stimulate an interaction between the electronic excited states of two dye molecules in which the excitation energy is transferred from the donor molecule to the acceptor molecule. The efficiency of the energy transfer process varies in proportion to the inverse sixth power of the distance between the donor and acceptor molecules (Figure 7.10). Boiling that down, *a small change in the distance between dye molecules produces a large change in the transfer efficiency.*

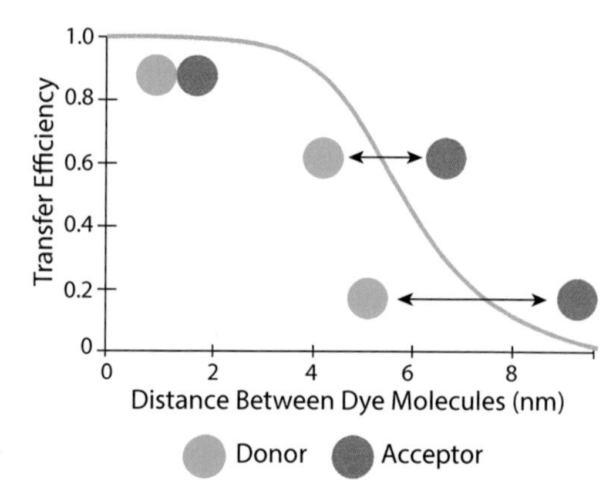

Figure 7.10 The efficiency of fluorescence resonance energy transfer (FRET) between two dye molecules is inversely proportional to the sixth power of the distance between the donor and acceptor dye molecules. *Note*: the transfer efficiency is defined in such a way that it is equal to 1.0 when the distance between dye molecules is equal to 0. *(Reproduced with permission from Jörg Langowski, http://www.dkfz.de/Macromol/research/smfret.html)*

One can measure the donor emission spectrum and the acceptor absorbance spectrum. If the dye molecules are close enough so that the fluorescence intensity (plotted as a function of wavelength) of these two spectra overlap, one can calculate a mathematical quantity known as the spectral overlap integral (Figure 7.11). Using this information, one can then calculate the transfer efficiency and therefore the distance between the dye molecules.

The FRET process was deemed a spectroscopic ruler in the experiments that pioneered its use in revealing proximity relationships in biological macromolecules and has since been called a molecular ruler and a spectroscopic nanoruler.[43] *Note*: Readers may worry that if the energy transfer efficiency varies as the inverse sixth power of the separation distance, it would blow up at small distances. But the transfer efficiency is defined such that it is 1.0 when the separation distance is equal to 0.[44] The famous FRET equation was formulated by German physical chemist Theodor Förster (and the acronym is often construed as Förster

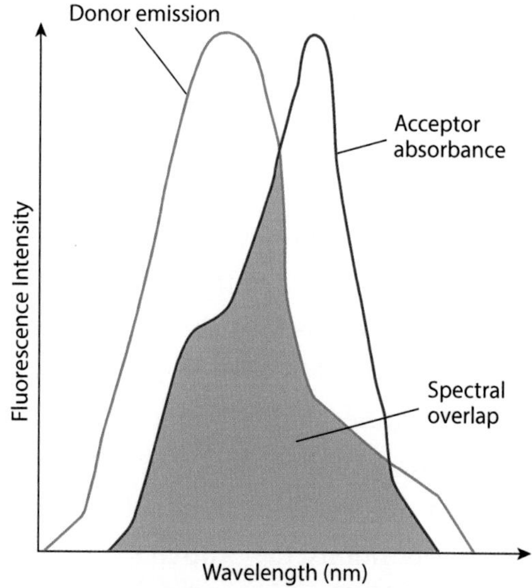

Figure 7.11 The region in which the two spectra overlap can be evaluated mathematically and the result is used to calculate the transfer efficiency and hence the distance between the dye molecules.

resonance energy transfer rather than fluorescence resonance energy transfer).

For readers interested in the exercises at the end of this chapter, we present the FRET equation for energy transfer efficiency as $E = \frac{1}{1+\left(\frac{r}{R_0}\right)^6}$, where "r" is the distance between the dye molecules and R_0 is the value of "r" when 50% of the energy is transferred. The value of R_0 can be readily obtained from experimental data. To repeat, R_0 is given by the value of "r" at E=0.5, that is, at 50% energy transfer efficiency.

FRET (Fluorescence resonance energy transfer) A distance-dependent transfer of energy from a donor molecule to an acceptor molecule. In our context, FRET is a technique that is used to measure the distance between fluorescent dye molecules (applicable only when they are separated by 10 nm or less).

Going back to Seeman's work, how did he verify the operation of his nanomechanical device? The transition between B-DNA and Z-DNA and vice versa consists (largely) of a rotary motion as suggested by the cartoons in Figure 7.9. (The structural transition also causes a small change in the separation of the base pairs along the helix axis, which is why we say it is largely, but not exclusively, a rotary motion.) Seeman prepared two sets of B-DNA molecules: those with a proto-Z region and those *without* a proto-Z region. The latter were used in a set of *control* experiments.

Figure 7.12 shows results of his FRET measurements that detect nanomechanical motion. In panel (a), the height variation in the blue vertical bars shows the donor energy transfer (%) in transitions between

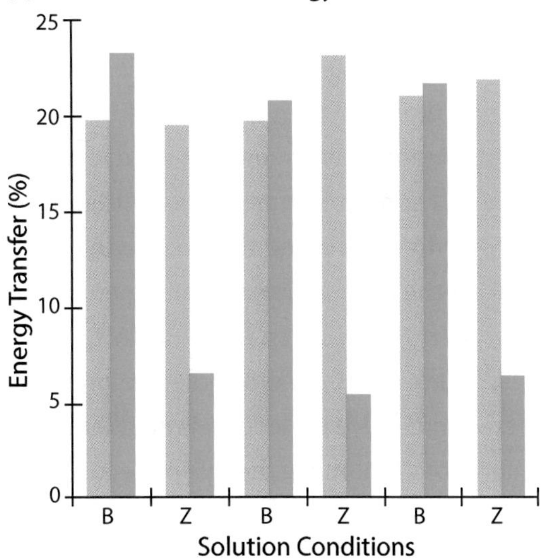

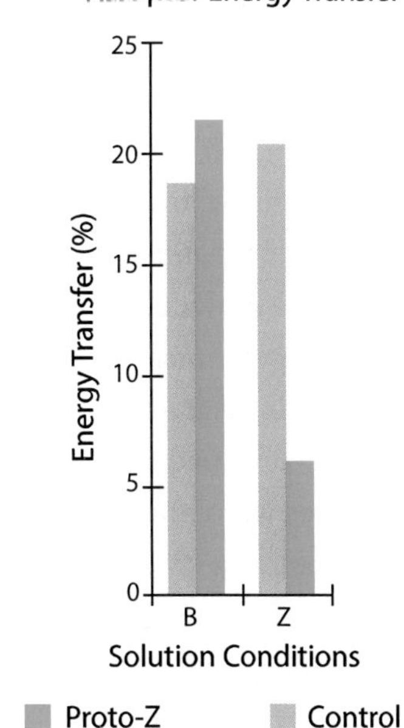

Figure 7.12 FRET demonstration of nanomechanical motion. Only the B-DNA molecule that contains the proto-Z nucleotide region shows a change in energy transfer when a particular salt is present or absent in the solution. *(Adapted with permission from Chengde Mao, Weiqiong Sun, Zhiyong Shen, and Nadrian C. Seeman, "A Nanomechanical Device Based on the B-Z Transition of DNA," Nature 397 (14 January 1999), 144–146.)*

B-DNA and Z-DNA. The cycling between B and Z is accomplished by the addition or removal of the salt hexaamminecobalt(III) chloride in the solution (and additional less dramatic changes in the composition of the solution). B-DNA is the stable form in the absence of the salt while Z-DNA is the stable form in the presence of the salt. However, the energy transfer (%) remains essentially constant (shown by the orange vertical bars) when the B form without the proto-Z region—i.e., the control molecule—is examined. That's because the control form of the B molecule is designed to be incapable of undergoing the B-Z transition. Panel (b) of Figure 7.12 shows a similar result when the acceptor energy transfer (%) is measured both for B-DNA with a proto-Z region and for B-DNA without a proto-Z region (the control results).

If this isn't clear, let's try again. Focus on panel (a) of Figure 7.12. Look at the vertical blue bars reading from left to right and ignore the orange bars. The height of the blue bar is high then low then high then low . . . This height fluctuation gives a measure of the energy transfer between the two dye molecules in response to an external stimulus. The external stimulus consists of Seeman changing the solution buffer—in essence, he is repeatedly adding and then removing the cobalt salt in the solution. When the blue bar is high the energy transfer between the dye molecules is large because they are close together. That is when the B-DNA form shown in the cartoon in panel (a) of Figure 7.9 will be the predominant molecule present. When the blue bar is low the energy transfer between the dye molecules is smaller because they are farther apart. Providing solution conditions that move the dye molecules farther apart induces the rotation of B-DNA into the Z-DNA form shown in the cartoon in panel (b) of Figure 7.9. As you look at the alternating height of the blue bars—high, low, high, low, high, low—visualize in synch with this change in height the reversible twist around the long central axis in either panel of the cartoon. This rotary motion puts the dye molecules closer together, then farther apart, then closer together, then farther apart . . . Seeman measures the tiny reversible distance change with his FRET equipment.

Our discussion of the first robust nanomechanical device has been intense. *Here is the most important point*: the device is predicated on the assumption that the distances between the dye molecules in the two states (B- and Z-DNA conformations) are well defined. *Well-defined distances can exist only if all parts of the DNA molecules maintain their structural integrity.* Seeman found that his initial attempts to demonstrate motion in devices that contained three-arm DNA branched junctions instead of double-crossover (DX) molecules were unsuccessful, owing to the flexibility of those components.[40] *He solved this problem with his creation of the double-crossover molecule (DX) because the DX motif behaves as a rigid unit.* It took Seeman 12 years from the conception of the device to the publication of the device in 1999.[45] Only with his invention of the DX molecule was he able to get a FRET signal that he believed; the other device designs were not stiff enough to give a reliable FRET signal.

Germans have a penchant for long words. If you've ever studied the language you may find that it has the feel of LEGO® about it: many little pieces are joined together to create a *complex* compound noun. They sprout up everywhere, like bright yellow dandelions in a lush green lawn. And Nadrian Seeman's LEGO®-like molecular constructions illustrate an aspect of his scientific talent: *Fingerspitzengefühl.* Literally, this word means, finger-tips feeling. In a newspaper opinion article, William Safire wrote, "Fingerspitzengefühl... that combination of sure-footedness on slippery slopes and sensitivity to nuance familiar to mountain goats, safecrackers and statesmen."[46] In working his way out of gnarly situations, Seeman has shown his intuitive flair—his fingertips feeling—for visual-spatial processing. His invention of the double-crossover molecule is a classic example of this.

Regardless of the ingenuity of the B-Z device and the difficult route to its realization, perhaps the reader is a little bemused. What's all the hubbub about? It's only a minuscule rotation of a few molecules. Puzzlement and even skepticism are understandable. But when someone wants to erect a tall building they begin modestly. Indeed, they set out in the opposite direction and dig a big hole in the ground. The significance of the first robust nanomechanical device was that it was a starting place. In later chapters we'll explore nanomechanical devices of far greater complexity and whose potential impact is more transparently obvious.

The Age of Discovery began in the early fifteenth century. Prince Henry the Navigator, a Portuguese royal prince, sent many sailing expeditions down Africa's west coast, always within sight of land. None had sailed past legendary Cape Bojador, just south of the Canary Islands. The area had treacherous reefs and unmanageable currents. Sailors had not dared to go beyond and inevitably turned back. Prince Henry sent more than a dozen expeditions over the course of a decade, trying to go farther than Cape Bojador. Finally, in 1434, Gil Eanes, the commander of one of the expeditions, succeeded in rounding the Cape. He found a route by sailing a few miles out to sea and then heading south for several miles before returning to the coastline. The discovery of a passable route marked the beginning of the Portuguese exploration of Africa. Eanes and his crew surmounted an obstacle by means of creativity and fortitude.[47,48] Over half a millennium later, Nadrian Seeman advanced the adventure that is nanotechnology. He realized that DNA could be induced to take forms other than the double helix, that it could be used to make junctions.

But his invention of structural DNA nanotechnology was stymied by a lack of rigidity. He had to take a different tack, to find an alternative branched motif. A decade after he constructed the first immobile nucleic acid junction he found a way to make such a junction rigid. And this he did with the DNA crossover molecule. Seeman was tenacious and he found a creative solution to his quandary. He did not merely grapple with a problem. He solved it.

Exercises for Chapter Seven: Exercises 7.1–7.2

Exercise 7.1

In Figure 7.13 we show some of the original experimental data from the classic FRET paper by Lubert Stryer and Richard P. Haugland (endnote 43). They used oligomers of poly-L-proline as spacers of defined length to separate an energy donor and an energy acceptor by distances ranging from 1.2 nm to 4.6 nm. (Poly-L-proline is what is known as a poly(amino acid), a synthetic polymer made up of many repeating units of the amino acid proline. Amino acids comprise the building blocks of proteins.) The experiments of Stryer and Haugland demonstrated agreement with the r^{-6} dependence predicted by Theodor Förster. What is the value of R_0 for the energy donor and the energy acceptor used in their experiment?

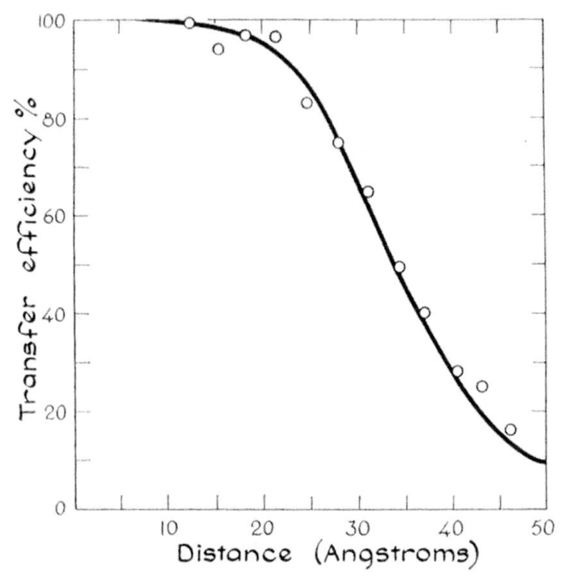

Figure 7.13 FRET distance dependence. *(Lubert Stryer and Richard P. Haugland, "Energy Transfer: A Spectroscopic Ruler," PNAS 58(1967), 719–726. Figure courtesy of Lubert Stryer.)*

Exercise 7.2

Two fluorescent molecules A and B are chemically attached to the end of the poly(amino acid) poly-L-proline; "A" is at one end of the molecule and "B" at the other end. In this particular poly(amino acid), the bond joining the prolines is not a peptide bond (the typical bond between amino acids) but a more *rigid* bond. We write A-(proline)$_n$-B to denote this polymer where n denotes the number of prolines in the chain. If A absorbs light that causes it to be in an electronically excited state, B fluoresces. If solutions of A-(proline)$_8$-B and A-(proline)$_{11}$-B are separately illuminated with light that excites A, for which solution will the fluorescence be brighter?

ENDNOTES

[1] Eric Smalley, "NYU's Nadrian Seeman," *TRN's View From the High Ground* (May 4–11 2005), http://www.trnmag.com/Stories/2005/050405/View_Nadrian_Seeman_050405.html

[2] Karen Hopkin, "3-D Seer," *The Scientist* (August 2011, Profile), 52–55.

[3] Ann Finkbeiner, "Crystal Method," *University of Chicago Magazine* (Sept–Oct/2011), http://mag.uchicago.edu/science-medicine/crystal-method

[4] Rong-Ine Ma, Neville R. Kallenbach, Richard D. Sheardy, Mary L. Petrillo, and Nadrian C. Seeman, "Three-Arm Nucleic Acid Junctions Are Flexible," *Nucleic Acids Research 14, No. 24* (1986), 9751.

[5] Mary L. Petrillo, Colin J. Newton, Richard D. Sheardy, and Nadrian C. Seeman, "Ligation and Flexibility of 3-Arm and 4-Arm Nucleic Acid Branched Junctions," *Biophysical Journal 51, No. 2* (February 1987), A569.

[6] Mary L. Petrillo, Colin J. Newton, Richard P. Cunningham, Rong-Ine Ma, Neville R. Kallenbach, and Nadrian C. Seeman, "The Ligation and Flexibility of Four-Arm DNA Junctions," *Biopolymers 27, No. 9* (1988), 1337–1352.

[7] David M.J. Lilley, "Structures of Helical Junctions in Nucleic Acids," *Quarterly Reviews of Biophysics 33, No. 2* (May 2000), 109–159.

[8] Robert Pool, "Dr. Tinkertoy," *Discover Magazine* (February 1997), http://discovermagazine.com/1997/feb#.URa6ijUjz6I

[9] Nadrian C. Seeman, "Nanotechnology and the Double Helix," *Scientific American 290, No. 6* (2004), 69.

[10] Robert Leighton, *The New Yorker* (4 February 2013), http://michaelmaslin.com/inkspill/anatomy-cartoon-robert-leighton-his-escher-cartoon-this-weeks-new-yorker/

[11] Tsu-Ju Fu and Nadrian C. Seeman, "DNA Double-Crossover Molecules," *Biochemistry 32* (1993), 3211–3220.

[12] Chengde Mao, "The Emergence of Complexity: Lessons from DNA," *PLoS Biology 2, No. 12* (December 2004), 2036-2038. http://www.ncbi.nlm.nih.gov/pmc/articles/PMC535573/pdf/pbio.0020431.pdf

[13] Nadrian C. Seeman, http://seemanlab4.chem.nyu.edu/cross.html

[14] Fu and Seeman, "DNA Double-Crossover," 3212.

[15] Masayuki Endo and Hiroshi Sugiyama, "Chemical Approaches to DNA Nanotechnology," *ChemBioChem 10* (2009), 2420–2443.

[16] Nadrian C. Seeman, "DNA Nicks and Nodes and Nanotechnology," *Nano Letters 1, No.1* (2001), 22–26.

[17] Erik Winfree, Furong Liu, Lisa A. Wenzler, and Nadrian C. Seeman, "Design and Self-Assembly of Two-Dimensional DNA Crystals," *Nature 394* (6 August 1998), 539–544.

[18] Xiaojun Li, Xiaoping Yang, Jing Qi, and Nadrian C. Seeman, "Antiparallel DNA Double Crossover Molecules as Components for Nanoconstruction," *J. Am. Chem. Soc. 118* (1996), 6131–6140.

[19] Erik Winfree, "On the Computational Power of DNA Annealing and Ligation," *DNA Based Computers: Proceedings of a DIMACS Workshop 27*, ed. Richard J. Lipton and Eric B. Baum (Providence, RI: American Mathematical Society, 1996), 199–221.

[20] Leonard M. Adleman, "Molecular Computation of Solutions to Combinatorial Problems," *Science 266* (1994), 1021–1024.

[21] Nadrian C. Seeman, "From Genes to Machines: DNA Nanomechanical Devices," *Trends in Biochemical Sciences 30, No. 3* (March 2005), 119–125.

[22] E. Winfree, X. Yang, and N.C. Seeman, "Universal Computation via Self-Assembly of DNA: Some Theory and Experiments," *DNA Based Computers II: Proceedings of a DIMACS Workshop 44*, ed. Laura Landweber and Eric Baum (Providence, RI: American Mathematical Society, 1999), 191–214.

[23] J.H. Reif, "Local Parallel Biomolecular Computation," *DNA Based Computers III: Proceedings of a DIMACS Workshop 48*, ed. Harvey Rubin and David Harlan Wood (Providence, RI: American Mathematical Society, 1999), 217–254.

[24] Mao, "The Emergence of Complexity," 2037.

[25] Winfree et al., "Design and Self-Assembly," 539.

[26] Nadrian C. Seeman, "In the Nick of Space: Generalized Nucleic Acid Complementarity and DNA Nanotechnology," *Synlett, No. 11* (2000), 1536–1548.

[27] Jiwen Zheng, Pamela E. Constantinou, Christine Micheel, A. Paul Alivisatos, Richard A. Kiehl, and Nadrian C. Seeman, "Two-Dimensional Nanoparticle Arrays Show the Organizational Power of Robust DNA Motifs," *Nano Letters 6, No. 7* (2006), 1502–1504.

[28] Ibid., 1503.

[29] Nadrian C. Seeman, "Nanomaterials Based on DNA," *Annu. Rev. Biochem. 79* (2010), 65–87.

[30] Nadrian C. Seeman and Philip S. Lukeman, "Nucleic Acid Nanostructures: Bottom-Up Control of Geometry on the Nanoscale," *Rep. Prog. Phys. 68* (2005), 237–270.

[31] Seeman, "Nanomaterials," 70.

[32] Banani Chakraborty, Ruojie Sha, and Nadrian C. Seeman, "A DNA-Based Nanomechanical Device With Three Robust States," *PNAS 105, No. 45* (11 November 2008), 17245–17249.

[33] Bruce H. Robinson and Nadrian C. Seeman, "The Design of a Biochip: a Self-Assembling Molecular-Scale Memory Device," *Protein Engineering 1, No. 4* (1987), 295–300.

[34] Xiaoping Yang, Alexander V. Vologodskii, Bing Liu, Börries Kemper, and Nadrian C. Seeman, "Torsional Control of Double-Stranded DNA Branch Migration," *Biopolymers 45* (1998), 69–83.

[35] Seeman, "From Genes to Machines," 120.

[36] Yang et al., "Torsional Control," 82.

[37] Seeman, "Nanomaterials," 80.

[38] Bernard Yurke, Andrew J. Turberfield, Allen P. Mills, Jr., Friedrich C. Simmel, and Jennifer L. Neumann, "A DNA-Fuelled Molecular Machine Made of DNA," *Nature 406, No. 6796* (10 August 2000), 605–608.

[39] Hao Yan, Xiaoping Zhang, Zhiyong Shen, and Nadrian C. Seeman, "A Robust DNA Mechanical Device Controlled by Hybridization Topology," *Nature 415, No. 6867* (3 January 2002), 62–65.

[40] Chengde Mao, Weiqiong Sun, Zhiyong Shen, and Nadrian C. Seeman, "A Nanomechanical Device Based on the B-Z Transition of DNA," *Nature 397* (14 January 1999), 144–146.

[41] Alexander Rich and Shuguang Zhang, "Z-DNA: The Long Road to Biological Function," *Nature Reviews Genetics 4, No. 7* (July 2003), 566–573.

[42] http://seemanlab4.chem.nyu.edu/BZ.Device.html

[43] Lubert Stryer and Richard P. Haugland, "Energy Transfer: A Spectroscopic Ruler," *PNAS 58* (1967), 719–726.

[44] http://bio.physics.illinois.edu/techniques.asp

[45] Paul S. Weiss, "A Conversation With Prof. Ned Seeman: Founder of DNA Nanotechnology," *ACS Nano 2, No. 6* (2008), 1091.

[46] William Safire, "Essay; Where's the Fingerspitzengefuhl?" *The New York Times* (9 March 1995).

[47] Daniel J. Boorstin, *The Discoverers* (New York: Random House, 1983), 165–167.

[48] Craig Loehle, "A Guide to Increased Creativity in Research—Inspiration or Perspiration?" *BioScience 40, No. 2* (February 1990), 123–129.

CHAPTER EIGHT

DNA Origami and DNA Bricks

Paul Rothemund is giving a lecture for a lay audience in Monterey, California, in 2007.[1] Rothemund is a computer scientist and bioengineer from Caltech. He's a tall young man, clean-shaven and bespectacled in a patterned long sleeve shirt, khaki trousers, and running shoes. Impassioned about his subject he speaks rapidly, a bundle of energy, his hands gesticulating as he roams the front of the stage. "There's an ancient and universal concept that words have power, that spells exist, and that if we could only pronounce the right words, then—whoosh. . . . There are many ways of casting molecular spells using DNA. . . . We think we can actually write programming languages for DNA. What we really want to do in the end is learn how to program self-assembly so that we can build anything." How can you make an arbitrary shape or pattern out of DNA? He tells the audience that he decided to write a molecular program to achieve a type of DNA origami where you take a long strand of DNA and cause it to fold into whatever shape or pattern you might want. Pressing the wireless clicker in his hand a new image appears on the screen behind him. It is a geometrical shape. Rothemund says that to get DNA to fold into this particular shape, to make a molecular spell using DNA, "I actually spent about a year in my home, in my underwear, coding. . . ."[1] Whatever works.

We're going to take on a subject, scaffolded DNA origami, that's quite different—both in the development of DNA nanotechnology and also in this book. One way in which it's different is that Nadrian Seeman did not invent it. Also, scaffolded DNA origami did not bother with optimization of sequences to avoid undesired secondary structures or binding interactions and generally ignored the normal, careful practices of DNA nanotechnology. Further, origami flung the door wide open to aperiodic structures.

DNA origami is a simple method for folding long, single-stranded DNA molecules into arbitrary two-dimensional shapes. A long, single strand of naturally occurring DNA—its sequence accepted as a given—is used to scaffold a couple of hundred shorter strands (designed and synthesized) to produce a two-dimensional or three-dimensional shape, either with straight or bent features. Rothemund's creation of origami draws on Seeman's crossovers: double-crossovers, triple-crossovers—all in all, multi-crossovers. Rothemund also recruits a design theme launched by William Shih, Gerald Joyce, and colleagues. It manipulates a scaffold strand with a small number of helper strands to direct folding of the scaffold.[2] Shih built a nanoscale octahedron, the first ever three-dimensional DNA wireframe object—a skeletal shape represented with only lines and vertices. Shih's work was itself foreshadowed by another folding strategy.[3] Rothemund vastly grew the generality of these methods.[4]

Scaffolded DNA Origami

The design of scaffolded DNA origami is performed in five steps, the first two by hand and the last three aided by computer. The first step creates a geometric model of a DNA structure to approximate the desired shape. Figure 8.1 (a) shows an example outlined in red (33 nm wide and 35 nm tall). One fills the shape in from top to bottom by an even number of parallel double helices idealized as cylinders. A periodic array of crossovers (shown as small blue crosses) holds the

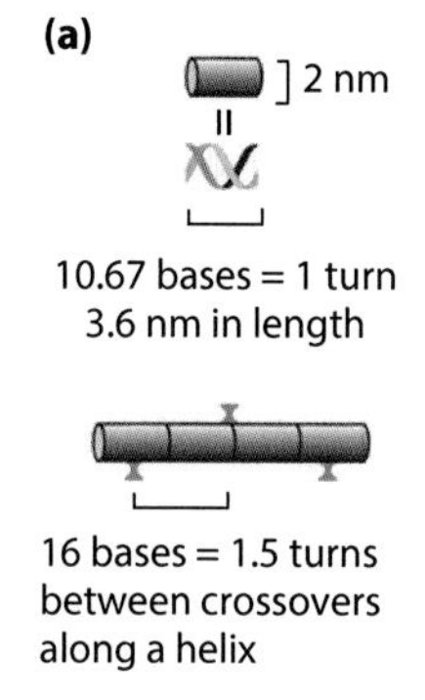

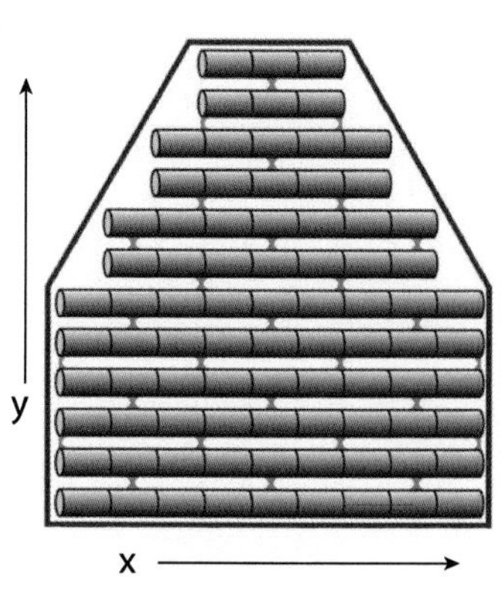

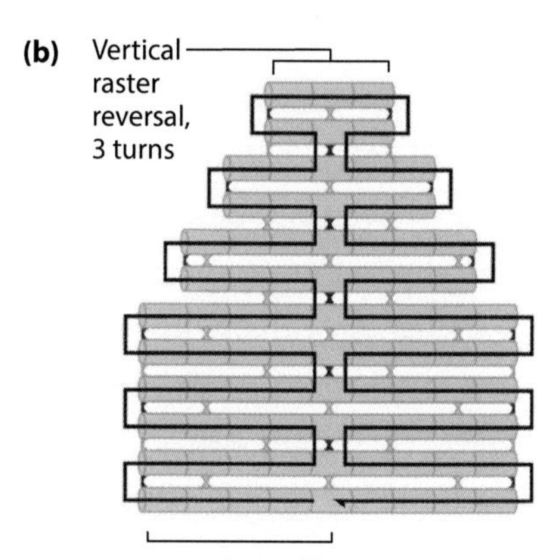

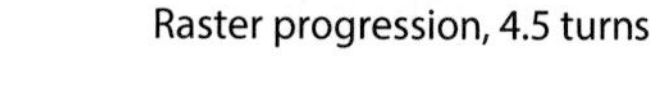

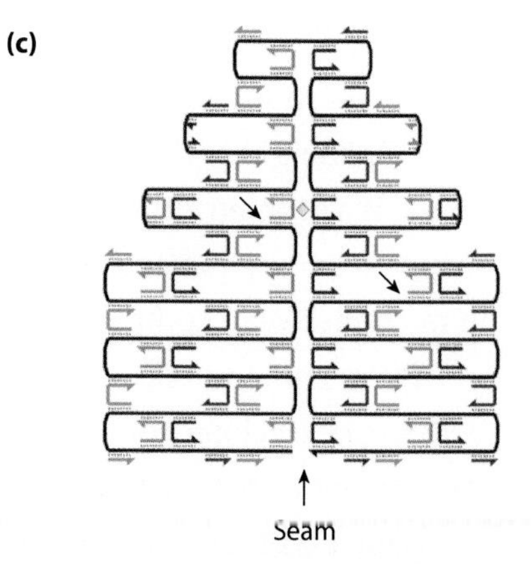

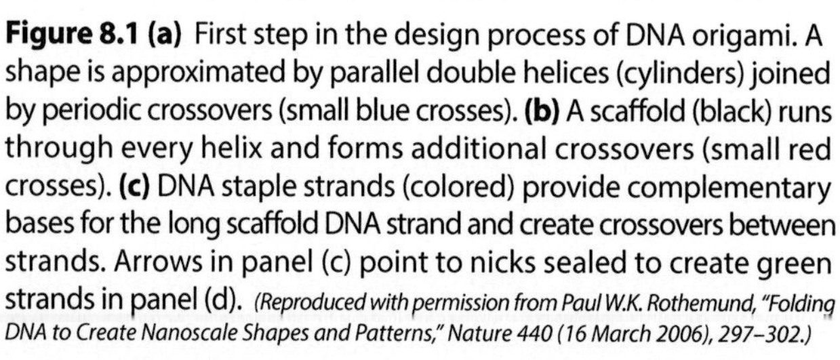
Figure 8.1 (a) First step in the design process of DNA origami. A shape is approximated by parallel double helices (cylinders) joined by periodic crossovers (small blue crosses). **(b)** A scaffold (black) runs through every helix and forms additional crossovers (small red crosses). **(c)** DNA staple strands (colored) provide complementary bases for the long scaffold DNA strand and create crossovers between strands. Arrows in panel (c) point to nicks sealed to create green strands in panel (d). *(Reproduced with permission from Paul W.K. Rothemund, "Folding DNA to Create Nanoscale Shapes and Patterns," Nature 440 (16 March 2006), 297–302.)*

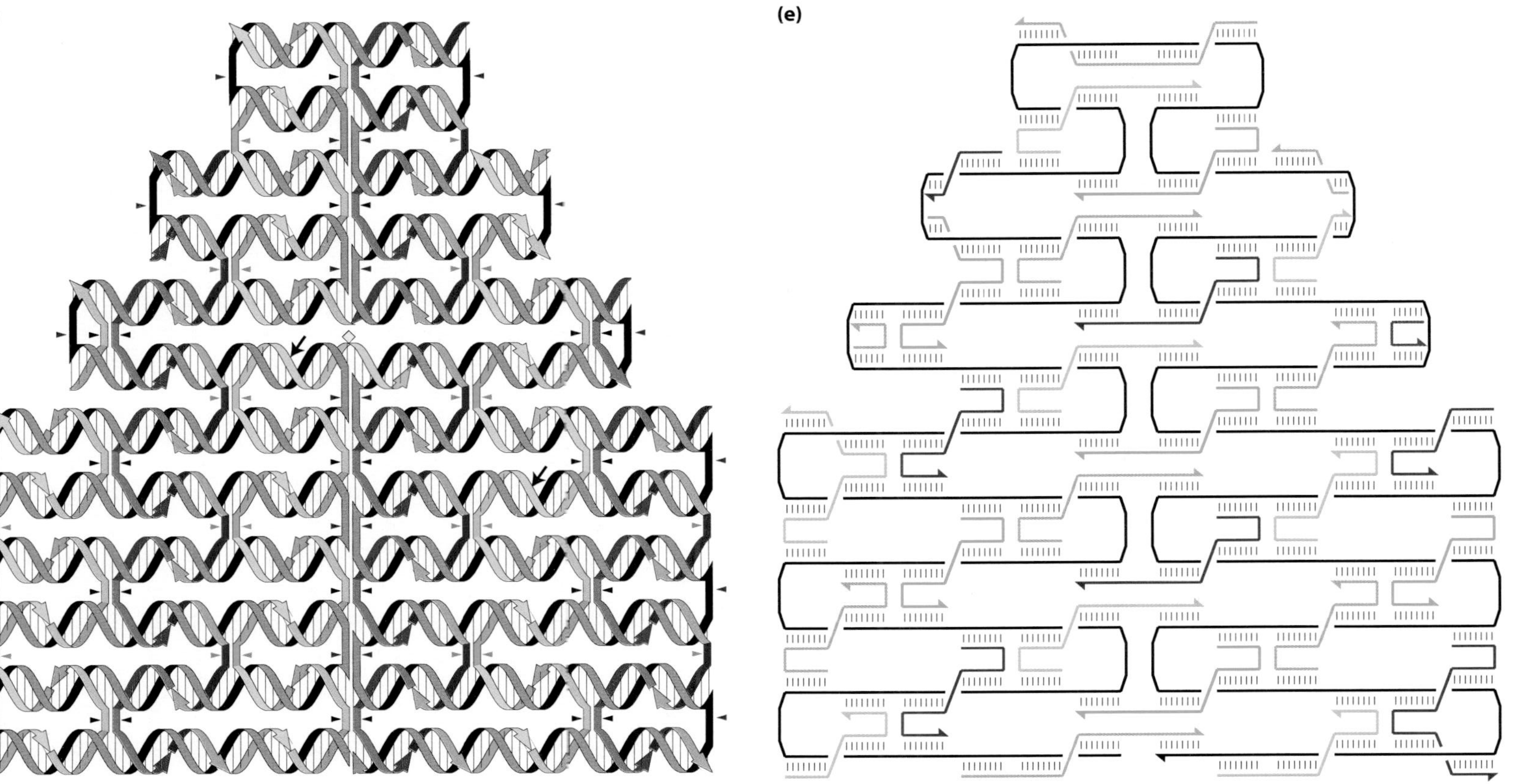

Figure 8.1 (d) Similar to the previous panel with strands drawn as helices. **(e)** A finished design after merges and rearrangements along the seam.
(Reproduced with permission from Paul W.K. Rothemund, "Folding DNA to Create Nanoscale Shapes and Patterns," Nature 440 (16 March 2006), 297–302.)

helices together. The crossovers designate positions where the strands running along one helix switch to an adjacent helix and continue there, just as we saw in Figure 7.3 of Chapter Seven (Seeman's DX molecules). Conceptually, the second step is the folding of a single long scaffold strand back and forth so that it comprises one of the two strands in every helix as shown in Figure 8.1 (b). Progression of the scaffold from one helix to another creates an additional set of crossovers shown as red crosses (for additional rigidity).[5] Note that the folding path leaves a seam in the middle of panel (b), i.e., a contour that the path does not cross. The use of the term *raster* in the figure refers to the way the area is filled out—from side to side in lines and then line-by-line from top to bottom. In folding a long scaffold back and forth in a raster fill pattern, there are technical details (we will skip) regarding the distance between scaffold crossovers as measured by the number of helix half-turns (see panel (b) labels).

Raster fill pattern A type of digital image that uses small rectangular pixels (derived from the phrase picture elements), arranged in a grid formation to represent an image. The raster, or scanning pattern, typically adds the pixels in a series of straight parallel lines progressing from left to right and then line-by-line from top to bottom.

In the third step a computer program generates the sequences of many short DNA staple strands—shown as colored strands in panel (c)—just as William Shih used what he called helper strands. These snippets of DNA are designed to provide *complementary bases* for the long scaffold DNA (*look how far we've come from Chargaff's base ratios in Chapter Two*) and create the periodic crossovers between strands. (The staple strands both direct the folding of the scaffold and also create duplexes with the scaffold in double-helical domains.[6]) Panel (d) presents the information contained in panel (c) but in a format that shows the crossovers with the strands now drawn as helices. The red triangles point to scaffold crossovers; both the black and blue triangles point to periodic crossovers (two different colors because of a technical point we won't discuss).

In the two final steps, computer programs examine and refine the design to minimize and balance twist strain between crossovers. Additional staple strands may be used to bridge seams and provide strengthening of the pattern. Observe that the staples shown in panel (d) reverse direction at the crossover points; thus these are antiparallel configurations of double-stranded DNA (one strand from the scaffold, one from the staple), and we know from Chapter Seven that these antiparallel duplexes are stable.

Panel (e) shows a finished design after the computer programs have completed merges and rearrangements along the seam. We've used Figure 8.1 to systematically

Scaffolded DNA origami A method of directed molecular self-assembly to build custom-shaped nanoscale objects composed of DNA. In general, long, naturally occurring DNA is used as the single-stranded scaffold DNA molecule. Hundreds of short staple strands of DNA are designed so they can connect—via hybridization—distant locations on the scaffold strand and fold it into the desired shape. Then the staples are synthesized and mixed in solution with the scaffold. The strands collectively self-assemble when annealed; that is, the temperature is lowered to allow hydrogen bonds to form between complementary base sequences.

present the steps of the design process because the creation of DNA origami was such a dramatic and unexpected advance that it merits our careful attention. (Rendering careful attention is a relative term. Rothemund augmented his six-page paper in *Nature* with 91 pages of online Supplementary Material.) Some years after Rothemund's publication, Seeman commented on the growing capabilities of the field that he founded: "A key unanticipated development was the advent of [scaffolded] DNA origami, which has vastly expanded the scale of addressable DNA structures."[7] We will shortly see that these structures are indeed addressable, meaning that one can write patterns on DNA origami constructs.

Annealing of DNA Using temperature changes to cause complementary sequences of single-stranded DNA to pair by the formation of hydrogen bonds.

DNA Origami Patterns

It's time to learn how Rothemund tested his method. He chose as the scaffold the genome of the bacteria-destroying virus known as M13mp18. The genome is the entirety of an organism's hereditary information. In the case of this virus it is a single strand of DNA about 7,000 nucleotides in length. M13mp18 was a good choice for this application because it's readily available and its entire base sequence is known. To fold the genomic DNA into shapes, the scaffold strand was mixed with several hundred staple strands in solution and annealed from 90 °C to 20 °C over the course of only two hours—a single laboratory step. To stimulate a mental picture of what happens when the staple strands enable the scaffold strand to fold into a predetermined shape, we present Figure 8.2. To grasp the simplicity of this figure, it's important to remember that all the design that was done in Figure 8.1 serves as input to a computer program: the model, seam positions, and folding path are input as lists of helix lengths in units of turns or bases. The folding path requires an additional list of orientations specifying its direction of travel to the left or right of any adjacent seams. Using the folding path as a guide, the design program applies the scaffold sequence to the model, and generates the appropriate set of staple strands.[8] Little wonder

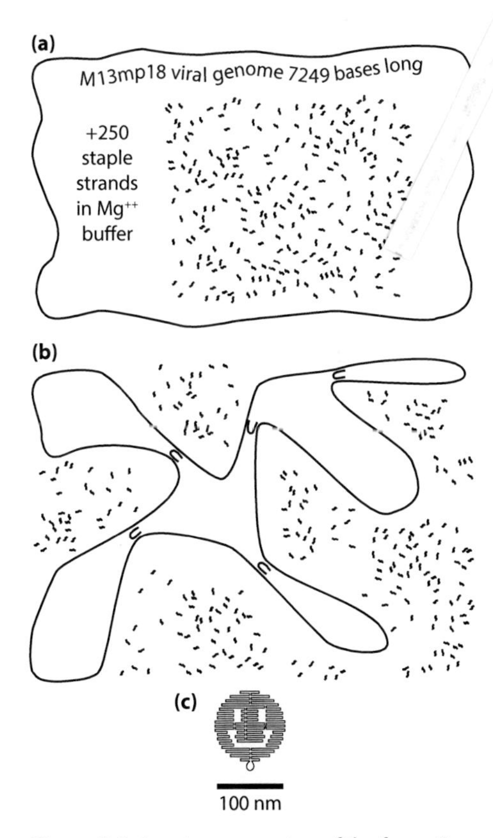

Figure 8.2 A cartoon overview of the formation of a DNA origami pattern. As we proceed from **(a)** to **(c)**, hundreds of staple strands enable a scaffold strand to fold into a predetermined shape as the temperature is lowered during the annealing process. *(Reproduced with permission from P.W.K. Rothemund, "Design of DNA Origami," Proceedings of the International Conference on Computer Aided Design (ICCAD), San Jose, CA (6–10 November 2005), IEEE Conference Publications, 470–477.)*

that Rothemund spent a full year writing computer code to achieve a folded shape, whether in his undies or fully clothed.

After the computer is finished, the mixture of strands is heated to 90 °C and cooled back to 20 °C over the course of about two hours (this is the annealing process). In another presentation Rothemund gave to a lay audience in 2008, one can actually watch how the design steps are implemented in solution. In that lecture, an enlightening animation—made by Shawn Douglas of the Wyss Institute at Harvard—shows the entire self-assembly process taking place.[9] Looking at the animated version of panel (d), Figure 8.1, you may goggle at the beautiful helical crossovers: they materialize as the staple strands massage the scaffold strand, like harbor tugboats gently nudging a cargo ship into port.

Rothemund checked out different folds beginning with a 26-helix square (panel (a) of Figure 8.3). This simple image is helpful in showing how colors indicate the raster fill pattern: red/orange is the first base and purple is the 7,000th base. The far left shows the folding path (the dangling curve represents unfolded sequence). Panel (b) demonstrated a more arbitrary shape—a five-pointed star. For panels, (a)–(c), there are no scale bars. All panels measure 165 nm x 165 nm. So far, so good.

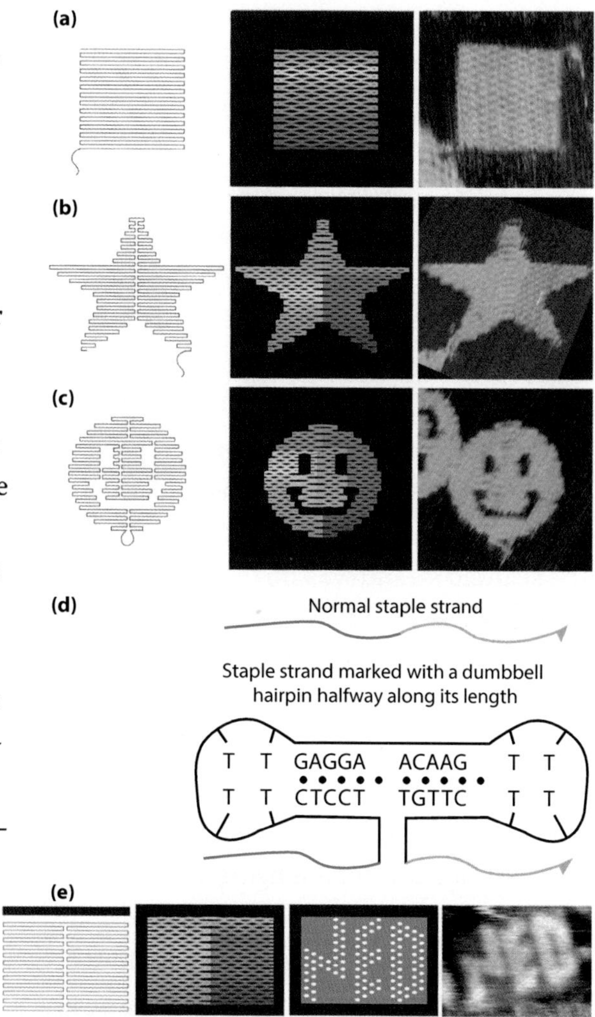

Figure 8.3 (a) A simple square: (left to right) the folding path, the colors of the raster fill pattern, and an AFM image. **(b)** A five-pointed star: folding path, raster fill pattern, and AFM image. **(c)** A three-hole disk—a smiley face. This origami design is topologically different from (a) and (b). Images (a)–(c) measure 165 nm x 165 nm. **(d)** Dumbbell hairpin nucleic acid sequence used to decorate an origami. The height difference between normal staple strands and labeled staple strands provides contrast when viewed in an AFM. **(e)** From left to right, the folding path, the color of the raster fill pattern, the decorated model for the pattern "NED" rendered using dumbbell hairpins on the underlying origami, the AFM image of the decorated origami pattern. Letters are 60 nm high. The black scale bar in the far left panel is 100 nm and applies to all panels. *(Panels **(a)–(c)** reproduced with permission from Paul W.K. Rothemund, "Folding DNA to Create Nanoscale Shapes and Patterns," Nature 440 (16 March 2006), 297–302; Panel **(d)** reproduced with permission from Online Supplementary Material for Paul W.K. Rothemund, "Folding DNA to Create Nanoscale Shapes and Patterns," Nature 440 (16 March 2006), 297–302; Panel **(e)** reproduced with permission from Paul W.K. Rothemund, "Scaffolded DNA Origami: From Generalized Multicrossovers to Polygonal Networks," Nanotechnology: Science and Computation, ed. Junghuei Chen, Natasa Jonoska, and Grzegorz Rozenberg (Berlin Heidelberg: Springer-Verlag, 2006), 3–21.)*

Now cast your mind back to our discussion of topology in Chapter Six. Remember the lame joke about the equivalence of a coffee cup and a donut? Rothemund knows that joke too. He anticipated criticism from all you topologists out there. He wanted to show that DNA origami need not only be about topological disks that are all equivalent to one another. Scaffolds can be routed arbitrarily through shapes. So he designed a three-hole disk—a smiley face. Panel (c) shows this design.

In addition to binding the DNA scaffold and maintaining its form, staple strands provide a method of decorating shapes with arbitrary patterns layered on top of the underlying shape. (This is the addressability that Seeman praised in the previous section.) To accomplish this, Rothemund used a second set of staple strands referred to as labeled strands. He called the resulting creations patterned origami. There are a variety of ways to label strands, and a simple one is to add a tiny nucleic acid sequence on the top of a normal/unlabeled strand. Rothemund synthesized what are called dumbbell hairpins—a variation of a naturally occurring phenomenon called a DNA hairpin—in which DNA exhibits *intra*strand base pairing. (Seeman used DNA hairpins in the previous chapter.)

We see both an unlabeled and a labeled staple strand in panel (d) of Figure 8.3. The second is marked halfway along its length with a dumbbell hairpin. Labeled strands give higher AFM height contrast than unlabeled (3 nm versus 1.5 nm). With dumbbell hairpins, Rothemund wrote arbitrary patterns on top of the substrate origami, to gain decorations. That is, height variations that show up as light (labeled staple) and dark (unlabeled staple) areas when viewed with an AFM.[10,11]

DNA hairpin When a single strand of DNA curls back on itself to become self-complementary it results in a partial double helix with a bend in it; several bases remain unpaired in the bend before the strand loops back on itself. In the classic case, the structure resembles a real hairpin and hence the name. These structures occur naturally and can also be synthesized to perform various functions in DNA nanotechnology. A dumbbell hairpin is a variation of the simpler structure and also occurs naturally.

Addressable DNA structures DNA nanostructures that can be labeled. Such labeling provides uniquely identifiable positions so that, for example, the initial structure can act as a substrate and additional patterns can be formed on top of it.

Panel (e) of Figure 8.3 (left to right) shows the folding path, the colors of the raster fill pattern, a model for the letters "NED" rendered using hairpins decorating the underlying rectangular origami, and an AFM image of the decorated origami pattern. This is a tribute to Nadrian Seeman. Ned is Nadrian Seeman's nickname. In the *Festschrift* (celebration publication) honoring Seeman on the occasion of his 60th birthday, Rothemund said that the pattern clearly showed Ned's influence on DNA nanotechnology.[12]

Consider the scale: letters are 60 nm high. Think back to "IBM" written with the STM that we saw in Chapter Three. Donald Eigler wrote that pattern using an ultrahigh-vacuum instrument. His sample was cooled down to 4K. That is, −452 °F or −269 °C. And the construction was done one atom at a time (serial fabrication). Rothemund's letters are twelve times larger (although he formed letters only 30 nm high in his publication cited in endnote 4). However, patterned origami requires no high-tech equipment and the work was done using parallel fabrication: 50 billion copies of the pattern—in contrast to just one copy—were formed in a single experiment.

The artwork does bear some resemblance to a child's crayon drawing held onto a refrigerator with magnets. But the reason for this is technical and has to do with the way staple strands are merged (panel (e), Figure 8.1). While any merge pattern creates the same shape, the pattern of merges dictates the types of decorations that can later be applied to the shape. One can employ a rectilinear merge pattern, that is, a square framework, in the underlying origami that gives a rectilinear arrangement of hairpin decorations. Alternatively, one can use a staggered merge pattern that is nearly hexagonal, giving a nearly hexagonal pattern of hairpin decorations. It is this latter type of decoration we see in panel (e) of Figure 8.3 (most evident in the dots in the third frame).

Strand Invasion Also Called Strand Displacement

Rothemund's success was surprising because his method ignored the normal, careful practices of DNA technology employed previously such as optimization of sequences to avoid undesired binding interactions (Chapter Seven). He suggested that several factors contributed to the success of scaffolded DNA origami; we'll discuss one that is something of a counterpoint to Seeman's attentive sequence selection. Rothemund considered any undesirable secondary structure that the scaffold DNA might assume and argued that it was unlikely that this secondary structure perfectly blocked the binding sites for all the staple strands that should bind to its sequence. (Secondary structures might be thought of as similar to the cyclization forms that plagued Seeman's early work.) Thus, Rothemund reasoned, staple strands may bind by partial matches at first (to gain a toehold) and then participate in a branch migration that displaces the secondary structure. (In this context, branch migration means that a new single DNA strand extends its partial pairing with its complementary strand as it displaces the resident strand from a DNA duplex.) A longer region of complementarity between the staple and the scaffold stabilizes the staple-scaffold interaction and—in terms of free energy—wins out over the scaffold secondary structure.[13]

Rothemund used an excess (by a factor of 100) of staple strands in his technique and speculated that this might help to drive the process. The process we've described, from toehold to takeover, is called *strand invasion* or *strand displacement.* Figure 8.4 illustrates the opening of scaffold secondary structure by strand displacement. In panel (a), five bases of undesired secondary structure in the scaffold occur in the middle of the binding site for the red staple strand. In (b), the red staple strand can still bind by 10 bases adjacent to the hairpin stem and gain a toehold via these partial matches. In (c), thermally driven fluctuations at the junction between the staple and the hairpin allow the staple strand to gain three more base pairs. Eventually, the thermally driven fluctuations result in the hairpin opening and that allows the remainder of the staple strand to bind as shown in (d).

In the previous chapter we cited the work of Bernard Yurke, Andrew Turberfield, Friedrich Simmel, and colleagues.[14,15] They made explicit use of strand displacement to actuate DNA nanomachines. For example, in creating the first sequence-dependent nanodevice. Rothemund remarked that it was Yurke's work on DNA motors that convinced him that strand invasion might make his scaffolded DNA origami possible.[13]

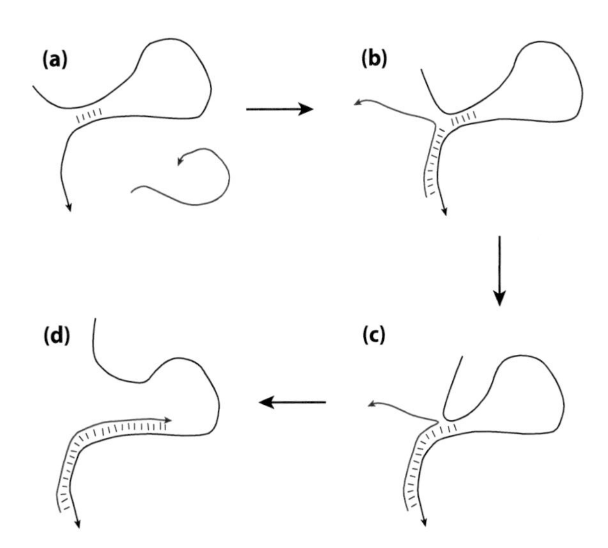

Figure 8.4 Opening of scaffold DNA secondary structure by strand displacement. From **(a)** to **(d)**, the staple structure gains a toehold and then thermal fluctuations enable the rest of the staple to bind to the scaffold, displacing the secondary structure. *(Reproduced with permission from Online Supplementary Material for Paul W.K. Rothemund, "Folding DNA to Create Nanoscale Shapes and Patterns," Nature 440 (16 March 2006), 297–302.)*

DNA Origami With Complex Curvatures in Three Dimensions

The unveiling of Rothemund's origami technique produced an avalanche of new work by others. A group led by Hao Yan at Arizona State University's Biodesign Institute has produced several exceptional results in recent years. (Yan is one of more than 40 students who earned his Ph.D. in Nadrian Seeman's lab.) One of the group's goals is to mimic geometries found in nature. Most biological molecules have globular shapes that contain intricate three-dimensional curves. Yan refined DNA origami folding to form two- and three-dimensional nanostructures that have substantial intrinsic curvature.[16] They bent double-helical DNA so it would follow the rounded contours of target objects, achieving a combination of flexibility and stability by careful consideration of factors such as the placement of crossovers (Figure 8.5).

Strand invasion also called Strand displacement In DNA nanotechnology, strand displacement refers to the hybridization of two partially or fully complementary DNA strands by means of the displacement of one or more pre-hybridized strands. Strand displacement can be initiated at so-called toeholds (complementary single-stranded domains) and then progress through a branch migration process.

One of the limitations of the conventional, block-based DNA origami designs is the level of detail that can be achieved. As originally devised, DNA origami structures are organized in a raster grid and each building block corresponds to a certain length of double-helical DNA (Figure 8.1). With the original technique, rounded elements are

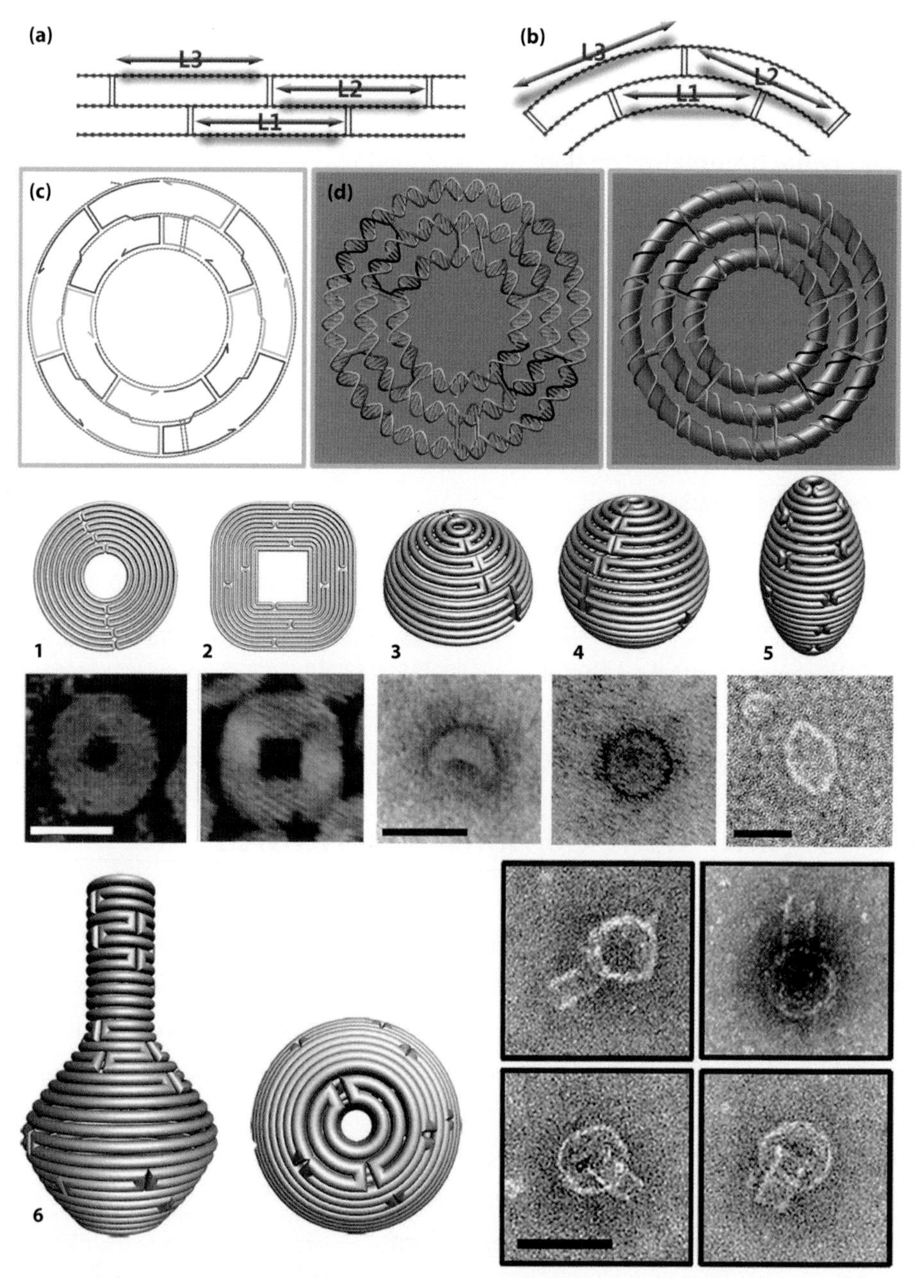

Figure 8.5 Complex curvature design. **(a)** Parallel DNA duplexes give multihelical structures. Constant distance between consecutive crossovers connecting adjacent helices (L1, L2, and L3). **(b)** Bending DNA into concentric rings gives in-plane curvature. Distance between crossovers in outer rings is greater than in inner rings (L3 > L2 > L1). **(c)** Three-ring concentric structure. Single-stranded scaffold is cyan, short staple strands are various colors. Two scaffold crossovers between adjacent rings make three-ring arrangement. **(d)** Helical/cylindrical views. The results: curved two- and three-dimensional nanostructures. Viewed top to bottom beginning with the third row of images: schematic designs, AFM images (total of 2), and TEM images (total of 7). **1.** nine-layer concentric ring structure; **2.** eleven-layer concentric square with rounded outer corners and sharp inner corners; **3.** hemisphere; **4.** sphere; **5.** ellipsoid; **6.** nanoflask with cylindrical neck and rounded bottom. Scale bars are 50 nm in both AFM images. Scale bars are 50 nm in all seven TEM images. *(Reproduced with permission from Dongran Han, Suchetan Pal, Jeanette Nangreave, Zhengtao Deng, Yan Liu, and Hao Yan, "DNA Origami With Complex Curvatures in Three-Dimensional Space," Science 332 (15 April 2011), 342--346.)*

only crudely approximated and intricate details are often lost. To tackle this issue, Yan began by defining the desired surface features of a target object with the scaffold, followed by manipulation of DNA's conformation and shaping of crossover networks to complete the design. Remarkably, the conformational changes were accomplished by deviating from the natural B-form twist by a slight overtwist or undertwist of the DNA helices to produce different bending angles.

We'll now consider in more detail how they achieved control over the degree of surface curvature. Yan's group began by making simple two-dimensional concentric ring structures, each ring formed from a DNA double helix (panels (a) and (b) of Figure 8.5). The concentric rings are bound together by means of strategically placed crossover points that bridge the gap between concentric helices. Such carefully thought-out crossovers help maintain the structure of concentric rings and this prevents the DNA from extending. The network of crossover points can also be designed to produce combinations of both in-plane and out-of-plane curvature. This attention to meticulous composition allows for the creation of curved three-dimensional structures.

DNA origami with complex curvature
Several methods can exploit DNA base pairing to create complex, closed shapes. The method we have discussed first decomposes the shape into circular contour lines. Then DNA double helices are designed that can bend and follow these contours while crossovers join DNA along adjacent contours. The method succeeds by a delicate balance of slightly overtwisted and slightly undertwisted helices. This produces different bending angles resulting in a combination of structural flexibility and stability.

However, if Yan only used this method, the range of curvature would still be limited for standard B-form DNA that will not tolerate large deviations from its preferred configuration of 10.5 base pairs/turn. But he recognized that if you could slightly overtwist or undertwist these helices, you could produce different bending angles. (We won't show these illustrations.) By combining the method of concentric helices with such non-B-form DNA (having 12 base pairs/turn), the group produced sophisticated forms, including spheres, hemispheres, ellipsoid shells, and even a round-bottomed nanoflask.

The bottom portion of panel of Figure 8.5 shows an example of the results: DNA nanostructures with complex three-dimensional curvatures. The fourth row of the figure shows (left to right) two AFM images and three TEM images of the forms in the third row of the figure. The lower right of the figure shows four TEM images of the *nanoflask* that is shown schematically on the lower left. The cylindrical neck and rounded bottom of the flask are clearly visible in the TEM images.

DNA Tiles in Two Dimensions

After its launch in 2006, DNA origami became the dominant player in the design and synthesis of complex DNA structures. In 2012, an alternative appeared. It was the product of a group led by Peng Yin of the Wyss Institute at Harvard. They demonstrated a modular style in two dimensions that they termed *DNA tiles.* [17] They then quickly and dramatically expanded the concept to three dimensions in the form of construction with DNA bricks. When Yin's two-dimensional version was published, Rothemund and collaborator Ebbe Sloth wrote a News and Views perspective in the journal *Nature.*[18] They reminded us that carpenters have been turning trees into furniture and dwellings for thousands of years, and so the discipline of woodworking uses well-established techniques for joining pieces of wood to make a desired form. Then there are those nanotechnologists of the school that Seeman founded, who try to use DNA as a material for crafting nanoscale shapes. But DNA-working, as Rothemund phrased it, has only been in development for three decades; DNA-working techniques are still evolving.

DNA tiles also called single-stranded tiles (SSTs) A tile in the present context is a single-stranded DNA molecule with four different sticky-ended binding domains that specify which four other tiles can bind to it as neighbors. (*Note*: there are no crossover molecules used in this approach.) SSTs are used as a modular style of nanostructure design in two dimensions.

There may be circumstances in which DNA origami is the best bet or where DNA bricks are better suited. Or it may be that some combination of DNA bricks and DNA origami will prevail. As Rothemund said, DNA nanotechnologists are still just apprentice DNA carpenters. Future results will probably surprise us. While awaiting further developments, we conclude this chapter with a description of the basics of Yin's method of DNA-working.

The most fundamental limitation of DNA origami is that for each new shape one wishes to create, one must design a new fold for the long scaffold strand and devise a new set of staple strands. Yin's modular style dispensed with scaffolds and staples and revisited a construction paradigm that we saw Seeman use in the previous chapter—DNA tiles. However, instead of using crossover molecules as the building block, each tile is a single DNA strand with four different binding domains that specify which four other tiles can bind to it as neighbors.[19] A single-stranded tile, referred to as an SST, consists of a 42-base strand of DNA composed entirely of linked sticky ends. The main design feature is a self-assembled rectangle with each of its SST strands folded into a 3 nm x 7 nm tile and attached to four neighboring tiles.

Consider Figure 8.6. Panel (a) shows the single-stranded tile motif, a short length of DNA bristling with sticky ends. The familiar 5′→ 3′ directionality is shown by the arrow-head at the 3′ end in domain 4. In panel (b) we see three equivalent representations of a self-assembled rectangle. This particular rectangle contains six parallel helices, each measuring about eight helical turns, but the basic strategy can be adapted to design rectangles with different dimensions. The tiny vertical bars represent base pairing. The left and middle figures in panel (b) are two different views of the same SST rectangle structure. Each standard (full) tile has 42 bases (these tiles are labeled with a "U"), and each top and bottom boundary (half) tile has 21 bases (labeled "L"). The reason for having

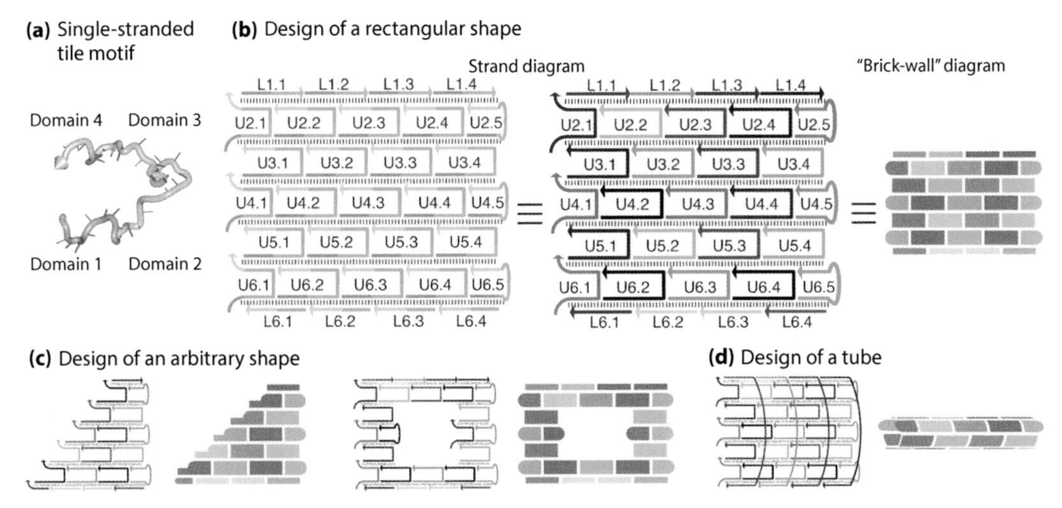

Figure 8.6 Self-assembly of molecular shapes using single-stranded tiles (SSTs). **(a)** The canonical SST motif. **(b)** Three representations of the main design feature: there are full/standard tiles (labeled "U") and boundary/half tiles (labeled "L") in the left and middle figures. Colors distinguish domains in the left figure and distinguish strands in the middle and right figures. The right figure is a simplified brick-wall diagram. **(c)** Selecting an appropriate subset of SSTs from the common pool in (b), one can design a desired target shape, e.g., a triangle (left) or a rectangular ring (right). **(d)** By linking pairs of half-tiles on the top and bottom boundaries into full tiles, one can transform the rectangle in (b) into a tube with prescribed width and length. *(Reproduced with permission from Bryan Wei, Mingjie Dai, and Peng Yin, "Complex Shapes Self-Assembled From Single-Stranded DNA Tiles," Nature 485 (31 May 2012), 623–627.)*

both standard and boundary tiles (i.e., full and half-tiles) is that the DNA strands on the edges of each shape have free binding domains that can lead to the shapes clumping together. This is addressed by adding these boundary tiles (called edge-protector strands) where necessary.

A pre-designed rectangular SST lattice can also be thought of as a molecular canvas where each SST serves as a 3 nm x 7 nm elemental molecular component. Designing a shape amounts to selecting its constituent strands/tiles on the canvas. One produces the desired form by annealing all those SSTs that contribute to the target shape and the remaining strands are excluded. Design of arbitrary shapes from a molecular canvas is depicted in Figure 8.7.

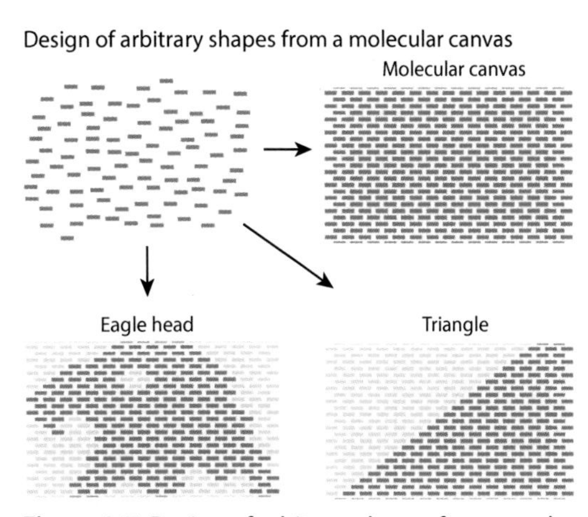

Figure 8.7 Design of arbitrary shapes from a molecular canvas. *(Reproduced with permission from Bryan Wei, Mingjie Dai, and Peng Yin, "Complex Shapes Self-Assembled From Single-Stranded DNA Tiles," Nature 485 (31 May 2012), 623–627.)*

This is a significant departure from DNA origami; small DNA strands self-assemble into large structures without the need for a scaffold. It also departs from traditional thinking about tile-based assembly. Yin did employ the techniques pioneered by Seeman to design SST strand sequences that minimized sequence symmetry for most of the structures. But he demonstrated the generalization of his technique (near the conclusion of 89 pages of online Supplementary Information) by successfully making structures from random sequences. Also, many scientists assumed that small strands would need to be mixed in precise ratios to avoid making fused or half-finished structures. But, unlike Seeman, Yin mixed strands together without careful adjustment of strand stoichiometry, i.e., the relative ratio of the strands. (Note that Rothemund's origami did not use equal concentrations of different staple strands, thus anticipating Yin's practice.) Yin thinks that the SST technique works because the strands are slow to assemble but grow quickly once they start. Thus, the shapes have a low probability of touching one another and fusing incorrectly as they begin to form.[20]

Using a master strand collection of 310 elemental molecular components, Yin used appropriate subsets to construct over 100 distinct and complex two-dimensional shapes composed of hundreds—in some cases more than one thousand—individual tiles (SSTs). These are shown in Figure 8.8. (Figure 8.7 eagle's head is in the second row from the bottom, and the third column in from the right.) The results point to another difference between SSTs and DNA origami. Rothemund's method requires a new set of staple strands for every design. Yin's tile technique is more versatile since millions of shapes can be crafted from the same set of tiles simply by leaving some out. As Yin said, "Once you have a pre-synthesized library [of tiles], you don't need any new DNA designs. You just pick your molecules."[20]

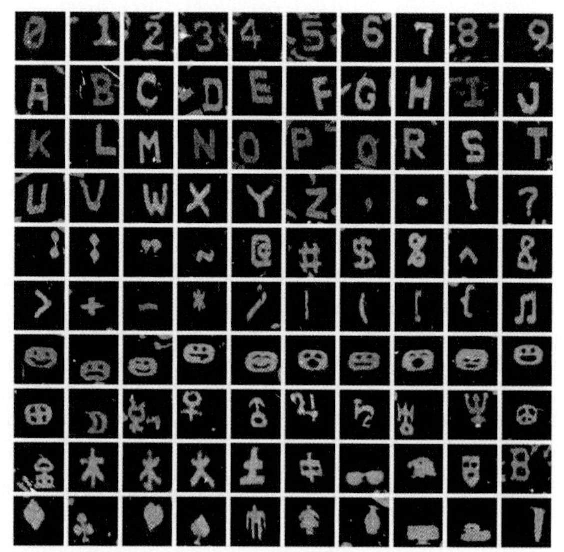

Figure 8.8 Complex shapes designed using a molecular canvas. AFM images of 100 distinct shapes, including the 26 capital letters of the English alphabet, 10 Arabic numerals, 23 punctuation marks and other standard keyboard symbols, 10 emoticons, 9 astrological symbols, 6 Chinese characters and other miscellaneous symbols. Each image is 150 nm x 150 nm. *(Reproduced with permission from Bryan Wei, Mingjie Dai, and Peng Yin, "Complex Shapes Self-Assembled From Single-Stranded DNA Tiles," Nature 485 (31 May 2012), 623–627.)*

DNA Bricks in Three Dimensions

It was only six months after he published his two-dimensional results that Yin and colleagues published their three-dimensional generalization. This blew scientists away. Henrik Dietz is a Principal Investigator in the Laboratory for Biomolecular Nanotechnology at the Technische Universität München (TUM) in Munich, Germany, and a major player in DNA nanotechnology. "The three-dimensional thing is so awesome, I almost got tears in my eyes because of the joy."[21] He expressed what a lot of others were thinking. To extend his modular-assembly method from two dimensions to three dimensions, Yin constructed what he called single-stranded bricks.[22] A canonical DNA brick is a 32-nucleotide single strand composed of four 8-nucleotide binding domains (sticky ends) as shown in Figure 8.9 (a). Notice the arrowhead in domain 4 indicating the 3′ end. There is a wealth of information in Figure 8.9 and we will provide an overview. For those do-it-yourselfers with intrepidity, there are helpful details in the 165 pages of Supplementary Materials available online.[23]

The bricks have modular simplicity. They're composed of only a short string of nucleotides—32 nucleotides (brick)—and self-assemble with great precision. Each DNA brick has a distinct nucleotide sequence, but all bricks adopt an identical shape when incorporated into the target structure: two 16-nucleotide antiparallel helices linked together. The two domains adjacent to the linkage are called the head domains and the other two are called the tail domains.

As a result of sequence design, a DNA brick with a tail domain whose sequence is abbreviated as "a" (panel (b) of Figure 8.9) can bind with a neighboring brick that has a head domain with complementary sequence "a*." The strands have been devised so that each pairing between bricks produces a 90° angle and this permits construction in three dimensions (more on this shortly). Figure 8.9 (c) is a molecular model that shows the helical makeup of a cuboid three-dimensional DNA structure; each strand has a particular sequence as indicated by a distinct color. (A cuboid is a solid composed of three pairs of rectangular faces placed opposite each other and joined at right angles.) Yin introduced a LEGO®-like model to depict the design in a simple manner, shown in panel (b). The two protruding round plugs, pointing in the same direction as the helical axes, represent the two tail domains; the two connected cubes with recessed round holes represent the two head domains.

Each brick must adopt one of two classes of orientation, horizontal or vertical. The two bricks connect to form a 90° angle by the familiar rules of hybridization (base pairing), represented in LEGO® terms as the insertion of a plug into a hole. An insertion is only allowed between a plug and a hole that carry complementary sequences with matching polarities. (Recall that the 5′→ 3′ direction defines the polarity.) A more detailed LEGO®-like model depicting polarity and other features can be found on page three of the Supplementary Materials online.

DNA bricks A modular-assembly method in three dimensions. One can use DNA bricks to construct complex three-dimensional nanostructures by using only short, synthetic DNA strands as modular components. Fabrication with DNA bricks is mediated solely by local binding interactions (sticky ends).

Before continuing, we share a beautiful insight from the Wyss Institute authors who evolved the two-dimensional DNA tiles into three-dimensional bricks. The authors are Yonggang Ke, Luvena L. Ong, William M. Shih, and Peng Yin. To appreciate their cleverness we revisit our terse remark, "The strands have been devised so that each pairing between bricks produces a 90° angle and this permits construction in

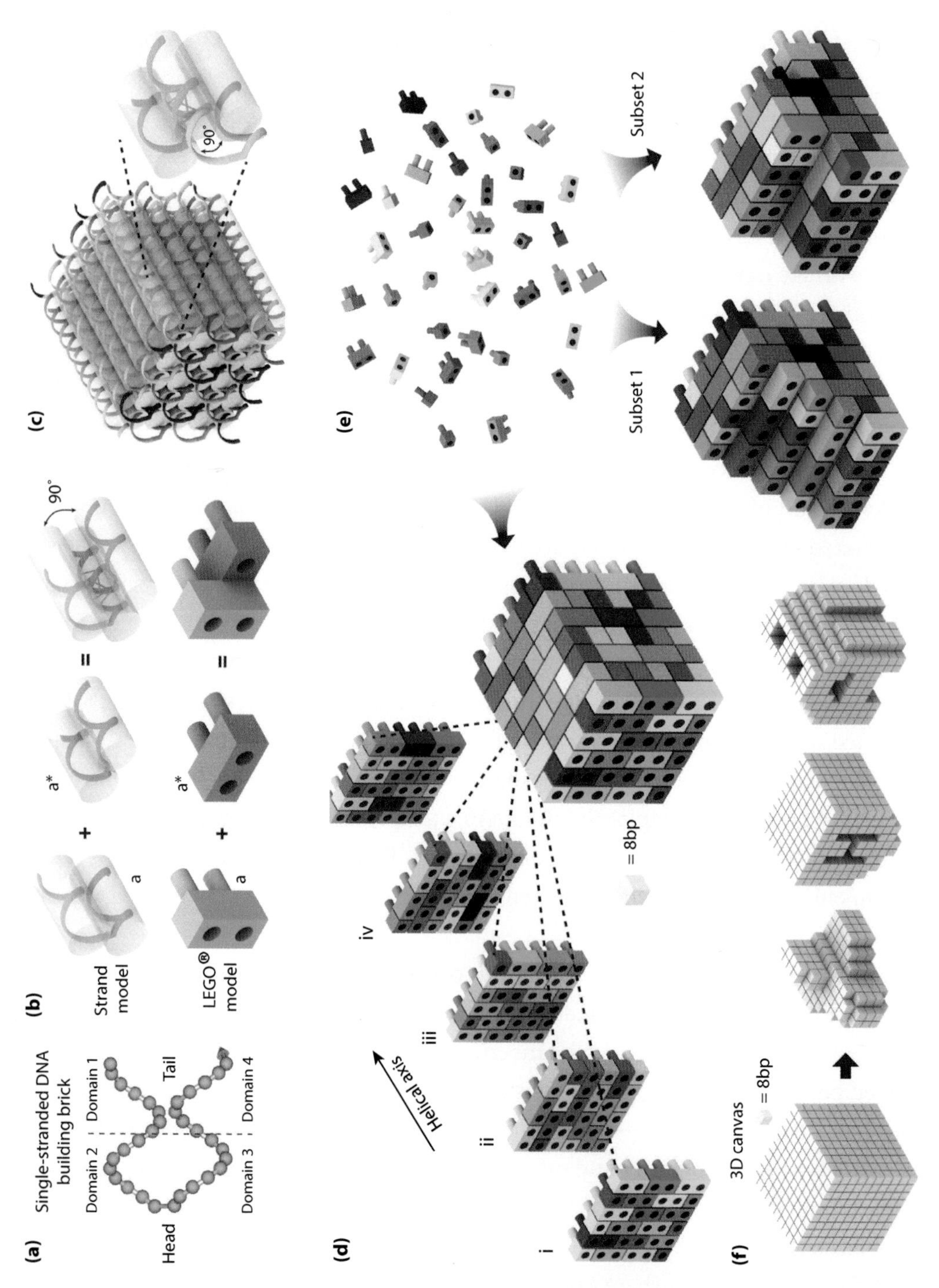

Figure 8.9 Design of DNA brick structures analogous to structures built of LEGO® bricks. **(a)** A 32-nucleotide, four-domain, single-stranded DNA brick; connected domains 2 and 3 are head domains while domains 1 and 4 are tail domains. **(b)** Each two brick assembly forms a 90° angle via nucleotide hybridization of two 8-nucleotide domains "a" and "a*". **(c)** Molecular model of the helical structure of a cuboid; the inset shows a pair of bricks. **(d)** A LEGO®-like model of the same cuboid; half-bricks are present on the boundary of each layer. **(e)** The cuboid is self-assembled from DNA bricks; using the cuboid as a molecular canvas, a smaller shape can be designed using a subset of the bricks. **(f)** Three-dimensional shapes designed from a molecular canvas of volume elements. *(Reproduced with permission from Yonggang Ke, Luvena L. Ong, William M. Shih, and Peng Yin, "Three-Dimensional Structures Self-Assembled From DNA Bricks," Science 338 (30 November 2012), 1177–1183.)*

three dimensions." If you return to the Nanotechnology Starter Kit of Chapter One, you'll see that the helical periodicity—the repeat distance—of the DNA double helix is about 10–10.5 nucleotide pairs. To make the leap into three dimensions, the authors reckoned that if they used a DNA strand that was only 8 nucleotides long, the base complementarity between strands of a double helix would produce a 90° angle (a dihedral angle, that is, an angle between planes). The angle derives from the approximate 3/4 right-handed helical twist of 8 base pairs of DNA. This is in contrast to a full helical turn that would arise from strands of about 10–10.5 nucleotide pairs.[24] When we conceptualize the helices as bricks, this means that a brick must adopt one of two possible classes of orientation: horizontal or vertical (panel (b) of Figure 8.9). The two bricks connect via hybridization (base complementarity) to form the 90° angle that is represented as the insertion of a plug into a hole. But such coupling is only allowed between a plug and a hole that carry complementary base sequences with matching polarity. This is not graphically depicted in the figure for the sake of simplicity.

Returning to Figure 8.9, panels (c) and (d) illustrate structural periodicities of the design by using, as an example, a cuboid structure that measures 6 helices x 6 helices x 48 base pairs. Note that each brick in the cuboid has a particular sequence and the color use is consistent with panel (b). The small whitish cube in panel (d) is an elemental molecular volume component (a voxel, by analogy to a pixel). It measures approximately 2.5 x 2.5 x 2.7 nm and 8 base pairs (bp) fit inside this volume element. The volume element is the joining together of two bricks, where 8 bases from one brick bind to a complementary set of 8 bases from a second brick to form an 8-bp unit. The binding together of two bricks is illustrated in the strand model and the LEGO®-like model in the far right of panel (b) and also in the inset to panel (c).

Bricks can be grouped into 8-bp layers that contain their head domains. Bricks follow a 90° counterclockwise rotation along successive 8-bp layers, resulting in a repeating unit that has consistent brick orientation and arrangement every four layers. The exploded view of panel (d) shows an example of this, where the first and fifth 8-bp layers share the same arrangement of bricks. Within an 8-bp layer, all bricks share the same orientation (all horizontal or all vertical) and form a staggered arrangement to cover the layer. On the boundary of each layer, some DNA bricks are bisected to half-bricks; they represent a single helix with two domains. The cuboid is self-assembled from DNA bricks in a one-step reaction—suggested by the arrow from panel (e) pointing to panel (d).

Each brick carries a particular sequence that directs it to fit only to its predesigned position, so the bricks are not interchangeable during self-assembly. Because of its modular architecture, a predesigned DNA brick structure can be used for construction of smaller custom shapes assembled from subsets of DNA bricks as implied by the other two arrows from panel (e) going to Subset 1 and Subset 2.

Last, panel (f) shows a molecular canvas. In panel (f) the LEGO®-like model is further abstracted to a three-dimensional model that contains only positional information of each 8-bp double strand. A cuboid structure that measures 10 helices x 10 helices x 80 base pairs is conceptualized as a three-dimensional molecular canvas that contains 10 x 10 x 10 volume elements or voxels (those whitish squares each of which contains an 8-bp double strand). Based on the three-dimensional canvas, a computer program first generates a full set of DNA bricks, including full-bricks and half-bricks, that can be used to build a prescribed custom shape. Then, using three-dimensional modeling software, a designer defines the target shapes by removing unwanted volume elements from the three-dimensional canvas, a process Yin likens to three-dimensional sculpting. The computer program analyzes the shape and automatically selects the correct subset of bricks for self-assembly of the shape. Characterization of the results is performed with agarose gel electrophoresis (a process similar to PAGE but using the gel agarose instead of polyacrylamide) and TEM.

Like Rothemund, Yin has conjured up a molecular spell using DNA. Erik Winfree (Seeman's collaborator in Chapter Seven) said, "It's astounding work."[25]

Molecular canvas Starting with a thousand-brick block, this master brick collection defines a cuboid molecular canvas in three dimensions. (A cuboid is a box-like structure composed of six rectangular faces. A cube is a special case in which all six faces are squares.) Each volume element serves as a three-dimensional molecular pixel—a voxel. One designs a shape by selecting only its constituent voxels—a subset of the entire brick collection—from the molecular canvas and anneals them to form the target shape.

DNA Brick Shapes in Three Dimensions

From the three-dimensional cuboid molecular canvas consisting of 10 x 10 x 10 = 1000 elemental molecular volume components (panel (f) of Figure 8.9), Yin constructed 102 different structures. These included solid shapes with sophisticated geometries, surface patterns, and hollow shapes with intricate tunnels and enclosed cavities. Among the most impressive of these are the closed-cavity and open-cavity shapes. Figure 8.10 shows TEM images of a sampling of these structures including a series of "empty boxes" with different sizes of cuboid cavities and open cavities with turning, branching, and crossing tunnels. The top row of each numbered column depicts a three-dimensional model; the second row is a computer-generated projection view; the third row is a TEM image computer-averaged from six different products with the same target nanoshape; the bottom row is a raw TEM image.

The shapes we've selected to present have additional transparent three-dimensional views that highlight the deleted volume elements (colored dark gray). For example, the top right model of shape 32 shows the enclosed cuboid cavity. In the projection model (i.e., second row) of Figure 8.10, views along the helical axis depict each individual helix as a dot, while views perpendicular to the helical axis show the helical bundles as lines.

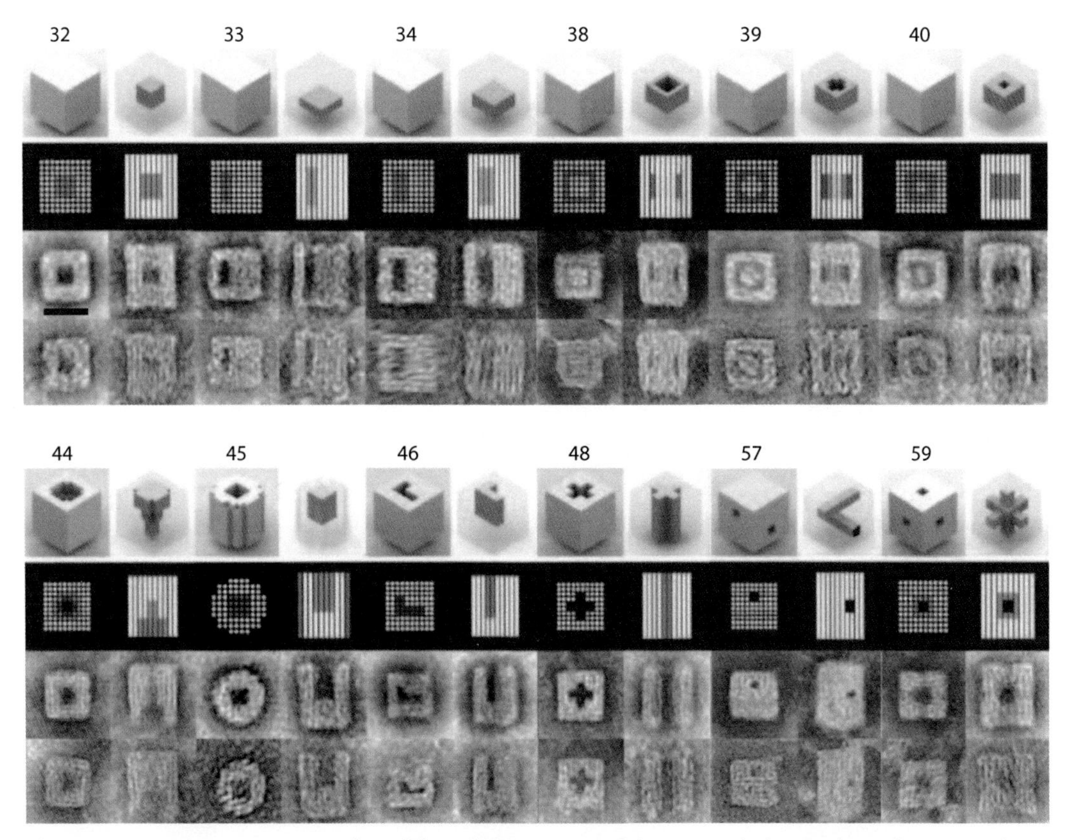

Figure 8.10 Computer-generated models and TEM images of shapes made from a three-dimensional molecular canvas. The top row for each numbered shape depicts a three-dimensional model, followed by a computer-generated projection view (second row), an image averaged from six different objects visualized by TEM (third row), and a representative raw TEM image (fourth row). This sampling (from a set of 102 shapes) includes a series of "empty boxes" with different sizes of cuboid cavities and open cavities with turning, branching, and crossing tunnels. The shapes have additional transparent three-dimensional views that highlight the deleted volume elements (colored dark gray)—e.g., the top right model of shape 32 shows the enclosed cuboid cavity. The scale bar shown in the third row below shape 32 is 20 nm and is the same for all shapes. *(Adapted with permission from Yonggang Ke, Luvena L. Ong, William M. Shih, and Peng Yin, "Three-Dimensional Structures Self Assembled From DNA Bricks," Science 338 (30 November 2012), 1177–1183.)*

The brightness of a dot is proportional to the length of a helix, and the brightness of a line is proportional to the number of corresponding helices that contribute to its projection.

Of course there is room for improvement. Yin remarked that low yields were observed for larger designs. Yield is essentially the ratio of the number of desired target structures produced to the total number of structures produced. Solving this challenge may require improvements in structure and sequence design, enzymatic synthesis for higher-quality strands, optimized annealing conditions, and a detailed understanding and perhaps explicit engineering of the kinetic assembly pathways of DNA brick structures.

Yield This is the ratio of the number of desired target structures produced to the total number of structures produced.

When mentioning Yin's two-dimensional modular construction technique, the forerunner of DNA bricks, we cited Rothemund's comments on this alternative to

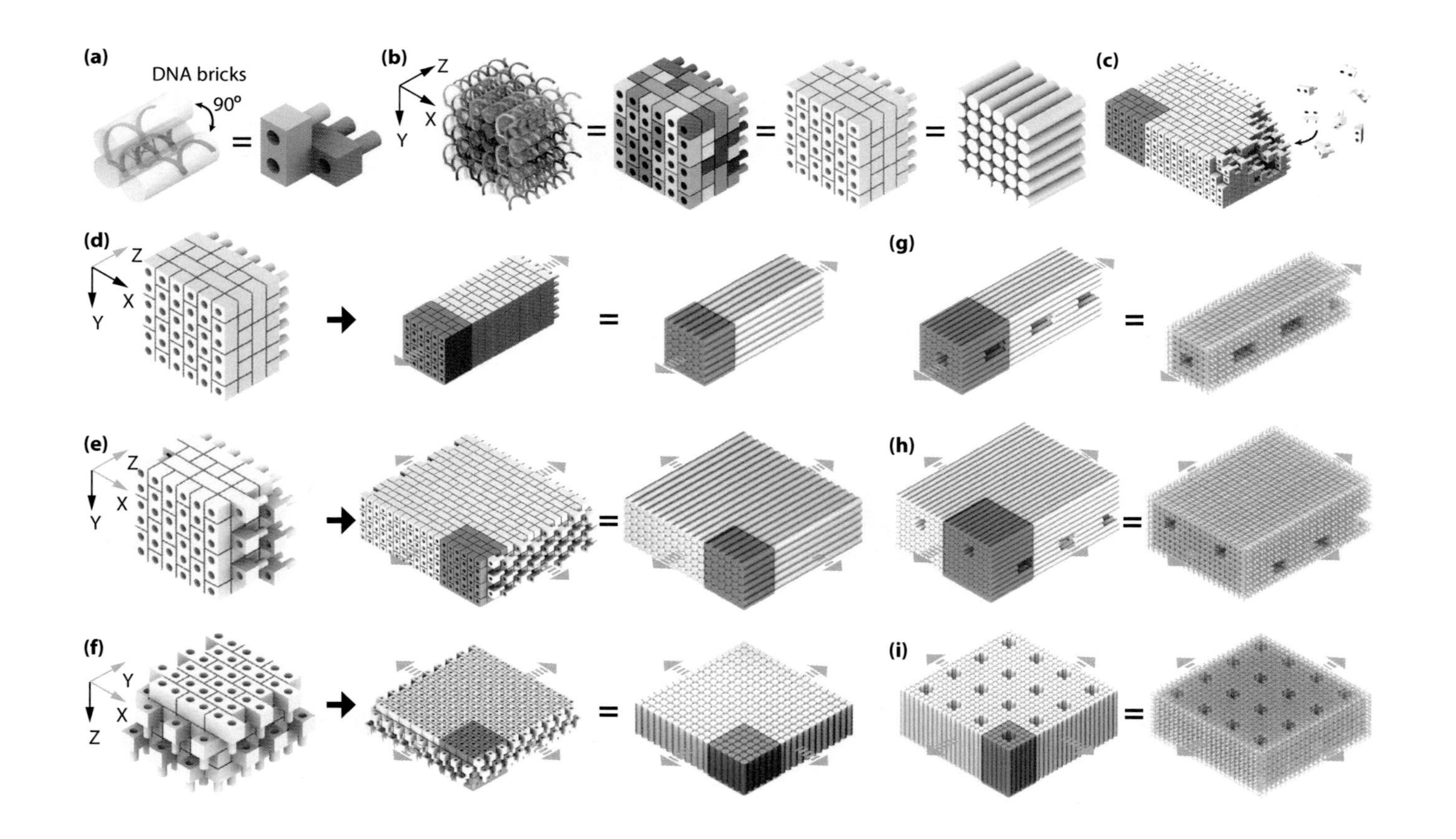
(a)
DNA bricks
90°
(b)
Z
X
Y
(c)
(d)
Z
X
Y
(g)
(e)
Z
X
Y
(h)
(f)
Y
X
Z
(i)

DNA origami. In his paper on DNA bricks, Yin echoed Rothemund's sentiments on the uncharted road ahead when he said, "The successes of constructions that use only short strands (as in bricks) and those that include a long scaffold (as in origami) together suggest a full spectrum of motif possibilities with strands of diverse lengths."[26] He observed that longer strands might provide better structural support while shorter ones might provide finer modularity and features. In the end, the use of both may give the most rapid progress toward greater complexity.

DNA Brick Crystals

Finally, we briefly describe a report from Yin's group on the application of DNA bricks in constructing DNA brick crystals that can grow to micron size in their lateral dimensions with precisely controlled depths up to 80 nm.[27] Using single-stranded DNA bricks, the group reported the creation of crystals with complex three-dimensional features at nanometer resolution. Figure 8.11 shows the design strategy for DNA brick crystals. Yin and colleagues constructed four groups of crystals: (1) Z-crystals that are one-dimensional "DNA-bundle" crystals extending along the z axis (panel (d) of Figure 8.11); (2) X-crystals that are one-dimensional crystals extending along the x axis; (3) ZX-crystals that are two-dimensional "multilayer" crystals extending along the z axis and the x axis as shown in panel (e); (4) XY-crystals that are two-dimensional "DNA-forest" crystals extending along the x axis and the y axis as seen in panel (f).

In their paper, the group also commented on the work of Seeman in making DNA crystals using the tensegrity triangle motif, a subject that we'll cover in the next chapter. They pointed out that Seeman's approach grows a three-dimensional crystal in all three directions with "no control in depth."[28] They contend that their modular strategy has advantages over alternative methods. Approaches like that of Seeman are hierarchical—individual strands first assemble into a discrete building block and then the building blocks assemble into a crystal. Yin stresses that in this case the subunits are incorporated into the crystal as preformed monomers. But, the homogeneity of complex monomers is often difficult to ensure and the addition of a defective monomer can compromise the growth of a well-ordered crystal. By contrast, brick crystals grow non-hierarchically.

Figure 8.11 Design of DNA brick crystals. **(a)** Strand (left) and brick (right) models showing two 32-nucleotide DNA bricks that form a 90° angle. **(b)** Models of a 6 helix x 6 helix x 24 base pair cuboid with increasing levels of abstraction: (left to right) a strand model, a brick model (in which colors distinguish brick species), a brick model with all bricks colored gray, and a model where cylinders represent DNA double helices. **(c)** Individual DNA strands, rather than pre-assembled multi-brick blocks, are directly incorporated into the growing crystal. **(d)–(f)** Brick and cylinder models of a one-dimensional Z-crystal (d), a two-dimensional ZX-crystal (e), and a two-dimensional XY-crystal (f), designed from the 6 helix x 6 helix x 24 base pair cuboid. **(g)–(i)** Cylinder and DNA-helix models of crystals with pores and tunnels: Z-crystal with a tunnel and periodic pores (g); ZX-crystal with two groups of parallel tunnels (h); XY-crystal with periodic pores (i). Repeating units of the crystals are denoted by blue boxes. Pink arrows indicate the directions of crystal growth. *(Reproduced with permission from Yonggang Ke, Luvena L. Ong, Wei Sun, Jie Song, Mingdong Dong, William M. Shih, and Peng Yin, "DNA Brick Crystals With Prescribed Depths," Nature Chemistry 6 (November 2014), 994–1002.)*

Yin's growth of DNA crystals from short, single-stranded DNA bricks does not involve the assembly of preformed, multi-stranded building blocks with well-defined shapes. Rather, Yin's crystals are designed to form via non-hierarchical growth with individual bricks (rather than preformed multi-brick components) directly incorporated into the crystal—as shown in panel (c) of Figure 8.11.

Seeman, Rothemund, and Yin

If we step back and mentally juxtapose the work of Seeman with that of Rothemund and Yin, it's confirming of Seeman's observation that for the first twenty years of DNA nanotechnology his group made the discoveries and the mistakes. For example, Rothemund employed Seeman's invention of crossover molecules as a means of making constructions rigid. In his original two-dimensional work, Yin elaborated (to great advantage) on Seeman's tile paradigm and used Seeman's software (Chapter Six) to design and optimize his DNA sequences.[17] However, Seeman believed that strand concentrations must be stoichiometric (i.e., equal concentrations of reactant strands) and that strands must be highly purified. Both Rothemund and Yin found this to be a misconception (as Rothemund phrased it).[29] Rothemund's origami did not use equal concentrations of different staple strands, thus anticipating Yin's practice of not carefully adjusting stoichiometry. Regarding sequence design or lack thereof, we recall that Rothemund had used the viral scaffold strand with its sequence taken as a given. (Rothemund had called out the optimization of sequences as another of Seeman's misconceptions.) And while Yin did initially generate DNA sequences by minimizing sequence symmetry, he also used randomly generated sequences and found that the results from these were comparable to the carefully designed sequences. After he found that the self-assembly yields were similar, Yin then used random sequences for all subsequent designs.[30]

But make no mistake about it. Paul Rothemund has written with fondness of "Ned and his academic children."[31] In his *Nature* perspective article, Rothemund said that the greatest advance in the field of nanotechnology was Nadrian Seeman's realization that one could use DNA the way a carpenter uses wood; that DNA doesn't just have to assume the form of a linear double helix. Seeman recognized that DNA can branch and form joints and that DNA joints can be used to create a variety of nanosized architectures. Rothemund said, "My bet is that if we succeed in building a mature nanotechnology, one with injectable surgical nanorobots and automobiles grown from tiny seeds, Ned Seeman's invention of structural DNA nanotechnology will have played a major role."[32]

Paul Rothemund wrote a paper for a book dedicated to Nadrian Seeman on the occasion of his 60th birthday.[12] In that paper he spoke of Seeman's DNA cube and "wondered what kind of mad and twisted genius had *conjured* it"[33] [italics added]. When Rothemund spoke of "casting molecular spells" to his lay audience in Monterey, he also employed a magic idiom. He may be onto something. Such superstitious talk is tongue-in-cheek, but it does allude to the enchantment of the discipline of DNA nanotechnology. It's a venture that seems so removed from everyday life and yet springs from DNA, the engine of life. If Seeman is a modern sorcerer, then Rothemund and Yin are two of the sorcerer's apprentices. They are stirring a seething, moiling cauldron. Just what the outcome will be and how it will come to pass no one knows.

Exercises for Chapter Eight: Exercises 8.1–8.2

Exercise 8.1

In this exercise we'll have another look at computer-generated models and TEM images of shapes made from Yin's three-dimensional molecular canvas. In Figure 8.12 we show six shapes that we didn't show in Figure 8.10—but they were made from the same molecular canvas as those in Figure 8.10.[22] The top row for each numbered shape depicts a 3D model, followed by a computer-generated projection view (second row). In the third row we have an image averaged from multiple different objects and visualized by TEM for each of the shapes shown in the top two rows. Using image-averaging makes for a clearer TEM image than if we just used a single representative raw image. However, the order of the TEM images is scrambled. Your job is to look carefully at the TEM images and match them to the corresponding 3D model/computer-generated projection view in the top two rows.

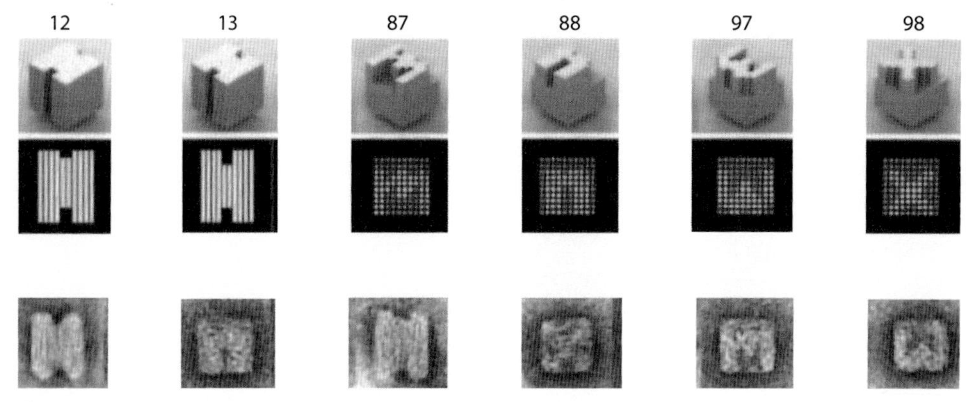

Figure 8.12 *(Adapted with permission from Yonggang Ke, Luvena L. Ong, William M. Shih, and Peng Yin, "Three-Dimensional Structures Self-Assembled From DNA Bricks," Science 338 (30 November 2012), 1177–1183.)*

Exercise 8.2

A polynucleotide, poly(dA-dT) is called a single-stranded, alternating copolymer—that is, the sequence of bases along the chain is ATATATA . . . What kind of structure would you expect this to assume in solution?

(Exercise reproduced with permission from Physical Biochemistry, *2/e by David Freifelder, p. 35, Copyright 1982 by W.H. Freeman and Company, New York. Used with permission of the publisher.)*

ENDNOTES

[1] "Paul Rothemund Casts a Spell with DNA," http://www.ted.com/talks/paul_rothemund_casts_a_spell_with_dna.html

[2] William M. Shih, Joel D. Quispe, and Gerald F. Joyce, "A 1.7-Kilobase Single-Stranded DNA That Folds Into a Nanoscale Octahedron," *Nature 427* (12 February 2004), 618–621.

[3] Hao Yan, Thomas H. LaBean, Liping Feng, and John H. Reif, "Directed Nucleation Assembly of DNA Tile Complexes for Barcode-Patterned Lattices," *PNAS 100, No. 14*, (July 8, 2003), 8103–8108.

[4] Paul W.K. Rothemund, "Folding DNA to Create Nanoscale Shapes and Patterns," *Nature 440* (16 March 2006), 297–302.

[5] Lloyd M. Smith, "The Manifold Faces of DNA," *Nature 440* (16 March 2006), 283–284.

[6] Carlos Ernesto Castro, Fabian Kilchherr, Do-Nyun Kim, Enrique Lin Shiao, Tobias Wauer, Philipp Wortmann, Mark Bathe, and Hendrik Dietz, "A Primer to Scaffolded DNA Origami," *Nature Methods 8, No. 3* (March 2011), 221–229.

[7] Nadrian C. Seeman, "Structural DNA Nanotechnology: Growing Along With *Nanoletters*," *Nano Letters 10, No. 6* (9 June 2010), 1971–1978.

[8] Paul W.K. Rothemund, "Design of DNA Origami," *Proceedings of the International Conference on Computer Aided Design (ICCAD)*, San Jose, CA (6-10 November 2005), IEEE Conference Publications, 470–477, http://ieeexplore.ieee.org/stamp/stamp.jsp?tp=&arnumber=1560114.

[9] Paul Rothemund, "Details DNA Folding," http://www.ted.com/talks/paul_rothemund_details_dna_folding.html

[10] Rothemund, "Folding DNA," 301.

[11] Rothemund, "Folding DNA to Create Nanoscale Shapes and Patterns," *Nature 440* (16 March 2006), Supplementary Note S6: "Hairpins for Creating Patterns," 63–69 (linked to the online version of the paper as a PDF download).

[12] Paul W.K. Rothemund, "Scaffolded DNA Origami: From Generalized Multicrossovers to Polygonal Networks," *Nanotechnology: Science and Computation*, ed. Junghuei Chen, Nataša Jonoska, and Grzegorz Rozenberg (Berlin, Heidelberg: Springer-Verlag, 2006), 3–21.

[13] Rothemund, "Folding DNA to Create Nanoscale Shapes and Patterns," *Nature 440* (16 March 2006), Supplementary Note S9: "Why Does Scaffolded DNA Origami Work?," 79–80 (linked to the online version of the paper as a PDF download).

[14] Bernard Yurke, Andrew J. Turberfield, Allen P. Mills Jr., Friedrich C. Simmel, and Jennifer L. Neumann, "A DNA-Fuelled Molecular Machine Made of DNA," *Nature 406* (10 August 2000), 605–608.

[15] Friedrich C. Simmel and Bernard Yurke, "Using DNA to Construct and Power a Nanoactuator," *Physical Review, E 63* (29 March 2001), 041913-1–041913-5.

[16] Dongran Han, Suchetan Pal, Jeanette Nangreave, Zhengtao Deng, Yan Liu, and Hao Yan, "DNA Origami With Complex Curvatures in Three-Dimensional Space," *Science 332* (15 April 2011), 342–346.

[17] Bryan Wei, Mingjie Dai, and Peng Yin, "Complex Shapes Self-Assembled From Single-Stranded DNA Tiles," *Nature 485* (31 May 2012), 623–627.

[18] Paul W.K. Rothemund and Ebbe Sloth Andersen, "The Importance of Being Modular," *Nature 485* (31 May 2012), 584–585.

[19] Bryan Wei, "Complex Shapes," 623.

[20] Ed Wong, "DNA Drawing With an Old Twist," *Nature News and Comments* (30 May 2012), http://www.nature.com/news/dna-drawing-with-an-old-twist-1.10742

[21] David Bradley, "Lego-Like DNA Bricks Are Child's Play," *Chemistry World* (29 November 2012), http://www.rsc.org/chemistryworld/2012/11/dna-lego-bricks-origami

[22] Yonggang Ke, Luvena L. Ong, William M. Shih, and Peng Yin, "Three-Dimensional Structures Self-Assembled From DNA Bricks," *Science 338* (30 November 2012), 1177–1183.

[23] Yonggang Ke, Luvena L. Ong, William M. Shih, and Peng Yin, "Three-Dimensional Structures Self-Assembled From DNA Bricks," *Science 338* (30 November 2012), Supplementary Materials I (linked to the online version of the paper as a PDF download).

[24] Ke et al., "Three-Dimensional Structures," 1177.

[25] Ruth Williams, "DNA Bricks," *The Scientist* (29 November 2012), http://www.the-scientist.com/?articles.view/articleNo/33501/title/DNA-Bricks/

[26] Ke et al., "Three-Dimensional Structures," 1182.

[27] Yonggang Ke, Luvena L. Ong, Wei Sun, Jie Song, Mingdong Dong, William M. Shih, and Peng Yin, "DNA Brick Crystals With Prescribed Depths," *Nature Chemistry* 6 (November 2014), 994–1002.

[28] Ibid., 1.

[29] Rothemund, "Folding DNA to Create," 301.

[30] Ke et al., "Three-Dimensional Structures," 1179.

[31] Rothemund, "Scaffolded DNA Origami," 3.

[32] Paul W.K. Rothemund, "What's the Greatest Innovation?" http://www.spiked-online.com/index.php?/innovationsurvey/article/3271/

[33] Ibid., 3.

CHAPTER NINE

DNA Assembly Line and the Triumph of Tensegrity Triangles

In mid-summer of 2009, Nadrian Seeman is one of many distinguished speakers at the 33rd Steenbock Symposium held at the University of Wisconsin.[1] This four-day gathering is in honor of Har Gobind Khorana, the Nobel Prize–winning biochemist from MIT. Seeman is dressed with characteristic informality in an unbuttoned, brown, Eisenhower-style jacket over a red shirt open at the neck. He's an engaging speaker and warms up the crowd by recalling his time as a postdoc at MIT. He tells them that he first knew Gobind when he (Seeman) was "a very, very minor player" at a crystallographic lab at MIT. Seeman left to set up his own lab at the University at Albany-SUNY and, "about four years later Gobind came to the university and I was *astounded* that he remembered somebody as inconsequential as myself in that era." The lights now dim and Seeman begins his technical talk. He's about two-thirds of the way through his lecture titled "DNA: Not Merely the Secret of Life." He has explained the premise of structural DNA nanotechnology going back to his epiphanic moment at the Albany pub and presented many slides of two- and three-dimensional constructions. Now he is ready to move from the static to the dynamic, from microscopic trellises to machines that move: nanomechanical devices.[2] By way of introducing this topic he shows a Rube Goldberg slide (Figure 9.1) on the big screen behind him. Seeman explains to the audience, "This shows a mechanical device that's not nano. This is an automatic napkin and the way it works is, the guy spoons his soup, he pulls on the string that pulls on the ladle, the cracker goes up, the parrot goes

Figure 9.1 Rube Goldberg's Self-Operating Napkin. A non-nanomechanical device for which Seeman has high praise. *(Artwork Copyright © and ™ Rube Goldberg Inc. All Rights Reserved. RUBE GOLDBERG ® is a registered trademark of Rube Goldberg Inc. All materials used with permission. rubegoldberg.com)*

after the cracker, the birdseed then falls down into the bucket that pulls on the string that opens the lighter that lights the rocket, the rocket takes off, the sickle cuts the string, this goes back and forth and wipes the guy's mustache. Now this is *far* more efficient and *effective* than anything we've yet made on the nanometer scale. [Audience laughter] But we're working."

DNA Nanoscale Assembly Line (Overview)

Seeman was indeed working. In the following year he demonstrated a way to get discernable work out of a complex nanomechanical architecture. (We mean *work* in the scientific sense. For example, in physics it's defined as the energy transfer that occurs when an object is moved over a distance by an external force at least part of which is applied in the direction of the displacement.)[3] We're going to look at Seeman's creation, a DNA nanoscale assembly line that he compared to an automobile assembly line. Analogous to its macroscopic analog, the nanoscale production line links individually selected components. It is a prototype molecular factory that can be programmed to manufacture eight different products.

The assembly line combined a number of recent developments in DNA nanotechnology. One is motile molecules, known as DNA walkers.[4] These molecules are powered by energy derived from DNA hybridization, the familiar process in which complementary single strands of DNA form a double-stranded helix, a duplex. DNA walkers use oligonucleotides as fuel to move from one binding site to another on a DNA-modified surface. The surface, a second assembly line component, is none other than a sheet of DNA origami, an origami track. A third component are DNA machines that serve as cargo-donating devices, holding gold nanoparticles as cargo.

First the big picture and then we'll have a closer look at individual components. Here's what happens: a DNA walker travels along a path fueled by the energy released when single strands of DNA bind to complementary strands to form duplexes. The origami track is lined with three cargo-donating devices—each bearing a different nanoparticle cargo—at fixed intervals in an assembly line arrangement. The devices are individually programmed to either donate or keep their cargo (that is, they are two-state devices). This means that as the walker sequentially encounters the devices, it can be loaded with cargo resulting in eight different possible end products (three two-state devices produce $2^3 = 8$ possible outcomes). Seeman said, "I think of these DNA walkers as being similar to a chassis of a car rolling down the assembly line and we are adding components to it."[5]

The walkers have seven limbs: four DNA strands serve as feet and three as hands that carry the donated cargo. The cargo devices are anchored in series to an origami tile that acts as the walker's track. The walker's feet are controlled by the fuel strands that are added from solution. The fuel strands displace the feet from the origami track enabling them to move to other positions on the track—to walk. As we saw way back in Chapter Two when presenting a flowchart for DNA synthesis, creating molecules is a challenging process that involves adding and removing protective groups at various stages to avoid undesired interactions among unfinished molecules. The hope is that a new chemistry will come out of Seeman's effort—new in the sense of control. Perhaps researchers will (eventually) be able to manufacture drugs and perform other types of organic synthesis more easily and cleanly. Figure 9.2 introduces the overall picture of the assembly line.[6] We'll wind our way back to it by looking at its individual components. We begin with a chat about DNA walkers.

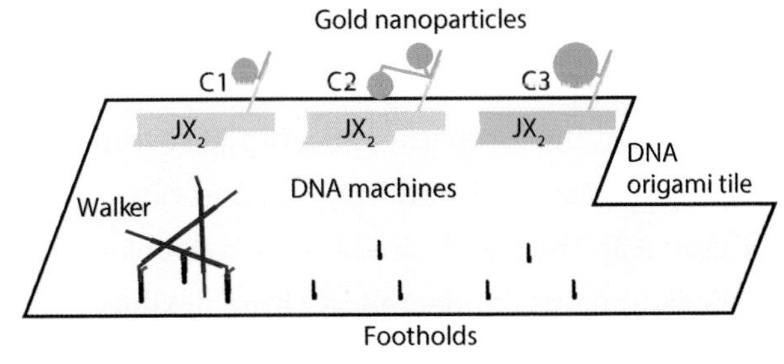

Figure 9.2 Overview of DNA nanoscale assembly line. *(Reproduced with permission from Max von Delius and David A. Leigh, "Walking Molecules," Chem. Soc. Rev. 40 (2011), 3656–3676.)*

DNA Walkers

The idea behind making walkers is biomimetic: to emulate motor proteins such as kinesin that carry large molecules around a cell. (By the way, it may have come to your attention that we've been pretty casual throughout this book in our use of words like biomimetic. Whether one calls research biomimetic, bioinspired, or biokleptic is somewhat arbitrary.)[7] Kinesin molecules use the energy of ATP hydrolysis as fuel to move in discrete steps along protein filaments called microtubules that are ubiquitous in cells. ATP hydrolysis is the cleaving of the molecule **a**denosine **tri**phosphate by reaction with water to produce ADP—**a**denosine **di**phosphate—and to release energy that can be used to produce work.

The first non-autonomous DNA walker—that is, one that relies on the sequential addition of chemical fuels in the form of fuel strands—was invented in Nadrian Seeman's group.[8] Two classes of fuel strands are set and unset strands. The former join single-stranded segments by hybridizing with them but can be removed by the addition of unset strands for which they have greater affinity. And we know why. There are more paired bases when the unset strands combine with the set strands and so this combination has lower free energy.

Seeman's walker was a two-component device where the two parts are connected by hydrogen bonds that are relatively easily broken chemical bonds. The two-components are a triple-crossover (TX) molecule, shaded blue, called the footpath, and a biped region (that performs the walking), shaded tan (Figure 9.3). The biped consists of two double-helical domains connected by three flexible DNA linker strands each nine nucleotides long. The three strands in the biped region are flexible because they are not complementary to any of the other strands in the system. They remain single-stranded throughout the operation of the system.

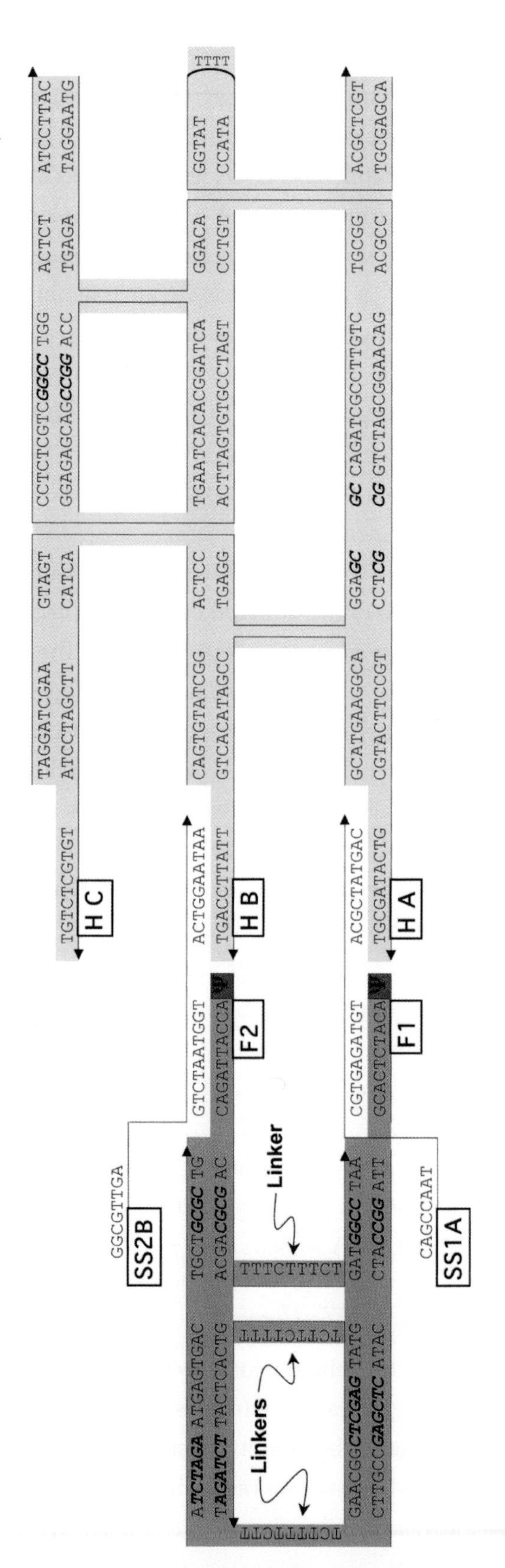

Figure 9.3 Cartoon of the biped system with DNA base sequences included. The biped is shown as tan and the footpath is shown as blue. *(Reproduced with permission from William B. Sherman and Nadrian C. Seeman, "A Precisely Controlled DNA Biped Walking Device," Nano Letters 4, No. 7 (2004), 1203–1207.)*

SS2B	≡	GGCGTTGAGTCTAATGGTACTGGAATAA

Figure 9.4 Detail of the set strand SS2B.

Each helical domain in the biped has a single-stranded portion called a foot on its end. The foot can pair with complementary strands of DNA. Likewise, each domain in the footpath has a single-stranded region—a foothold—that can hybridize with complementary strands. A foot attaches to a foothold when a set strand complementary to both is added to the solution. Follow this example closely: when the black and white strand labeled SS2B is present, it connects the biped Foot 2 (labeled F2) to the Foothold B (labeled H B).

DNA walkers Molecules composed of DNA that can walk along tracks also composed of DNA. In the work presented here we only consider non-autonomous walkers. Single-stranded DNA feet move along a DNA track in response to the sequential addition to the aqueous solution of single-stranded DNA fuel strands. Energy for motion is derived from DNA hybridization (the binding together of strands having complementary base sequences).

To further clarify, the set strand called SS2B has the complete sequence shown in Figure 9.4, where the first set of eight bases (black) are *un*paired, the second set of ten bases (tan) are hybridized to Foot 2 (F2), and the third set of ten bases (blue) are hybridized to Foothold B (H B). The eight *un*paired bases in the set strand are referred to as a toehold; they allow the set strand to be removed by the addition of an *un*set strand. This is the strand displacement phenomenon (also called competitive hybridization) that we saw in the previous chapter. The free energy gained through the formation of the new base pairs powers the walker.

Figure 9.5 cartoons the operation of the biped. The sequences of each foot, foothold, and toehold are represented by unique colors. Complementary strand sequences are shown with the same color. For instance, the magenta and aqua sections of Set Strand 1A complement Foot 1 and Foothold A, respectively. Panel (a) shows the state of the system called 1A, 2B: this means that Foot 1 is set to Foothold A and Foot 2 is set to Foothold B. For the biped to take a step, Foot 2 must be released from Foothold B, as seen in panels (b) and (c). To make this happen, Unset Strand 2B is introduced into the solution. It binds to the toehold of Set Strand 2B, as shown in panel (b). This is followed by branch migration (Chapter Eight) that results in the complete hybridization (all bases paired) of Set Strand 2B and Unset Strand 2B. That's the state of the system in panel (c). In panel (c)—State 1A—we find Foot 2 connected to the footpath only through the flexible linkers to Foot 1 and Set Strand 1A. We need to get on to other matters so we won't walk you through the whole procedure.

Cartoons aside, how does one verify the state of the biped system? This is done by taking a small sample of the solution and exposing it to ultraviolet light. Here's why that works: if you look closely at the bottom portion of Foot 1 and Foot 2, you'll see two small red rectangles at the ends of each of the feet of the biped. Now go back to

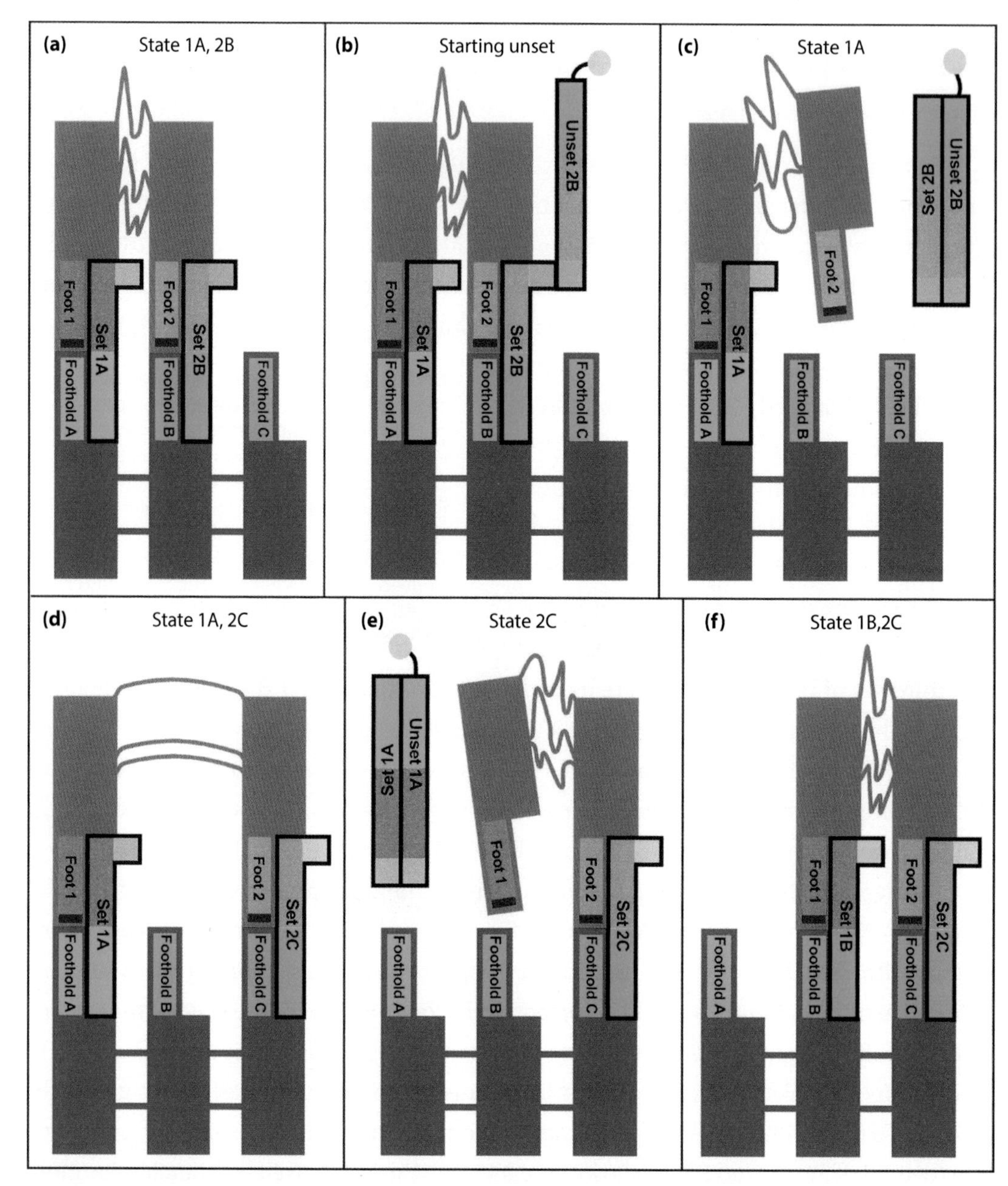

Figure 9.5 Cartoon depiction of biped system taking a full step. Note: the system is rotated 90° clockwise relative to Figure 9.3. *(Reproduced with permission from William B. Sherman and Nadrian C. Seeman, "A Precisely Controlled DNA Biped Walking Device," Nano Letters 4, No. 7 (2004), 1203–1207.)*

Figure 9.3 and look at the red boxes on the topmost end of the biped; inside each box is the Greek symbol Ψ (pronounced "psi"). These are chemical groups called psoralens. Psoralen is a cross-linker used to attach the foot strands to their respective set and foothold strands by strong chemical bonds (covalent bonds). A carefully performed quantitative analysis has shown that to optimize the psoralen cross-linking efficiency, the 5′ two bases of each foot must be 5′-AC-3′, and the 3′ two bases of each foothold must be 5′-GT-3′.[9]

To verify the state of the biped walker system, Seeman took a small sample of the solution and exposed it to ultraviolet light. If a foot is attached to a foothold, then the psoralen on the end of the foot can form covalent bonds by linking to the T on the end of the foothold, the T in the set strand, or to both (Figure 9.6). So by running the sample on a denaturing gel he could see three possible structures as single bands.

The target product that he was looking for was the largest, with all three strands covalently linked, shown in panel (b) of Figure 9.6. The two failure complexes, panels (c) and (d), have only two of the three strands linked. The target product can form with a high yield (a large percentage of the three-strand complex) for an attached foot. But it cannot form at all for an unattached foot. Because each foot and foothold has a unique length, denaturing PAGE of the ultraviolet-treated sample reveals which feet were detached as well as which feet were attached to which footholds.

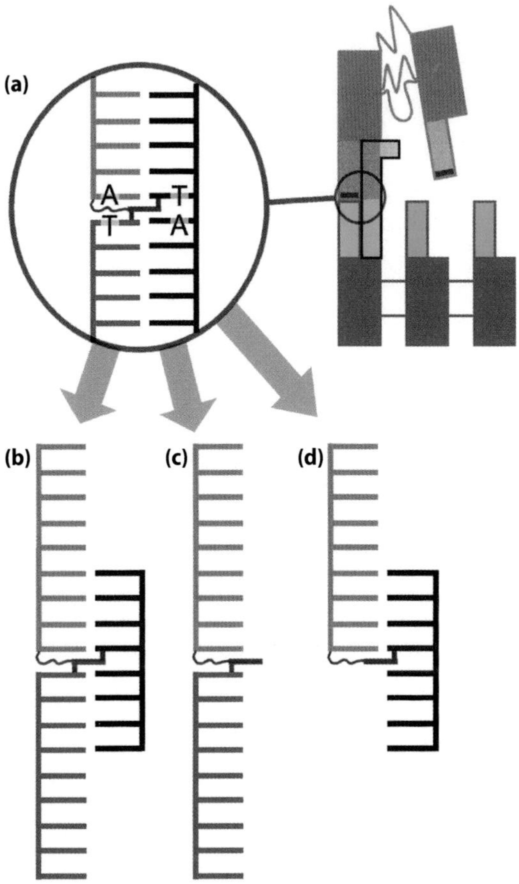

Figure 9.6 The inset circle **(a)** shows a detailed picture of a foot strand (tan), a set strand (black), and a foothold strand (blue). The psoralen (red) is covalently attached to the foot. But after ultraviolet radiation, it can form covalent links to either the set strand, the foothold, or both. This results in three possible structures that show up on a denaturing gel. The target complex **(b)** is the largest, with all three strands covalently linked. The two failure complexes **(c** and **d)** have only two of the three strands linked. *(Reproduced with permission from William B. Sherman and Nadrian C. Seeman, "A Precisely Controlled DNA Biped Walking Device," Nano Letters 4, No. 7 (2004), 1203–1207.)*

In Seeman's publication, he shows a nifty record of a biped walk: a stained PAGE with a cartoon above each lane to show the state of the system.[10] You can now appreciate that the verification of the state of the system was done with exacting care. It was no walk in the park.

The walker we've described has an inchworm-type gait instead of a foot-passing gait, but the latter has been demonstrated to have a whole host of designs including many autonomous walkers (ones that don't rely on sequential addition of fuel strands). This is a field rich with clever constructions and the interested reader will find the critical review by Max von Delius

Fuel strand A single strand of DNA that acts as a chemical fuel for a DNA device, e.g., a DNA walker. The fuel strand hybridizes with another DNA single strand thus lowering the free energy of the system. This provides energy to do work, e.g., powering structural change or motion.

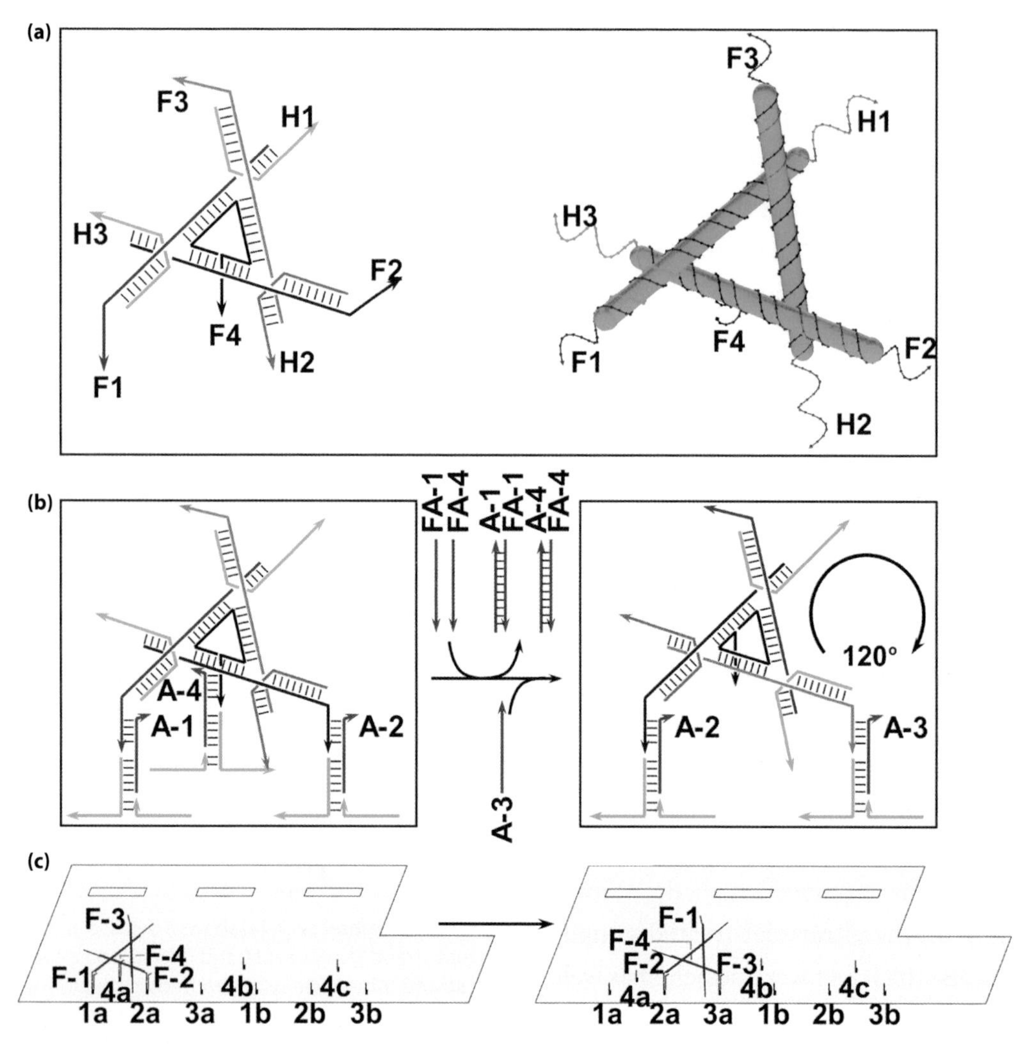

Figure 9.7 DNA walker based on a tensegrity triangle. **(a)** Walker structure. The drawing on the left is a stick figure indicating three hands (H1-H3) and four feet (F1-F4) all consisting of single-stranded DNA segments. The image on the right shows the strand structure. **(b)** One stride of the walker's movement. The walker is positioned at its starting point by anchoring two corner feet (F1 and F2) and the middle foot F4 on the DNA origami track (which is not shown in this panel). Fuel strands are added to the system to remove anchor strands A1 and A3, releasing foot F1 and F4. Then new anchor strand A4 is added to bind foot F3 to the track. The walker thus walks one step forward by rotating its body 120°. Panel **(c)** shows the walker's first step. *(Reproduced with permission from Hongzhou Gu, Jie Chao, Shou-Jun Xiao, and Nadrian C. Seeman, "A Proximity-Based Programmable DNA Nanoscale Assembly Line," Nature 465 (13 May 2010), 202–205.)*

and David A. Leigh (in the endnotes) to be rewarding reading. However, we will forge ahead to the walker used by Seeman in his DNA nanoscale assembly line.

The construction is based on tensegrity triangles (Figure 9.7). The principle of tensegrity was first demonstrated by sculptor Kenneth Snelson and the word *tensegrity*—a compound word combining tensile and integrity—was coined by R. Buckminster Fuller.[11,12] Tensegrity is a construction principle involving rigid struts and flexible tendons; struts push outward and tendons pull inward and the balance between these counteracting forces of tension and compression lead to stable, rigid structures.[13] In Chapter Seven we saw Seeman use a double-crossover triangular motif that is based on tensegrity triangles. In the case of tensegrity-based walkers, Seeman did not employ the double-crossover. Instead, he chose three four-arm DNA junctions that make use of the tensegrity strategy. They comprise the walker shown in Figure 9.7. The walker has three hands and four feet, all consisting of single-stranded DNA segments. As we've seen before, movement of the walker is controlled by the addition of fuel strands. Because the design choice is an equilateral triangle, each step of the triangular walker entails a 120° rotation.

We've devoted a lot of attention to walkers because they're fun and because they use concepts such as fuel strands that give us a head start on other parts of the DNA assembly line. It's time for us to take a look at the other components of the DNA nanoscale assembly line and then put them all together. The two other components are the DNA machines and the DNA origami track. We have the background to tackle both of these and the intriguing way that Seeman coupled them. We'll start with the DNA machines that draw on our knowledge of crossover molecules.

DNA Machines and Paranemic Crossover Molecules

The new crossover motif we introduce is called a paranemic crossover. (If you're a fan of words, the *nem* comes from the same ancient Greek word for *thread* that inspired the term *nematic* liquid crystal.) A paranemic crossover DNA molecule, or PX DNA, is the result of a crossover at every possible position in two strands of double-stranded DNA with the same polarity.[14,15] That is, the crossovers occur between a strand of one duplex and a strand of the other duplex, both of whose 5′→3′ directions have the same relative orientation (Figure 9.8).

In the paranemic molecules, two strands are drawn in red and two in blue, where the arrowheads indicate the 3′ ends of the strands. The thin horizontal lines within the two double-helical domains indicate the complementary base pairing (in which every nucleotide participates). A paranemic crossover molecule in which two adjacent sites are not crossed over but rather remain juxtaposed is called a JX_2 molecule. Building on our earlier topology lesson, PX and JX_2 are topoisomers of one another: they have distinct topologies but the same chemical formula and bond connectivities.

Figure 9.9 shows the color coding of the PX and JX_2 strands. Labels indicate that the top ends, A and B, are the same in both molecules, but the bottom ends, C and D, are rotated 180°. This rotation is the basis for the operation of a device in which Seeman and colleagues used strand replacement to interconvert the PX and JX_2 motifs.[16] This device is integral to the DNA machines of the assembly line.

The principles of operation of the device are illustrated in part (b). To begin, several modifications are made to the four-stranded molecule, PX. Similar modifications are made to the four-stranded molecule, JX_2. One strand of each of the blue and red strand pairs

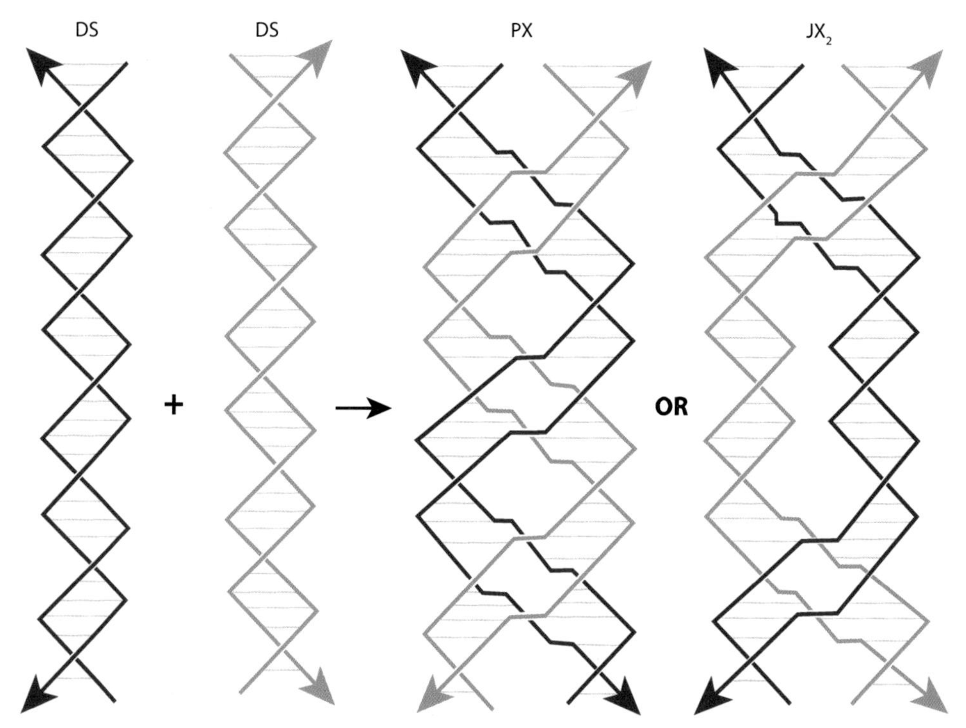

Figure 9.8 Two double-stranded helices (DS) form crossovers to create two types of paranemic molecules, PX DNA and JX_2 DNA. *(Reproduced with permission from Xiaoping Zhang, Hao Yan, Zhiyong Shen, and Nadrian C. Seeman, "Paranemic Cohesion of Topologically-Closed DNA Molecules," J. Am. Chem. Soc. 124 (2002), 12940–12941.)*

in PX is broken into multiple strands and the red and blue strands of opposite polarity are connected by hairpin loops. (Recall from the previous chapter that a hairpin loop is an unpaired loop of DNA created when a strand folds and forms base pairs with another section of the same strand.)[17] The net effect is that the PX molecule shown in (b) now consists of one red strand, one blue strand, and two green strands.

The green strands in PX are set strands (because they set the state of the device to be in the PX conformation). Similarly, the JX_2 molecule has purple set strands. The set strand associated with the red strand has a 5′ single-stranded extension, and the set strand associated with the blue strand has a similar 3′ extension. Extensions like these are used to initiate branch migration that leads to removal of the strand from the branched motif, because it is paired with a complementary strand along its entire length. So a complement to the entire length of the fuel strand will pair with that strand in preference to the partially paired set strand in the PX (or JX_2) motif.

If you're wondering, the black circles on the (unset) fuel strands show that these strands have been biotinylated. This means that a small

Paranemic molecule A type of DNA crossover molecule that occurs in two different forms, PX and JX_2. Conversion from one form to the other takes place in a two-step process mediated by short DNA fuel strands of two classes, set and unset. Such conversion produces measurable work.

DNA machine By using strand replacement to convert from one form to another, the paranemic motif provides a device that constitutes a DNA machine. Such DNA machines facilitate the donation of cargo to DNA walkers in a DNA assembly line.

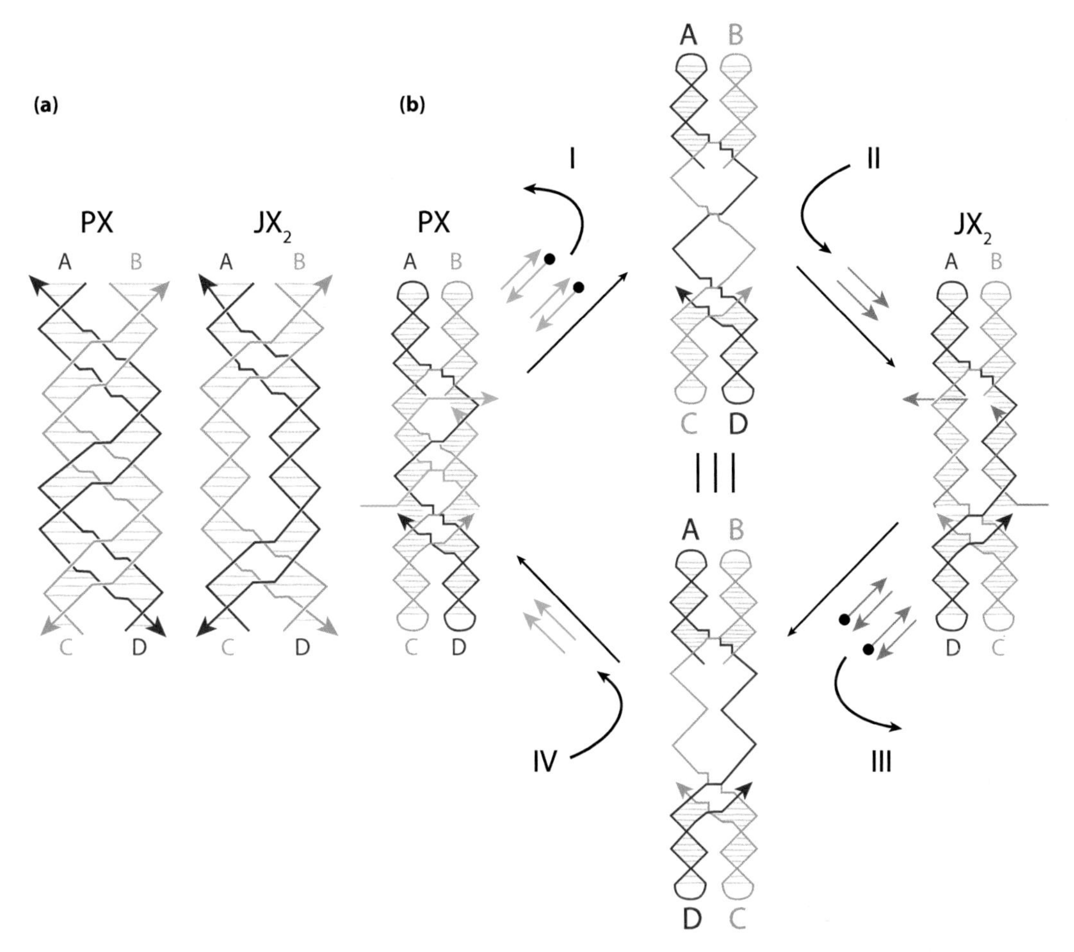

Figure 9.9 (a) The color coding of the strands and labels indicate that the top ends are the same in both molecules, but the bottom ends are rotated by 180°. This rotation is the basis for the operation of the device, shown in **(b)**. The green set strands in PX are removed in process "I" by the addition of fuel strands (green). The same technique is used with the purple fuel strands. Note that the identity symbol, ≡, between the two intermediate states (flanked by PX and JX_2), indicates that they are identical. *(Reproduced with permission from Hao Yan, Xiaoping Zhang, Zhiyong Shen, and Nadrian C. Seeman, "A Robust DNA Mechanical Device Controlled by Hybridization Topology," Nature 415 (3 January 2002), 62–64.)*

molecule called biotin has been covalently attached to the strand. Biotin has a very strong affinity for the protein streptavidin. Seeman was able to exploit this by introducing streptavidin into the solution. This removes the biotinylated strands and, because the complementary set strands are hybridized to the unset strands, they're flushed out as well.

Process I in Figure 9.9 part (b) shows the addition of fuel strand complements to the two green set strands of the PX device, producing the unstructured intermediate at the top of the drawing. Process II shows the addition of pale-purple set strands that convert the intermediate to the JX_2 conformation. Process III shows the addition of fuel strands that convert the JX_2 molecule to the unstructured intermediate, and process IV shows the addition of the green set strands to produce the PX conformation again.

The upshot of these maneuvers is a device with a four-step cycle that leads to two robust endpoints, the PX state and the JX_2 state. Seeman used both electrophoresis and

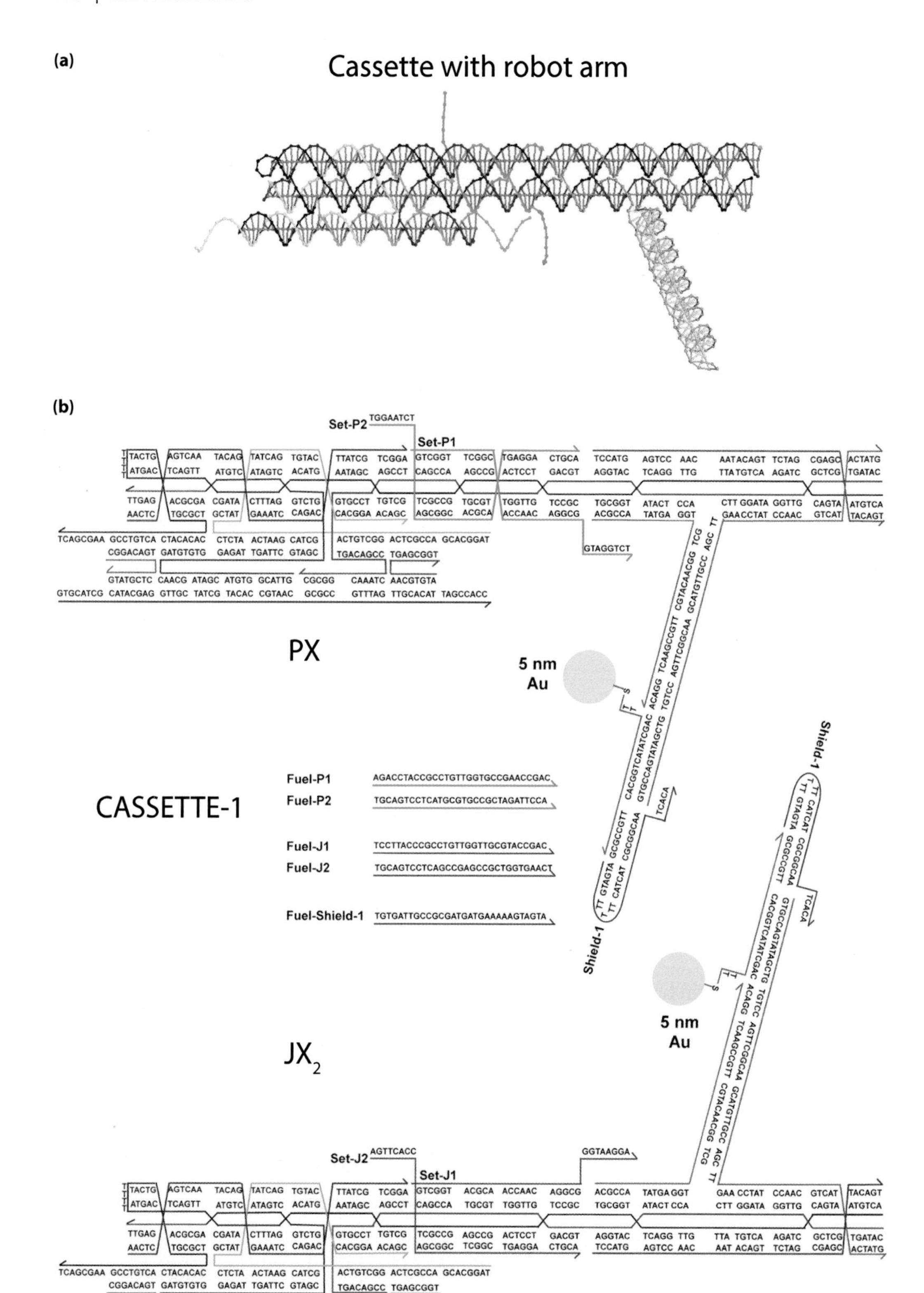
(a)
Cassette with robot arm
(b)
Set-P2
Set-P1
PX
5 nm Au
CASSETTE-1
Fuel-P1 AGACCTACCGCCTGTTGGTGCCGAACCGAC
Fuel-P2 TGCAGTCCTCATGCGTGCCGCTAGATTCCA
Fuel-J1 TCCTTACCCGCCTGTTGGTTGCGTACCGAC
Fuel-J2 TGCAGTCCTCAGCCGAGCCGCTGGTGAACT
Fuel-Shield-1 TGTGATTGCCGCGATGATGAAAAAGTAGTA
Shield-1
Shield-1
5 nm Au
JX2
Set-J2
Set-J1

AFM to provide gel evidence and images that demonstrate the success of the device.[16] We won't show these results.

DNA Cassette With Robot Arm and DNA Origami Track

It's time now to describe how Seeman placed the device at a precise location in a DNA array thus enabling its operation at a specific site, that is, within a fixed frame of reference.[18] He did this by adding additional features to the basic PX-JX_2 device, notably, a cassette. The original incarnation of the cassette predated the assembly line. It consists of three double-helical domains of DNA side-by-side, one of which is much shorter than the other two (panel (a) of Figure 9.10). The two longer helical domains compose the basic rotary PX - JX_2 device we described in the previous section. The shorter DNA duplex is called an insertion domain. The insertion domain contains sticky ends that bind with their complementary strands in a two-dimensional DNA array and thus allow for the precise, specific placement of the cassette. (When he first invented the cassette, he placed it in a two-dimensional DNA crystalline array rather than an origami-derived base plane.)

Another feature he added to the basic PX-JX_2 device is a long robot arm. This is the extension protruding down and to the right in panel (a) of Figure 9.10. It is a double helix that points in opposite directions in the PX and in JX_2 state. So the arm flips (it rotates by 180°—check out panel (b) of Figure 9.10). Using AFM one can see the before and after positions, and this enables differentiation of the two states. Seeman designed this cassette with an eye to nanorobotics: it's crucial to be able to insert controllable devices into a substrate, leading to a diversity of structural states. And then he went even further.

DNA cassette A structure that enables the insertion of a DNA nanomechanical device at a specific site in a two-dimensional crystalline DNA array.

Instead of a two-dimensional crystalline DNA array, he merged the cassettes with a DNA origami track.[19] His origami substrate included slots to allow for the precise insertion of multiple cassettes. The slots in the origami—imagine them as the *eyes* of Rothemund's *smiley face* origami—contain sticky ends that can hybridize with sticky ends on the insertion domain of the cassettes. The notch in the upper right of the track in Figure 9.2 is there to enable recognition of orientation during AFM.

To see the cassette with robot arm in action, panel (b) of Figure 9.10 shows the first cassette in each of its two possible states, PX and JX_2.[20] In the present incarnation (the one used for the assembly line), there are now four double-helical domains of DNA

Figure 9.10 (a) Cassette consisting of three helical domains. The insertion domain is much shorter than the other two. It contains sticky ends that facilitate its placement into a specific site in a two-dimensional DNA base plane (not shown). **(b)** The nitty-gritty details of a cassette showing its robot arm carrying a gold nanoparticle cargo. When the PX-JX_2 device switches from one state to the other, the robot arm rotates by 180°. *(Panel **(a)** reproduced with permission from Baoquan Ding and Nadrian C. Seeman, "Operation of a DNA Robot Arm Inserted Into a 2D DNA Crystalline Substrate," Science 314 (8 December 2006), 1583–1585; panel **(b)** portion reproduced with permission from Online Supplementary Material for Hongzhou Gu, Jie Chao, Shou-Jun Xiao, and Nadrian C. Seeman, "A Proximity-Based Programmable DNA Nanoscale Assembly Line," Nature 465 (13 May 2010), 202–205.)*

instead of three. The bottom two (shorter) duplexes compose the insertion domain that fits into the origami track. The top two duplex domains form the PX/JX_2 device motif with the robot arm sticking out. Earlier we used the word "cargo" quite a lot. What we see here is the cargo strand with a gold nanoparticle, 5 nm Au, attached. The robot arm rotates by 180° (the robot arm flips) when the device changes from one state to the other.

DNA Assembly Line

As all rivers run to the sea, at last we are ready for the DNA assembly line.[21] The origami track has three slots for cassettes and nine single-strand DNA extensions (Figure 9.2). The triangular walker uses the cargo-donating cassettes as stations. In a serial manner, the walker moves into the proximity of each of the DNA machines (observe the black stems in Figure 9.2). Each DNA machine is initially in the JX_2 state (so the robot arm points *up* in each case in Figure 9.2) and each holds a different cargo of gold nanoparticles. To see this, look closely at Figure 9.2 (yet again) to notice the thin line protruding above the main part of each cassette (color-coded to match the cassette colors). These are the robot arms. Each arm is tethered to a gold nanoparticle cargo. C2 is tethered to a pair of gold nanoparticles.

The three DNA machines are independently programmed either to donate cargo or not, so the assembly line can produce eight distinct products. Each machine carries a different type of cargo: a 5-nm gold particle, a coupled pair of 5-nm particles, and a 10-nm particle. The hands of the walker accept and bind the cargo species that are placed for pick up. The feet of the walker bind to single strands on the origami surface and allow locomotion. Figure 9.11 allows you to scrutinize bit by bit the proximity-based cargo attachment process for cassette 1.

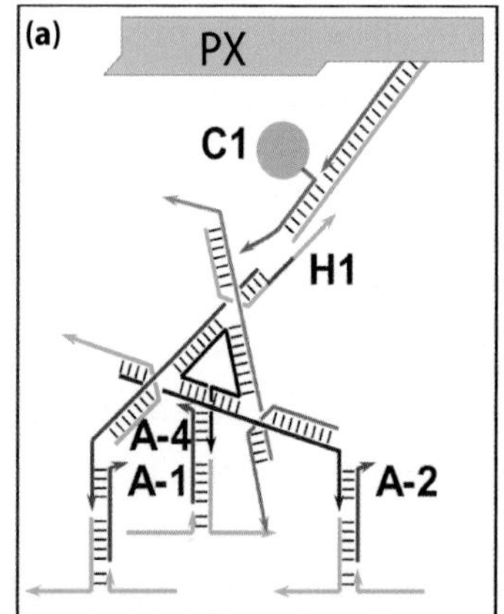

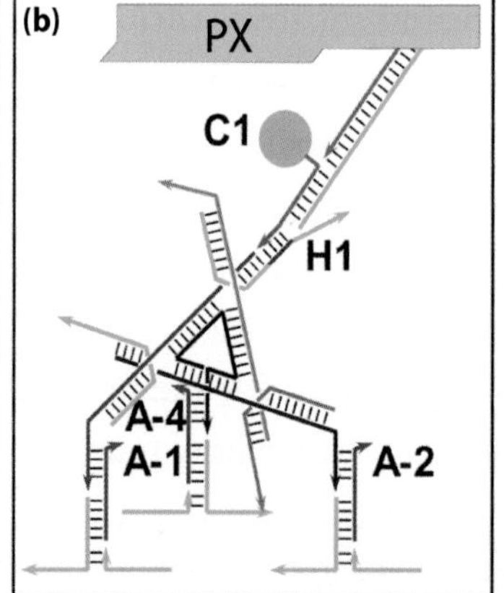

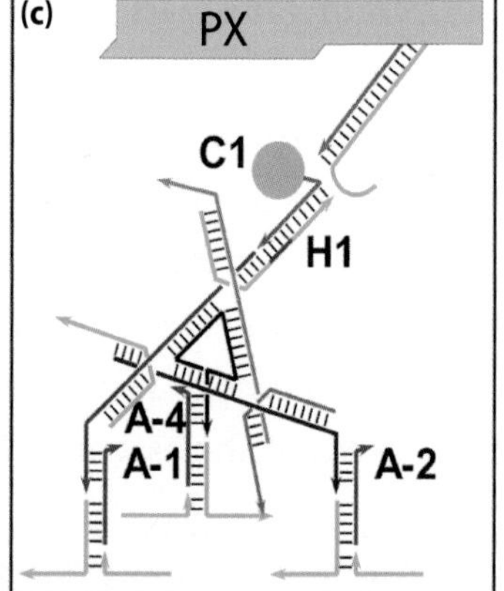

Figure 9.11 Details of the cargo transfer from the cassette to the walker. The PX state brings the arm of cassette one close to the first hand, H1 **(a)**, the brown toehold binds its complement (red, **b**) and branch migration transfers the cargo strand to H1 **(c)**. *(Reproduced with permission from Hongzhou Gu, Jie Chao, Shou-Jun Xiao, and Nadrian C. Seeman, "A Proximity-Based Programmable DNA Nanoscale Assembly Line," Nature 465 (13 May 2010), 202–205.)*

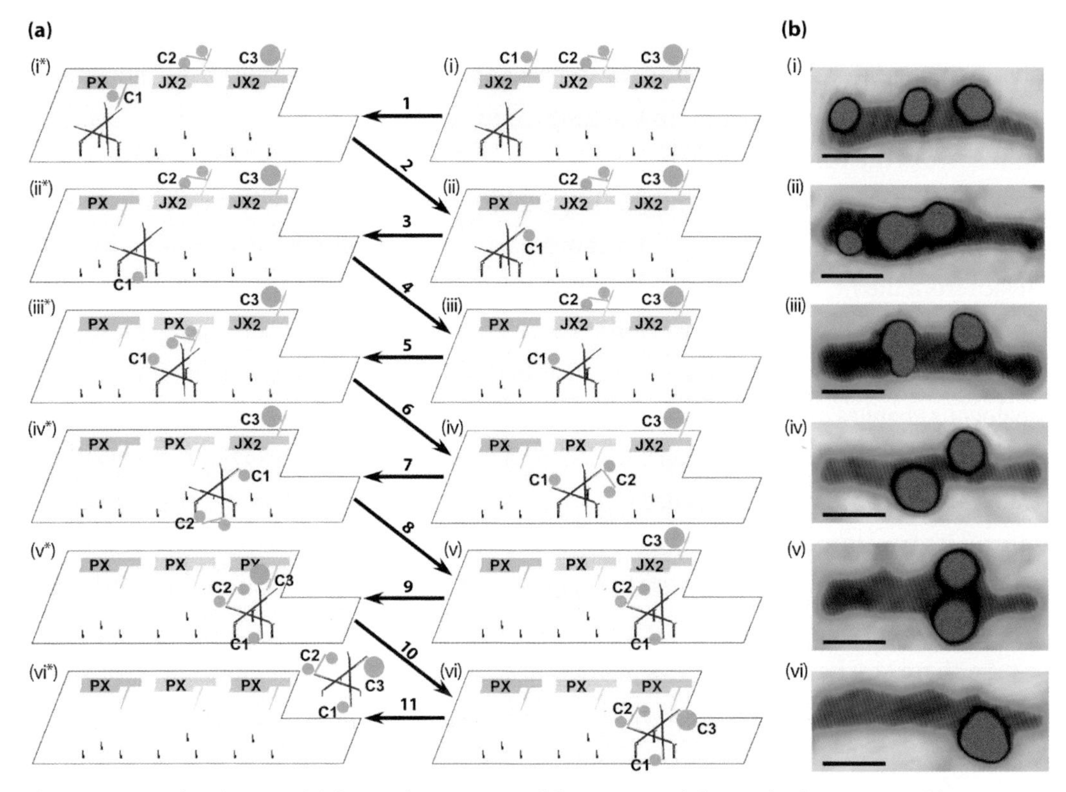

Figure 9.12 Molecular assembly line and its operation. **(a)** Origami track (tan outline), programmable two-state machines inserted in series into the substrate (blue, purple, and green) and walker (red triangular arrangement of DNA double helices). Machines cargoes (greenish-brown dots) are a 5 nm gold particle (C1), a coupled pair of 5 nm particles (C2), and a 10 nm particle (C3). Machines can be in two states: PX (ON or donate cargo) or JX$_2$ (OFF or do not donate cargo). In this example, the walker collects cargo from each machine. **(b)** AFM imaging corresponding to the process steps in states (i)–(vi). In this AFM mode, only gold nanoparticle cargoes and the origami are visible; nanoparticles attached to the walker are not individually resolved. (TEM images, not shown, resolve the individual gold nanoparticles.) The scale bar is 50 nm. *(Reproduced with permission from Hongzhou Gu, Jie Chao, Shou-Jun Xiao, and Nadrian C. Seeman, "A Proximity-Based Programmable DNA Nanoscale Assembly Line," Nature 465 (13 May 2010), 202–205.)*

A touch of irony: when Seeman started out (ca. 1980), he was trying to suppress branch migration and create immobile branch junctions. As Figure 9.11 illustrates, thirty years later he's making good use of branch migration for cargo transfer.

We illustrate the operation of the entire assembly line with the example shown in Figure 9.12. In this particular case, all machines donate cargo to the walker. The 11 separate processing steps are sketched in Figure 9.12, panel (a) and the state of the system is visualized using AFM in panel (b). (Note that the gold nanoparticle cargoes and the origami are the only features visible in the images, and that the nanoparticles attached to the walker are not *individually* resolved. TEM images *do* resolve the individual gold nanoparticles but we won't show these images.)

Steps 1, 5, and 9, respectively, involve the transitions of the first, second, and third two-state devices from the JX$_2$ (OFF) state to the PX (ON) state, which allows cargo donation. The actual transfers of the cargo particles to the walker, respectively, occur in steps 2, 6, and 10. Steps 3 and 7 entail the movement of the walker from a cargo-donating station to an intermediate position. In steps 4 and 8 the walker completes its movement from the intermediate position to the next cargo-donating station. Step 11 removes the walker from the origami track.

The changes in the AFM images (in panel (b) of Figure 9.12) when going from state (i) to state (ii) show the motion of a cargo particle associated with switching its DNA device from the JX_2 state to the PX state (thus allowing particle transfer). When the walker and its first cargo particle move from the first particle-donating station to the second, this is reflected in changes to the AFM images that correspond to states (ii) and (iii). The analogous changes involving the second particle-donating station are shown in the transition from state (iii) to state (iv), where the second particle has been moved to the walker track. In the same way, the transition from state (iv) to state (v) shows the walker has moved its two cargoes to the third cargo-donating station. Finally, the changes in the AFM images corresponding to states (v) and (vi) show the addition of the third cargo to the walker.

So there you have it. Seeman's nanoscale automobile-style assembly line. Not yet up to the sophistication of a Rube Goldberg contraption. But getting there.

The Triumph of Tensegrity Triangles

Seeman has also made headway toward the goal closest to his heart. He shared his thoughts with Paul Weiss, Editor-in-Chief of *ACS NANO* (a publication of the American Chemical Society on nanoscience and nanotechnology). "What we'd like to be able to do with it, the original idea I had at that pub back in 1980, is to use the DNA as a host lattice for macromolecular-scale guests—basically, to solve the crystallization problem of macromolecular crystallography in a general way. Make . . . box-like things that are connected by sticky ends; make that lattice. If the boxes are filled with an ordered material, then the whole system will be ordered and you can do your crystallography and be happy."[22]

And what is the potential for society? One hope is to determine the crystal structures of potential drug targets to better design drugs precisely tailored to their targets. "The program to facilitate the structural basis of drug target crystallization could enable scientists to develop new cures for various diseases. The HIV protease inhibitors were developed from crystal structures of the protease, but there was a lot of luck involved in getting those crystals. We would like to eliminate the dependence on luck."[23]

Seeman has taken a big step closer to his thirty-year old dream.[24] He reported the crystal structure at 0.4 nm resolution (more commonly referred to as 4 Å resolution; 1 nm = 10 Å) of a designed, self-assembled, three-dimensional crystal based on the DNA tensegrity triangle. This means he can determine the position of each atom with a precision of 0.4 nm. He was elated with this result, as well he should be. At the Steenbock Symposium he presented pre-publication results. Projecting a photograph of the crystals on the screen (panel (b) of Figure 9.13), he said, "This is the most exciting slide I can show because this shows macroscopic objects and we know where every atom is in these things. . . ."[1] The scale bar in Figure 9.13 (b) is 200 μm (1000 nm = 1 μm).

The tensegrity triangle in three dimensions is a rigid DNA motif with three-fold rotational symmetry. It consists of three helices such that their helix axis directions do not all share the same plane. (This may not be obvious in a simple two-dimensional schematic such as Figure 9.13 (a), but in a tick we'll show stereoscopic images.) This directionality is an important design feature. It's essential that the directions of propagation associated with the sticky ends—the bases colored red in (a)—do not share the same plane but rather extend to form a three-dimensional arrangement of matter. The three helices of the tensegrity triangle are connected pair-wise by three four-arm branched junctions—at each corner of (a)—that produce the stiff, alternating, over-and-under motif shown schematically in panel (a). As a result we have three helical domains, each containing two double helical turns of 21 nucleotide pairs, including sticky ends. To see one of these domains, follow a green strand along its length, count the nucleotide pairs, and be sure to include as pairs the sticky ends TC and GA. The six sticky ends form three identical complementary pairs.

In total, there are seven strands in the molecule, three that participate in a crossover near the corners—magenta in (a)—three that extend for the length of each helix (green), and a final nicked strand at the center (blue) that completes the crossovers and also forms the double

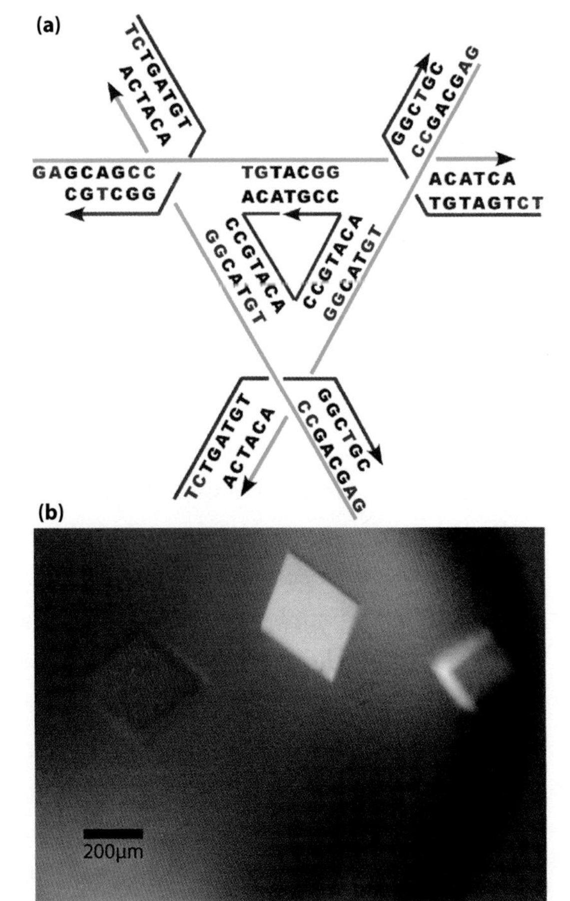

Figure 9.13 Schematic design, sequence, and crystal pictures of a three-dimensional crystal based on the tensegrity triangle. **(a)** The three unique strands are shown in magenta (strands restricted to a single four-arm junction), green (strands that extend over each edge of the tensegrity triangle), and dark blue (a nicked strand at the center passing through all three junctions). Arrowheads indicate the 3′ ends of strands. Sticky ends are shown in red letters. **(b)** Optical image of crystals of the tensegrity triangle. The scale bar is 200 μm (or 200,000 nm). *(Reproduced with permission from Jianping Zheng, Jens J. Birktoft, Yi Chen, Tong Wang, Ruojie Sha, Pamela E. Constantinou, Stephan L. Ginell, Chengde Mao, and Nadrian C. Seeman, "From Molecular to Macroscopic via the Rational Design of a Self-Assembled 3D Crystal," Nature 461 (3 September 2009), 74–77.)*

helices between the crossovers. The nicked site of the center strand is only three-fold rotationally *averaged*, occurring with one-third occupancy in each edge. In crystallography, the occupancy of a particular atom is a measure of the fraction of molecules in the crystal in which that atom occupies the position specified in the model of the crystal. (We'll return to this point later.) The green and magenta strands indicate an over-and-under motif. By tailing the three helices with short single-stranded sticky ends to make them cohesive, the helices can be directed to connect with helices belonging to six other molecules in six different directions. This results in a three-dimensional, periodic lattice—that is, *a crystal.*

DNA tensegrity triangles Three self-assembling double helices of DNA that point in three independent directions so as to define a three-dimensional structure having three-fold rotational symmetry. Each of the ends of the helices has two unpaired bases (sticky ends) that bind to complementary pairs of bases on other triangles. This enables the construction of macroscopic three-dimensional crystals. These crystal lattices contain periodically arranged cavities that might be used to host biomolecules in a three-dimensional periodic arrangement. In that event, X-ray diffraction could be used to determine the structure of the guest molecules.

An historical side point: notice the nucleotides in Figure 9.13 (a) that are in light blue lettering. By studying electron density maps of the crystals (Chapter Five), Seeman found that these nucleotides had A-DNA characteristics. Rosalind Franklin first observed the transformation from B-DNA to a dehydrated form that she called A-DNA. The existence of these two hydration-dependent forms of DNA (with concomitant differences in the X-ray patterns of the two forms) was a significant complicating factor in deciphering the structure of DNA—one that we did *not* discuss in Chapter Two.[25]

When we looked at X-ray diffraction in Chapter Five we saw patterns of dark spots. Each spot corresponds to a type of variation in the electron density map of the molecule under investigation. The crystallographer is said to *index the crystal* by relating which variation corresponds to which spot and this is done in the context of a lattice (one is said to "index the crystal in a such-and-such lattice"), that is, an infinite array of points each of which has an identical environment. When Seeman did this, he indexed the crystal in a rhombohedral lattice (rhombohedral shapes are evident in the images of panel (b) of Figure 9.13). He determined the elemental repeating unit of the lattice, the unit cell, to have an edge measuring about 7 nm and an angle between the edges of about 101°.

Figure 9.14 presents two stereoscopic images of the lattice formed by tensegrity triangles. Some people have no trouble at all in viewing images such as these that are designed to give a three-dimensional effect without 3D glasses. This is called free-viewing. There are tricks to getting the three-dimensional effect to happen if it doesn't come easily.[26] The three directions that define this lattice are indicated by the red, green, and yellow color-coding in Figure 9.14 (a) which shows the surroundings of a given tensegrity triangle.

In Figure 9.14 (b), we see the open nature of this stick-like lattice, which illustrates the rhombohedron that is flanked by eight of the triangles. The red triangle is at the rear, bonded by sticky ends to the three yellow triangles that flank it lying in a plane closer to the reader. The yellow triangles are bonded to the green triangles lying in a plane even closer to the reader. An eighth triangle lying closest to the reader and directly above the red triangle has been excluded for clarity. The volume of this rhombohedral cavity is about 103 nm^3.

In our discussion of Figure 9.13 (a), we remarked that the nicked site of the center strand is only three-fold rotationally *averaged*, occurring with one-third occupancy in each edge. This is something of a concern. As Seeman explained in the Supplementary Information to his paper, "The nick and the missing phosphate group in the central

Figure 9.14 Lattice formed by tensegrity triangles. **(a)** Surroundings of a triangle. This is a stereoscopic image that distinguishes three independent directions by base-pair color. The central triangle is flanked by six other triangles. **(b)** Rhombohedral cavity formed by tensegrity triangles. This stereoscopic image shows seven of the eight triangles that comprise the rhombohedron's corners. The cavity outline is drawn in white. The rear red triangle connects through one edge each to the three yellow triangles in a plane closer to the reader. The yellow triangles are connected through two edges each to two different green triangles that are even nearer to the reader. *(Reproduced with permission from Jianping Zheng, Jens J. Birktoft, Yi Chen, Tong Wang, Ruojie Sha, Pamela E. Constantinou, Stephan L. Ginell, Chengde Mao, and Nadrian C. Seeman, "From Molecular to Macroscopic via the Rational Design of a Self-Assembled 3D Crystal," Nature 461 (3 September 2009), 74–77.)*

strand [dark blue inverted triangle in panel (a) of Figure 9.13] in principle destroy the 3-fold rotational symmetry of the triangle." However, he pointed out that the identity of the three sticky end pairs was expected to restore this symmetry by placing the nick in three different sites. He concluded that, "This averaging is borne out in the data processing of the data at 4 Å resolution."[27] So, once again, his ingenuity carried the day.

However, there is a caveat. The applications that have been suggested for designed three-dimensional nucleic acid crystalline systems include the scaffolding of biological systems for crystallographic structure determination, as well as the organization of nanoelectronics. As Seeman concedes, "Both of these applications will probably be most usefully realized with scaffolding that is not three-fold rotationally averaged. Nevertheless, following this beginning, the other steps apparently needed for these applications are likely to prove incremental and feasible." [28] And while we're tangentially on the subject of his X-ray data, let's observe that the diffraction images that gave him a wealth of data were collected at the National Synchrotron Light Source at Brookhaven National Laboratory and the Advanced Photon Source at Argonne National Laboratory. We mentioned both of these facilities in Chapter Five.

Before going further, we owe you a couple of anaglyphs (Figure 9.15). There are two popular ways of creating three-dimensional illusions. If you're a movie fan you may be familiar with the anaglyph technique. Anaglyph 3D encodes an image for each eye using filters of chromatically opposite colors (e.g., red/blue). The two differently filtered images provide different light information for each eye and by doing so, trick your typical perception and cause the foreground to jump out at you. On the other hand, a side-by-side pair of stereoscopic images makes you do all the work yourself.[29] To view anaglyphs you'll need to have a pair of anaglyph glasses. You may find that plastic glasses (about $5) perform better than cardboard gel glasses (about 10¢). Figure 9.15 shows anaglyphs of the stereoscopic images shown in Figure 9.14: the surroundings of a tensegrity triangle (top) and the rhombohedral cavity formed by tensegrity triangles (bottom).

The construction of a three-dimensional crystal in which the position of every atom is known—by design—is a landmark in structural DNA nanotechnology. However, some of the issues that remain may surprise you. Seeman was able to achieve 0.4 nm (= 4 Å) resolution. That sounds terrific. For the purposes he has in mind, it's not. Here's what he said: "2-3 Å is sort of what you should get from a good macromolecular crystal. Not necessarily nucleic acid crystals, they tend to be about an Angstrom worse than protein crystals, but 2-3 [Å], in that ballpark."[30]

Recently, he and his colleagues have reported progress in resolution by varying the lengths of the sticky ends used in constructing the crystals.[31] Their results indicate that not only do the lengths and sequences of the sticky ends define the interactions between the motifs, but they also have an impact on the resulting resolution. Seeman's team anticipates that redesigned assemblies will form three-dimensional crystals with better resolution. This will be all-important in the precise determination of the crystallographic structure of scaffolded biological macromolecules (recall Figure 1.1 in Chapter One).

Seeman was a speaker at the annual meeting of the American Association for the Advancement of Science (AAAS) in February 2013 and he touched on this very point. Here is his abstract on the subject of controlling the structure of DNA in three-dimensions: "However, this is a milestone, not an end point. We must now advance to controlling the structure of other components as guests in DNA structures. We have achieved this object in 1D with amyloid fibrils, and in 1D, 2D . . . with gold nanoparticles. However, this is still a highly problematic enterprise, both synthetically and analytically. When control of non-DNA species can be achieved routinely, a new era in nanoscale control will await us."[32] Regardless of the uncertainties ahead, one writer observed, "While Seeman admits that there are still a lot of steps needed to reach these goals, this is a man who sticks with an idea until it works."[33]

Figure 9.15 Anaglyphs of the surroundings of a tensegrity triangle (top) and the rhombohedral cavity formed by tensegrity triangles (bottom).

Since this is the last chapter in which we'll focus on Seeman's contributions, it's appropriate to take a moment to say a few words about the people whose creativity and hard work were indispensable: Seeman's exceptional students and postdocs. Over the years Seeman has given hundreds of presentations throughout the world and his slides invariably call out in large letters his collaborators from other universities and his student and postdoc collaborators within his own lab. A number of his former students and postdocs have gone on to establish research programs of their own at other universities with notable success. Among his many

Anaglyph A type of stereoscopic illustration—i.e., an illustration that appears to have three-dimensional depth and solidity. This is achieved with a pair of images of the same subject that are superimposed. The images are printed in different colors (usually red and green) and viewed with appropriate filters over each eye.

accomplished lab members whose work appears in this book are Jens Birktoft, Jie Chao, Junghuei Chen, Pamela Constantinou, Baoquan Ding, Tsu-Ju Fu, Hongzhou Gu, Xiaojun Li, Shiping Liao, Furong Liu, Philip S. Lukeman, Chengde Mao, Mary L. Petrillo, Jing Qi, Lisa Wenzler Savin, Ruojie Sha, Zhiyong Shen, Tong Wang, Xing Wang, Shou-Jun Xiao, Hao Yan, Xiaoping Yang, Xiaoping Zhang, Xinshuai Zhao, and Jianping Zheng.

We've seen Seeman's tenacity throughout the chapters of this book devoted to his research. It is one of his distinguishing characteristics. Regardless of wrong turns and culs-de-sac, he finds a way back to the main road—and success. Seeman's nanoscale version of an automobile assembly line calls to mind a simile used by the author E.L. Doctorow. The Winter 1986 issue of *the Paris Review* contained the transcript of an interview with the author.[34] The interviewer, George Plimpton, conducted the session in public in the main auditorium of New York City's 92nd Street YMHA. An audience of about five hundred was on hand. The 92nd Street Y is in the Central Park neighborhood of Manhattan—only five miles north of the Washington Square neighborhood, home to the NYU campus where Seeman would later become a member of the chemistry faculty. Not far into the interview, Plimpton questioned Doctorow about his approach to writing a book. He asked, "Do you have any idea how a project is going to end?" The author's response seems to describe the journey Seeman is on. Doctorow said that in the main he did not. "It's hard to explain. I have found one explanation that seems to satisfy people. I tell them it's like driving a car at night: you never see further than your headlights, but you can make the whole trip that way."

Exercise for Chapter Nine: Exercise 9.1

Exercise 9.1

Biological molecular walkers such as the motor protein kinesin have inspired DNA walkers. The movements of naturally occurring nanomachines such as single motor-protein molecules can be analyzed directly using optical techniques (optical tweezers) that allow the experimenter to position them as desired. [35] A remarkable experiment reported in 1993 was the first ever glimpse of a biological engine turning over.[36] You are the experimenter and you've positioned a single kinesin molecule that is on a microtubule (as mentioned in the subsection called DNA Walkers in the present chapter). You've tagged the kinesin molecule with a silica bead that makes optical trapping and hence, observation, feasible. (Optical trapping, also referred to as optical tweezers, is a technique based on the transfer of momentum between a beam of radiation—typically a beam emitted by a laser—and the object that it is passing through, e.g., a silica bead.) Panel (a) of Figure 9.16 cartoons the position of an optically trapped kinesin molecule moving along a microtubule.[37] The traces of position versus time in panel (b) of the figure show the movement of a single kinesin molecule as a function of the particular mechanical and chemical experimental conditions chosen.[38] For the traces shown, what is the average rate of movement of the kinesin molecule along the microtubule?

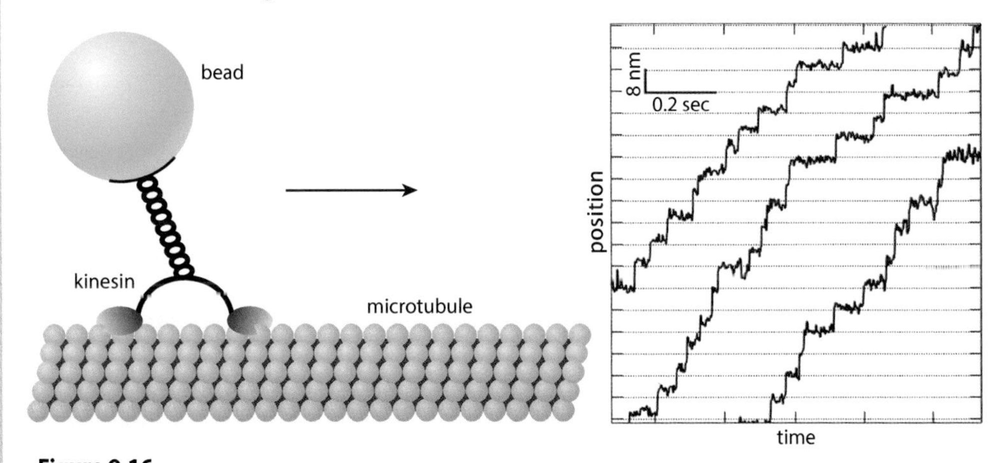

Figure 9.16

*((**a**) Reproduced with permission from Hajin Kim and Taekjip Ha, "Single-Molecule Nanometry for Biological Physics," Rep. Prog. Phys. 76 (2013), 1–16. (**b**) Unpublished kinesin stepping traces courtesy of Steven M. Block, Departments of Biology and Applied Physics, Stanford University.)*

ENDNOTES

[1] Nadrian C. Seeman, "DNA: Not Merely the Secret of Life," *Synthetic Genes to Synthetic Life, 33rd Steenbock Symposium* (July 30th—August 2nd 2009), University of Wisconsin-Madison, http://www.biochem.wisc.edu/seminars/steenbock/symposium33/ and linked to http://www.youtube.com/watch?v=kDTqbKqJRPw&feature=youtu.be

[2] Hao Yan and Yan Liu, "DNA Nanotechnology: An Evolving Field," *Nanotechnology: Science and Computation*, ed. Junghuei Chen, Nataša Jonoska, and Grzegorz Rozenberg (Berlin, Heidelberg: Springer-Verlag, 2006), 35–53.

[3] http://www.britannica.com/science/work-physics

[4] Lloyd M. Smith, "Molecular Robots on the Move," *Nature* 465 (13 May 2010), 167–168.

[5] Mike Brown, "Molecular Robots on Nano-Assembly Lines," *Chemistry World* (12 May 2010), http://www.rsc.org/chemistryworld/News/2010/May/12051003.asp

[6] Max von Delius and David A. Leigh, "Walking Molecules," *Chem. Soc. Rev. 40* (2011), 3656–3676.

[7] Andrea E. Rawlings, Jonathan P. Bramble, and Sarah S. Staniland, "Innovation Through Imitation: Biomimetic, Bioinspired and Biokleptic Research," *Soft Matter 8* (2012), 6675–6679.

[8] William B. Sherman and Nadrian C. Seeman, "A Precisely Controlled DNA Biped Walking Device," *Nano Letters 4, No. 7* (2004), 1203–1207.

[9] Ornella Gia, Sebastiano Marciani Magno, Anna Garbesi, Francesco Paolo Colonna, and Manlio Palumbo, "Sequence Specificity of Psoralen Photobinding to DNA: A Quantitative Approach," *Biochemistry 31* (1992), 11818–11822.

[10] Sherman and Seeman, "A Precisely Controlled," 1206.

[11] Kenneth Snelson, "Snelson on the Tensegrity Invention," *International Journal of Space Structures 11, Nos. 1 and 2* (1996), 43–48.

[12] Richard Buckminster Fuller, "Tensile-Integrity Structures," United States Patent Office, Patent No. 3,063,521, November 13, 1962.

[13] Dage Liu, Mingsheng Wang, Zhaoxiang Deng, Richard Walulu, and Chengde Mao, "Tensegrity: Construction of Rigid DNA Triangles With Flexible Four-Arm DNA Junctions," *J. Am. Chem. Soc. 126* (2004), 2324–2325.

[14] Xiaoping Zhang, Hao Yan, Zhiyong Shen, and Nadrian C. Seeman, "Paranemic Cohesion of Topologically-Closed DNA Molecules," *J. Am. Chem. Soc. 124* (2002), 12940–12941.

[15] Zhiyong Shen, Hao Yan, Tong Wang, and Nadrian C. Seeman, "Paranemic Crossover DNA: A Generalized Holliday Structure With Applications in Nanotechnology," *J. Am. Chem. Soc. 126* (2004), 1666–1674.

[16] Hao Yan, Xiaoping Zhang, Zhiyong Shen, and Nadrian C. Seeman, "A Robust DNA Mechanical Device Controlled by Hybridization Topology," *Nature 415* (3 January 2002), 62–64.

[17] http://www.nature.com/scitable/definition/hairpin-loop-mrna-314

[18] Baoquan Ding and Nadrian C. Seeman, "Operation of a DNA Robot Arm Inserted Into a 2D DNA Crystalline Substrate," *Science 314* (8 December 2006), 1583–1585.

[19] Hongzhou Gu, Jie Chao, Shou-Jun Xiao, and Nadrian C. Seeman, "Dynamic Patterning Programmed by DNA Tiles Captured on a DNA Origami Substrate," *Nature Nanotechnology 4* (April 2009), 245–248.

[20] Hongzhou Gu, Jie Chao, Shou-Jun Xiao, and Nadrian C. Seeman, "A Proximity-Based Programmable DNA Nanoscale Assembly Line," *Nature 465* (13 May 2010), Supplementary Information Figure S2: "The Sequence of Cassette 1 in the PX and JX_2 States," 10 (linked to the online version of the paper as a PDF download).

[21] Hongzhou Gu, Jie Chao, Shou-Jun Xiao, and Nadrian C. Seeman, "A Proximity-Based Programmable DNA Nanoscale Assembly Line," *Nature 465* (13 May 2010), 202–205.

[22] Paul S. Weiss, "A Conversation With Prof. Ned Seeman: Founder of DNA Nanotechnology," *ACS NANO 2, No. 6* (2008), 1089–1096.

[23] Nadrian C. Seeman, "Why You Should Care About Molecular Nanotechnology," http://www.foresight.org/nano/NedSeeman.html

[24] Jianping Zheng, Jens J. Birktoft, Yi Chen, Tong Wang, Ruojie Sha, Pamela E. Constantinou, Stephan L. Ginell, Chengde Mao, and Nadrian C. Seeman, "From Molecular to Macroscopic via the Rational Design of a Self-Assembled 3D Crystal," *Nature 461* (3 September 2009), 74–77.

[25] Jeffrey M. Vargason, Keith Henderson, and P. Shing Ho, "A Crystallographic Map of the Transition From B-DNA to A-DNA," *PNAS 98, No. 13* (June 19, 2001), 7265–7270.

[26] Magic Eye, Inc. and Rachel Cooper, "How to See 3D: Magic Eye 3D and More," http://www.vision3d.com/3views.html

[27] Jianping Zheng, Jens J. Birktoft, Yi Chen, Tong Wang, Ruojie Sha, Pamela E. Constantinou, Stephan L. Ginell, Chengde Mao, and Nadrian C. Seeman, "From Molecular to Macroscopic via the Rational Design of a Self-Assembled 3D Crystal," *Nature 461* (3 September 2009), Supplementary Information: "X-Ray Diffraction Data Processing," 1 (linked to the online version of the paper as a PDF download).

[28] Zheng et al., "From Molecular to Macroscopic," 76–77.

[29] Gedalyah Reback, "The Difference Between 3D Stereograms and Anaglyphs," November 14, 2012, http://www.snapily.com/blog/the-difference-between-3d-stereograms-and-anaglyphs/

[30] Weiss, "A Conversation with Prof. Ned Seeman," 1092.

[31] Yoel P. Ohayon, Arun Richard Chandrasekaran, Esra Demirel, Sabrine I. Obbad, Rutu C. Shah, Victoria T. Adesoba, Matthew Lehmann, Jens J. Birktoft, Ruojie Sha, Paul M. Chaikin, and Nadrian C. Seeman, "Impact of Sticky End Length on the Diffraction of Self-Assembled DNA Crystals," *Journal of Biomolecular Structure and Dynamics 31, Supplement* (2013), 84–85.

[32] Nadrian C. Seeman, "Controlling the Structure of Matter Using the Information in DNA," AAAS 2013 Annual Meeting, http://aaas.confex.com/aaas/2013/webprogram/Paper8533.html

[33] Nina Notman, "Designing 3D DNA Crystals," *Chemistry World* (2 September 2009), http://www.rsc.org/chemistryworld/news/2009/september/02090904.asp

[34] E.L. Doctorow, The Art of Fiction No. 94, *The Paris Review, No. 101* (Winter 1986), http://www.theparisreview.org/interviews/2718/the-art-of-fiction-no-94-e-l-doctorow

[35] Karel Svoboda, Christoph F. Schmidt, Bruce J. Schnapp, and Steven M. Block, "Direct Observation of Kinesin Stepping by Optical Trapping Interferometry," *Nature 365* (21 October 1993), 721–727.

[36] Jonathon Howard, "One Giant Step for Kinesin," *Nature 365* (21 October 1993), 696–697.

[37] Hajin Kim and Taekjip Ha, "Single-Molecule Nanometry for Biological Physics," *Rep. Prog. Phys. 76* (2013), 1–16.

[38] Unpublished kinesin stepping data courtesy of Steven M. Block, Departments of Biology and Applied Physics, Stanford University.

BRIEF INTERLUDE

Back to Methuselah

Nadrian Seeman often begins his invited talks with a slide showing a naked lady. She's *The Dream* by French Post-Impressionist painter Henri Rousseau (Figure I1.1.). Seeman tells his audience that she represents biology, which "is *not* what we're talking about."[1,2] As we know, in the hands of Seeman and others, DNA is even more than the molecule that makes life work, its genetic material. It is a programmable construction tool—programmable in that one can specify the sequence of bases in the DNA and those sequences then determine the structure one creates.

Figure I1.1 Henri Rousseau's painting, The Dream. *(Reproduced with permission from The Museum of Modern Art, Digital Image © The Museum of Modern Art/Licensed by SCALA/Art Resource, New York.)*

Seeman makes no claim to be an artist or an expert in art. But he looks for inspiration in visual art, particularly in paintings and mosaics.[3] He quotes the serpent speaking to Eve in George Bernard Shaw's play *Back to Methuselah*, "You see things; and you say 'Why?' But I dream things that never were; and I say 'Why not?' "[4] Seeman believes that this characterizes much of the scientific enterprise. Most practicing scientists are either asking "Why?" about phenomena in nature or are dreaming up new things and asking "Why not?"[5]

Seeman's imagination was sparked by his recollection of M.C. Escher's woodcut *Depth*. When he went to the campus pub in Albany he was thinking of 6-arm junctions as planar objects with 6-fold symmetry, something along the lines of a snowflake.[6] But when *Depth* flashed into his mind he recognized that the fish in the picture were analogous in their branching to a 6-arm junction: starting from the middle of each fish there is a head, a tail, a top fin, a right fin, a bottom fin, and a left fin. So there are a total of six protrusions that are not planar but rather three-dimensional. Moreover, the fish are organized like the molecules in a molecular crystal—that is, they are arranged in repeating arrays from front to back, from left to right, and from top to bottom. As Seeman recalled, "When I realized that by analogy I could think of the extremities of the fish as nucleic acid double helices, it was a short step to imagine their intermolecular associations being directed by 'sticky ends.'"[7] This was central to the genesis of his subsequent research program: to make interesting and useful molecular structures and topologies from DNA, using its chemical information to control its structure in three dimensions.

Molecular-Scale Weaving

Seeman finds it easy to be inspired by art in his work with DNA. Ideas arise from any depiction of lines that wrap around each other or that form braided and woven patterns. Roman mosaics are among his favorite sources for design concepts. Seeman took the photograph of a Roman mosaic in Conimbriga, Portugal, shown in Figure I1.2. (a).[8] There are double-helical images formed into branched molecules at the upper left. But Seeman was most intrigued by the braided weave on the bottom and the right. Figure I1.2. (b) shows a section of the mosaic highlighting the braided structure. Mosaics such as this led him to wonder if he could make woven structures out of DNA. In fact, he and his colleagues constructed a portion of the woven pattern shown in Figure I1.2, the first step in what Seeman described as "automatic molecular-scale weaving."[9]

To do justice to the discussion in Seeman's paper would require the rigorous definition of a large number of new terms, many of which have their roots in mathematics. For the intended purpose of this *brief* interlude that would not be appropriate. One alternative is to bypass the material without further comment. Another is to ask for the reader's tolerance and touch lightly on Seeman's goal. His purpose was to demonstrate the nanoscale organization of matter into strong molecular structures that display the advantageous properties of analogous macroscopic structures. For example, deliberately braided woven molecules (nanoscopic) that emulate a woven fabric (macroscopic).

So . . . we begin by observing that a half-turn of DNA, about six nucleotide pairs, corresponds to a node or a crossing point in a knot or a catenane.[10] A mathematical knot is slightly different than the knots we're all familiar with. If you take a shoelace and tie a knot in it *and* fasten the ends together, say, with tape, then you've created a mathematical knot. (Elsewhere, Seeman shows the close relationship between knots and catenanes.[11])

Figure I1.2 (a) A Roman mosaic in Conimbriga, Portugal, photographed by Nadrian C. Seeman. **(b)** A section of the mosaic illustrating a braided structure. *(Reproduced with permission from Nadrian C. Seeman, © Nadrian C. Seeman.)*

Mathematical knots are always formed on a closed loop with no loose ends. And if you think of the knot as casting a shadow on the floor (that is, you project the three-dimensional object onto a plane), then the place(s) where you see the shadow of the knot crosses itself is called a crossing or a node. One speaks of positive and negative nodes—in Seeman's case the signs are generated relative to a group of vertical helix axes—and refer to the nature of the crossing, i.e., whether the crossing is an overcrossing or an undercrossing. (Don't fret over this too much. We're trying to navigate Seeman's work while not getting all knotted up in the rigging.)

Take a look at Figure I1.3. Panel (a) is the section of the mosaic we saw in panel (b) of the previous figure. In panel (b) of Figure I1.3 you see that, relative to a vertical helix axis, Seeman assigned a sign to nodes: + (green) or – (yellow) depending on the nature of the crossing. In panel (c), Seeman emphasized the equivalence between a half-turn of DNA (6 nucleotide pairs) and a node in a knot. He replaced node signs by a set of 6 horizontal lines that correspond to base pairs using the same color coding as in the previous panel.

Now consider Figure I1.4. Panel (a) is a schematic example of a deliberately braided structure. Seeman refers to it as a prototype of a braided catenane.[12] Each molecule consists of two cyclic strands, one red and one blue. Each of the crossings corresponds to a half-turn of DNA, similar to the representations in panel (c) of Figure I1.3. The molecule has three vertical double-helix-containing domains (top, center, and bottom). Seeman also furnishes strand polarity indicators: the bow-tie structure on the left represents 3′-3′ linkages while the filled circles on the right represent 5′-5′ linkages. The need to maintain antiparallelism in the DNA strands forming the double-helical segments resulted in the need for these unusual linkages. Such linkages make the synthesis of the strands containing them more involved than the synthesis of strands made up exclusively of 5′-3′ linkages. Panel (b) of Figure I1.4 shows the base sequence and design of one of the braided molecules synthesized for his study.

(a)

(b)

(c)

Figure I1.3 Braided topologies. **(a)** Identical to panel (b) of Figure I.2. **(b)** Shows nodes—green (+) and yellow (–) depending on the nature of the crossing—relative to a vertical axis. **(c)** The equivalence between a half-turn of DNA (6 nucleotide pairs) and a node in a knot is emphasized by replacing node signs by 6 horizontal lines that correspond to base pairs. *(Reproduced with permission from Tanashaya Ciengshin, Ruojie Sha, and Nadrian C. Seeman, "Automatic Molecular Weaving Prototyped by Using Single-Stranded DNA," Angew. Chem. Int. Ed. 50 (2011), 4419–4422.)*

Panel (b) subtly incorporates a handy molecular biology technique that we learned about in Chapter Six. Recall that in the analysis of his DNA cube Seeman used restriction enzymes that recognize specific sequences of DNA called restriction sites. The labels PvuII (in cyan) and HindIII (in magenta) are restriction enzymes that cut the braided molecule at the color-matched restriction sites.

Moors and Crossover Molecules

Since one of Seeman's goals is designing periodic matter, Moorish art has been another favorite resource because it contains examples of interesting topologies in periodic or locally periodic patterns. The Alhambra in Granada, Spain, is a palace and fortress that contains exceptional art and architecture from the centuries during which Moorish monarchs ruled. The pattern from the Alhambra mosaic shown in Figure I1.5 (a) is a

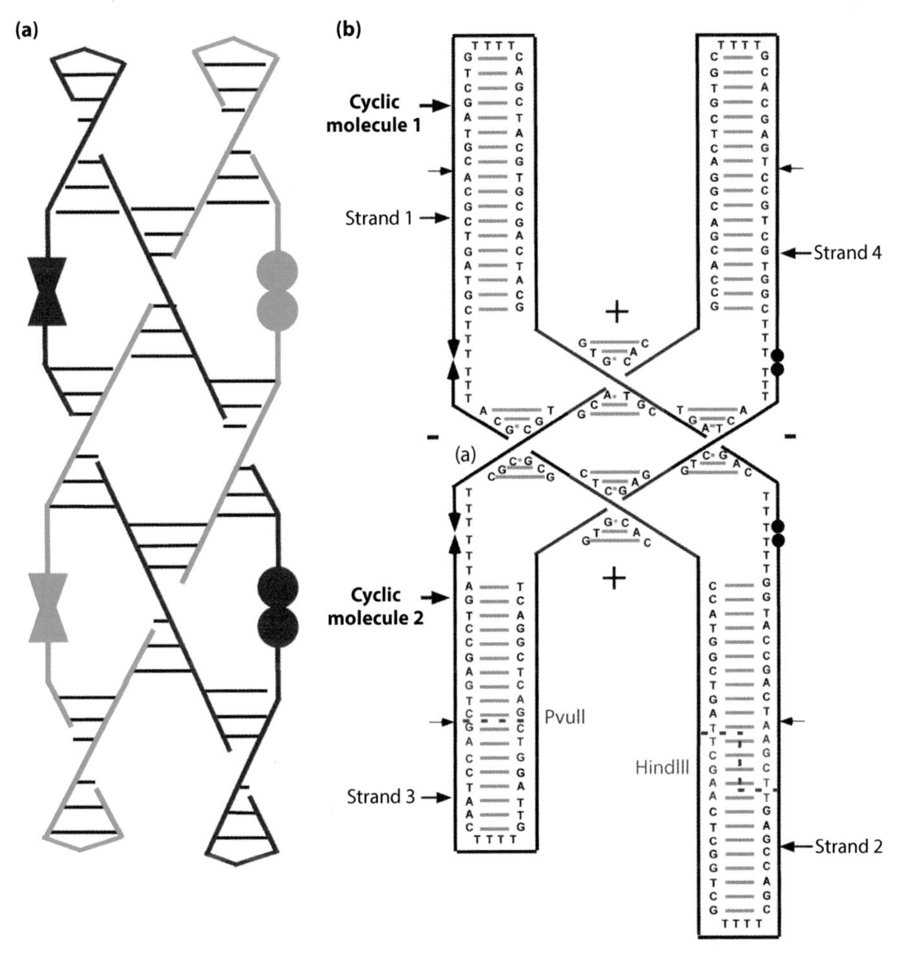

Figure I1.4 (a) Schematic for a braided molecule. **(b)** Base sequence and design for a braided molecule. *(Reproduced with permission from Tanashaya Ciengshin, Ruojie Sha, and Nadrian C. Seeman, "Automatic Molecular Weaving Prototyped by Using Single-Stranded DNA," Angew. Chem. Int. Ed. 50 (2011), 4419–4422.)*

complex catenane with locally periodic features. The pattern shown in Figure I1.5 (b) is the strand structure of a two-dimensional DNA crystalline network built from DNA double crossover molecules, specifically the DAE-based systems we discussed in Chapter Seven. Seeman had this to say, "Both patterns suggest that different species of nucleic acids could be mixed to build novel networks with distinct and possibly useful properties."[13] Now we know of the inspiration the Alhambra provided Seeman as well as M.C. Escher's influence on him. To close the loop, we note that Escher's visits to the Alhambra in the 1920s and 1930s had a strong impact on his own work.[14]

Tensegrity Sculpting

We mentioned in Chapter Nine that sculptor Kenneth Snelson originated the concept of tensegrity. Figure I1.6 (a) is a photograph taken by Seeman of a Snelson creation called *Free Ride Home* located in the Storm King sculpture park in Mountainville, New York.[15]

(a)

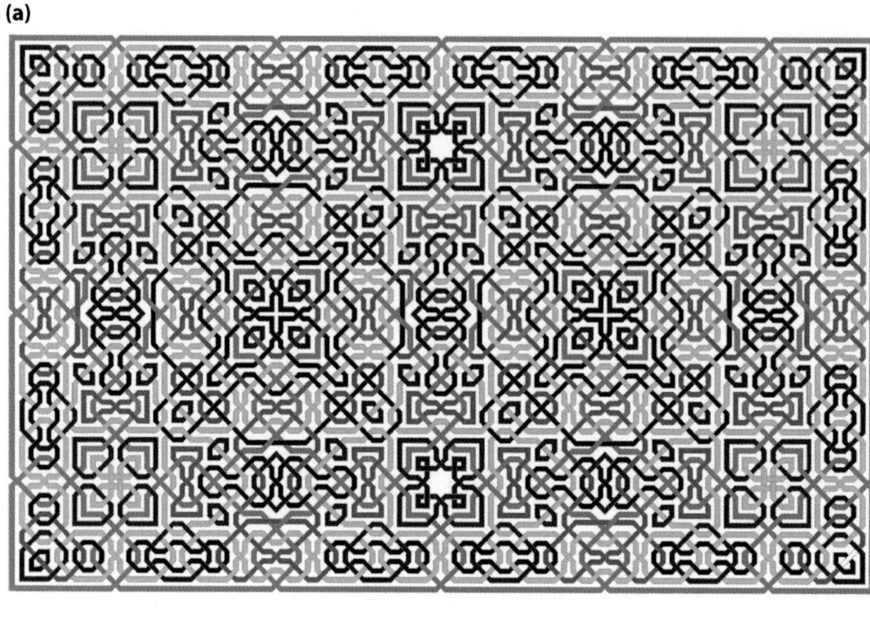

(b)

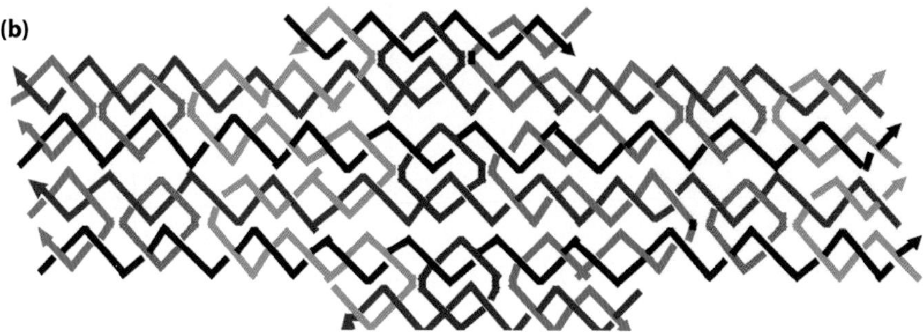

Figure I1.5 Periodic catenanes. **(a)** A locally periodic catenane found in a mosaic from the Alhambra, consisting of numerous linked cycles. **(b)** The periodic DNA pattern called DAE-O, one of the DX-based systems discussed in Chapter Seven. The strands in the DNA pattern are shaded differently to differentiate them for ease of viewing. *(Reproduced with permission from Nadrian C. Seeman, © Nadrian C. Seeman.)*

Snelson uses his idea in many of his artworks, and nanotechnologists other than Seeman have also drawn inspiration from the tensegrity design principle. William Shih of the Wyss Institute at Harvard, along with his colleagues, is one of them. Shih has formed nanoscale three-dimensional tensegrity structures from rigid bundles of DNA double helices and segments of single-stranded DNA that act as tension-bearing cables.[16] In doing so, Shih moved beyond the standard DNA origami methods that rely only on paired bases to provide structural integrity. Instead, he incorporated stretched single-stranded DNA segments as nanoconstruction elements that act in solution as *entropic springs*. The term *entropic spring* refers to the fact that there are fewer ways to arrange the single-stranded DNA the more it is extended. This leads to a decrease in entropy and thus an increase in free energy. The result is that the restoring force on the strand increases as the length increases.[17] Figure I1.6 (b) shows four schematic views of a tensegrity structure (known as a tensegrity prism) and transmission electron micrographs of those structures constructed from DNA by Shih and his colleagues.[18] Assembly is a one-pot reaction as is true for other styles of DNA origami.

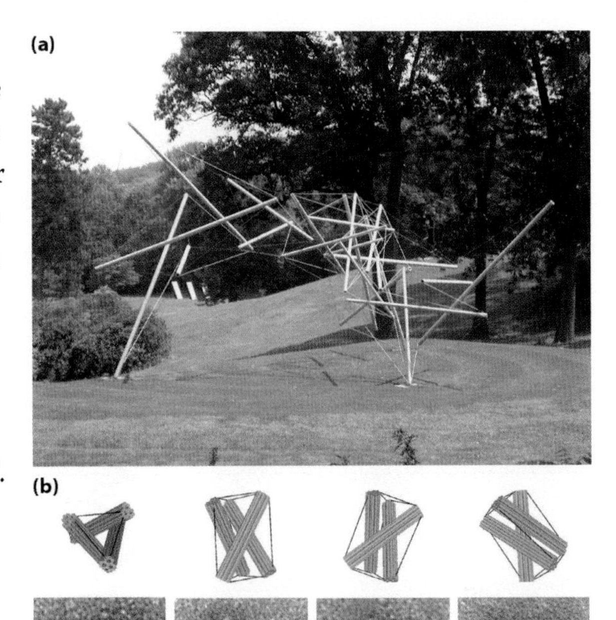

Figure I1.6 (a) Sculpture incorporating tensegrity principles. **(b)** Schematic models of DNA tensegrity prisms and transmission electron micrographs of those structures constructed from DNA. The scale bars are all 20 nm. *(Part **(a)** reproduced with permission from Nadrian C. Seeman, photo © Nadrian C. Seeman; part **(b)** reproduced with permission from Tim Liedl, Björn Högberg, Jessica Tytell, Donald E. Ingber, and William M. Shih, "Self-Assembly of Three-Dimensional Prestressed Tensegrity Structures From DNA," Nature Nanotechnology 5 (July 2010), 520–524.)*

Mayan Pottery, Chirality, and the Handedness of Life

One of the most notable DNA motifs that Seeman has worked with is the PX motif. As we saw in Chapter Nine, this motif looks like two double helices wrapped around each other (although somewhat more complex) as illustrated in Figure I1.7 (b). Figure I1.7 (a) shows a Mayan vessel that contains motifs reminiscent of this system (white rectangle markups). Seeman remarked that the Mayan pottery showed both left-handed and right-handed chiralities. Chirality means handedness. (The word chirality derives from the Greek *kheir* meaning "hand.") Chiral objects are not superposable with their mirror images. If you hold your hands in front of you with the palms facing together you see that your hands are mirror images of one another. If you now turn your hands so that both palms face the same direction, your thumbs will point in opposite directions. Alternatively, when you point your thumbs in the same direction, your palms face opposite directions. Thus your hands are mirror images of one another, but not superposable on one another—each hand is a chiral object.[19] (The word *superimposable* is often used in this context instead of *superposable*. There are differences in the meaning of these two words, but we won't belabor the point.)

Returning to Figure I1.7, the white rectangles in panel (a) highlight motifs of different chiralities. Panel I1.7 (c) details the structure in the rightmost of the two white rectangles and rotates the image by 90° counterclockwise. Chirality is an essential characteristic of

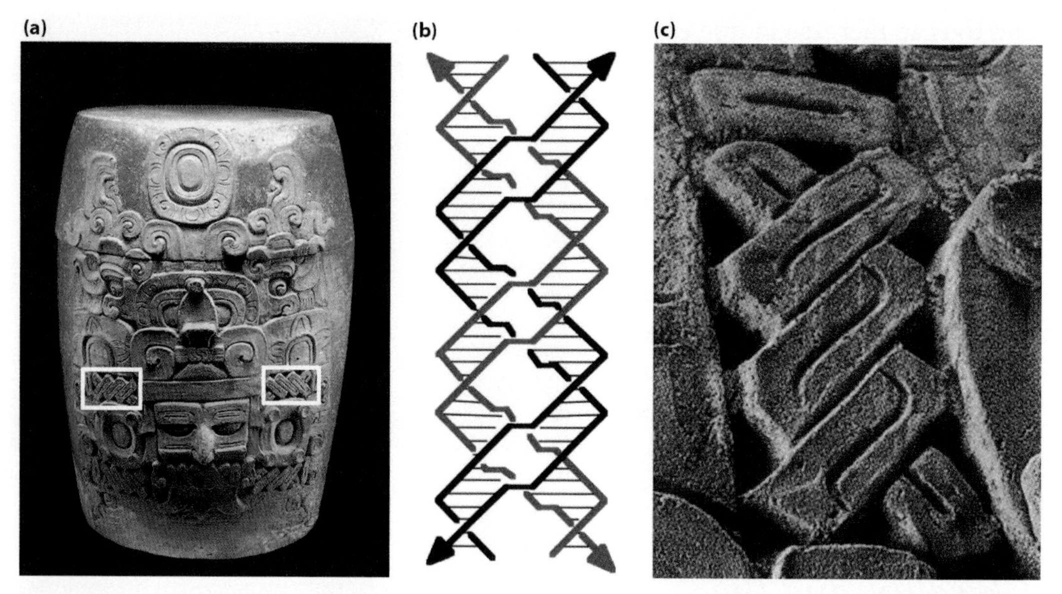

Figure I1.7 (a) Mayan pottery. **(b)** The PX motif in structural DNA nanotechnology. **(c)** Detail of the structure in the rightmost white rectangle of panel (a) but rotated counterclockwise by 90°. *(Adapted with permission from Nadrian C. Seeman, © Nadrian C. Seeman.)*

cellular life, but how a handedness preference emerged in the prebiotic world is unknown. A few exceptions aside, it is a commonplace to say that the nucleic acids RNA and DNA are made up of only right-handed molecules—the right-handed sugars ribose and deoxyribose—and proteins are made up of only left-handed molecules—the left-handed amino acids. This means that the biomolecules of life (DNA, RNA, and proteins) each possess only a single handedness. Thus it is said that life is homochiral. The question is: why? Why should all amino acids be left-handed and all sugars be right-handed? Why not the other way around? An awkward situation in origins research is that life is homochiral and no one is certain why this is so. In the next Interlude, in Part III of this book, we return to the subject of chirality.

Henry Rousseau's painting *The Dream* is about biology. No, that's not what Seeman's structural DNA nanotechnology is about. Nor is the work of Clark and Bellini concerned with biology, *per se*. However, we've seen that disparate aesthetic patterns in art can be linked to branched DNA motifs and constructions. And through art we've come upon the realm of chirality, and chirality is inextricably intertwined with biology and the mystery of life's origins. Seeman commented that not only has DNA structural chemistry been influenced by art, but that in the broadest sense, science and art can be thought of as manifestations of similar types of thinking.[20] In fact, many of the experiments presented in this book suggest that contemporary scientists are doing art, in the old meaning of that word. That is, they are making artifacts, or things that do not naturally exist.[21]

ENDNOTES

[1] Nadrian C. Seeman, "DNA: Not Merely the Secret of Life," Arthur M. Sackler Symposium, National Academy of Sciences, April 11, 2007, http://sackler.nasmediaonline.org/2007/nbm/nadrianseeman/nadrianseeman.html

[2] Ann Finkbeiner, "Crystal Method," *University of Chicago Magazine* (Sept–Oct/2011), http://mag.uchicago.edu/science-medicine/crystal-method

[3] Nadrian C. Seeman, "Art as a Stimulus for Structural DNA Nanotechnology," *Leonardo* 47, No. 2 (2014), 142–149.

[4] George Bernard Shaw, *Back to Methuselah: A Metabiological Pentateuch* (New York: Oxford University Press, 1947), 5.

[5] Seeman, "Art as a Stimulus," 143.

[6] Ibid., 145.

[7] Ibid., 145.

[8] Ibid., 146–147.

[9] Tanashaya Ciengshin, Ruojie Sha, and Nadrian C. Seeman, "Automatic Molecular Weaving Prototyped by Using Single-Stranded DNA," *Angew. Chem. Int. Ed. 50* (2011), 4419–4422.

[10] Nadrian C. Seeman, "The Design of Single-Stranded Nucleic Acid Knots," *Molecular Engineering 2, Issue 3* (September 1992), 297–307.

[11] http://seemanlab4.chem.nyu.edu/knot-cate.html

[12] Tanashaya Ciengshin, "Automatic Molecular Weaving," 4419.

[13] Seeman, "Art as a Stimulus," 147.

[14] Doris Schattschneider, "The Mathematical Side of M.C. Escher," *Notices of the American Mathematical Society 57, No. 6* (June/July 2010), 706–718.

[15] Seeman, "Art as a Stimulus," 149.

[16] Tim Liedl, Björn Högberg, Jessica Tytell, Donald E. Ingber, and William M. Shih, "Self-Assembly of Three-Dimensional Prestressed Tensegrity Structures From DNA," *Nature Nanotechnology 5* (July 2010), 520–524.

[17] Richard Berry, "Biological Molecules 2: Modelling DNA and RNA" (15 January 2013) and http://biologicalphysics.iop.org/cws/article/lectures/48662

[18] Liedl, et al., "Self-Assembly," 521.

[19] http://www.chem.ucla.edu/harding/tutorials/stereochem/id_mole_chiral.html

[20] Seeman, "Art as a Stimulus," 148.

[21] Paolo Garbolino, "What the Scientist's Eye Tells the Artist's Brain," *Art as a Thinking Process*, ed. Mara Ambrožič and Angela Vettese (Berlin: Sternberg Press, 2013), 74.

CHAPTER TEN

DNA Nanotechnology Meets the Real World

We revisit Nadrian Seeman's email interview. Seated at the computer in his comfortably chaotic New York office, he is asked, "Where is DNA likely to show up first in terms of useful technology? How soon could this happen?" Seeman's short answer: "I don't know." Then he adds, "Perhaps in the applications of crystal self-assembly for structural purposes." The interviewer submits several more questions while making reference to the "many, well-hyped promises of nanotechnology." Then he takes another head-on run at the prediction question. He asks, "In terms of technology and anything affected by technology, what will be different about our world in 5 years? In 10? In 50? What will have surprised us in ten years, in fifty?" Seeman does not rise to the bait. His answer is, "I am a scientist, not a futurist. I don't make predictions."[1]

Scientists building the edifice of DNA nanotechnology are optimistic and passionate over the outlook, but they dislike prognosticating. This reticence toward speaking as futurists serves them particularly well when proof-of-concept experiments provide tantalizing hints at real-world applications. In this chapter we present two such experiments, both embryonic efforts to connect—quite literally—with the real world. The first is the construction of synthetic lipid membrane channels using DNA nanostructures.[2] Synthetic membrane channels can punch holes into lipid vesicles and might find a use in gene therapy to inject material through a lipid bilayer membrane and into a cell's interior. This could be an alternative way of accomplishing transfection (getting nucleic acids into cells), a subject we spoke about in connection with Chad Mirkin's research in Chapter Three.

The second experiment uses a robotic device to target cells so that specific cell surface proteins can trigger the release of encapsulated drug payloads.[3] The device is based on DNA origami and is called a molecular nanorobot. So far, it has succeeded in delivering biochemical instructions encoded in antibody fragments that target both lymphoma cell lines and leukemia cell lines in tissue cultures (*in vitro*). The nanorobot has achieved cancer cell growth arrest while leaving healthy cells alone.

Cell Membrane Channels

To begin our examination of synthetic lipid membrane channels, we'll first learn about naturally occurring channels. These pores penetrate lipid bilayer membranes and enable the transport of ions such as sodium (Na^+), potassium (K^+), calcium (Ca^{++}), and chlorine (Cl^-) through the otherwise impermeable membranes. (An ion is an atom or molecule having a net positive or negative charge.) To get a feeling for how important these channels are to life, we look back at a commentary written by eminent physiologist Clay M. Armstrong, M.D., of the University of Pennsylvania. Armstrong wrote, "Ion channels are involved in every thought, every perception, every movement, every heartbeat. They developed early in evolution, probably in the service of basic cellular tasks like energy production and osmotic stabilization of cells, and evolved to underlie the elaborate electrical system that provides rapid perception and control."[4] On a sobering note, mutations in the proteins that form the channel can lead to diseases such as multiple sclerosis and cystic fibrosis. Channel-related diseases are major drug targets.

We first saw a cross section of a cell's membrane structure in the Liquid Crystals Starter Kit in Chapter One. It showed not only the lipid bilayer but also permanently attached proteins. Then in Chapter Four we elaborated on the fluid mosaic model of the cell membrane. There is a certain irony in this understanding of how a cell membrane is constructed. As Roderick MacKinnon of The Rockefeller University, a Nobel Laureate in Chemistry wrote, "The evolution of the lipid cell membrane solved one problem and created another. It enabled compartmentalization of life's essential ingredients, but it made it almost impossible for charged atoms, the ions, to move into and out of cells. . . . Nature had to fashion specific mechanisms to get ions across the membrane. . . . Membrane-spanning proteins called ion channels were nature's solution."[5]

One of the most well-studied channels is the potassium channel.[6] This consists of proteins called K^+ channels that promote rapid diffusion of potassium ions across cell membranes. This movement underlies many fundamental biological processes including electrical signaling in the nervous system. Potassium channels can use different methods of gating—the processes by which the pore opens and closes. Some K^+ channels are ligand-gated, which means that pore-opening transitions are coupled

Cell membrane channels Pores composed of proteins that form passageways to enable ions to pass through lipid bilayer membranes.

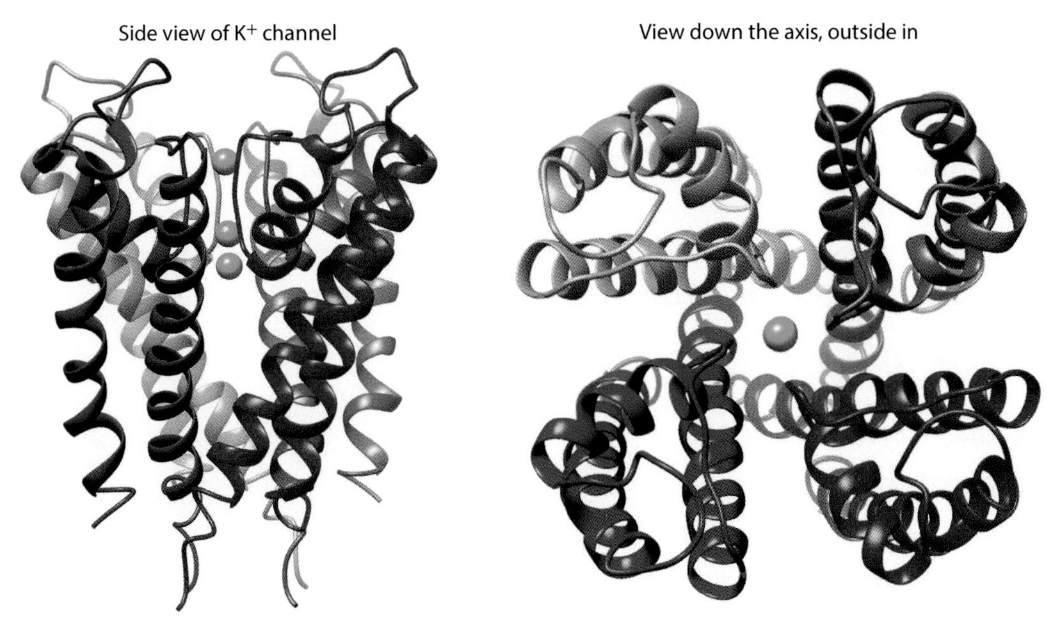

Figure 10.1 Potassium channel (K+) from the bacterium *Streptomyces lividans*. The channel, a transmembrane protein, is a tetramer with four-fold symmetry about the central axis. Each subunit consists of a pair of alpha helices, one lining the central pore and one interacting with the lipid membrane. *(Courtesy of Ray Fort Jr. and Rachel Kramer Green. PDB ID 1BL8 from Doyle, D.A., Cabral, J.M., Pfuetzner, R.A., Kuo, A., Gulbis, J.M., Cohen, S.L., Chait, B.T., Mackinnon, R., "The Structure of the Potassium Channel: Molecular Basis of K+ Conduction and Selectivity," Science 280 (1998), 69–77 and created with Eric F. Pettersen, Thomas D. Goddard, Conrad C. Huang, Gregory S. Couch, Daniel M. Greenblatt, Elaine C. Meng, Thomas E. Ferrin, "UCSF Chimera—A Visualization System for Exploratory Research and Analysis," J Comput Chem 25, No. 13 (25 October 2004), 1605–1612.)*

to the binding of an ion, a small organic molecule, or even in some cases to a protein. Others are voltage-gated by the electrical potential that exists across the membrane.

Figure 10.1 shows a side view and a view down the axis of a K^+ channel.[7] Note that the helices comprising the proteins are called alpha helices and are composed of amino acids (these helices should not be mistaken for single-stranded DNA helices). One alpha helix acts as a voltage sensor that detects the electrical potential across the membrane. Figure 10.2 shows in cross section the potassium channel embedded in the lipid membrane. For a concise, illuminating discussion of pores and channels, specifically potassium channels, the reader can visit the site provided in the endnote.[8]

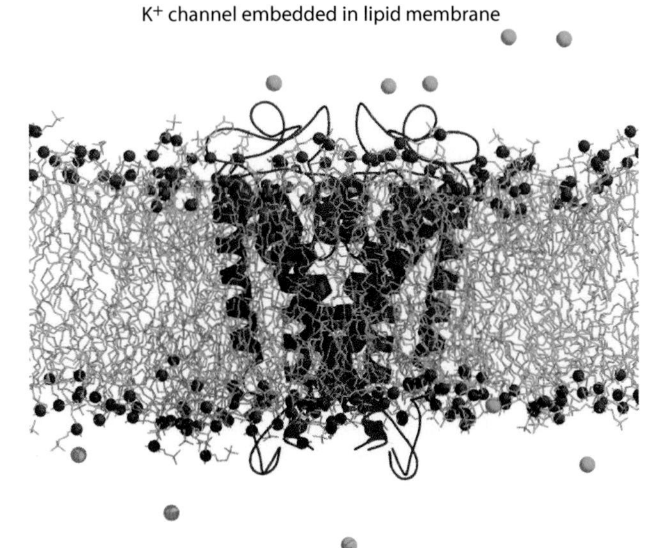

Figure 10.2 Potassium channel is embedded in a lipid membrane. The channel excludes ions smaller than potassium but admits larger ones. This is accomplished by a so-called selectivity filter at the external entrance to the channel. *(Reproduced with permission from Mark S.P. Sansom, Indira H. Shrivastava, Kishani M. Ranatunga, and Graham R. Smith, "Simulations of Ion Channels Watching Ions and Water Move," TIBS 25 (August 2000), 368–374.)*

Included there is an explanation of how the channel's selectivity filter operates and the answer to the question of how it keeps out smaller ions, the "little guys." Conformational changes within the membrane are responsible for pore opening in potassium channels. Inner helices obstruct the pore in the closed state and expand its intracellular diameter in the opened state.[9]

Finally, we must acknowledge that new computer simulation studies have challenged the established view of the details of how the potassium channel functions.[10]

Synthetic Membrane Channels via DNA Nanotechnology

We now turn to synthetic channels fabricated by structural DNA nanotechnology and show how they attempt to mimic the real thing (Figure 10.3).[2] Hendrik Dietz, Friedrich C. Simmel, and colleagues at the Technische Universität München (TUM) in Munich, Germany, reported a synthetic membrane channel constructed entirely from DNA and anchored to a lipid membrane by cholesterol side chains. The shape of their synthetic channel was bio-inspired by the channel protein alpha-hemolysin, a pore-forming toxin from the bacteria *Staphylococcus aureus*. The bacteria secrete alpha-hemolysin monomers that bind to the outer membrane of susceptible cells. The monomers form oligomers (molecular complexes consisting of a small number of

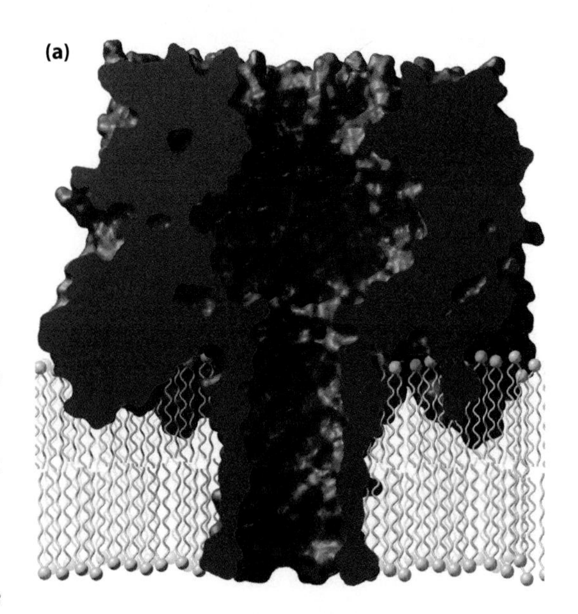

Figure 10.3 (a) Alpha-hemolysin a bacterial toxin that can self-assemble to form a transmembrane channel. **(b)** Synthetic DNA membrane channel made by scaffolded DNA origami and bio-inspired by the alpha-hemolysin transmembrane channel. *(Panel **(a)** adapted with permission from Aleksei Aksimentiev, "Alpha-Hemolysin: Self-Assembling Transmembrane Pore," Theoretical and Computational Biophysics Group, University of Illinois at Urbana-Champaign, http://www.ks.uiuc.edu/Research/hemolysin/; **(b)** reproduced with permission from Martin Langecker, Vera Arnaut, Thomas G. Martin, Jonathan List, Stephan Renner, Michael Mayer, Hendrik Dietz, and Friedrich C. Simmel, "Synthetic Lipid Membrane Channels Formed by Designed DNA Nanostructures," Science 338 (16 November 2012), 932–936.)*

monomers) to create a channel that can damage or destroy the host cell. Figure 10.3 (a) shows a cross-sectional view of a naturally occurring alpha-hemolysin transmembrane channel.[11]

The group at TUM used scaffolded DNA origami to build their membrane channel (panel (b) of Figure 10.3). In its entirety, the channel is composed of 54 double-helical DNA domains packed on a honeycomb lattice. The channel consists of two modules. One is a red stem that penetrates the membrane and extends from the membrane's outer side to its inner side. The other module is a barrel-shaped cap that adheres to the outside of the membrane. Cylinders (both silver and red) represent double-helical DNA domains. Orange strands with orange ellipsoids depict cholesterol-modified oligonucleotides. These modified DNA strands hybridize to other single-stranded DNA called adaptor strands to form duplex DNA. Adhesion of the channel to the lipid bilayer is promoted by 26 hydrophobic cholesterol moieties that protrude from the portion of the barrel adhering to the outside of the membrane. (A moiety is a distinct part of a larger molecule.) The stem protrudes centrally from the barrel and consists of six, red double-helical DNA domains that, taken together, form a hollow tube. Overall, the appearance of the naturally occurring channel to the synthetic channel is striking.

It is the interior of this tube that acts as a transmembrane channel. The tube has an interior diameter of about 2 nm and a length of about 42 nm that runs through both stem and barrel (parts (a) and (b) of Figure 10.4). The TEMs in Figure 10.4 (c) are taken from purified structures and confirm that the intended shape is realized. As shown in Figure 10.5, the group used small unilamellar lipid vesicles (known as SUVs) to show that the synthetic DNA channels bind to lipid bilayer membranes in the desired orientation. The cholesterol-modified face of the barrel forms a tight contact with the membrane; the stem appears to protrude into the lipid bilayer (panel (b) of Figure 10.5). Regarding the small

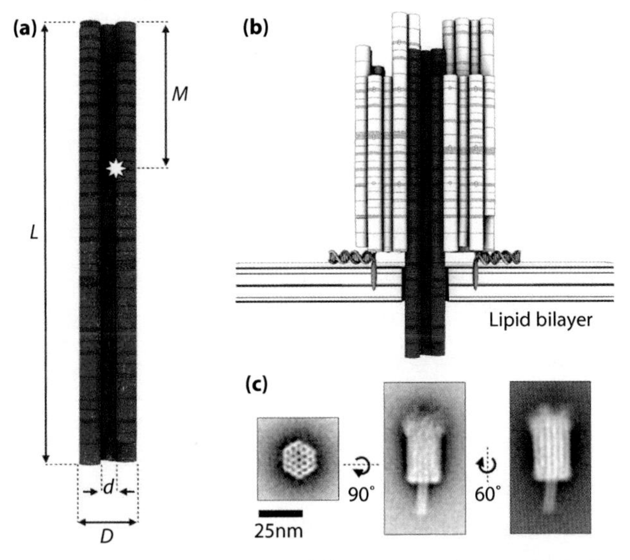

Figure 10.4 (a) Dimensions of the transmembrane channel. Length L = 47 nm, tube diameter D = 6 nm, inner diameter d = 2 nm. The length of the central channel fully surrounded by DNA helices is 42 nm. The star symbol shows the position of a 7-base strand extension acting as a defect in channel mutants, as discussed in the text. **(b)** Cross-sectional view through the channel when incorporated in a lipid bilayer. **(c)** Averaged negatively-stained TEM images obtained from purified DNA channel structures. *(Reproduced with permission from Martin Langecker, Vera Arnaut, Thomas G. Martin, Jonathan List, Stephan Renner, Michael Mayer, Hendrik Dietz, and Friedrich C. Simmel, "Synthetic Lipid Membrane Channels Formed by Designed DNA Nanostructures," Science 338 (16 November 2012), 932–936.)*

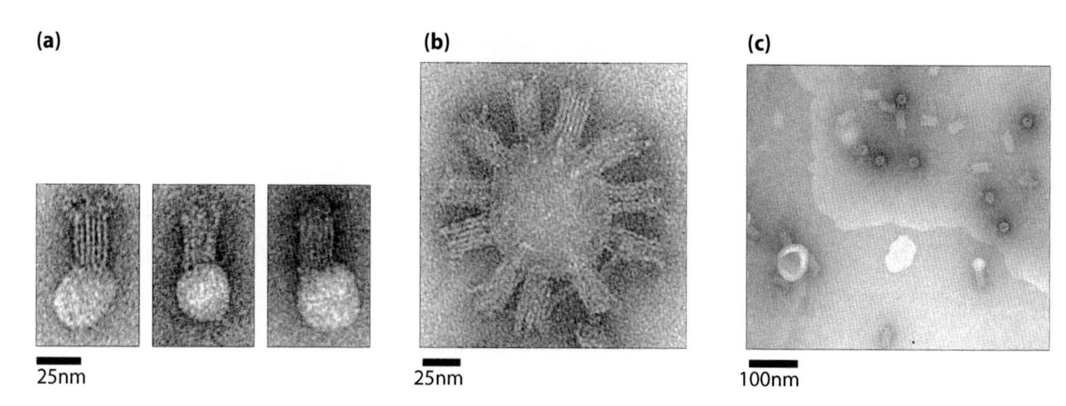

Figure 10.5 (a) TEM images of a DNA channel adhering to a single small unilamellar vesicle (SUV). **(b)** Multiple channels adhering to a single SUV and **(c)** multiple channels binding to an extended lipid bilayer. *(Reproduced with permission from Martin Langecker, Vera Arnaut, Thomas G. Martin, Jonathan List, Stephan Renner, Michael Mayer, Hendrik Dietz, and Friedrich C. Simmel, "Synthetic Lipid Membrane Channels Formed by Designed DNA Nanostructures," Science 338 (16 November 2012), 932–936.)*

unilamellar lipid vesicles, they're artificially prepared vesicles (cavities) composed of a lipid bilayer. They are said to be *uni*lamellar because they have only one lipid bilayer. This is distinct from a micelle (Figure 4.7 of Chapter Four) that has only a lipid *mono*layer.

Current Gating

To demonstrate the electrical conductivity of the synthetic membrane pores, the group used a technique called single-channel electrophysiology, a description of which would lead us too far astray. However, we note that this procedure measures currents in the picoampere range. A picoamp (pA) is 10^{-12} amps (one trillionth of an ampere). By comparison, a typical household circuit may carry 15–20 amperes. Having established that the channel was electrically functional, the TUM group then looked for the current gating behavior that is characteristic of natural ion channels, where it is caused by switching between distinct channel conformations with different conductances. (Conductance is the degree to which an object conducts electricity. It is the reciprocal of the object's electrical resistance.)

> **Ion Channel Current Gating** Gating describes conformational changes (i.e., changes in shape or structure) in an ion channel. For example, a voltage can induce these changes. Gating currents result from the movement of small electrically charged molecules in the ion channel. The current will change as the channel conformation changes.

We provide an example of (voltage-controlled) gating in naturally occurring channels in the multiple current traces in panel (a) of Figure 10.6. These are single-channel measurements of calcium channels (work that does originate with the group at TUM).[12] The pattern of electrical activity is typified by brief openings of the channel occurring in clusters with occasional measurements dominated by long-lasting openings. The channel openings produce current pulses of about 1.2 pA owing to the passage of small ions.

The synthetic channels do not have this sophisticated natural gating mechanism. However, the Dietz group did find gating behavior in the synthetic channels that they

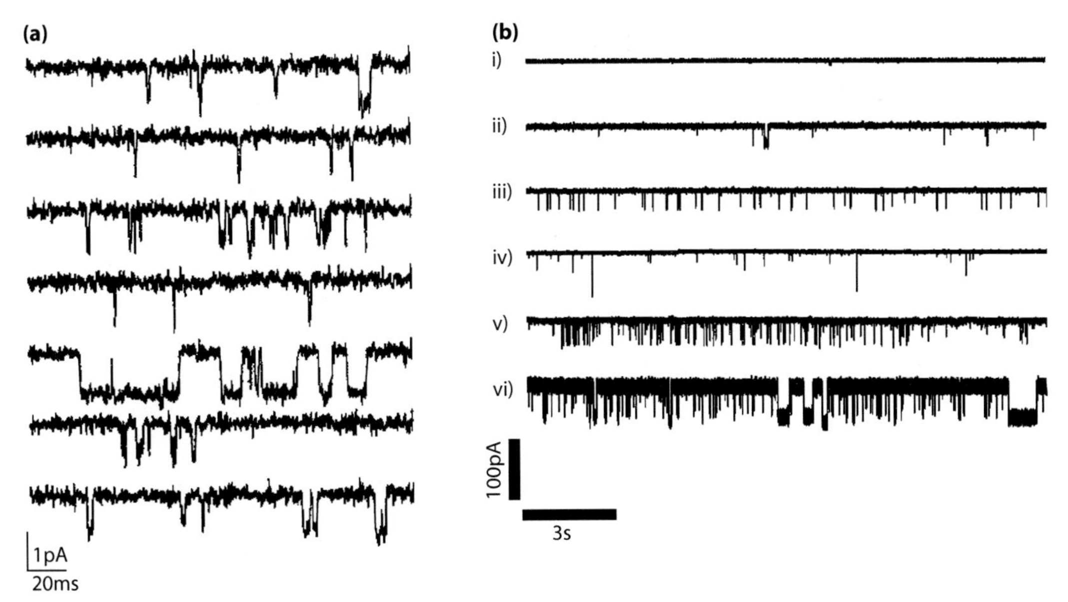

Figure 10.6 (a) Current gating behavior in naturally occurring calcium ion channels (not from the Dietz group). The data are plotted as current (in picoamps) versus time (in milliseconds). **(b)** Synthetic DNA channels display gating behavior as shown by these traces of current (vertical axis) versus time (horizontal axis). The vertical displacements in the current from the steady state are the signature of channel gating events. Traces (i)–(iii) are from original, unmodified channels while (iv)–(vi) are from "mutant" or modified channels, as discussed in the text. *(Panel **(a)** reproduced with permission from B. Nilius, P. Hess, J.B. Lansman, R.W. Tsien, "A Novel Type of Cardiac Calcium Channel in Ventricular Cells," Nature 316 (1 August 1985), 443–446; **(b)** adapted with permission from Martin Langecker, Vera Arnaut, Thomas G. Martin, Jonathan List, Stephan Renner, Michael Mayer, Hendrik Dietz, and Friedrich C. Simmel, "Synthetic Lipid Membrane Channels Formed by Designed DNA Nanostructures," Science 338 (16 November 2012), 932–936.)*

believe might be caused by thermal fluctuations of the structure. This is shown in panel (b) of Figure 10.6 where the traces were recorded at a bias voltage of 100 mV.[13]

They hypothesized that stochastic (that is, random or probabilistic) unzipping and rezipping of short double-helical domains in the channel might also contribute to the current gating they observed. To test this idea, they designed what they called channel mutants or modified channels. These differed from the originals by a single-stranded heptanucleotide (seven nucleotides) that acted like a defect and protruded from the central transmembrane tube. Supportive of their hypothesis, these mutant channels showed more pronounced gating than did the originals as seen in Figure 10.6 (iv)–(vi). They also significantly differed in their gating time statistics (not shown) and every mutant channel displayed gating whereas some of the originals showed none at all as seen in panel (i). Thus the group concluded that the transmembrane current depended on fine structural details of the synthetic DNA channel.

Channels as Single-Molecule Sensors

We're going to consider one more line of work from the TUM group. They performed a series of experiments on synthetic pores to show their potential use as sensing devices of single molecules. Single-molecule sensing with natural pores is a technology that is presently coming of age. A specific example of such sensing is the sequencing of DNA using nanopores together with DNA polymerase. We'll provide some background on this line of research before considering how the synthetic channel may offer advantages. Figure 10.7 shows the idea.[14]

Pause for a moment. Think back to Chapter Two. Frederick Sanger's gene sequencing technique was bio-inspired by DNA polymerase, the enzyme that replicates DNA. Now we've come back to DNA polymerase—this time as a tool for sequencing DNA.

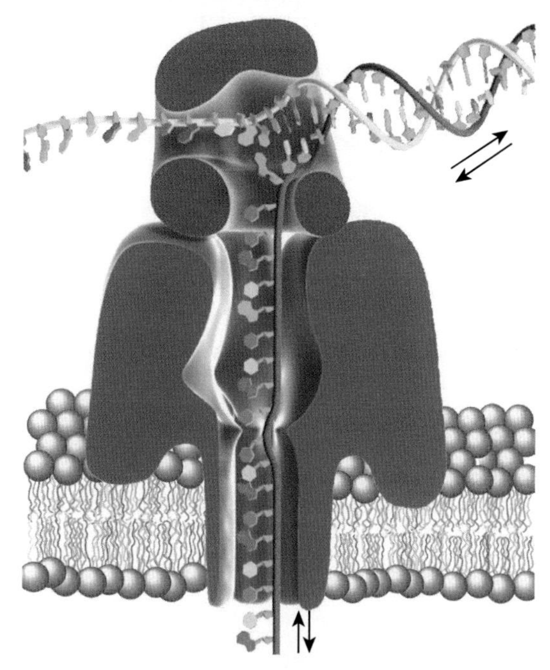

Figure 10.7 Nanopores for DNA sequencing. DNA inserted in a nanopore, with speed control provided by a DNA polymerase (brown). An alpha-hemolysin nanopore (gray) is embedded in a lipid bilayer. The DNA template (red backbone) is inserted into the pore by an applied electric field. Its motion into or out of the pore (arrows) can be controlled by the applied electric field and the polymerase activity. DNA sequence information is obtained by changes in the ionic current running along the DNA through the nanopore, which occur as the DNA is ratcheted through the pore by the polymerase. *(Reproduced with permission from Grégory F. Schneider and Cees Dekker, "DNA Sequencing with Nanopores," NatureBiotechnology 30, No. 4 (April 2012), 326–328.)*

The goal in using nanopores to help identify bases is to sequence long strands of DNA and to do so rapidly. The basic concept is to force a DNA molecule through a nanoscale pore in a membrane (using an applied voltage) and read off each base when it reaches the narrowest constriction of the pore. This is done by using modulations of the ion current passing through the pore to identify the base. The principle of operation is the transient blockage of ionic current by the molecules that are translocated (moved from one place to another). Researchers employ an imaginative strategy in order to slow the translocation speed down (from about one-millionth of a second per base) to be able to reliably identify individual bases. The trick is to use DNA polymerase to ratchet a DNA template strand through a nanopore in single-nucleotide steps while complementary DNA is being synthesized by the polymerase. Blocking oligomers provide further control of the process.[15,16] These blocking oligomers come into play because the polymerase is bound to the single-stranded DNA template strand (containing the sequence to be read). And the strand itself is bound by hybridization to a primer. Figure 10.8 tries to break down this sophisticated, hypothetical manipulation into simpler steps.

As the polymerase extends the primer to synthesize double-stranded DNA, it acts like a motor that ratchets the DNA template through the pore. (Check out Figure 2.7 in Chapter Two to reacquaint yourself with primers.) The blocking oligomer protects the

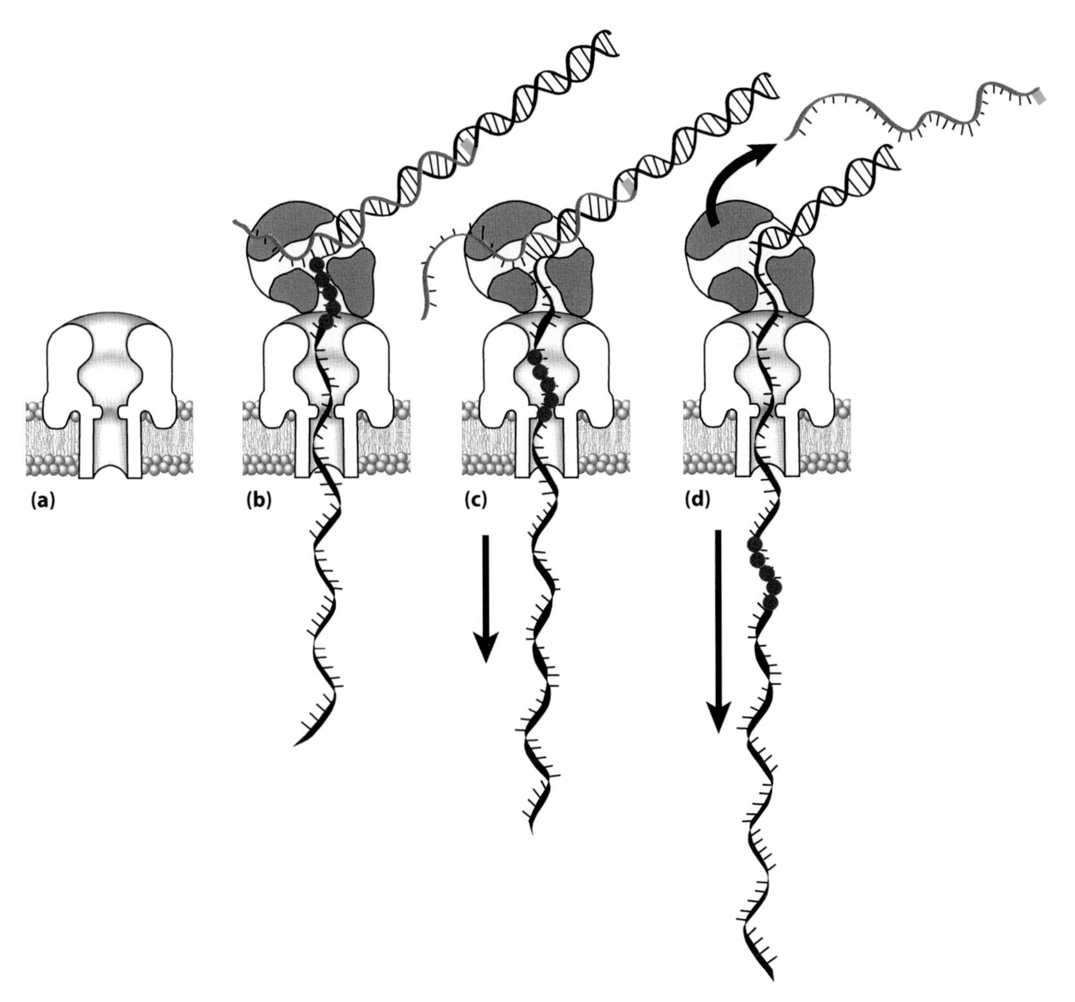

Figure 10.8 Forward ratcheting of DNA through a nanopore (conjectured process): **(a)** the open channel. **(b)** Nanopore capture of a polymerase-DNA complex with a blocking oligomer (red line) bound to it. **(c)** Mechanical unzipping of the blocking oligomer promoted by the applied voltage, which ratchets the DNA template forward through the nanopore. **(d)** Release of the blocking oligomer, which exposes the 3′ end of the DNA primer within the polymerase active site. (Dots are added for clarity.) *(Adapted with permission from Gerald M. Cherf, Kate R. Lieberman, Hytham Rashid, Christopher E. Lam, Kevin Karplus, and Mark Akeson, "Automated Forward and Reverse Ratcheting of DNA in a Nanopore at 5-Å Precision," Nature Biotechnology 30, No. 4 (April 2012), 344–348.)*

3′ end of the primer in solution and allows the template to be positioned in the pore. After the single-stranded end of the template goes through the pore, the blocking oligomer is mechanically unzipped by the applied voltage, thereby allowing DNA synthesis to proceed (Figure 10.8).[16] Ionic current traces have been made revealing current blockades that are consistent with the conjectured ratcheted passage of the DNA through the pore (we will not show these current traces). A current blockade is a transient reduction of the ionic current through the pore that is produced when large molecules, e.g., DNA, impede the flow of small ions.[17]

We call the reader's attention to the fact that endnote 16, a publication on DNA sequencing using naturally occurring nanopores, employs none other than the channel protein alpha-hemolysin that the TUM group used for bio-inspiration.

Oligomers Molecular complexes that consist of a small number of monomers.

DNA sequencing with nanopores A developing technique using naturally occurring protein nanopores to read DNA base sequences. This approach may offer advantages in speed, cost, and length of strands read—compared to present methods. Synthetic nanopores constructed by DNA nanotechnology might possibly be used for this purpose.

Current blockade A transient reduction of the ionic current through a nanopore that is produced when large molecules such as DNA impede the flow of small ions.

The synthetic lipid membrane channel/pore constructed by scaffolded DNA origami might be used for the same purpose as the natural channels. The TUM group pointed out that altering the geometry of biological pores and introducing chemical functions through genetic engineering or chemical conjugation are challenges that confront the users of naturally occurring membrane pores as nanopore sensors. By contrast, the geometry of synthetic DNA objects as well as their chemical properties can be tailored for custom nanopore sensing applications.

To establish the relevance of the synthetic channels, the group did two sets of experiments that showed current blockades similar to those observed with the natural pores. We will discuss one of these experiments in which they added hairpins to single-stranded DNA and studied current blockades attributable to hairpin unzipping (Figure 10.9). Single-stranded DNA is expected to fit through the 2 nm central pore of the synthetic channel, but secondary structures such as hairpins are not. To translocate through the channel, the secondary structures must unzip. This should show up as a characteristic time delay in the current blockades. They used a hairpin with a stem made of nine base pairs flanked by 50 thymidine molecules on the 3′ end and 6 thymidines on the 5′ end. (Thymidine is the familiar base thymine (T) combined with deoxyribose.) The DNA with a hairpin is shown in the center of a channel in the drawings in panel (a).

The hairpin molecules were initially added to the outside of a lipid membrane containing a single synthetic DNA channel that displayed a stable current baseline

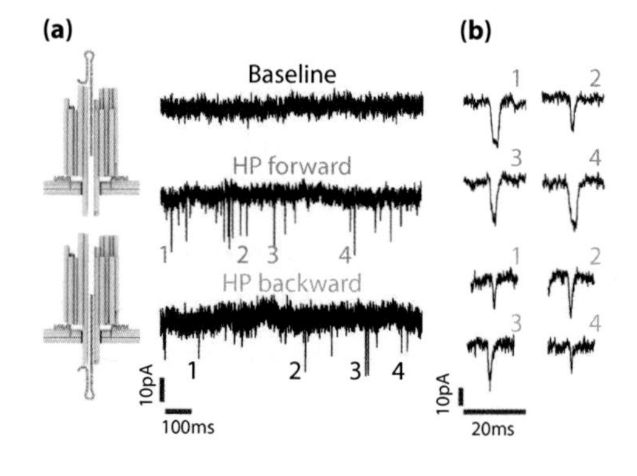

Figure 10.9 DNA translocation studies using synthetic lipid membrane channels. The addition of hairpins to a single strand of DNA results in the appearance of current blockades attributed to the unzipping and translocation of hairpin molecules from one side of the membrane to the other. This process can be performed in either direction (hairpin forward or hairpin backward), as is also observed in natural channels. The transient current blockades are shown in panel **(a)**. Panel **(b)** shows representative blockade events on an expanded time scale for forward (top, 1–4) and backward (bottom, 1–4) translocation of DNA hairpins. *(Reproduced with permission from Martin Langecker, Vera Arnaut, Thomas G. Martin, Jonathan List, Stephan Renner, Michael Mayer, Hendrik Dietz, and Friedrich C. Simmel, "Synthetic Lipid Membrane Channels Formed by Designed DNA Nanostructures," Science 338 (16 November 2012), 932–936.)*

without gating (current trace labeled Baseline in Figure 10.9). Then a voltage was applied leading to the capture, unzipping, and translocation of the hairpin structures. This was detected as transient current blockades as shown in the middle current trace of Figure 10.9. After about 30 minutes the direction of the applied voltage was reversed, again leading to transient current blockades from unzipping and translocations in the opposite direction. That is, molecules that had accumulated inside the lipid membrane by previous translocations through the channel were now going backward through the channel. Panel (b) shows representative blockade events for forward (top) and backward (bottom) translocation of DNA hairpins.

So there you have it. One attempt to marry constructions made by structural DNA nanotechnology to the real world. The authors of the study believe that fully synthetic lipid membrane channels are only a first step. They anticipate harnessing ion transfer for driving complex nanodevices, bio-inspired by the rich diversity of natural membrane machines such as ion pumps, rotary motors, and transport proteins.

Molecular Nanorobots Built by DNA Origami: Cell-Targeted Drug Delivery

Next up: molecular nanorobots for cell-targeted drug delivery. A group at the Wyss Institute for Biologically Inspired Engineering at Harvard University built an autonomous DNA nanorobot using scaffolded DNA origami.[3] In contrast to the non-autonomous walkers we saw in the previous chapter, this nanorobot is autonomous. Thus, once the nanorobot is programmed it carries out its tasks without further operator intervention. The nanorobot is capable of transporting molecular payloads to cells. It then senses cell surface inputs and determines if the inputs are appropriate. If so, they trigger activation of the robot which then reconfigures its structure for payload delivery (Figure 10.10). The device can carry a variety of materials and is controlled by an aptamer-encoded logic gate that enables it to respond to an array of cues.

Aptamers are short amino acid polymers (or they can be single-stranded nucleic acids) that recognize and bind to targets with high affinity and selectivity. The word comes from the Latin *aptus* meaning "to fit."[18]

The logic gate used in this study is known as a logical AND. Think of it in terms of locks and keys—specifically, a lock that has places for two keys. The AND means that both keys must fit in the lock for the lock to open. The design of the nanorobot incorporated complementary aptamer DNA duplexes on the left (small dotted box in panel (a)) and right sides of the barrel-shaped nanorobot. The aptamer strands are attached to one domain ("domain" refers to either the top or bottom of the barrel) and partially complementary aptamer strands are attached to the other domain. When both aptamers recognize their targets, the lock duplexes dissociate, and the nanorobot

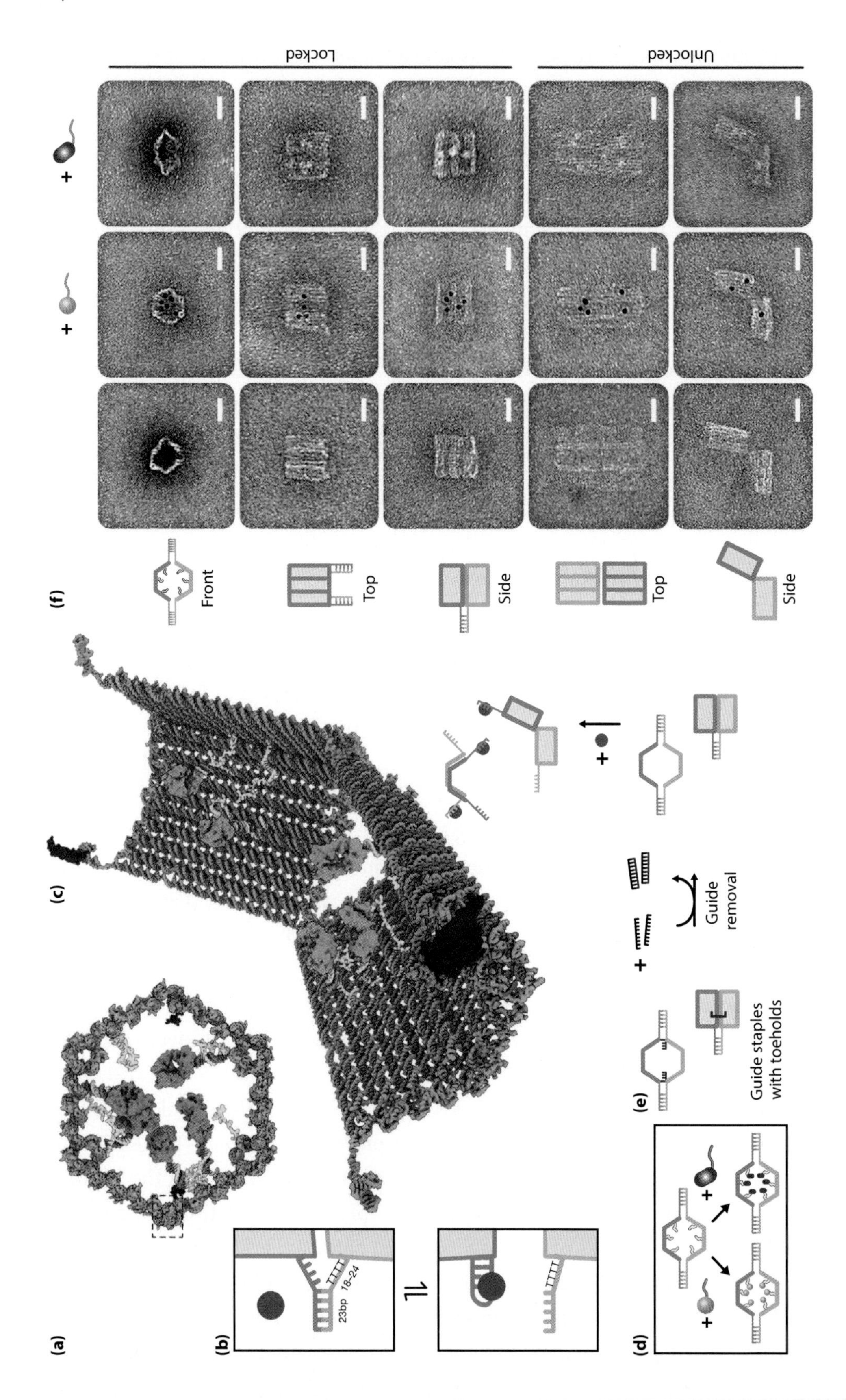
(a)
(b)
23bp
18–24
TTTT
(c)
(d)
(e)
Guide staples
with toeholds
Guide
removal
(f)
Front
Top
Side
Top
Side
Locked
Unlocked

undergoes a dramatic reconfiguration to expose its previously sequestered surfaces. The change is evident by comparing the closed nanorobot as seen in panel (a) to the perspective view of the opened nanorobot in panel (c). This is the AND gate we spoke about, now in the context of the aptamers.

Aptamers Short single-stranded nucleic acids (or amino acid polymers) that recognize and bind to molecular targets with high affinity and selectivity.

AND gate Thinking in terms of locks and keys, both aptamer locks must be open simultaneously to activate the robot. The robot remains inactive when only one of the two aptamer locks is open. The keys to these locks are the cell surface antigens. When the locks open in response to the keys, the nanorobot can deliver its antibody (or gold nanoparticle) cargo.

Researchers Ido Bachelet, George M. Church, and Shawn M. Douglas at Harvard's Wyss Institute used scaffolded DNA origami to create the nanorobot in the form of a hexagonal barrel measuring 35 nm x 35 nm x 45 nm. The top and bottom domains of the barrel (blue and orange) are covalently attached in the rear by single-stranded scaffold hinges. The domains can be non-covalently fastened in the front by single-strand staples modified with DNA aptamer-based locks as we discussed.

Harvard's team performed a one-pot reaction: 196 oligonucleotide staple strands directed a 7,308-base scaffold strand (derived from a virus) into its target shape. To operate the device in response to proteins, the group designed a DNA aptamer-based lock mechanism that opens in response to the binding of antigen keys, panel (b) of Figure 10.10. The cartoon in panel (c) shows the nanorobot opened by protein displacement of aptamer locks while the two domains remain constrained in the rear by scaffold hinges.

Twelve payload attachment sites were arranged in an inward-facing ring in the middle of the barrel to enable different payload orientations and spacings. The attachment sites are oligonucleotide staple strands with 3′ extensions that are complementary to a linker sequence coupled to the intended cargo (panel (d)).

Two types of cargo were loaded (by passive diffusion from solution): 5-nm gold nanoparticles covalently attached to linker molecules and various Fab′ antibody fragments, also attached to linkers. Antibodies are proteins produced by the body's immune system

---

Figure 10.10 (a) The front of a closed nanorobot loaded with a protein payload. Two DNA-aptamer locks fasten the front of the device on the left (shown in dotted box) and right. Payload attachment sites are shown in yellow and the payload in pink. **(b)** Aptamer lock mechanism, consisting of a DNA aptamer (blue) and a partially complementary strand (orange). The lock can be stabilized in a dissociated state by its antigen key (red), which is a protein. **(c)** Perspective view of the nanorobot opened by protein displacement of aptamer locks. The two domains of the opened nanorobot (blue, top and orange, bottom) are constrained in the rear by scaffold hinges. **(d)** Payloads such as gold nanoparticles (gold) and antibody Fab′ fragments (magenta) loaded in the nanorobot. **(e)** Front and side views show guide staples (red) having 8-base toeholds that aid assembly of nanorobot; after folding, guide staples are removed by addition of fully complementary oligos (black) and nanorobots can then be activated by interaction with antigen keys (red). **(f)** Negatively stained TEM images of the nanorobot in both open and closed configurations, with and without cargo (note the cartoons to the left and the top). Left column, unloaded; center column, robots loaded with 5-nm gold nanoparticles; right column, robots loaded with Fab′ fragments. The tiny black dots are gold nanoparticles; the small white spots are Fab′ fragments. Scale bars are 20 nm. *(Reproduced with permission from Shawn M. Douglas, Ido Bachelet, and George M. Church, "A Logic-Gated Nanorobot for Targeted Transport of Molecular Payloads," Science 335 (17 February 2012), 831–834.)*

in response to the presence of a foreign substance called an antigen. The meaning of the term Fab′ is this: the F stands for fragment and the ab stands for antigen. Thus, Fab denotes a portion of a protein called an antibody that binds to an antigen (another type of protein). The prime (′) at the end of the word means that this fragment contains more amino acids than the Fab alone.

Antibody Proteins produced by the body's immune system in response to the presence of a foreign substance called an antigen.

Antigen Any substance that causes the body's immune system to produce antibodies against it. Antigens may have entered the body from the outside or may have been generated within the body.

Panel (f) of Figure 10.10 shows negatively stained TEM images of the nanorobot in both the closed and open configurations, with and without cargo. Look at the cartoons to the left and the top of the figure for guidance. Manual counting showed that on average four attachment sites within the nanorobot were populated when loading gold nanoparticles and three sites were populated when loading antibody fragments. In the left column the nanorobot is not loaded with any cargo. The cargo is visible in the second and third columns of the figure. The gold nanoparticles are electron-dense and show up as black spots in the middle column of TEM images. The antibody fragments are much less electron dense and negative staining is used so they show up as white spots (right column of panel (f)).

When we spoke of DNA bricks the issue of yield came up. Yield has been important from the inception of the field (beginning with Seeman's research) although we did not draw attention to it. In the present case, the researchers wanted to ensure assembly of the device to high yield in its closed state. One obstacle they needed to overcome was that the robot is essentially a spring-loaded device, one that has a bias to be open.[19] To address this, they incorporated two guide staples adjacent to the lock sites that span the top and bottom domains of the nanorobot. (We refer the reader to the online Supporting Material, Figure S15, for beautiful illustrations of the guide staples.)[20] The guide staples could be readily removed after origami folding as shown in panel (d). As assessed by manual counting of nanorobot images by TEM, folding with the aid of guide staples was hugely successful and increased the yield of closed robots from 48% to 97.5%. So the researchers were able to more than double the number of (desired) closed robots as a percentage of the total number of robots produced when they used guide staples.

Tests of Nanorobot Function

We've presented the design of this DNA origami structure and the way in which it's supposed to function. Now we look at tests of robot function. The team selected a payload such that robot activation would be coupled to labeling of an activating cell (Figure 10.11). Fluorescent molecules play a reporter role here as we've seen in other contexts. FITC (fluorescein isothiocyanate) is the fluorescent molecule used in this study. Robots loaded with fluorescently labeled antibody fragments against human leukocyte antigen (HLA)-A/B/C (an essential element in immune function) were mixed with different cell types expressing human HLA-A/B/C and various "key" combinations (to be described). The nanorobots remain inactive in the presence of key$^-$ cells (middle), but activate on engaging key$^+$ cells (bottom). The results were analyzed by flow cytometry.

Flow cytometry Flow cytometry is a technology used to measure and analyze multiple physical characteristics of cells as they flow in a fluid stream through a laser beam. The light that is scattered in several directions is captured and analyzed by software to uncover cellular properties such as size and internal complexity.

Flow cytometry is a widely used method to measure and analyze multiple physical characteristics of individual cells. The combining form *cyto-* is derived from the Greek *kytos* meaning "hollow," as a cell or container. In flow cytometry, thousands of cells per second in a fluid stream move through a laser beam in an ordered fashion. The light that's scattered in several directions is captured and analyzed by software to uncover cellular properties such as size and internal complexity.[21]

As shown in Figure 10.11 (row b), in the absence of the correct combination of keys, the robot remained inactive. In the inactive state, the sequestered antibody fragments were not able to bind to the cell surface, resulting in a baseline fluorescence signal (far right of the row (b) of the figure). However, when the robot encountered the proper combination of antigen keys, it was free to open and bind to the cell surface via its

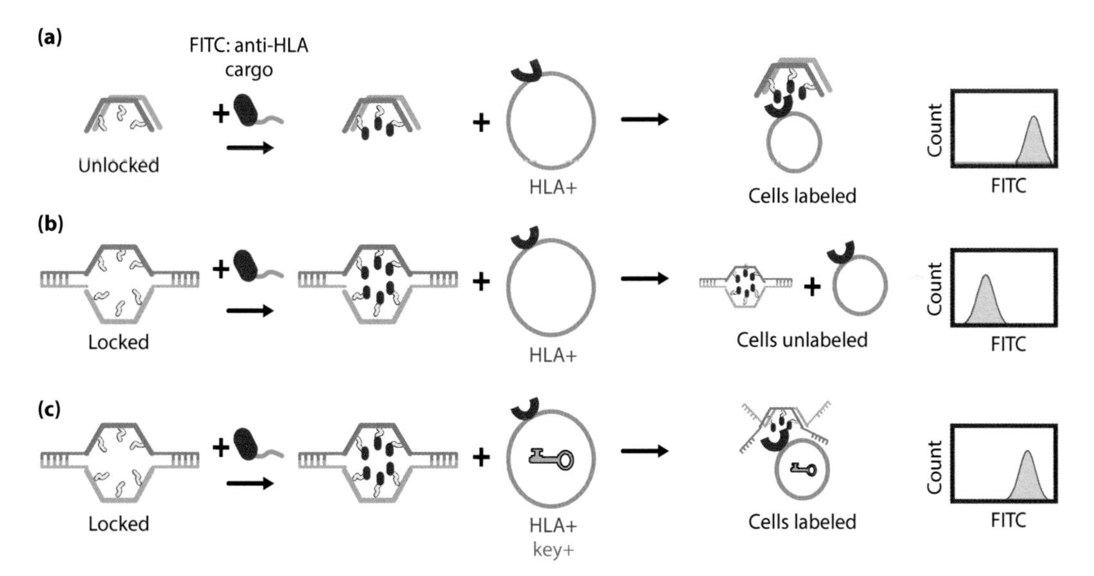

Figure 10.11 Robots were loaded with fluorescently labeled antibody Fab′ fragments to human HLA-A/B/C. **(a)** In their unlocked state, robots will bind to any cell expressing the HLA-A/B/C antigen. **(b)** Robots remain inactive in the presence of key$^-$ cells. **(c)** They activate upon engaging key$^+$ cells. *(Reproduced with permission from Shawn M. Douglas, Ido Bachelet, and George M. Church, "A Logic-Gated Nanorobot for Targeted Transport of Molecular Payloads," Science 335 (17 February 2012), 831–834.)*

antibody payload, causing an increase in fluorescence (cartoon in the far right of the top and bottom rows of the figure).

The Harvard group upped the ante by testing the robots against six different cell types expressing various combinations of antigen keys. To understand Figure 10.12 it's important to keep in mind the AND gate mechanism. The robot could be programmed to activate in response to a single type of key by using the same aptamer sequence in both lock sites. Alternatively, different aptamer sequences could be encoded in the locks to recognize two different inputs. Both locks needed to be opened simultaneously to activate the robot. The robot remained inactive when only one of the two locks was opened. Thus, the lock mechanism is equivalent to a logical AND gate, with possible inputs of cell surface antigens (the keys) either not binding or binding to aptamer locks. Possible outputs are that the nanorobot remains closed or it performs a conformational

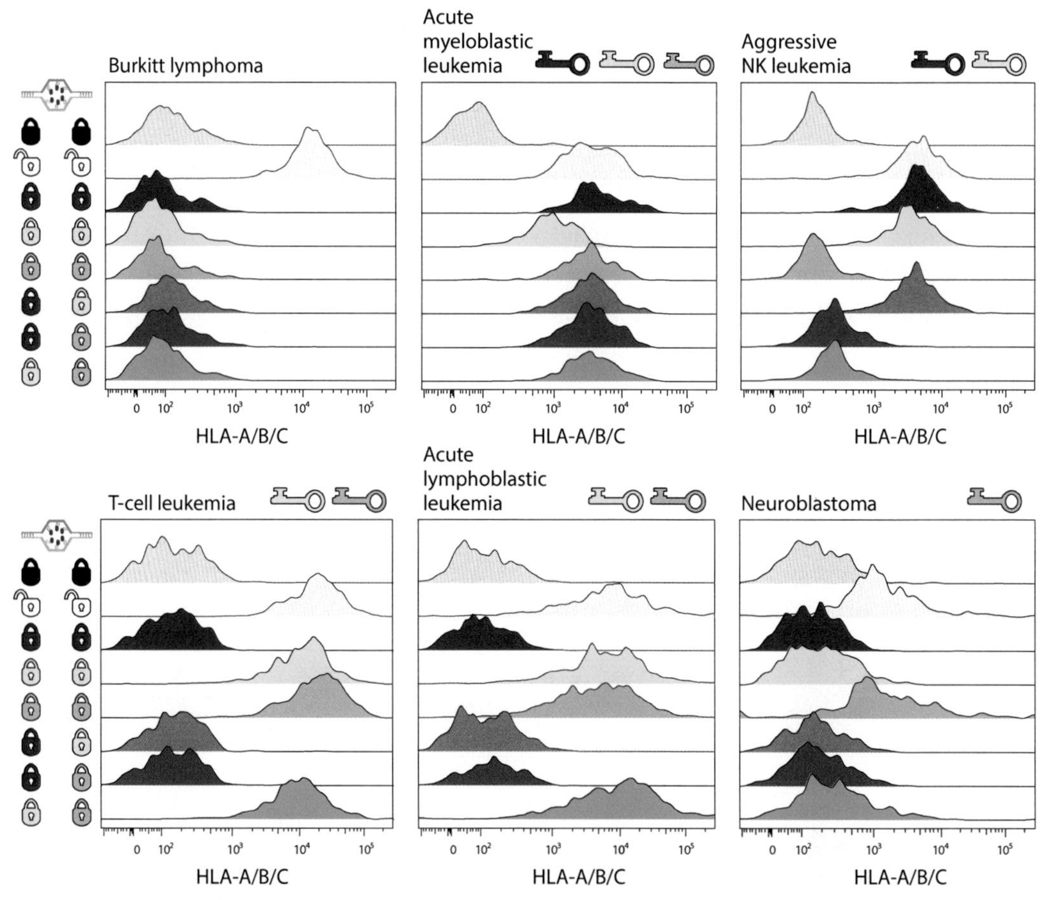

Figure 10.12 Generality and robustness of the aptamer-encoded logic gating. Eight different nanorobot versions were tested with six different cell types expressing various combinations of antigen keys. The robot could be programmed to activate in response to a single type of key by using the same aptamer sequence in both lock sites. Alternatively, different aptamer sequences could be encoded in the locks to recognize two inputs. Both locks needed to be opened simultaneously to activate the robot. The robot remained inactive when only one of the two locks was opened. Each histogram displays the count of cells versus fluorescence due to anti–HLA-A/B/C labeling. *(Reproduced with permission from Shawn M. Douglas, Ido Bachelet, and George M.Church, "A Logic-Gated Nanorobot for Targeted Transport of Molecular Payloads," Science 335 (17 February 2012), 831–834.)*

rearrangement to expose its payload. Consider Figure 10.12 comprising six histograms of cell count versus fluorescence due to anti-HLA-A/B/C labeling.

To test generality and robustness of the aptamer-encoded gating, Douglas and colleagues designed six different robots using combinations of aptamer locks drawn from a set of three well-characterized aptamer sequences. Color match indicates lock-and-key match. The three aptamer sequences have the names: 41t, active against platelet-derived growth factor (PDGF), shown in red; TE17, shown in yellow; and sgc8c, shown in blue. The six robot versions—plus a permanently locked negative control and a no-lock positive control—were loaded with fluorescently labeled antibody to human HLA-A/B/C Fab′ and used to probe six different cell lines, each expressing different profiles of the key antigens recognized by the three chosen aptamers. Each histogram displays the count of cells versus fluorescence due to anti-HLA-A/B/C labeling. Reading from left to right, we begin with Burkitt lymphoma cell line. This activated none of the robots. Thus, none of the cell counts changed—so none of the colored peaks moved along the axis labeled HLA-A/B/C in Figure 10.12

By sharp contrast, an acute myeloid leukemia cell line activated all robots (with the exception of the permanently locked control). A third cell line, isolated from a patient with large granular lymphocytic leukemia, aggressive NK type (NKL), activated robots with two 41t locks (two red locks), two TE17 locks (two yellow locks), and with one 41t and one TE17 lock. A fourth cell line, isolated from acute T cell leukemia, activated robots bearing two TE17 locks, two sgc8c locks, as well as one TE17 lock and one sgc8c lock. The fifth cell line was isolated from acute lymphoblastic leukemia and had a similar profile to the T cell leukemia. Finally, a neuroblastoma cell line activated robots with two sgc8c locks. Figure 10.12 confirms that the aptamer-encoded logic-gating (that is, the nanorobot's lock and key mechanism) worked with all six different cell types.

To summarize, the six different robots—plus a permanently locked negative control and a no-lock positive control—performed as desired. Permanently locked nanorobots did not exhibit significant binding to any cell population. Unlocked nanorobots labeled each subpopulation. The other six nanorobots were effective and selective in targeting six combinations of protein aptamer locks each of which had been designed to target different types of cancer cells in culture (i.e., *in vitro*).

Test of Binding Discrimination: Healthy Cells Versus Leukemia Cells (NK Cells)

The Harvard team further examined the performance of the logic-gating mechanism in several ways. We present just one of these experiments—a notable result that showed the ability of the nanorobot to discriminate between healthy cells and cells expressing an aggressive NK type of leukemia. To simulate physiological conditions, they mixed granular lymphocytic leukemia cells (NK cells) with healthy whole-blood cells and found that robots gated with 41t locks (two red locks) discriminated NK cells with high precision as shown in Figure 10.13. Off-target binding occurred in only 0.6% of sampled cells; that is, healthy cells were virtually ignored by the nanorobot. This absence of collateral damage is enormously important. Secondary harm to normal tissue is well known to anyone whose life has been touched by the ravages of our current modality of chemotherapy to treat cancer. However, we must add that whether or not nanorobots will work in a living organism has yet to be demonstrated.

To understand Figure 10.13 we must say a few words about forward- versus side-scatter dot plots, a staple of flow cytometric analysis. In flow cytometry, forward scatter refers to the laser light reflected by cells at angles less than 90° and provides information about the size of the cell. The correlation of forward scatter and cell size is a complicated issue and we will take it as a given. Side scatter refers to laser light reflected by cells at right angles (90°) and provides information about the granularity or complexity of the cells. Again, we will take as a given that side scatter correlates with cell structure or cell complexity.

In a plot of forward- scatter versus side-scatter—Figure 10.13—each dot on the graph gives the forward- and side-scatter values from a single cell. As was true in Figure 10.12, the 41t aptamer lock on the nanorobot is shown as a red lock icon, and the robot was programmed to activate in response to a single type of key by using the same aptamer sequence in both lock sites. Hence, two red locks are shown along with the robot in the cartoon above the scatter dot plot. The healthy whole-blood cells are spelled out in

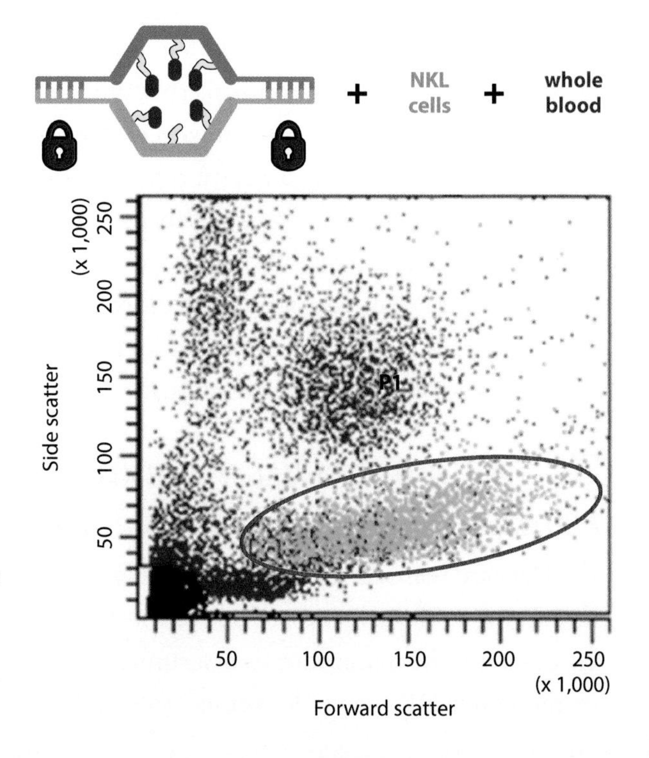

Figure 10.13 The scatter dot plot shows the discrimination between healthy cells and NKL leukemia cells. Forward- versus side-scatter dot plot of 4:1 mixture of healthy human whole-blood cells and aggressive NKL leukemia cells. The ellipse shows the position on the scatter dot plot of the subpopulation of all cells that are NKL cells (but includes some "colocalized" healthy cells). The green dots represent cells that are engaged by the antibody-loaded nanorobots. The green dots outside the ellipse are likely healthy cells mistakenly engaged by the nanorobots off-target binding. Off-target binding occurred in only 0.6% of sampled cells: there is virtually no collateral damage to healthy cells.
(Reproduced with permission from Shawn M. Douglas, Ido Bachelet, and George M. Church, "A Logic-Gated Nanorobot for Targeted Transport of Molecular Payloads," Science 335 (17 February 2012), 831–834.)

a red font in the cartoon and are shown as red dots in the plot. The green dots represent cells that are engaged by the antibody-loaded robots. The ellipse shows the position on the scatter dot plot of the subpopulation of all cells that are NKL cells—but the ellipse includes some healthy cells. The red dots inside the ellipse are healthy cells that happen to "colocalize" with the NKL cells in terms of their physical parameters—size and granularity.[22,23] (Some of the red dots inside the ellipse could be false negatives.) The green dots outside the ellipse are likely to be healthy cells mistakenly engaged by the nanorobots—false-positive selections (off-target binding).[22] We repeat a most notable finding: Off-target binding occurred in only 0.6% of sampled cells. That is, there is virtually no collateral damage to healthy cells. (P1 in the center of the plot stands for Population 1; it is a label generated by the data analysis software that designates all cells within the ellipse—both the green and red dots.)

Forward- versus side-scatter dot plot In a scatter dot plot, each cell is represented by a dot that is positioned on the horizontal and vertical axes according to the intensities of scattered light detected for that cell. Light scattering information is collected in the forward direction (light reflected by cells at angles less than 90° from the direction of the incident light) and in the side direction (light reflected at 90°). The forward- versus side-scatter plot provides information about both the size of the cells and the granularity, that is, the complexity of the cells. Scatter dot plots reveal quantitative percentages of cells with various properties and show discrete subpopulations of cells with different intensities.

We conclude with one last line of investigation. Douglas, Bachelet, and Church examined the ability of an activated nanorobot to interface with cells and induce tunable changes in cell behavior. They performed several such experiments and we report one of them. They chose to target NKL cells using a pair of 41t aptamer locks and loaded the robots with a combination of antibody to human CD33 and antibody to human CDw328 Fab′ fragments. This antibody combination has been shown to induce growth arrest in leukemic cells. This much is shown in the cartoons on the left of Figure 10.14. To appreciate the significance of the findings summarized in the accompanying graph, we introduce the terms *cell cycle* and *apoptosis*.

The growth, replication, and division of cells consist of a series of processes collectively known as the cell cycle. The cell cycle can be divided into two major stages: interphase

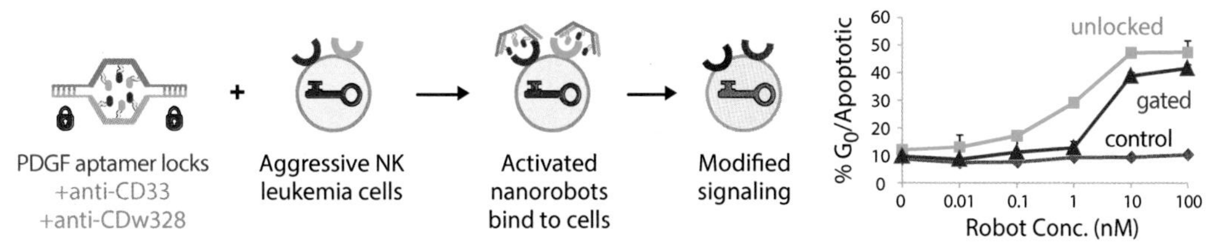

Figure 10.14 Cartoons: Nanorobots manipulate target cell signaling. A single dose of nanorobots loaded with an equal mixture of antibody to human CD33 and antibody to human CDw328/Siglec-7 Fab′ fragments (cyan and magenta, respectively) and gated by 41t locks recognizing platelet-derived growth factor (PDGF) were used to treat NKL cells at various concentrations. Graph: NKL cells treated with nanorobots were analyzed after 72 hours for cell cycle distribution by propidium iodide (a fluorescent molecule used to stain cells). Note the percent of cells at sub-G_1 stage (G_0/Apoptotic = apoptotic and G_0). *(Reproduced with permission from Shawn M. Douglas, Ido Bachelet, and George M. Church, "A Logic-Gated Nanorobot for Targeted Transport of Molecular Payloads," Science 335 (17 February 2012), 831–834.)*

(a phase between mitotic events) and mitosis. Mitosis is cell division, and mitosis can be further broken down to different stages that are not relevant to our discussion. Stages in interphase, however, are germane to the graph in Figure 10.14. The G_1 phase is the first stage of interphase and is a period of cell growth during which the cell produces proteins and cytoplasmic organelles (specialized subunits within a cell). There is a point during the G_1 phase when a cell either proceeds to the next stage of interphase (the S phase during which DNA is replicated where S stands for synthesis of DNA) or it enters a state of dormancy called the G_0 phase. In the G_0 phase the cell is in a non-dividing state. In cancer cells, dormancy is a state in which the cells may survive, but growth is arrested. Make a mental bookmark of this last statement while we familiarize ourselves with the word *apoptosis.*

Cell cycle The growth, replication, and division of cells consists of a series of processes collectively known as the cell cycle.

Dormancy During the cell cycle, a cell may enter a state of dormancy (called the G_0 phase). When a cell is dormant it is in a non-dividing state. In cancer cells, dormancy is a state in which the cells may survive but growth is arrested.

Apoptosis A genetically determined process of cell self-destruction (sometimes called programmed cell death). It is a process that eliminates damaged or unneeded cells. If apoptosis is halted, e.g., by genetic mutation, the result may be uncontrolled cell growth and tumor formation.

Apoptosis describes inherently programmed cell death, a basic biological phenomenon in the cycle of the cell.[24] The name is derived from a Greek word to describe the "dropping off or falling off" of petals from flowers, or leaves from trees. (The authors who proposed this term suggested that the stress should be on the penultimate syllable and that the second half of the word should be pronounced like "ptosis" with the "p" silent.) As pointed out in endnote 24, apoptosis plays a complementary but opposite role to mitosis and, in the case of cancer cells, it can be the result of therapeutically induced tumor regression. Thus, in the G_0 phase, cells are not dividing nor are they dead; they are in a sub-G_1 stage, sandwiched between proliferation via cell division and death via apoptosis.

Now check the graph in Figure 10.14. The vertical axis shows the percent of NKL cells that are either in the G_0 phase or are apoptotic (G_0/Apoptotic) after exposure to nanorobots that have docked to the cells' receptor sites and discharged their antibody cargo. We see on the horizontal axis that as the nanorobot concentration in the solution was increased, the robots induced growth arrest in NKL cells in a dose-dependent fashion. That is, the robots caused the NKL cells to be in a sub-G_1 phase: the NKL cells were either dormant (G_0) or apoptotic (dead). This was experimentally determined by flow cytometry by examination of the cell cycle distribution among the NKL cells using a fluorescent dye (propidium iodide) that stains DNA. The fluorescence intensity of the stained cells at certain wavelengths correlates with the amount of DNA they contain, and this DNA profile reveals whether the cells are in the G_1 stage or a sub-G_1 stage (that is, either G_0 or apoptotic).

Remember that just like synthetic ion channels, this study had its roots in DNA origami. The nanorobots are a hybrid of structural DNA nanotechnology, antibodies, aptamers, and metal atomic clusters in the form of gold nanoparticles. The nanorobots integrate sensing and logical computing functions (the AND gate) and are aimed at specific targeting of human cancers. No, they are not a cure for cancer. Nonetheless, as Paul Rothemund said, the experiment "takes us one more step along the path from the smartest drugs of today to the kind of medical nanobots we might imagine."[25]

A DNA origami box with a controllable lid was one of the forerunners of the logic-gated nanorobot. Kurt Gothelf is a chemist and Director of the Center for DNA Nanotechnology at Aarhus University in Denmark. He was part of the team that made the DNA origami box with controllable lid. When the logic-gated nanorobot paper appeared, he said, "This is one of the things the field has needed, something to show that, hey, this can actually be useful. People have been talking a lot about robots that enter your body, and go to a place where something is wrong and fix it. This is the first example that this might come true one day."[25] Shawn Douglas, first author of the nanorobot paper, put it this way, "My dream is for one of these devices to ultimately go through clinical trials and become an actual therapeutic that would be a novel treatment for some type of cancer."[26]

Nadrian Seeman declared that he was not a futurist; he doesn't make predictions. Neither do the other players presented in Chapters Six through Ten. Like the rest of us, their wish is that structural DNA nanotechnology will benefit society. The field has come a long way in its first 35 years but what it will deliver in the next 35 years or the 35 years after that is anybody's guess. The pertinent "h" word is not hype. It's hope.

Exercises for Chapter Ten: Exercises 10.1–10.2

Exercise 10.1

Figure 10.15 shows current trace (vi) from panel (b) of Figure 10.6—current gating events from a synthetic ion channel at an applied voltage of 100 mV. In the Supplementary Material to their publication, Dietz and his colleagues point out that the distinct gating behavior shown in this trace is actually channel conductance switching between several *sub*conductance levels. **(a)** Estimate the average current increment in picoamperes (10^{-12} amp) with channel opening. **(b)** Estimate the single channel conductance (based on this switching between *sub*conductance levels) in picosiemens (10^{-12} siemens). Conductance is the reciprocal of resistance and is measured in units of siemens when resistance is measured in ohms.

Figure 10.15
(Adapted with permission from Martin Langecker, Vera Arnaut, Thomas G. Martin, Jonathan List, Stephan Renner, Michael Mayer, Hendrik Dietz, and Friedrich C. Simmel, "Synthetic Lipid Membrane Channels Formed by Designed DNA Nanostructures," Science 338 (16 November 2012), 932–936.)

Exercise 10.2

You're operating a flow cytometer and will analyze cells using forward- versus side-scatter dot plots. Going into the experiment you have a very general idea of the structure of the cells you're examining (Figure 10.16). Suppose the light from the laser is incident on the sample from the left side. Draw a simple diagram to show what features of the cells are likely to contribute to the forward-scatter and side-scatter components of the scatter dot plots you will be collecting. That is, show how the laser light is scattered by all the features that you see in Figure 10.16.

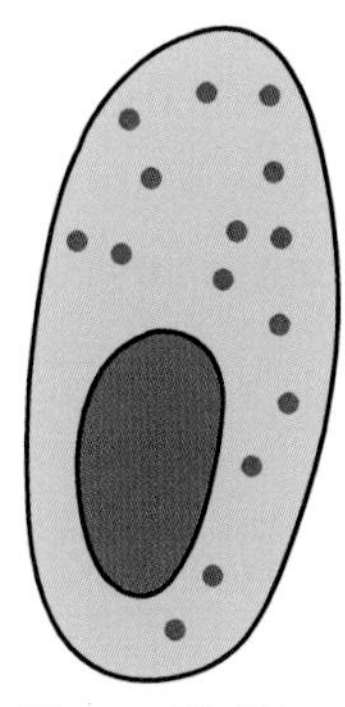

Figure 10.16

ENDNOTES

[1] Eric Smalley, "NYU's Nadrian Seeman," *TRN's View From the High Ground* (4–11 May 2005), http://www.trnmag.com/Stories/2005/050405/View_Nadrian_Seeman_050405.html

[2] Martin Langecker, Vera Arnaut, Thomas G. Martin, Jonathan List, Stephan Renner, Michael Mayer, Hendrik Dietz, and Friedrich C. Simmel, "Synthetic Lipid Membrane Channels Formed by Designed DNA Nanostructures," *Science 338* (16 November 2012), 932–936.

[3] Shawn M. Douglas, Ido Bachelet, and George M. Church, "A Logic-Gated Nanorobot for Targeted Transport of Molecular Payloads," *Science 335* (17 February 2012), 831–834.

[4] Clay M. Armstrong, "Early Views of Channels and Gates," in "Ion Channels: From Idea to Reality," *Nature Medicine 5, No. 10* (October 1999), 1105–1109.

[5] Roderick MacKinnon, "Potassium Channel's Secret," in "Ion Channels: From Idea to Reality," *Nature Medicine 5, No. 10* (October 1999), 1105–1109.

[6] Declan A. Doyle, João Morais Cabral, Richard A. Pfuetzner, Anling Kuo, Jacqueline M. Gulbis, Steven L. Cohen, Brian T. Chait, and Roderick MacKinnon, "The Structure of the Potassium Channel: Molecular Basis of K^+ Conduction and Selectivity," *Science 280* (3 April 1998), 69–77.

[7] Images courtesy of Professor Ray Fort Jr. and Dr. Rachel Kramer Green.

[8] http://chemistry.umeche.maine.edu/CHY431/Channels3.html

[9] Roderick MacKinnon, "Potassium Channels," *FEBS Letters 555* (2003), 62–65.

[10] David A. Köpfer, Chen Song, Tim Gruene, George M. Sheldrick, Ulrich Zachariae, Bert L. de Groot, "Ion Permeation in K^+ Channels Occurs by Direct Coulomb Knock-On," *Science 346, No. 6207* (17 October 2014), 352–355.

[11] Aleksei Aksimentiev, "Alpha-Hemolysin: Self-Assembling Transmembrane Pore," Theoretical and Computational Biophysics Group, University of Illinois at Urbana-Champaign, http://www.ks.uiuc.edu/Research/hemolysin/

[12] B. Nilius, P. Hess, J.B. Lansman, and R.W. Tsien, "A Novel Type of Cardiac Calcium Channel in Ventricular Cells," *Nature 316* (1 August 1985), 443–446.

[13] Personal communication; email from Freidrich Simmel to K.D. on 24 July 2015.

[14] Grégory F. Schneider and Cees Dekker, "DNA Sequencing With Nanopores," *Nature Biotechnology 30, No. 4* (April 2012), 326–328.

[15] Elizabeth A. Manrao, Ian M. Derrington, Andrew H. Laszlo, Kyle W. Langford, Matthew K. Hopper, Nathaniel Gillgren, Mikhail Pavlenok, Michael Niederweis, and Jens H. Gundlach, "Reading DNA at Single-Nucleotide Resolution With Mutant MspA Nanopore and phi29 DNA Polymerase," *Nature Biotechnology 30, No. 4* (April 2012), 349–353.

[16] Gerald M. Cherf, Kate R. Lieberman, Hytham Rashid, Christopher E. Lam, Kevin Karplus, and Mark Akeson, "Automated Forward and Reverse Ratcheting of DNA in a Nanopore at 5-Å Precision," *Nature Biotechnology 30, No. 4* (April 2012), 344–348.

[17] Tom Z. Butler, Jens H. Gundlach, and Mark Troll, "Ionic Current Blockades From DNA and RNA Molecules in the α-Hemolysin Nanopore," *Biophysical Journal 93, No. 9* (1 November 2007), 3229–3240.

[18] Andrew D. Ellington and Jack W. Szostak, "*In Vitro* Selection of RNA Molecules That Bind Specific Ligands," *Nature 346* (30 August 1990), 818–822.

[19] Douglas, Bachelet, and Church, "A Logic-Gated Nanorobot," 832.

[20] Shawn M. Douglas, Ido Bachelet, and George M. Church, "A Logic-Gated Nanorobot for Targeted Transport of Molecular Payloads," *Science 335* (17 February 2012), Supporting Material, Figure S15, 24 (linked to the online version of the paper as a PDF download).

[21] Howard M. Shapiro, *Practical Flow Cytometry* (Hoboken, New Jersey: John Wiley & Sons, 2003).

[22] Personal communication; email from Ido Bachelet to K.D. on 13 May 2013.

[23] Kenneth W. Dunn, Malgorzata M. Kamocka, and John H. McDonald, "A Practical Guide to Evaluating Colocalization in Biological Microscopy," *Am J Physiol Cell Physiol. 300, No. 4* (April, 2011), C723–C742.

[24] J.F.R. Kerr, A.H. Wyllie, and A.R. Currie, "Apoptosis: A Basic Biological Phenomenon With Wide-Ranging Implications in Tissue Kinetics," *British Journal of Cancer 26, No. 4* (August 1972), 239–257.

[25] Alla Katsnelson, "DNA Robot Could Kill Cancer Cells," *Nature News and Comment* (16 February 2012), http://www.nature.com/news/dna-robot-could-kill-cancer-cells-1.10047

[26] Alyssa Danigelis, "DNA Robots Deliver Deadly Punch to Bad Cells," *Discovery News* (16 February 2012), http://news.discovery.com/tech/robotics/dna-robot-nanotechnology-cancer-cells-120216.htm

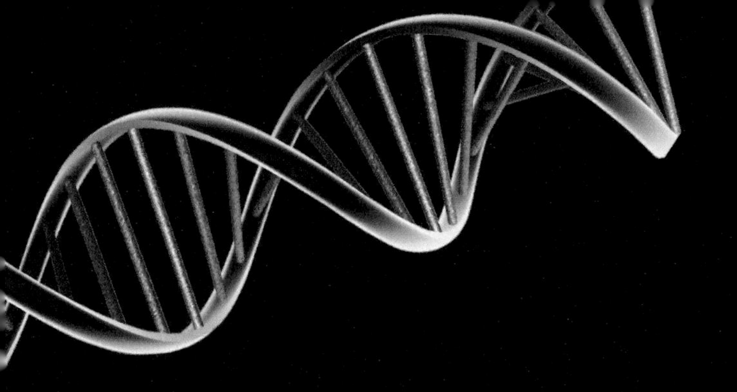

PART III

The Possible Origins of Life's Information Carrier

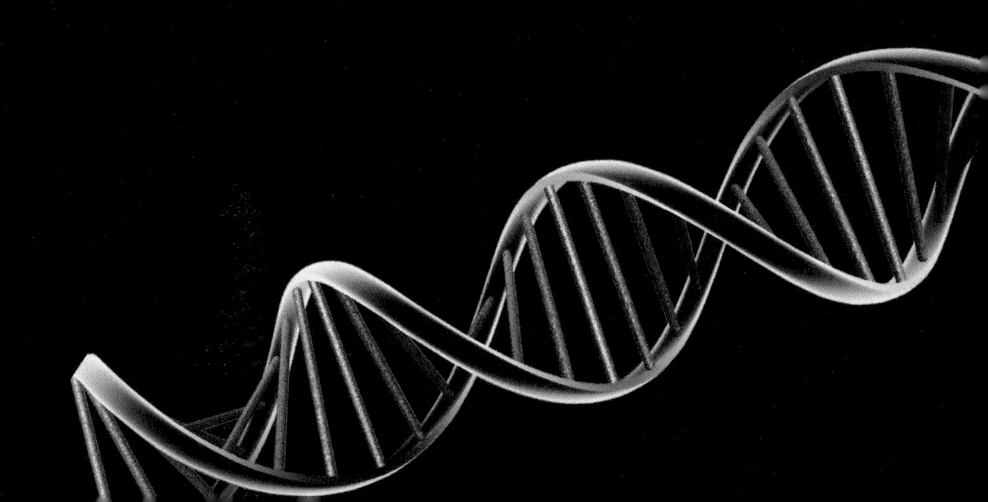

CHAPTER ELEVEN

Chance Findings

Michi Nakata, Noel Clark's extraordinary post doc, generally worked at night. There are no windows on the subterranean 2B level of the Duane Physical Laboratories on the University of Colorado campus in Boulder. Absent both blue skies and evening stars, hers was an austere basement world. Glazed cement floors, cinder block walls painted white, unsightly ventilation ductwork, exposed water pipes and electrical conduits, racks of test equipment, sturdy but unattractive tables with metal frames and wood surfaces crowded with instruments. Banks of long fluorescent bulbs cast their light on conjoined angular shapes, creating serrated shadows with a faint hint of menace. Large metal cabinets and cheerless I-beams captured the pallor and sterility of a warehouse. What passed for animation on the 2B level was provided by the atonal thrum and throb of scientific instruments. The buzz of lasers accompanied the resonant hum of high-voltage power supplies while the vacuum pumps' incessant pocketa-pocketa-pocketa provided the bass. The air was redolent of burnt microwave popcorn. It was here that Nakata tried to coax nature into revealing its secrets.

In the fall of 2004 she was given a new project: try to use short lengths of DNA to make molecules with tunable length and rigidity that could form liquid crystals. Long DNA molecules (when hydrated) had been known to form liquid crystal phases since the 1940s.[1] The fibers of DNA drawn out of a viscous gel (as Maurice Wilkins said, "It's just like snot!") for X-ray diffraction in the 1950s by Rosalind Franklin and by Wilkins were in liquid crystal phases.[2] Clark now asked the question: Could one obtain liquid crystals with shorter nucleic acid chains? How short? Nakata purchased single strands of two different lengths and compositions in September and the vials of powder waited for her to free up time from other experiments.

By December she found some time. She mixed a portion of each vial with water and placed the samples in standard glass cells that fit in the stage of her depolarizing light microscope. Now she was ready to mount the first cell in the temperature-controlled stage between crossed polarizers. The experiment was to heat the sample, let it cool down, and see if there was any discernible phase transition into a liquid crystalline state: a birefringence signature, a color or a pattern that suggested the formation of liquid crystals. Walking over to one of the metal cabinets, she got a new notebook with its cover labeled "Liquid Crystal Laboratory, Professor Noel A. Clark" and filled out the lines that read "Investigator," "Project," "Notebook Start Date," and "Book Number"—Michi Nakata, DNA, 12/21/2004 and number 261LC.

The first evening passed uneventfully. On the second evening she was amazed to see that snippets of DNA duplexes, ultrashort double-stranded DNA—nanoDNA—formed liquid crystals where none should be.[3] In a twinkling she left her drab basement world behind. She had stepped through a gateway to Elysium. For many decades during the Christmas season, the Japanese have embraced with fervor Beethoven's Ninth Symphony.[4] We don't know if Nakata was listening to the *Choral* at this time. But she might well have been singing her own *Ode to Joy*.

NanoDNA Short duplexes of B-form DNA of lengths less than 20 base pairs (bp).

Onsager's Criterion for an Isotropic-Nematic Liquid Crystal Phase Transition

To understand why Nakata's findings were so enchanting, we start with what we are already familiar with: duplex, B-form, long DNA. These molecules range from semiflexible polymers with chain lengths in the millions of base pairs (bp), down to approximately 100 bp rigid, rodlike objects. The appearance of liquid crystal phases in long DNA has been accounted for theoretically by modeling B-DNA as a rigid or semiflexible rod-shaped solute. (A solute is a substance that creates a solution when dissolved in a solvent.)

The starting point for such modeling is Lars Onsager's treatment of sterically repulsive hard rods (i.e., the rods cannot penetrate one another).[5] Onsager was a physical chemist and a Nobel Prize winner in Chemistry; he had been nominated for both the Physics and the Chemistry prize. He was also a colorful character. Born in Norway and educated there as a chemical engineer, he later emigrated to the United States. After brief stays at Johns Hopkins University and Brown University, he was appointed a Sterling and Gibbs Fellow—a postdoctoral fellowship—in the chemistry department at Yale University. The department was soon embarrassed to discover that he had no Ph.D.[6] He quickly remedied that situation in 1935, one year after he was appointed to the faculty as an assistant professor.

It has been suggested that the year 1949 was his *annus mirabilis*, his year of wonders.[7] It was then that he published his paper on anisotropic solutions of rod-shaped molecules and other major contributions. The rod-shaped molecules paper described the isotropic-nematic (I-N) phase transition of rod like polymers and put the statistical theory of liquid crystals on a firm mathematical basis, with certain simplifying assumptions.[5] (Recall from Chapter Five that the term *isotropic* describes properties that do *not* change with direction; there is no orientational order and no positional order; all liquids and gases are isotropic. A liquid crystalline material is said to be in its isotropic phase when it behaves like an ordinary liquid.)

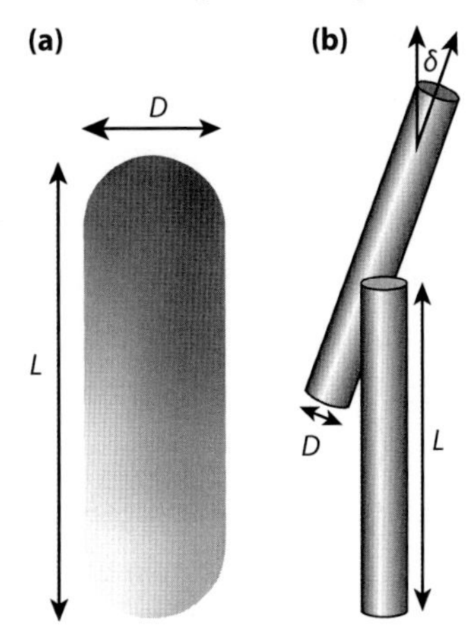

Figure 11.1 (a) A prolate rod of diameter *D* and length *L*. **(b)** The excluded volume of two rods depends on their orientation with respect to one another. *(Part **(a)** reproduced with permission from Giuliano Zanchetta.)*

For a sense of Onsager's mathematical presentation we look at a simple physical description of his arguments.[8] Consider a suspension of hard rods, that is, prolate objects that exert no forces whatsoever on one another until they touch, at which point they experience infinite repulsions. A prolate object is one that has a polar diameter, L, longer than its equatorial diameter, D (panel (a) of Figure 11.1). Think about the way hard rods can pack in solution. The favored alignment is one in which neighboring rods are more likely to be parallel. This is a result of the excluded volume. The excluded volume of a molecule is the volume that is impenetrable to other molecules in the solution because of the presence of the first molecule. This is a formal way of saying the obvious, but there is an important consequence to this observation. If the rods are oriented parallel to one another there is very little excluded volume—the rods can come very close to one another. But if the rods are at some angle, δ, with respect to one another, there is a large volume around each rod that another rod cannot enter because of the hard-rod repulsion between the two (panel (b) of Figure 11.1).[9,10] In fact, the excluded volume of one hard rod with respect to a second hard rod is a minimum when their axes lie along the same direction and a maximum when they are perpendicular. Speaking more informally, rodlike objects take up less space—they can pack more efficiently—when they are parallel. You can readily verify this if you try to fit spilled wooden matches back into their box without first aligning them.[11]

Onsager took all this on mathematically and found that at a sufficiently high concentration, hard rods in solution spontaneously undergo a phase transition from the isotropic phase (no preferential ordering) to the orientationally ordered nematic phase. To get a feeling for why this

Excluded volume The volume that is impenetrable to other molecules in the solution because of the presence of the first molecule.

is so, we use the term *entropy*. Entropy is a quantitative way to measure the dispersal of energy in a process. Onsager found that there is a competition between two forms of entropy that leads to the isotropic-nematic phase transition. On the one hand, there is orientational entropy corresponding to the tendency of rods to stay orientationally disordered. There is also positional or packing entropy that is higher for aligned rods because of the excluded volume interaction. These forms of entropy compete: parallel arrangements of anisotropic objects lead to a decrease in orientational entropy. But there is a concomitant increase in positional or packing entropy that is higher for aligned rods due to the excluded volume.[12] At sufficiently high concentrations, greater positional order will win. Saying it another way, at sufficiently high concentrations of hard rods in solution, the nematic phase is entropically favored.

Entropy A quantitative way to measure the dispersal of energy in a process.

Free energy Energy available to do work rather than just be dispersed as heat.

Work The energy transfer that occurs if an object is moved over a distance by an external force at least part of which is applied along a direction parallel to the displacement.

Connecting this with the term *free energy*, we can rephrase Onsager's conclusion to say that when the nematic phase is formed, the gain in free energy associated with aligning the rods outweighs the loss of orientational entropy. That is, at high enough concentrations the overall free energy of the system will be minimized for the orientationally aligned nematic state.

We can join Onsager's theory and Nakata's observation of nanoDNA liquid crystals by adding just a soupçon more vocabulary: volume fraction. Volume fraction, denoted by ϕ, is the volume of a constituent of a mixture divided by the sum of the volumes of all constituents prior to mixing. In our situation, the constituents consist of the repulsive hard rods.

Volume fraction The volume of a constituent of a mixture divided by the sum of the volumes of all constituents prior to mixing. Volume fraction is denoted by φ.

Shape anisotropy (also referred to as aspect ratio) The ratio of a molecules' length to its diameter, *L/D*.

Onsager's Criterion For molecules considered as hard rods, the shape anisotropy, L/D, must be such that $L/D > \sim 5$ for the molecules to exhibit liquid crystal phases.

Isotropic-Nematic phase transition (I-N) A liquid crystal transition from a phase in which the properties of the molecules do not change with direction (isotropic phase) to a phase in which there is a preferred direction of orientation of the molecules (nematic phase)—a degree of orientational order.

In his calculations, Onsager used known values of the relevant parameters for B-DNA. From the experimental literature he knew that $D \sim 2$ nm and $L \sim N/3$ nm for B-DNA molecules having a chain length N (measured in bp). Recall that L is the length and D is the diameter of a hard rod (i.e., a prolate rod). The ratio L/D is referred to as the shape anisotropy, a term familiar to us from Chapter Four. The term *aspect ratio* is also used to denote L/D. Onsager found that nematic order comes about if the volume fraction, ϕ, of anisotropic, hard rods is greater than $4D/L \approx 24/N$.

Almost 50 years later, two physical chemists, Peter Bolhuis and Daan Frenkel, made computer-simulated phase diagrams for hard rods. They quantitatively confirmed Onsager's results. For B-DNA molecules having a length $N > 28$ bp, the shape anisotropy is $L/D > 4.7$. The simulations of Bolhuis and Frenkel show that for an aspect ratio $L/D < 4.7$ there should be no liquid crystal phases.[13]

In case the conclusion went by too fast, let's return to it—because that is the punch line. According to Onsager's criterion, for $L/D < 4.7$ there should be no liquid crystal phases (other than the isotropic). Nakata's DNA samples were not at all anisotropic in shape. Their L/D ratio was far too small to allow a nematic phase to form. For a DNA chain 6 bp long, $L/D \sim 1$. Even for a DNA chain of 20 bp, $L/D \sim 3$—far too small for liquid crystal formation.[14] But Nakata saw liquid crystals where none should be.

Excited by her observations, Nakata brought them to Clark who asked her to pursue the matter further with additional short segments of DNA. She then burned through about $40,000 of DNA samples. At that time, commercially available short oligomers of DNA with chosen sequences cost about $1,000 per milligram.[15] But it was worth every penny. She found nematic liquid crystal ordering in short B-DNA duplexes—which they now referred to as nanoDNA—of length 6 bp $< N <$ 20 bp (Figure 11.2). The nematic phase appeared for N values that were an order of magnitude (that is, a factor of 10) smaller than those predicted from the volume fraction of hard rods that exists at the boundary of the isotropic-nematic phase transition. This could not be explained by the Onsager-Bolhuis-Frenkel (OBF) criterion. The DNA duplexes manifestly lacked the shape anisotropy required for liquid crystal ordering. Please be patient. In time we will see why this was not much ado about nothing.

The mystery was a good fit to the postulates of Nobelist Peter Medawar whose book came up in Nadrian Seeman's interview. Clark's curiosity had led to a problem that was not too simple to be useful or too difficult to be soluble.

NanoDNA Seems to Violate Onsager's Venerable Criterion

The data that Nakata obtained were striking. In addition to several forms of light microscopy, postdoc Nakata and graduate student Giuliano Zanchetta from the Bellini group performed X-ray diffraction experiments at the Advanced Photon Source, the synchrotron radiation facility at Argonne National Laboratory. This technique enabled them to evaluate columnar liquid crystal ordering of nanoDNA that appeared at high concentrations (Figure 5.2, Chapter Five). In the columnar liquid crystal phase the molecules exhibit a degree of translational order as well as a degree of orientational order. This form of molecular organization was startling because, like the nematic phase, the hard-rod models did not predict it.

Panel (a) of Figure 11.3 is a plot of DNA concentration, *c*, versus *N*, the length of the oligomers of DNA.[16] Observe that the horizontal axis is presented as a *logarithmic* scale. That is, a scale of measurement that uses intervals corresponding to orders of magnitude (i.e., factors of ten), rather than the more familiar linear scale.

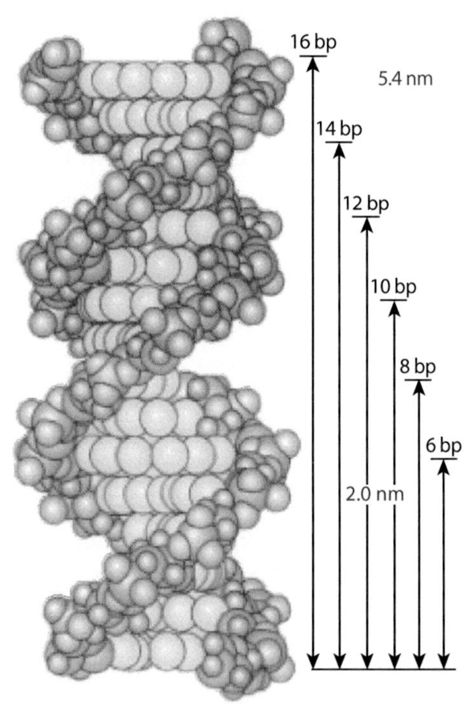

Figure 11.2 NanoDNA: length in nanometers (nm) and base pairs (bp). *Note*: the diameter of nanoDNA is about 2 nm. *(Adapted with permission from Solomon/Martin/Martin/Berg. Biology, 10E. © 2015 Brooks/Cole, a part of Cengage Learning, Inc. Reproduced by permission. www.cengage.com/permissions.)*

The plot includes the experimental data from the Bellini/Clark collaboration as well as data obtained on long DNA (lDNA) by other groups. The solid red, vertically oriented rectangle shows data from 1990 when another set of researchers reported short DNA ($N = 12$) spontaneously forming a liquid crystal-like phase and the rectangle shows the range of that liquid crystal phase.[17] The solid red triangles and solid red curves bound the measured isotropic-nematic (I-N) phase coexistence line for lDNA ($N > 100$ bp). The solid red circles and red dotted line give the measured nematic-columnar (N-C_U) phase boundary of lDNA. For $N < 20$, phase transitions are marked by red open symbols (I-N, triangles; N-C_U, circles; C_U–C_2, squares). C_2 is a crystalline phase. The range of each phase is indicated by colored columns (I, magenta; N, cyan; C_U, yellow), at $T = 20$ °C for $20 > N > 8$ and at $T = 10$ °C for $N = 6$. Note that the word "crystalline" is not a typo. At sufficiently high concentrations, nanoDNA forms a crystal, rather than a liquid crystal.[18]

Having had a look at experimental data, now consider phase diagrams of several theoretical models of interacting rod-like particles of axial ratio L/D filling a volume fraction ϕ (panel (b) of Figure 11.3).[19] Red lines are the boundaries of I-N coexistence according to Onsager's model. Dotted blue lines are the I-N and I-C_2 phase transitions according to the Bolhuis-Frenkel computer simulations. The only other point we'll touch on in this panel is the solid black Onsager-Bolhuis-Frenkel I-N line that incorporates several predictions for the I-N transition.

Bellini and Clark combined information from the two parts of Figure 11.3 to interpret their nanoDNA data in a seminal paper. To avoid being cowed by Figure 11.4 it may be necessary to take a deep breath and remember to exhale. Figure 11.4 is unquestionably a busy one. But it does the job of summarizing their analysis in a single panel.[20] We hasten to add that we will not run through all the bells and whistles packed into Figure 11.4. We will only discuss features that give us an informed understanding of how Bellini and Clark were able to reconcile their results with deeply rooted theoretical predictions.

The panel labeled (a) shows that liquid crystal phases are found in the sub-Onsager region of the phase diagram of nanoDNA where they are not expected to be on the basis of their shape. NanoDNA lacks the requisite shape anisotropy to satisfy Onsager's criterion: its *L/D* is too small. However, a combination of light and X-ray structural studies of nanoDNA provided the clues to explain the mystery.

The studies showed that end-to-end adhesion and subsequent stacking of the short duplex oligomers into anisotropic rod-shaped aggregates produce these phases. The extended duplex units are long and rigid enough to form liquid crystal phases consistent with Onsager's theory. *That is the take-home message of the upper right inset to Figure 11.4 that is labeled (b).*

Like a troupe of lithe circus acrobats who form a human pillar with each standing on the shoulders of the one below, the snips of nanoDNA stacked upon one another to form elongated lengths

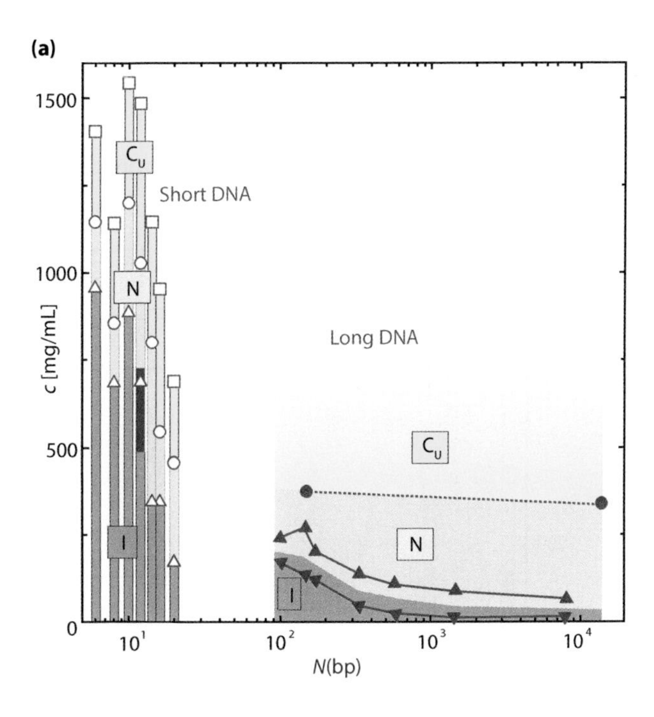

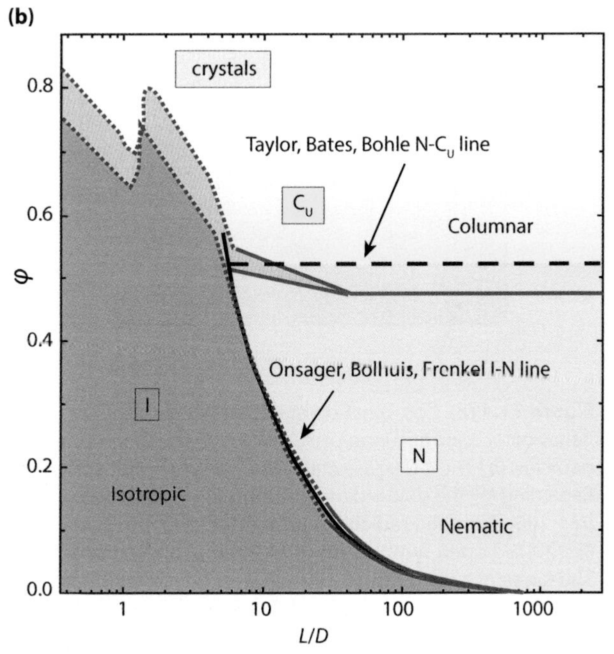

Figure 11.3 (a) Experimental data for DNA concentration, *c*, versus oligomer length, *N*, showing phase behavior for long and short (nano) DNA. **(b)** Phase diagrams from several theoretical models of interacting rodlike particles of axial ratio *L/D* (end-to-end length/diameter) filling a fraction φ of the total volume. *(Reproduced with permission from Online Supporting Material for Michi Nakata, Giuliano Zanchetta, Brandon D. Chapman, Christopher D. Jones, Julie O. Cross, Ronald Pindak, Tommaso Bellini, and Noel A. Clark, "End-to-End Stacking and Liquid Crystal Condensation of 6– to 20–Base Pair DNA Duplexes," Science 318 (23 November 2007), 1276–1279.)*

of DNA. This gave them the aspect ratio needed to satisfy Onsager's criterion. The stacking process was illustrated in Figure 1.7 (a) of Chapter One and we reproduce it here for convenience as Figure 11.5.

The end-to-end attraction of the duplexes results from the hydrophobicity of the faces of their terminal base pairs. We are familiar with this hydrophobic interaction

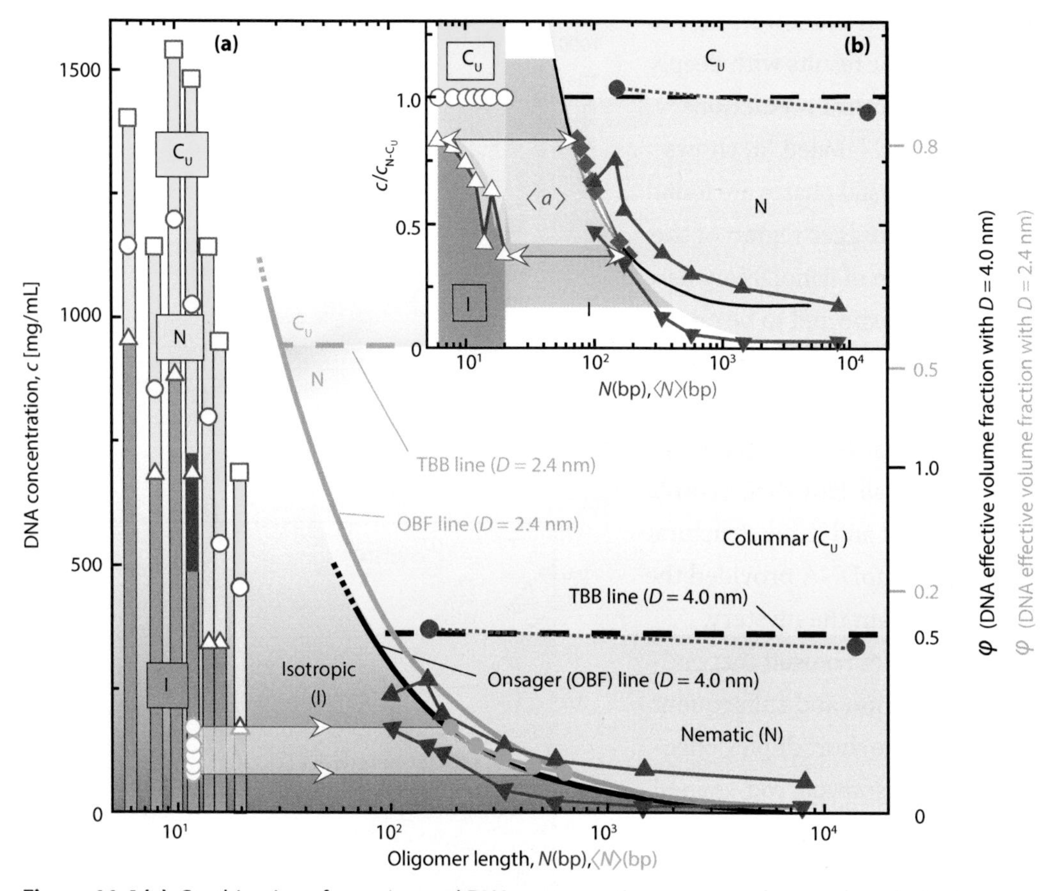

Figure 11.4 (a) Combination of experimental DNA concentration, c, versus oligomer length, N, phase behavior for nano and long DNA, along with the theoretical behavior from several models of interacting rodlike particles. **(b)** The c-N phase diagram of (a) but with c scaled (also called normalized) with respect to (c_{N-C_U}). Regarding (a), the solid red triangles and solid red curve bound the measured I-N phase coexistence for lDNA ($N > 100$). The solid red circles and red dotted line give the measured N-C_U phase boundary of lDNA. For $N < 20$, red open symbols mark the several phase transitions (I-N, triangles; N-C_U, circles; C_U- C_2, squares). The range of each phase is indicated by colored columns (I, magenta; N, cyan; C_U, yellow), at $T = 20$ °C for $20 > N > 8$ and $T = 10$ °C for $N = 6$. The solid red rectangle gives the range of the $N = 12$ liquid crystal phase obtained by another set of researchers. Theoretical phase boundaries for these transitions from model systems are shown for two choices of the volume fraction φ axis. One uses the DNA effective electrostatic diameter $D_{eff} = 4.0$ nm (heavy black lines/labels), applicable at low c. The other uses the DNA chemical diameter $D = 2.4$ nm (heavy orange lines/labels), applicable at high c (i.e., small N). The $D_{eff} = 4.0$ nm phase diagram (black OBF I-N line and dashed black TBB N-C_U line) accounts well for the lDNA I-N and N-C_U data. The open and closed green dots represent, respectively, the results of a simulation (where $L = D = 4.0$ nm), and their effective aggregate lengths ⟨N⟩ at the I-N transition. As to (b), it is the C_U-N phase diagram of (a), but with c scaled with respect to (c_{N-C_U}). This enables an estimate of the length ⟨N⟩(purple diamonds) and aggregation number ⟨a⟩ (blue arrows) in the sDNA aggregates. *(Reproduced with permission from Michi Nakata, Giuliano Zanchetta, Brandon D. Chapman, Christopher D. Jones, Julie O. Cross, Ronald Pindak, Tommaso Bellini, and Noel A. Clark, "End-to-End Stacking and Liquid Crystal Condensation of 6- to 20-Base Pair DNA Duplexes," Science 318 (23 November 2007), 1276–1279.)*

from the formation of lipid bilayers in membranes. Moreover, we had a sneak preview of how hydrophobicity drives the stacking of duplexes in Figure 5.2 of Chapter Five.

long DNA (lDNA) This denotes conventional DNA, as opposed to nanoDNA.

Columnar phase A liquid crystal phase consisting of molecules arranged in parallel columns. The columns are free to bend and to shift with respect to one another along the columnar direction.

Nematic-Columnar (N-C_U) phase transition A liquid crystal phase transition from a phase in which the molecules exhibit a degree of orientational order (nematic phase) to a phase in which the molecules exhibit a degree of translational order as well as a degree of orientational order (columnar phase).

Living polymerization systems provide a good model for the experimental observations of reversible end-to-end self-assembly of the oligomers into linear chains.[21] An example of such a system in nature is the protein actin that participates in important cellular functions such as muscle contraction and cell motility. Actin can form filament chains—biomolecular aggregations—without irreversible chain breaking. And filamentary actin is omnipresent in cell structure. As chemist Sandra Greer observed, "The thrill of science comes when we can find common features among disparate systems and are able to construct a common model to describe those various systems."[22]

If you are pleased to accept the take-home message from inset (b) in Figure 11.4 at face value, then the following subsection (explaining specifics of the figure) may be skipped with no loss of continuity. If you skip, please go to the subsection *Shifting Gears.*

The Details

There are several delicate issues at play in inset (b). If you have a thirst for such subtlety and want to understand how the take-home message was deduced, we present some of these fine points. One is the effective diameter of DNA duplexes. To begin, we return to the main panel, (a). We first observe that the symbol "N" is used in two different ways in this study: as the number of base pairs and also as the nematic phase in phase transition notation. We see that the lDNA I-N phase boundary measured for 100 bp < N < 8,000 bp is in reasonable agreement with the OBF (Onsager-Bolhuis-Frenkel) line. But this result requires that the effective double-helix diameter is taken to be D_{eff} = 4.0 nm to account for the electrostatic repulsion between chains at low concentration. What does this mean?

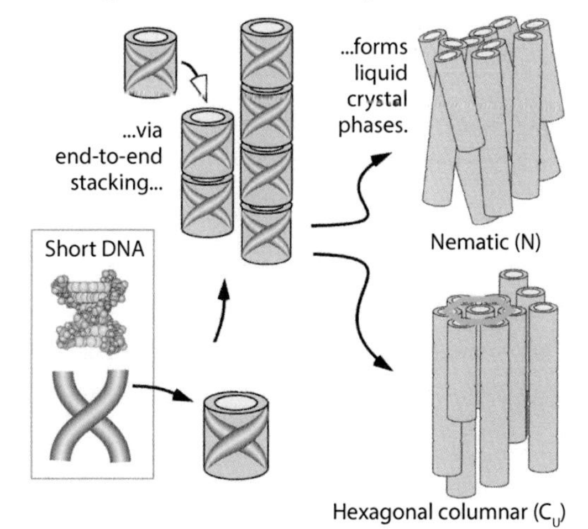

Figure 11.5 Short segments of DNA (nanoDNA) pair up end-to-end to form liquid crystal phases. *(Reproduced with permission from Michi Nakata, Giuliano Zanchetta, Brandon D. Chapman, Christopher D. Jones, Julie O. Cross, Ronald Pindak, Tommaso Bellini, and Noel A. Clark, "End-to-End Stacking and Liquid Crystal Condensation of 6- to 20-Base Pair DNA Duplexes," Science 318 (23 November 2007), 1276–1279.)*

Earlier we introduced the concept of excluded volume. The effective diameter is a parameter that characterizes the excluded volume of DNA by accounting for the conformational properties of actual electrically charged DNA, rather than DNA idealized as hard rods.[23] Because duplex DNA is a highly negatively charged molecule, electrostatic repulsion can make the effective diameter much greater than the geometrical diameter that is used when considering idealized hard rods. The negative charge of the DNA phosphate groups and base pairs attracts small cations from solution (usually Na^+). These cations—ions with a positive charge—create a positively charged cloud around the strands of the DNA duplex. The cation cloud partially screens the electrostatic interaction between strands, but nonetheless, the steric/hard excluded volume effects we spoke of earlier are now augmented by electrostatic interactions. The choice of $D_{eff} = 4.0$ nm for the DNA duplex also puts the lDNA experimental N-C_U phase boundary (solid red dots connected by dotted red line) at an excluded volume, $\phi = 0.55$. This agrees with the line derived from modeling long rods, the Taylor, Bates, Bohle (TBB) line in panel (a). (The intercept of the horizontal black dashes with the right-hand vertical axis is at $\phi = 0.55$.) Thus, the lDNA I-N phase boundary can be interpreted in terms of an electrostatically swollen DNA diameter, $D_{eff} = 4.0$ nm.

While the lDNA I-N phase boundary can be interpreted in terms of an electrostatically swollen DNA diameter, it is impossible to do so for the nanoDNA I-N data on the basis of shape factors alone. This is because any reasonable diameter yields axial ratios where there are no liquid crystal phases in any of the models. So how to proceed? Here's the path forged by Bellini and Clark.

In the case of nanoDNA, the liquid crystal phases appear in the $N = 20$ oligomer duplexes (i.e., 20 bp duplexes) at concentrations in the range of those of the lDNAs—in spite of lacking an L/D sufficiently large to enable liquid crystal ordering in the hard rod model. In fact, they noted that for $N = 20$, the N-C_U transition concentration for nanoDNA was nearly the same as the $N = 20$ N-C_U transition concentration for lDNA.

Effective diameter, D_{eff} A parameter that characterizes the excluded volume of DNA by accounting for the conformational properties of actual electrically charged DNA, rather than idealized hard rods.

Second, they observed that for lDNA, the DNA concentration at the N-C_U transition did not depend on L, the length of the DNA molecules. This suggested that the phase behavior of the $N = 20$ oligomers of nanoDNA might be understandable on the basis of the hard rod model, if their effective length, L_{eff}, was appropriately adjusted. In the simplest picture this assumes end-to-end aggregation into units of total length L_{eff} that are sufficient to increase the $N = 20$ L_{eff} to contact the Onsager line. This is shown in the "construction" indicated by the horizontal double-headed arrows in panel (b). For the $N = 20$ case this implies aggregates of average length $\langle N \rangle \sim 200$ bp, i.e., consisting of ~10 oligomer duplexes. (We note that the symbol $\langle a \rangle$ in panel (b) represents the mean number of oligomer duplexes in the aggregates.)

Now have a look at the open green dots at the bottom left of panel (a). This is data from a computer simulation performed by another group.[24] They used idealized spheres with $L = D = 4$ nm to model the aggregation of such spheres into semiflexible polymer chains of arbitrary length. The construction (the right-pointing arrows) with the closed green dots at the bottom center of panel (a) shows the effective lengths, N, now measured in base pairs, of the aggregates at which the nematic phase appears in their simulations. And these lengths match the OBF line. This indicates that the model aggregates behave effectively as hard rods and provides part of the rationale for a similar construction in panel (b).

Effective length L_{eff} A length introduced to make the phase behavior of nanoDNA obey the Onsager criterion. The simplest way to do this is to assume that the nanoDNA duplexes aggregate into units of total length L_{eff}.

At the high concentrations where liquid crystal phases are found for the shortest oligomers, the distance between duplexes lessens and approaches the so-called chemical diameter where simple steric repulsion dominates the electrostatic interactions. Thus, with decreasing N and increasing concentration, the effective chain diameter evolves from $D \sim 4$ nm to $D \sim 2.4$ nm. (That is, the effective diameter is an adjustable parameter in the models.)[25] This change shifts the model phase boundaries to higher concentrations accounting for the increased concentration necessary for liquid crystal phase formation. This is shown in panel (a) by shifting from the black ϕ scale to the orange ϕ scale (labels appearing on the far right of the figure). The black ϕ applies when $D \sim 4$ nm; the orange ϕ applies when $D \sim 2.4$ nm.

Scaling of data is a widely used means of analysis, for example, when variables span different ranges. Scaling transforms the data values to another range and this often entails both a shift and a change of the scale, i.e., a magnification or reduction.

The effective diameter variation is said to be scaled out in panel (b), where the concentration axis (the vertical axis on the left of the figure) is normalized by the number of mg/mL at the N-C_U transition, denoted by (c_{N-C_U}). This assumes that the DNA concentration at the N-C_U transition (c_{N-C_U}) corresponds, for each oligomer, to an effective volume fraction $\phi \approx 0.55$. That is, the diameter, D, appropriate for a given oligomer is that which makes its N-C_U transition occur at $\phi \sim 0.55$. This scaling yields $D = 4.0$ nm from the lDNA N-C_U data, comparable to that required to fit the lDNA I-N data.

The scenario of end-to-end stacking may be further analyzed in the frame of panel (b). Once the effective volume fraction is held by the strict requirement placed by the N-C_U transition, the amount of linear aggregation can be evaluated by horizontally projecting the I-N data points onto the expected I-N L/D for linear aggregation. The construction (horizontal double-headed arrows in panel (b)) thus gives an estimate of $\langle N \rangle$ (shown as purple diamonds), and $\langle a \rangle$. These are, respectively, the mean number of base pairs (= total aggregate lengths in base pairs) and the mean number of duplex oligomers in an aggregate necessary to generate an Onsager nematic. For the $N = 6, 8, 10, 12, 14, 16, 20$ bp oligomers, the mean numbers of oligomer duplexes in the aggregates, $\langle a \rangle$, and the total aggregate lengths in base pairs, $\langle N \rangle$, obtained from panel (b) are, respectively, $\langle a \rangle = 12, 10, 9, 8, 11, 6, 9$ and $\langle N \rangle = 75, 80, 87, 97, 160, 100, 180$ bp.

If you'd like to check your understanding of this, the way to obtain the $\langle N \rangle$ values is to guesstimate them from the horizontal axis (a logarithmic axis) based on the position of the purple diamonds. (Remember, for a logarithmic axis the intervals correspond to orders of magnitude (i.e., factors of ten), rather than the more familiar linear scale.) Then, to find the a values, divide $\langle N \rangle$ by N. For example, for the purple diamond representing $\langle N \rangle = 14$, you'll find that $\langle N \rangle = 160$, and $\langle a \rangle = 160/14 \approx 11$. In the foregoing, be aware that there is a small deviation from monotonic order in the values of $\langle N \rangle$ for the values of $\langle N \rangle = 14$ and $\langle N \rangle = 16$, as seen in the plot of the open red triangles on the left side of panel (b).

Shifting Gears

After dwelling on the niceties of nano versus long DNA for some time, we take a moment for reflection. Cellular DNA contains all of the information required for an organism to function and DNA molecules achieve formidable sizes. Nature packs DNA so it is secluded from ordinary cell activities while remaining accessible for replication or transcription (the process of making an RNA copy of a DNA molecule; the RNA transcript is then used to make proteins). Efficient packing requires DNA ordering; accessibility requires a fluid structure. *These are the defining characteristics of liquid crystals, as we explored in an earlier chapter, and speak to why DNA and liquid crystals are so inextricably linked together.*[26]

Now back to our story. Michi Nakata not only obtained data that were striking; she also took arresting photographs through the depolarizing microscope. In Giuliano Zanchetta's Ph.D. thesis he called her an artist of science. Before moving on to the second experimental surprise provided by nanoDNA, we present Figure 11.6. This shows the optical textures of the liquid crystal phases of a series of solutions of nanoDNA of increasing length.

Figure 11.6 Optical textures of the liquid crystal phases of a series of solutions of nanoDNA of increasing length obtained by depolarized light microscopy, viewed through crossed polarizers. Isotropic regions are black. N is the nematic phase, C_U is the columnar phase, and C_2 is a phase showing dendritic growth forms (meaning branching like a tree) indicative of more solid-like ordering at still higher DNA concentration. Labels show the number of base pairs (bp), the base sequences, and the length of a single duplex. The width of each image is 120 microns (=120,000 nm). *(Reproduced with permission from Michi Nakata, Giuliano Zanchetta, Brandon D. Chapman, Christopher D. Jones, Julie O. Cross, Ronald Pindak, Tommaso Bellini, and Noel A. Clark, "End-to-End Stacking and Liquid Crystal Condensation of 6– to 20–Base Pair DNA Duplexes," Science 318 (23 November 2007), 1276–1279.)*

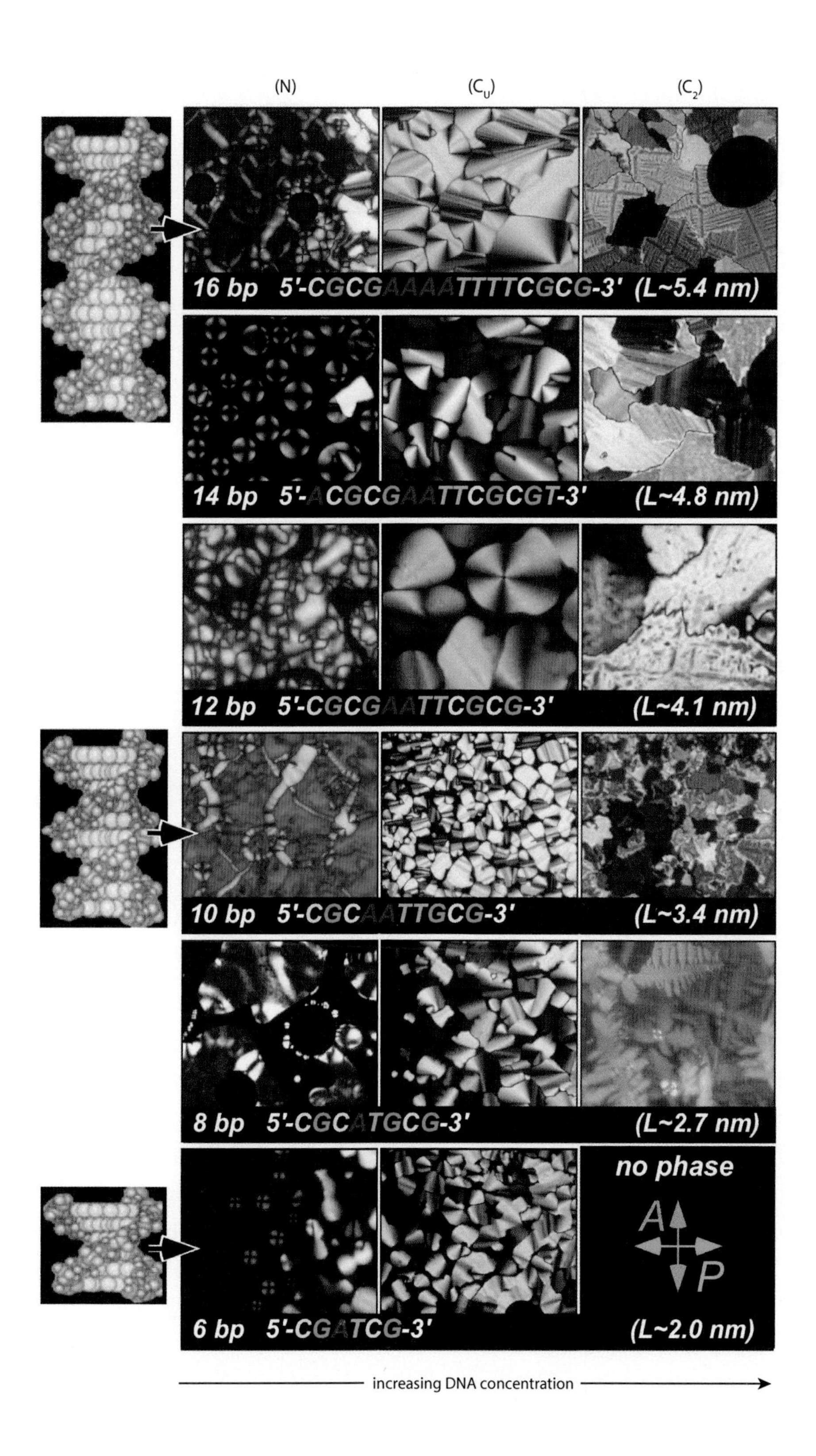
(N)
(C_U)
(C_2)
16 bp 5'-CGCGAAAATTTTCGCG-3' (L~5.4 nm)
14 bp 5'-ACGCGAATTCGCGT-3' (L~4.8 nm)
12 bp 5'-CGCGAATTCGCG-3' (L~4.1 nm)
10 bp 5'-CGCAATTGCG-3' (L~3.4 nm)
8 bp 5'-CGCATGCG-3' (L~2.7 nm)
no phase
A
P
6 bp 5'-CGATCG-3' (L~2.0 nm)
increasing DNA concentration

Phase Separation Into Liquid Crystal Droplets

After discovering nanoDNA liquid crystal phases, the team played with the temperature of the nanoDNA solution. They found, not surprisingly, that as the temperature was increased the DNA helices tended to break resulting in single-stranded disordered coils, a process known as denaturation (Figure 11.7). This led to their interest in examining the effect added impurities had on the liquid crystal phases. In particular, they wanted to explore the role of the denaturation of the helix itself, because after many duplexes became coils, the nanoDNA liquid crystal phases vanished. To make a system inherently made up of helices and coils, they combined the two 12 bp mutually complementary sequences, namely, CCTCAAAACTCC along with GGAGTTTTGAGG. They mixed them in unbalanced ratios.

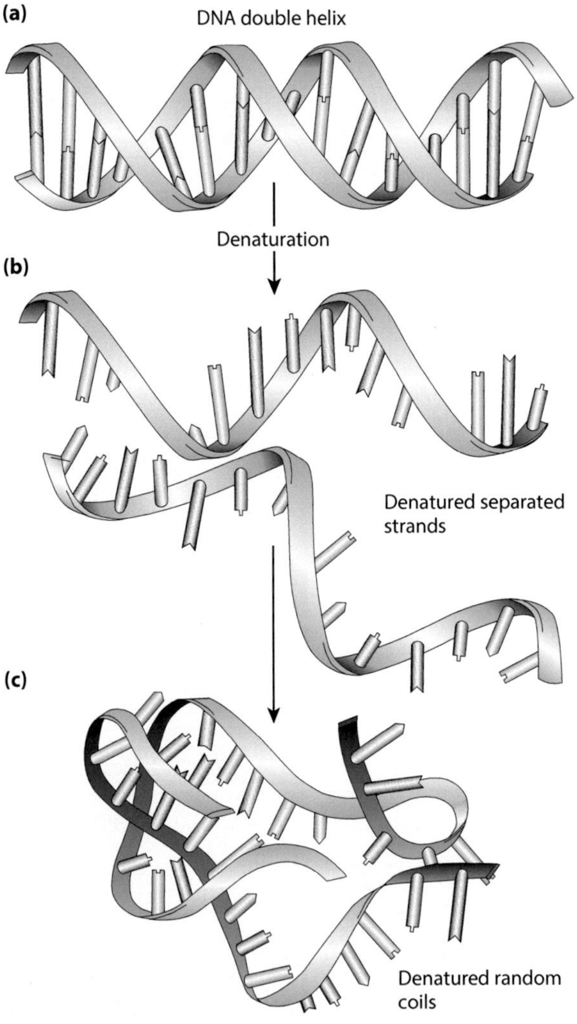

(a) DNA double helix

Denaturation

(b)

Denatured separated strands

(c)

Denatured random coils

Figure 11.7 DNA helix to coil transition.

This led to a stunning and unforeseen result. At some single-stranded DNA concentration, liquid crystal phases were expected to become disturbed and disappear. But this did not happen. Instead, a phase separation occurred between the helices and the pool of coils, with double-stranded DNA organizing into liquid crystal droplets.[27]

Consider this from another perspective. For liquid crystal phases of nanoDNA to form, the duplexes must stack to develop sufficient shape anisotropy (and rigidity). This puts constraints on the structure of the individual duplexes. So base sequence substitutions that reduce the ability of the nanoDNA duplexes to form linear rigid aggregates will reduce the stability of the liquid crystal phases.

For example, the addition of unpaired bases at the nanoDNA duplex ends, eliminates liquid crystal ordering by weakening end-to-end adhesion. It is the interplay between base sequence and liquid crystal ordering that leads to a means of condensation of complementary nanoDNA duplexes from mixed solutions of complementary and non-complementary oligomers.

This process is demonstrated in the cartoon of Figure 11.8 for a mixture of the mutually complementary, but not self-complementary, CCTCAAAACTCC (called "Z") + GGAGTTTTGAGG (called "Y") 12-nucleotide oligomers. In such a solution, at $T < T_m$, the DNA melting temperature, the complementary pairs form rigid duplexes that have a tendency to aggregate end-to-end. But the non-complementary oligomers remain as flexible single strands.

Liquid crystal (LC) ordering shows up as a phase transition. Along with the transition there is nearly complete phase separation of the duplexes: duplex DNA is sequestered into the columnar liquid crystal domains (for $T < T_{LC}$). The highly flexible single-stranded DNA remains in the isotropic phase. To summarize the content in Figure 11.8 (d) have a look at Figure 11.9.

Intrigued by the results, Bellini and Clark performed additional experiments using optical microscopy to study phase behavior in mixtures of 12- to 22-bp-long nanoDNA oligomers.[28] In each case they found that when the concentrations were high enough, mixtures phase-separated via the nucleation of duplex-rich liquid crystalline domains from an isotropic background rich in single strands. Then, depending on both concentration and temperature, they were able to observe either nematic or columnar phases that

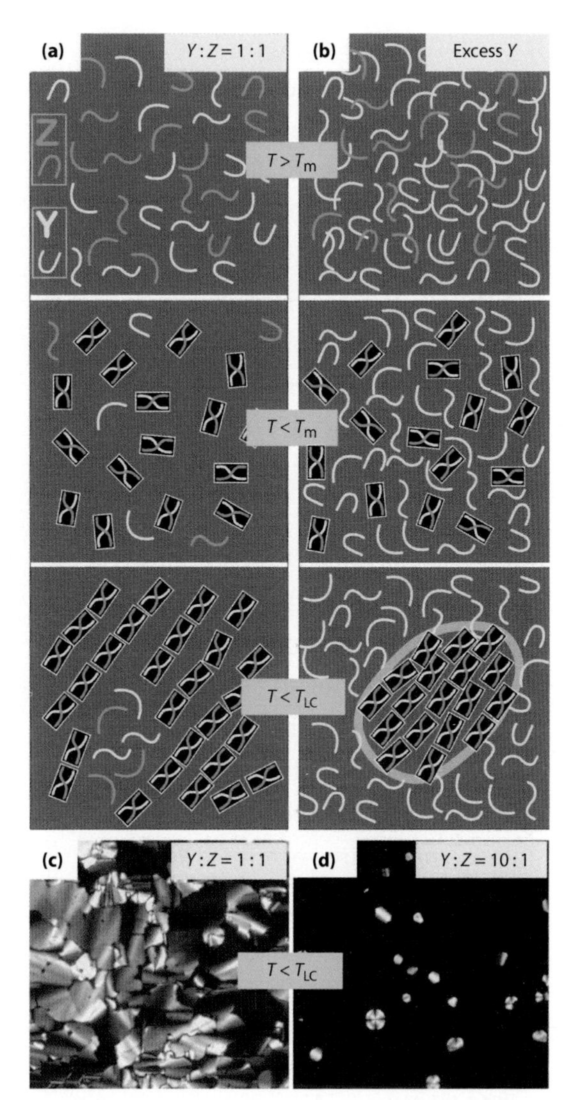

Figure 11.8 (a) In a 1:1 mixture of complementary but different nanoDNA oligomers, duplexes form upon cooling below their melting temperature, T_m and the liquid crystal phase appears below T_{LC} via a phase transition, filling nearly the whole area with liquid crystal domains, panel **(c)**. In this case the phase is C_U. **(b)** With one of the oligomer species in excess (Y, shown as yellow), the transition to the liquid crystal phase is marked by the appearance of isolated liquid crystal domains (C_U) that sequester all of the other complementary oligomer (Z, shown as green) into liquid crystal droplets. **(c)** and **(d)** Are depolarized light microscopy images of liquid crystal domains; the area occupied by the domains in the Y:Z = 10:1 case is consistent with an essentially complete condensation of Z oligomers into the liquid crystal phase. *(Reproduced with permission from Online Supporting Material for Michi Nakata, Giuliano Zanchetta, Brandon D. Chapman, Christopher D. Jones, Julie O. Cross, Ronald Pindak, Tommaso Bellini, and Noel A. Clark, "End-to-End Stacking and Liquid Crystal Condensation of 6- to 20-Base Pair DNA Duplexes," Science 318 (23 November 2007), 1276–1279.)*

sequestered these duplexes from the isotropic. The team determined that the phase separation was approximately complete, and corresponded to a spontaneous purification of duplexes from the single-strand oligomers.

The investigators interpreted the phase separation behavior in these experiments as the combined result of several factors. One was the lowering of free energy by the base pairing of single strands to form double strands (hybridization). Another was the further lowering of free energy from the end-to-end stacking of duplexes. A third factor was depletion-type attractive interactions (more on this in a moment) favoring the segregation of the more rigid duplexes from the flexible single strands, another free-energy-lowering reaction.

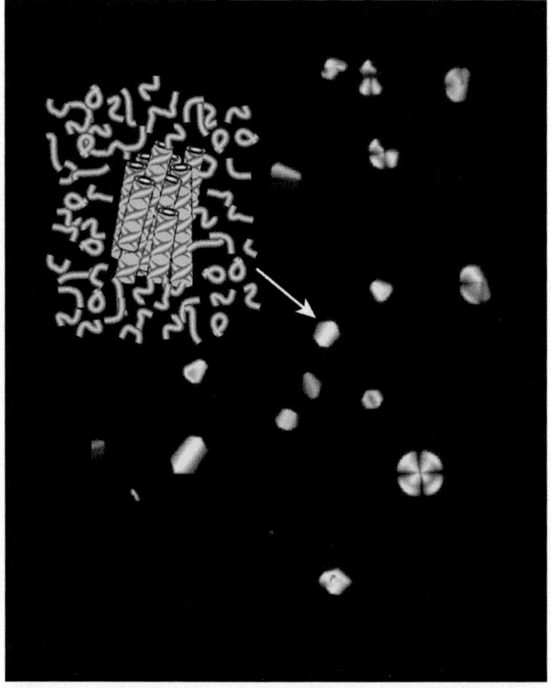

Figure 11.9 Self-selection of complementary oligomers by liquid crystal phase separation (bright domains). Single strands (orange) are expelled from the liquid crystal droplets composed of duplexes (green). The liquid crystal droplets scavenge duplexes from the isotropic phase (black), resulting in nearly complete phase separation. *(Adapted with permission from Michi Nakata, Giuliano Zanchetta, Brandon D. Chapman, Christopher D. Jones, Julie O. Cross, Ronald Pindak, Tommaso Bellini, and Noel A. Clark, "End-to-End Stacking and Liquid Crystal Condensation of 6– to 20–Base Pair DNA Duplexes," Science 318 (23 November 2007), 1276–1279.)*

Time for a note on thermodynamics. Equilibrium thermodynamics often comes down to competing processes: some lower the overall free energy and some raise the free energy. If the net result is a lowering, then the process is energetically favored. To keep the record straight, we are guilty of introducing the term *free energy* in Chapter Six in an *ad hoc* way. However, for those interested, we've cited in the endnotes a concise primer on equilibrium thermodynamics, written by Peter Atkins, an accomplished chemist at the University of Oxford.[29]

Bellini and Clark discussed two theoretical models for nanoDNA phase behavior in their papers.[30] Each theoretical approach shows that the origin of the phase condensation is the contrast in rigidity of single-strand DNA and elongated/stacked double-stranded DNA. We will provide both explanations. The first is a general mechanism for phase separation owing to what is known as the depletion interaction (also called depletion attraction or the depletion effect).[31] The second is a mathematical model that demonstrates the phase separation of a mixed solution of rigid and flexible rods promoted by the entropic gain (to be described) associated with the ordering of the rigid rods.

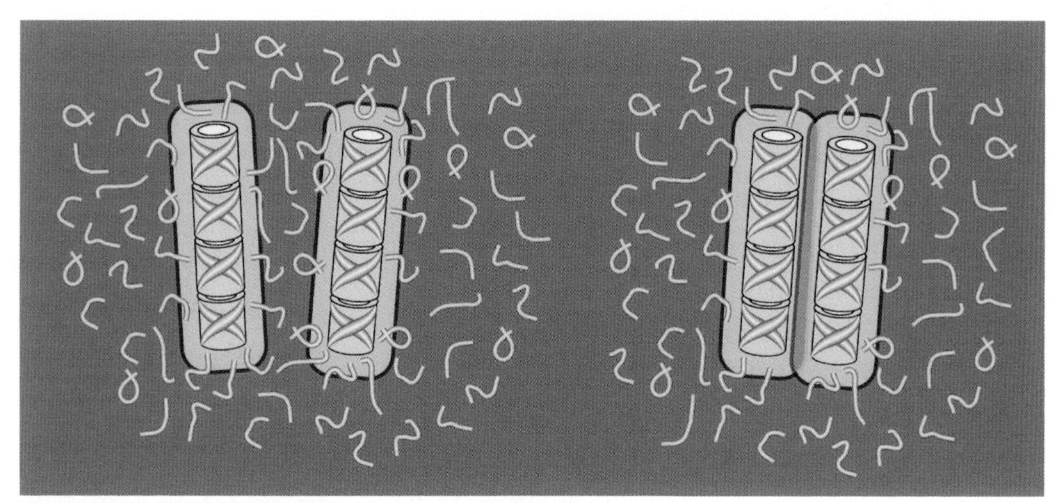

Figure 11.10 The depletion interaction. *(Courtesy of Noel A. Clark.)*

The Depletion Interaction

Imagine you have two large particles surrounded by a sea of smaller particles, shown on the left in Figure 11.10 as rods of two different sizes.[32] Although we've shown the larger rods as stacked DNA duplexes and the smaller ones as single strands of DNA, the depletion effect has broad applicability beyond DNA.

When the large particles are far apart they experience uniform pressure from the small particles over their whole surface. When the large particles come close together a region exists between the large particles where the small particles can no longer fit. The reduction of small particle density in the space between large particles that are sufficiently close together is referred to as depletion. The pressure exerted on the large particles is no longer uniform. The result is that the large particles are pushed together by an effective attraction (the depletion interaction).

The attractive force is usually described in terms of entropy, a subject we visited at the beginning of this chapter. Thus we speak of entropic forces and call upon one of our stalwarts, viz., minimization of free energy. The free energy is reduced due to what is called enhanced entropic freedom of the small particles. What this means is that the volume available to the small particles is increased by the aggregation of the large particles and thus the entropy of the small particles is increased.[33,34]

One way to understand the increase in volume available to the small, flexible particles is to characterize the large particles by a depletion zone around them. We show the

Depletion interaction In a solution that contains solutes of two distinct sizes—large and small—the small particles will exert pressure on the large particles from all sides. But when two large particles come close together, the small ones are excluded from the volume between the two. This produces a force on opposite sides of the two large particles that can be thought of as an effective attraction (the depletion interaction) between the two.

depletion zone by the black lines surrounding the stacked duplexes, Figure 11.10. The thickness of the depletion zone is known as the radius of gyration of the small particles, Figure 11.11. The radius of gyration, R_g, measures the dimension of a flexible polymer chain as its effective size changes dynamically due to its flexibility.[35]

On average, the small particles are excluded from the depletion zone.[36] When the large particles are so close that their depletion zones overlap, the net size of the depletion zone of the two large particles is decreased since the small particles are too big to enter the dark purple region as suggested on the right side of Figure 11.10. Thus, the volume available to the small particles is increased. This produces an entropy-driven phase separation of the small and large particles.

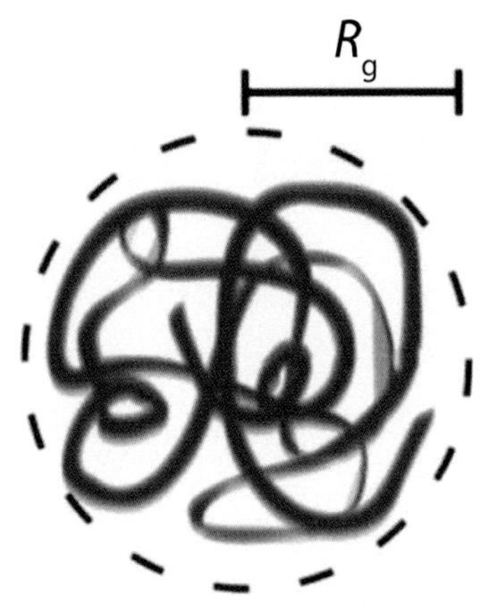

Figure 11.11 Radius of gyration. *(Reproduced with permission from Danielle J. Mai, Christopher Brockman, and Charles M. Schroeder, "Microfluidic Systems for Single DNA Dynamics," Soft Matter 8 (2012), 10560–10572.)*

Radius of gyration, R_g
A way to measure the dimension of a flexible polymer chain in which its effective size dynamically changes owing to its flexibility.

The phase separation occurs because the gain in entropy associated with the increase in the volume available to the small particles overcomes the loss of mixing entropy. The mixing entropy is lowered because there is a smaller volume available for the two differently sized particles to mix when the small particles can no longer enter the volume between the larger particles. The entropy-driven phase separation is what is referred to as an entropic force.[34,37,38]

Now that we're acquainted with the mechanism behind the depletion interaction, we return to Bellini and Clark. They were able to support their interpretation of the strong phase separation in their experiments by calling on a number of precedents in the literature. For example, the packing of rigid solutes by means of the depletion interaction promoted nematic ordering of rigid rodlike particles mixed with flexible polymers.[39] The physical chemists who authored this 1994 paper pointed out that the underlying phase separation was not only of fundamental interest but also of considerable biological and technological importance. By way of illustration, this phase separation was observed over two centuries earlier in patients with a variety of illnesses. The aggregation of their red blood cells caused a significantly increased sedimentation rate. This was likely induced by the depletion mechanism associated with increased concentrations of protein molecules in the blood.[40]

Bellini and Clark gave another example in the case of long DNA where compression into liquid crystal phases was obtained by adding a hydrophilic flexible polymer to a solution of long DNA.[41] Here too, the authors of the publication pointed to the practical relevance of their fundamental studies, specifically to overcoming the obstacles to making gene therapy work.[42]

As an additional instance, Bellini and Clark noted the case of protein crystallization resulting from protein separation in mixtures with polyethylene glycol solutions.[43] This experiment was performed in the laboratory of Clark's own Ph.D. thesis advisor, George Benedek, at MIT. The study is of particular relevance in the context of nanoDNA liquid crystal condensation since the size of the proteins (as small as 1.5 nm radius) and the flexible chains (as small as an average coil radius of 0.6 nm) closely approximates the size of double- and single-stranded nanoDNA.[30]

Flory's Model

The second theoretical explanation for the nanoDNA phase separation observed by Bellini and Clark is the work of chemist Paul J. Flory who published his result four years after he won the Nobel Prize for Chemistry.[44] Flory's classic paper showed that the origin of the phase condensation can be seen as the contrast in rigidity of single-stranded DNA and the stacked double-stranded DNA. To appreciate Flory's approach, think of the bending flexibility of DNA (either single- or double-stranded) in terms of the bending of a thin rod. While you do this, we'll enhance our lexicon with another morsel: bend persistence length. We won't spell out the formal definition but rather the import of the term.

Remember that we're looking at nanoDNA in solution and there is a lot of thermal energy present. All the solvent molecules (not just the solute DNA) in the solution are constantly in motion, colliding with each other and bouncing back and forth. If we consider the DNA as a thin, elastic rod immersed in a fluid, the rod will be randomly bent by thermal motion at different places along its length if that length is greater than its bend persistence length. That is, segments of the rod will point in random, uncorrelated directions if their separation is greater than the bend persistence length.

Alternatively, as biophysicist Philip Nelson wrote, it is only over separations less than the bend persistence length that the rod remembers which way it was pointing, hence the name persistence length.[45] A quantitative standard is long duplex DNA that typically has a persistence length of about 50 nm or about 150 base pairs (bp).

Bend persistence length A quantity that characterizes the bending rigidity, or alternatively, the flexibility, of DNA. Choose a segment of the DNA. If the segment is shorter than the bending persistence length, then along that segment the DNA can be thought of as a rigid rod with roughly a single direction. If the chosen segment of DNA is longer than the bend persistence length, then that segment can be thought of as a flexible rod with a direction that varies randomly from one location (within the chosen segment) to the next.

Solute, Solvent A solute is the component in a solution that is present in the lesser amount, while a solvent is the component in a solution that is present in the largest amount.

Bellini and Clark mixed flexible single-stranded DNA oligomers with a bend persistence length of about 1–2 nm with rigid duplexes that stack to form semi-rigid aggregates with a bend persistence length of about 20 nm.[30] It is this disparity in the bend persistence lengths of the two species—single strands and stacked duplex strands—that are at the heart of the Flory theory. As was true for the depletion interaction, Flory's calculations were not limited to DNA. He considered a very general system consisting of a solvent and two solutes, one a rigid rod, and the other a randomly coiled polymer chain. Flory found nematic/isotropic phase coexistence: the rigid species preferred the nematic phase and the flexible species preferred the isotropic phase. The segregation of the two species was essentially complete in the Bellini/Clark experiments.

Flory's paper was mathematically rigorous, but at the end of the paper he skillfully sketched his results in an informal way. It's pleasing to read his unadorned (and anthropomorphic) description of rodlike particles phase-separating in a mixture: "From a physical point of view, the isotropic phase offers to a rodlike solute none of the advantages of orientation. Hence, this component prefers the anisotropic [nematic] phase where obstruction by neighboring species is alleviated by mutual alignment. . . . The anisotropic phase approaches the selectivity of a pure crystal in its rejection of a foreign component, in this case the random coil."[46]

To conclude our discussion of phase separation we note that it is a collective effect of duplex and single-stranded nanoDNA. Whether one uses the depletion interaction argument or Flory's model, both double- and single-stranded nanoDNA are at play. In the solutions Bellini and Clark examined, duplexes alone would not form liquid crystals because their concentration is below that required for liquid crystal formation.[47]

Bellini and Clark's key finding from their phase separation experiments is that in a mixture of complementary and non-complementary nanoDNA, the complementary DNA is found only in the liquid crystal domains. Thus, in the nanoDNA liquid crystal phases, the hierarchy of coupled steps—duplexing, end-to-end stacking of duplexes, and liquid crystal phase ordering and separation—create a structural gatekeeper that enables only DNA that is capable of duplexing to enter. DNA molecules that do enter are then organized into a structure that, if stabilized by covalent links (a process known as ligation), should enhance their complementarity and the liquid crystal phase stability.

Bellini and Clark grabbed the measurable series of phenomena summarized above and fused them with a piercing, inventive guess. In the next chapter we begin to explore the weighty implications of their guess.

We opened this chapter with a look at Noel Clark's exceptional postdoc, Michi Nakata, at work in the lab (Figure 11.12). Her foundational contributions to the experimental study of nanoDNA have resulted in many publications. Sadly, all of them appeared posthumously as Nakata died tragically at age 30. Clark said of her, "She was a superstar—somebody that would come along once in a couple of generations."[48]

"To an Athlete Dying Young," is a poem in A.E. Housman's volume of poetry titled *A Shropshire Lad*. The fifth stanza of the poem reads:

"Now you will not swell the rout

Of lads that wore their honours out,

Runners whom renown outran

And the name died before the man."

Michi Nakata

November 21, 1975

September 30, 2006

Figure 11.12 *(Courtesy of Noel A. Clark.)*

Exercises for Chapter Eleven: Exercises 11.1–11.2

Exercise 11.1

How would the end-to-end distance of (a) a random coil and (b) a rodlike molecule vary with increasing temperature?

Exercise 11.2

Suppose we have two hard spherical gas molecules. What is the excluded volume of one molecule with respect to the other molecule?

ENDNOTES

[1] W.T. Astbury, "X-Ray Studies of Nucleic Acids," *Symposia Soc. Exp. Biol. 1* (1947), 66–76.

[2] J.E. Lydon, "The DNA Double Helix—the Untold Story," *Liquid Crystals Today 12, No. 2* (2003), 1–9.

[3] Unpublished.

[4] http://www.japantimes.co.jp/culture/2010/12/24/music/japan-makes-beethovens-ninth-no-1-for-the-holidays/#.VfehQL5fHW4

[5] Lars Onsager, "The Effects of Shape on the Interaction of Colloidal Particles," *Annals of the New York Academy of Science 51* (1949), 627–658.

[6] H. Christopher Longuet-Higgins and Michael E. Fisher, "Lars Onsager," *Biographical Memoirs, National Academy Press 60* (1991), 183–233.

[7] Ibid., 212.

[8] Giuliano Zanchetta, "Liquid Crystalline Phases in Oligonucleotide Solutions," Ph.D. dissertation, Università Degli Studi Di Milano (2007), 34.

[9] E. van den Pol, A. Lupascu, M.A. Diaconeasa, A.V. Petukhov, D.V. Byelov, and G.J. Vroege, "Onsager Revisited: Magnetic Field Induced Nematic-Nematic Phase Separation in Dispersions of Goethite Nanorods," *J. Phys. Chem. Lett. 1* (2010), 2174–2178.

[10] Daan Frenkel, "Order Through Entropy," *Nature Materials 14* (January 2015), 9–12.

[11] Zanchetta, "Liquid Crystalline," 34.

[12] Alessandro Speranza and Peter Sollich, "Simplified Onsager Theory for Isotropic-Nematic Phase Equilibria of Length Polydisperse Hard Rods," *J. Chem. Phys. 117, No.11* (15 September 2002), 5421–5436.

[13] Peter Bolhuis and Daan Frenkel, "Tracing the Phase Boundaries of Hard Spherocylinders," *J. Chem. Phys. 106, No. 2* (8 January 1997), 666–686.

[14] Tommaso Bellini, "Exploring Soft Matter With DNA," *Frontiers of Soft Matter 2012 Symposium* (18 May 2012), Boulder, CO, slide 14.

[15] Unpublished.

[16] Michi Nakata, Giuliano Zanchetta, Brandon D. Chapman, Christopher D. Jones, Julie O. Cross, Ronald Pindak, Tommaso Bellini, and Noel A. Clark, "End-to-End Stacking and Liquid Crystal Condensation of 6– to 20–Base Pair DNA Duplexes," *Science 318* (23 November 2007), Supporting Online Material, Figure S1, 16 (linked to the online version of the paper as a PDF download).

[17] Todd M. Alam and Gary Drobny, "Magnetic Ordering in Synthetic Oligonucletides. A Deuterium Nuclear Magnetic Resonance Investigation," *J. Chem. Phys. 92, No. 11* (1 June 1990), 6840–6846.

[18] Richard Wing, Horace Drew, Tsunehiro Takano, Chris Broka, Shoji Tanaka, Keiichi Itakura, and Richard E. Dickerson, "Crystal Structure Analysis of a Complete Turn of B-DNA," *Nature 287* (23 October 1980), 755–758.

[19] Nakata et al., "End-to-End," Supporting Online Material, Figure S2, 16.

[20] Michi Nakata, Giuliano Zanchetta, Brandon D. Chapman, Christopher D. Jones, Julie O. Cross, Ronald Pindak, Tommaso Bellini, and Noel A. Clark, "End-to-End Stacking and Liquid Crystal Condensation of 6– to 20–Base Pair DNA Duplexes," *Science 318* (23 November 2007), 1276–1279.

[21] M. Szwarc, " 'Living' Polymers," *Nature 178* (24 November 1956), 1168–1169.

[22] Sandra C. Greer, "Reversible Polymerizations and Aggregations," *Annu. Rev. Phys. Chem. 53* (2002), 173–200.

[23] Maxim D. Frank-Kamenetskii, "Biophysics of the DNA Molecule," *Physics Reports 288* (1997), 13–60.

[24] Xinjiang Lü and James T. Kindt, "Monte Carlo Simulation of the Self-Assembly and Phase Behavior of Semiflexible Equilibrium Polymers," *J. Chem. Phys. 120, No. 21* (1 June 2004), 10328–10338.

[25] Valentin V. Rybenkov, Nicholas R. Cozzarelli, and Alexander V. Vologodskii, "Probability of DNA Knotting and the Effective Diameter of the DNA Double Helix," *Proc. Natl. Acad. Sci. 90* (June 1993), 5307–5311.

[26] Kunal Merchant and Randolph L. Rill, "DNA Length and Concentration Dependencies of Anisotropic Phase Transitions of DNA Solutions," *Biophysical Journal 73* (December 1997), 3154–3163.

[27] Nakata et al., "End-to-End," Supporting Online Material, Figure S5, 20.

[28] Giuliano Zanchetta, Michi Nakata, Marco Buscaglia, Tommaso Bellini, and Noel A. Clark, "Phase Separation and Liquid Crystallization of Complementary Sequences in Mixtures of NanoDNA Oligomers," *PNAS 105, No. 4* (29 January 2008), 1111–1117.

[29] Peter Atkins, *The Laws of Thermodynamics: A Short Introduction* (Oxford: Oxford University Press, 2010).

[30] Nakata et al., "End-to-End," Supporting Online Material, "Condensation of Duplex Short DNA From a Duplex/Single Stranded Mixture," 13–14 (linked to the online version of the paper as a PDF download).

[31] Sho Asakura and Fumio Oosawa, "On Interaction Between Two Bodies Immersed in a Solution of Macromolecules," *Journal of Chemical Physics 22, No. 7* (1 July 1954) 1255–1256.

[32] Xiaohui Sun, Christophe Danumah, Yang Liu, and Yaman Boluk, "Flocculation of Bacteria by Depletion Interactions Due to Rod-Shaped Cellulose Nanocrystals," *Chemical Engineering Journal 198–199* (1 August 2012), 476–481.

[33] Shao-Tang Sun, Izumi Nishio, Gerald Swislow, and Toyoichi Tanaka, "The Coil-Globule Transition: Radius of Gyration of Polystyrene in Cyclohexane," *J. Chem. Phys. 73, No. 12* (15 December 1980), 5971–5975.

[34] Jeffrey J. Kuna, Kislon Voïtchovsky, Chetana Singh, Hao Jiang, Steve Mwenifumbo, Pradip K. Ghorai, Molly M. Stevens, Sharon C. Glotzer, and Francesco Stellacci, "The Effect of Nanometre-Scale Structure on Interfacial Energy," *Nature Materials 8, No. 10* (2009), 837–842.

[35] Danielle J. Mai, Christopher Brockman, and Charles M. Schroeder, "Microfluidic Systems for Single DNA Dynamics," *Soft Matter 8* (2012), 10560–10572.

[36] Vikram Prasad, "Weakly Interacting Colloid Polymer Mixtures" (Ph.D. diss., Harvard University, 2002), 60. http://weitzlab.seas.harvard.edu/links/thesis/VikramPrasad_Thesis.pdf

[37] Jacob N. Israelachvili and Håkan Wennerström, "Entropic Forces Between Amphiphilic Surfaces in Liquids," *J. Chem. Phys. 96* (1992), 520–531.

[38] L. Helden, R. Roth, G.H. Koenderink, P. Leiderer, and C. Bechinger, "Direct Measurement of Entropic Forces Induced by Rigid Rods," *Physical Review Letters 90, No. 4* (31 January 2003), 048301-1–048301-4.

[39] H.N.W. Lekkerkerker and A. Stroobants, "Phase Behavior of Rod-Like Colloid + Flexible Polymer Mixtures," *Il Nuovo Cimento 16D, No. 8* (August 1994), 949–962.

[40] Ibid., 949.

[41] Rudi Podgornik, Helmut H. Strey, and V. Adrian Parsegian, "Colloidal DNA," *Current Opinion in Colloid and Interface Science 3* (1998), 534–539.

[42] Ibid., 536.

[43] Onofrio Annunziata, Neer Asherie, Aleksey Lomakin, Jayanti Pande, Olutayo Ogun, and George B. Benedek, "Effect of Polyethylene Glycol on the Liquid-Liquid Phase Transition in Aqueous Protein Solutions," *PNAS 99, No. 22* (29 October 2002), 14165–14170.

[44] Paul J. Flory, "Statistical Thermodynamics of Mixtures of Rodlike Particles. 5. Mixtures With Random Coils," *Macromolecules 11, No. 6* (November-December 1978), 1138–1141.

[45] Philip Nelson, *Biological Physics: Energy, Information, Life* (New York: W.H. Freeman and Company, Updated First Edition, 2008), 347.

[46] Paul J. Flory, "Statistical Thermodynamics of Mixtures of Rodlike Particles. 5. Mixtures With Random Coils," *Macromolecules 11, No. 6* (November–December 1978), 1138–1141.

[47] Giuliano Zanchetta, "Phase Separation and Liquid Crystallization," 1111.

[48] John Aguilar, "Death of a Beautiful Mind," http://www.coloradodaily.com/ci_13124877

CHAPTER TWELVE

Unexpected Consequences

Scientists from three continents are gathered in Boulder, Colorado, for a three-day symposium, *Frontiers of Soft Matter 2012*. From May 16 to May 18 they've congregated to honor Noel Clark on the occasion of his 70th birthday. No one cares that this is a belated birthday celebration, tardy by about 1½ years. Speakers from North America, Europe, and Asia are in buoyant mood, interlacing their scientific talks with stories and photographic memories of their association with Clark. Lively lunches and dinners border the technical sessions. On Friday, the last day of the symposium, Tommaso Bellini is the final speaker. He's a burly man, sturdily built with a full head of black hair and a beard. His presentation is warm and personal but methodical, beginning with the large body of evidence for the liquid crystal ordering of DNA and RNA oligomers. Over two dozen colorful slides show the progress of the research through increasingly subtle experiments designed to tease out the fine points of the modes of assembly of nanoDNA (and nanoRNA) and the condensation of liquid crystal domains. At the beginning of the slide show Bellini displayed a succinct overview of what was to follow. One line, without further explanation, read: *Noel's vision*. Now the moment has come. Bellini taps his wireless clicker and a slide appears on the screen titled "night prophecy (3 a.m. Italian time)." On the left is a photograph of Clark giving a *University of Colorado Wizards Presentation* to educate and entertain several hundred children and their parents on a Saturday morning (Figure 12.1).[1] He's wearing the classic pointed wizard hat. Although his beard is a little short for your archetypal wizard, he attempts to make up for this with his zealous posturing. The main portion of the slide recounts a telephone call that

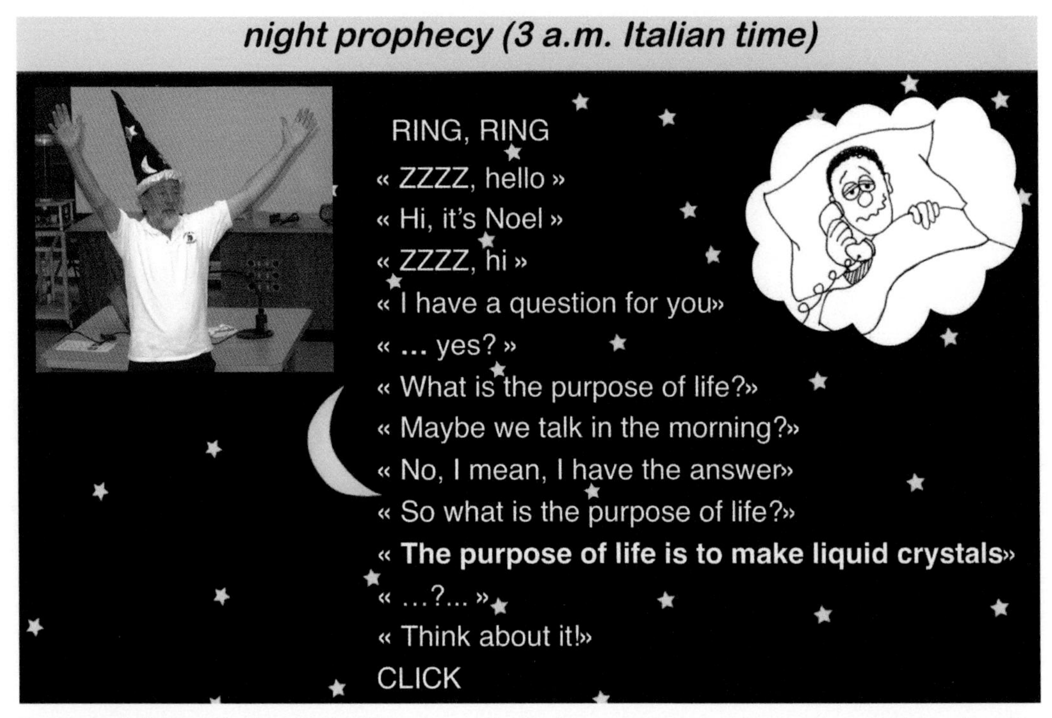

Figure 12.1 "Night Prophecy," also referred to as Noel's vision, as presented by Tommaso Bellini on May 18, 2012, at the symposium to honor Noel Clark. *(Courtesy of Tommaso Bellini.)*

Clark made to Bellini. After many months of meditative thought, Clark had a startling insight that he felt an urgent need to share. One year after the talk, Bellini affirmed that the slide was not a joke. It was an accurate representation of what had occurred.[2] It was Noel's vision.

Clark hit upon the answer through his interpretation of the macroscopic phase separation of complementary nanoDNA. Here's the way it happened.[3] In Milan, Giuliano Zanchetta bought some nanoDNA sequences that were not self-complementary but were mutually complementary. He used mixtures of the two sequences with unbalanced ratios and heated the solutions. Then he cooled them down and watched for liquid crystals to form. He gave an account of his results to the Boulder-Milan groups in his *Work Report Jul-Aug 06* (Figure 12.2 is a slide from his report).[4]

This was the first time that the researchers had gotten liquid crystals to form from non-self-complementary mixtures of DNA oligomers.

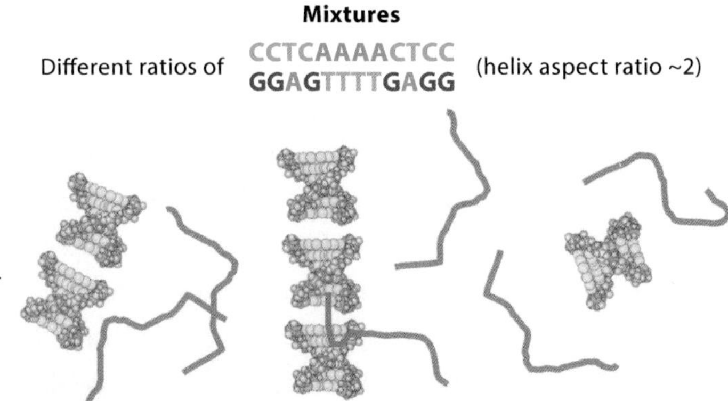

Figure 12.2 Mutually complementary (but not self-complementary) DNA oligomers used by Zanchetta. *(Courtesy of Giuliano Zanchetta.)*

And it showed the dramatic phase separation. It was a result that got their attention. However, Michi Nakata died soon after (in September 2006). Science was pushed to the side while everyone mourned, no one more so than Clark. Nakata had been a graduate student at the Tokyo Institute of Technology and had spent most of 2003 doing experiments for her thesis as a guest in Clark's lab. After receiving her degree in Tokyo, in 2004 she returned to Clark's lab as a postdoc for 2½ years until her tragic death in Boulder. Her talent, energy, and charisma had been a wonder to everyone.

Nakata had been selected to write the first draft of the team's first paper announcing their findings. That task now fell to Clark. Several months after Nakata's death, Clark began to work on the paper. He was struck anew by the Zanchetta data; the result was basting, bubbling, and burbling in the marinade of two years of experimental ingredients. In an epiphany he realized that the team's findings had profound implications for the possible origins of life on Earth. He decided to prepare a special colloquium to honor Nakata. He titled the talk, *Liquid Crystals and the Origin of Life: Michi's Story* and presented it on February 28, 2007, at the University of Colorado.[5]

Hierarchical Self-Assembly

Clark made sense of the body of evidence his team had accumulated as *hierarchical, staged self-assembly*—duplexing, end-to-end adhesion, liquid crystal phase formation and ultimately phase separation of liquid crystal-forming complementary oligomers; graduated steps one following from the previous; a matryoshka, a Russian nesting doll. With an intuitive leap, he reasoned that such a series of processes might have been instrumental on the prebiotic Earth as a means of templating the linear polymer structure of our bio-information carriers. Moreover, Clark understood—years in advance of the additional experimental evidence to support his premise—that in the presence of appropriate ligation chemistry, end-to-end aggregation and phase separation would strongly promote elongation of complementary oligomers. That is, he anticipated that ligation, the joining together of DNA fragments by strong covalent (electron-sharing) bonds, would be strongly favored in the liquid crystal phase relative to that in the isotropic phase. This is because within the liquid crystal drops, the terminal groups on neighboring oligomers are close to each other and thus their effective concentration is much higher than in the isotropic phase.

> **Template (verb)** To act as a template.
>
> **Ligation** In the context of nucleic acids, ligation is the joining together of DNA fragments by strong covalent (electron-sharing) bonds.
>
> **Hierarchical self-assembly** Stages of self-assembled organization in which one level of self-assembly guides the next.

Moreover, every ligation in the liquid crystal phase produces an extended complementary oligomer. So the process builds on itself. The formation of the liquid crystal phase by the complementary duplexes is said to display an autocatalytic effect. That is, the reaction product is itself the catalyst for that reaction. The liquid crystal phase is able to catalyze its own formation. This establishes conditions that would strongly promote the growth of complementary chains into longer complementary chains in those oligomers that form liquid crystals relative to those oligomers that do not. Complementarity promotes the extended polymerization of complementary oligomers. The fact that Bellini and Clark found the liquid crystal ordering to depend sensitively on complementarity introduces selectivity into this process. It means that the overall structure of the complementary assemblies generated will be templated by the liquid crystal geometry.

Autocatalysis A process in which the reaction product is itself the catalyst for that reaction. For example, in liquid crystal autocatalysis, the liquid crystal phase is able to catalyze its own formation.

Clark's vision, a process of hierarchical self-assembly with stages of duplexing, end-to-end stacking, phase segregation, and ligation, is a route to molecular selection, templating, and autocatalysis. He proposed that the liquid crystal phase of nanonucleic acids could autocatalytically select, template, and replicate its constituent molecules.[6,7]

Embracing the art of science, Clark transcended the limits of observation. "And as imagination bodies forth/The forms of things unknown, the poet's pen/Turns them to shapes and gives to airy nothing/A local habitation and a name," quoth Shakespeare.[8]

Clark's intuition is a lot to chew on. In case you balk at this line of reasoning, please know that it may take some time to sink in, as may also be true for the scenario it implies for the origins of life on the prebiotic Earth. Even when grasped, the argument raises all kinds of questions. For example, what about RNA? It's believed by many researchers studying the origins of life on Earth that life's first vestiges were systems of short (up to about 30 bases long) self-replicating molecules, emerging out of the primordial soup of small prebiotic organic molecules in solution. If RNA is an earlier form of nucleic acid than DNA, where does RNA fit in the Bellini/Clark proposition? Clearly, if they had experimental results on RNA it would strengthen the credibility of their hypothesis.

NanoRNA

From a structural point of view, RNA shares some basic features with DNA, but it usually adopts an A-conformation (panel (a) of Figure 12.3). As a result, terminal bases are tilted ~ 20° (called the inclination angle) with respect to the plane perpendicular to the helical axis. In B-form DNA the tilt is ~ 0°.[9] In spite of the large tilt angle, concentrated solutions of self-complementary RNA oligomers exhibited similar self-assembly into nematic and columnar liquid crystal phases as did B-form DNA (panel (b) of Figure 12.3). This indicated that the form of spontaneous ordering discovered by Bellini and Clark is universal in nucleic acids.[10]

The tilted base structure corresponds to a crucial difference between DNA and RNA. The straightness of the stacks required for liquid crystal ordering implies that for RNA the tilt of the terminals must be matched between adjoining lengths of nanoRNA. This in turn constrains the relative positions of the sugar–phosphate chains of RNA at their chemical discontinuities along the weakly aggregated physical polymer. And this touches on an important point that we have not yet spoken about.

NanoDNA and nanoRNA both spontaneously self-assemble into polymer aggregates. However, when the duplexes stack end-to-end they are only weakly bound. One speaks of this as being physically rather than chemically bound (and hence the term *physical polymer*). To underscore the point, look closely at panel (b) of Figure 12.3 and you'll see that the authors have shown gaps in the stacked nanoRNA duplexes. It is the chemical process of ligation—to which we devote Chapter Thirteen—that renders these aggregates chemically bound via strong covalent bonds. Bellini and Clark speculated that in the case of nanoRNA, the relative positions of the sugar–phosphate chains at these gaps (i.e., at these chemical discontinuities) might be particularly favorable for promoting ligation in a chemical environment suitable for such a ligation reaction.[11]

Panel (c) of Figure 12.3 shows both nematic and columnar phases of nanoRNA seen in depolarized transmission light microscopy. Bellini and Clark also found the same phase separation phenomenon in nanoRNA as they found in nanoDNA. Note, however,

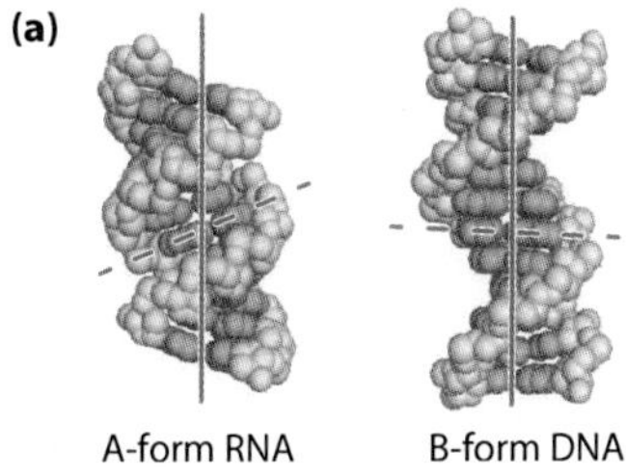

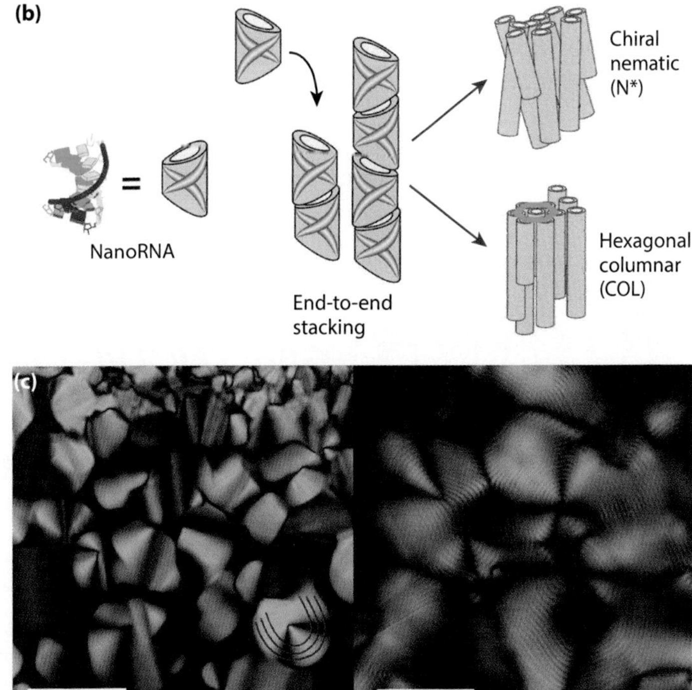

Figure 12.3 (a) A-form RNA and B-form DNA. We show the helical axis in blue and the inclination angles of the base pairs as dashed lines. **(b)** Sketch of end-to-end stacking showing liquid crystal ordering of nanoRNA. Both columnar and nematic phases are formed. **(c)** Depolarized transmission light microscopy images of liquid crystal textures observed in concentrated nanoRNA mixtures. Left image: columnar phase in a 10 bp sample; red lines indicate the orientation of columns of duplexes. Right image: nematic phase of 22 bp. The scale bars are 50 microns (=50,000 nm).

*(Panel **(a)** adapted with permission from Simon L. Bullock, Inbal Ringel, David Ish-Horowicz, and Peter J. Lukavsky, "A′-form RNA Helices Are Required for Cytoplasmic mRNA Transport in Drosophila,"* Nature Structural and Molecular *Biology 17, No. 6 (2010), 703–709; Panels **(b)** and **(c)** portions adapted with permission from Giuliano Zanchetta, Tommaso Bellini, Michi Nakata, and Noel A. Clark, "Physical Polymerization and Liquid Crystallization of RNA Oligomers," J. Am. Chem. Soc. 130 (2008), 12864–12865.)*

that the bases in RNA are not the same as in DNA. In RNA, the complementary base to adenine (A) is the base uracil (U), an unmethylated form of the thymine (T) found in DNA. Having said that, the phase-separated nanoRNA complementary sequences were GGAGUUUUGAGG and CCUCAAAACUCC. These nanoRNA sequences bring up another point that we didn't explore in the previous chapter when we had our hands full jousting with daunting phase diagrams.

Blunt Ends and Sticky Ends

We've stressed mutually complementary sequences in speaking of duplex self-assembly and subsequent stacking in both nanoDNA and nanoRNA. But when Bellini and Clark did their initial experiments they were quite surprised that they obtained liquid crystal phases with blunt-ended duplexes but not with duplexes having unpaired bases at their ends. To flesh out this description, the initial experiments on nanoDNA were done with two sequences. One was CGCAATTGCG, which is self-complementary (sometimes referred to as a duplex-forming palindromic oligomer). That is, it can hybridize with another strand of the same sequence and thus form a rigid, squat helix with blunt ends. The second sequence was CGCAATTGCGTTTTTTTTTT. This sequence is only partially self-complementary, forming a helix with two long, dangling tails (Figure 12.4).

This latter structure, a rigid core with flexible ends, is similar in appearance to a typical liquid crystal molecule (not necessarily one made from nucleic acids) and was expected to favor the formation of liquid crystal phases. A good model of a typical liquid crystal molecule is a short pencil with a short piece of cooked spaghetti attached to each end.[12] The flexible ends (cooked spaghetti) seem to allow the molecule to position itself more easily between other molecules as they all move about and thus help to form liquid crystal phases.

We pause for a brief digression. Eagle-eyed readers may have raised an eyebrow on seeing some of the base sequences in this chapter and the previous chapter. In the English language, palindromes are familiar to us in the form of words such as *radar* and *racecar*,

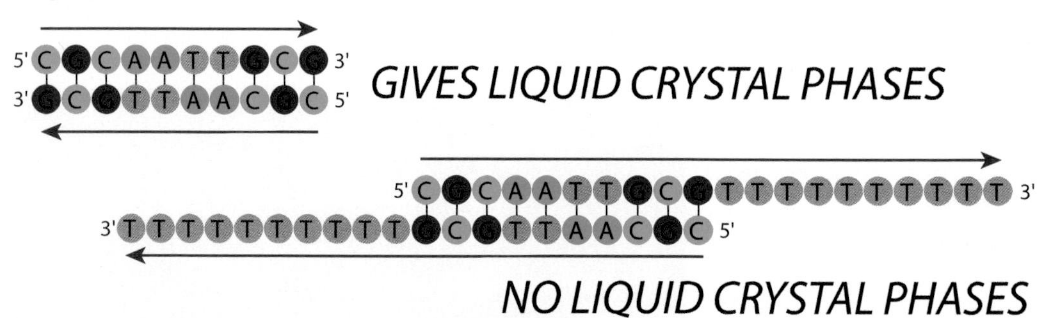

Figure 12.4 DNA sequences used in the initial Bellini/Clark experiments. The self-complementary sequence forms liquid crystal phases but the partially self-complementary sequence does not. *(Adapted with permission from Giuliano Zanchetta, "Spontaneous Self-Assembly of Nucleic Acids: Liquid Crystal Condensation of Complementary Sequences in Mixtures of DNA and RNA Oligomers," Liquid Crystals Today 18, No. 2 (December 2009), 40–49.)*

and, for Napoleon enthusiasts, phrases such as "*Able was I ere I saw Elba.*" In the case of nucleic acids, the word palindrome means that a sequence reads the same in the 5′→ 3′ direction as its complementary sequence reads in the 5′→ 3′ direction.[13] Consider the oligomer 5′-CGCAATTGCG-3′. The sequence that is complementary to this is 3′-GCGTTAACGC-5′. The sequence (and its complement) is palindromic, and is sometimes called a self-complementary sequence. We have wandered a little from our narrative and while we cannot dwell on this subject, we note that these types of sequences have come under close scrutiny. In the case of DNA, such structures can have profound effects, for example, in certain human diseases.[13]

Blunt end DNA A fragment of duplex DNA in which there are no unpaired bases (i.e., no overhangs) at either terminus. Both strands of the duplex are the same length.

Sticky end DNA A fragment of duplex DNA in which a terminal portion has a stretch of unpaired bases. The strands of the duplex are not the same length.

Now back to our story. In the Bellini/Clark experiments, the rigid helix without tails formed by CGCAATTGCG and its complement was the negative control because its aspect ratio, lower than two, should prevent spontaneous alignment. But what they found was just the opposite of what was expected. The oligonucleotide with tails, CGCAATTGCGTTTTTTTTTT, having a partially self-complementary sequence, showed no liquid crystal phases. But the control oligomer displayed wonderfully colored textures when observed in a depolarizing light microscope.

The explanation for the results summarized in Figure 12.4 comes in two parts. As we've seen in the previous chapter, the rigid helix without tails formed duplexes that stacked to form longer chains with an aspect ratio capable of forming liquid crystal phases. On the other hand, when the partially self-complementary strands formed duplexes, the presence of the dangling tails disturbed the end-to-end adhesion of these duplexes and thus suppressed the formation of liquid crystals.[14]

But what about those sticky ends that we know from our time with structural DNA nanotechnology? If the terminal sequences in the Bellini/Clark experiments were not boring strings of thymines (Ts) but instead were complementary sticky ends, would that enable liquid crystal formation? *Yes.*

The team studied the behavior of concentrated solutions of partially complementary sequences that formed duplexes with dangling unpaired bases chosen so as to favor stickiness among helices. And they found liquid crystalline structures.[15] For good measure, they got these results with both nanoDNA and nanoRNA. This finding widened the conditions for spontaneous long-range ordering in oligomeric nucleic acids, thus strengthening the notion that nucleic acids have remarkable self-assembly capability. We'll now chronicle the kernel of the investigation.

The study focused on the 12-base oligonucleotide, CGCGAATTCGCG, that they had used before (third row down from the top of Figure 11.6 in Chapter Eleven) but which we had not called out by its name. That's right. This particular self-complementary, duplex-forming oligomer has its own moniker. It is the well-studied Dickerson dodecamer also called the Dickerson-Drew dodecamer. Chemists Richard E. Dickerson and Horace R. Drew provided the first detailed description of a right-handed DNA double helix. They carried out a single-crystal structure analysis of B-DNA built from the dodecamer that now bears their names.[16] There have been over 750 papers published on this molecule.[5] Figure 12.5 shows the duplexes formed by the dodecamer.

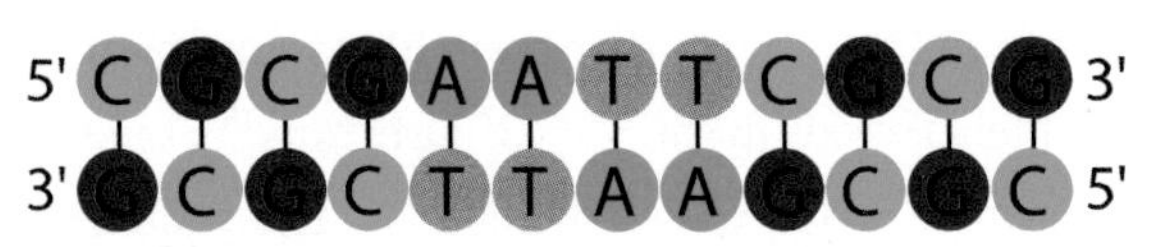

Figure 12.5 Duplexes formed by the Dickerson dodecamer DD12.

Dickerson dodecamer (DD12) The 12-base oligonucleotide, CGCGAATTCGCG, also referred to as the Dickerson-Drew dodecamer. This dodecamer was used to build B-DNA duplexes that provided the first detailed structural description of a right-handed DNA double helix.

Bellini and Clark reported the concentration-temperature phase behavior of both DNA and RNA 14-mers that they built by adding the sequence CG or AT to one terminal of the Dickerson dodecamer (referred to as DD12) or to the analogous sequence for RNA where uracil (U) takes the place of thymine (T). In this way they forged partially self-complementary sequences that, after hybridizing, provided exposed dangling, sticky nucleotides at their ends. These sticky ends provided attractive interactions between the duplexes enabling them to elongate by stacking.

To be exact, in the experiments they used the following sequences. For DNA: DD12 (5′-CGCGAATTCGCG-3′), DD12-AT (5′-CGCGAATTCGCGAT-3′), DD12-CG (5′-CGCGCGAATTCGCG-3′), and DD12-TT (5′-CGCGAATTCGCGTT-3′). For RNA: DD12 (5′-CGCGAAUUCGCG-3′), DD12-CG (5′-CGCGCGAAUUCGCG-3′), and DD12-UU (5′-CGCGAAUUCGCGUU-3′).

Bellini and Clark showed that by combining pairing (hybridizing) and stacking, the DNA DD12-AT and DD12-CG and the RNA DD12-CG samples all self-organize to form the same liquid phases as they had reported earlier using DD12 alone. They also analyzed various contributions to the free energy and found that the pairing of CG sticky ends increases the end-to-end adhesion relative to DD12. But the pairing of AT terminals decreases the adhesion. Last, for the DNA DD12-TT and the RNA DD12-UU samples they found no liquid crystal phase formation.

This brings us back to the distinction between physically bound aggregates and chemically bound aggregates that we called out in the previous section on nanoRNA. The presence of liquid crystal phases in the DNA DD12-AT and DD12-CG and in the RNA DD12-CG samples, but no ordering in the DNA DD12-TT and RNA DD12-UU samples is telling. It's a further indication that liquid crystallization of nanoDNA and nanoRNA proceeds from end-to-end adhesion through the formation of linear aggregates that are chemically discontinuous but physically bound.

Let's pause for a moment in our examination of these detailed points to return to the wider perspective. In the paper we've been discussing, the authors stress that the weak attractive interaction between duplexes due to the pairing of the two dangling nucleobases at their terminals provides sufficient free energy to promote the formation of long-range liquid crystalline ordering in solutions of oligomeric DNA and RNA. This adds on to the already observed supramolecular self-assembly capability in solutions of duplexes with blunt ends. Bellini and Clark concluded that the end-to-end adhesion of oligomeric duplexes promoted by a delicate combination of pairing and stacking may have been instrumental in promoting the initial prebiotic formation of linear polymers of nucleic acids.

Supramolecular A complex of molecules held together by noncovalent bonds.

To close our examination of the DNA and RNA 14-mers, Figure 12.6 shows depolarized light microscopy images and cartoons of the temperature dependence of the nematic-columnar phase boundaries in a DNA DD12-CG sample.[15] The figure shows the melting process for the nematic phase—right-side micrographs and cartoons.

Follow along as the temperature increases, reading from bottom to top in the figure. At 20 °C (panel (c)), the columnar and nematic phases coexist; columnar phase on the left, nematic on the right. The orange dashed line approximately indicates the location

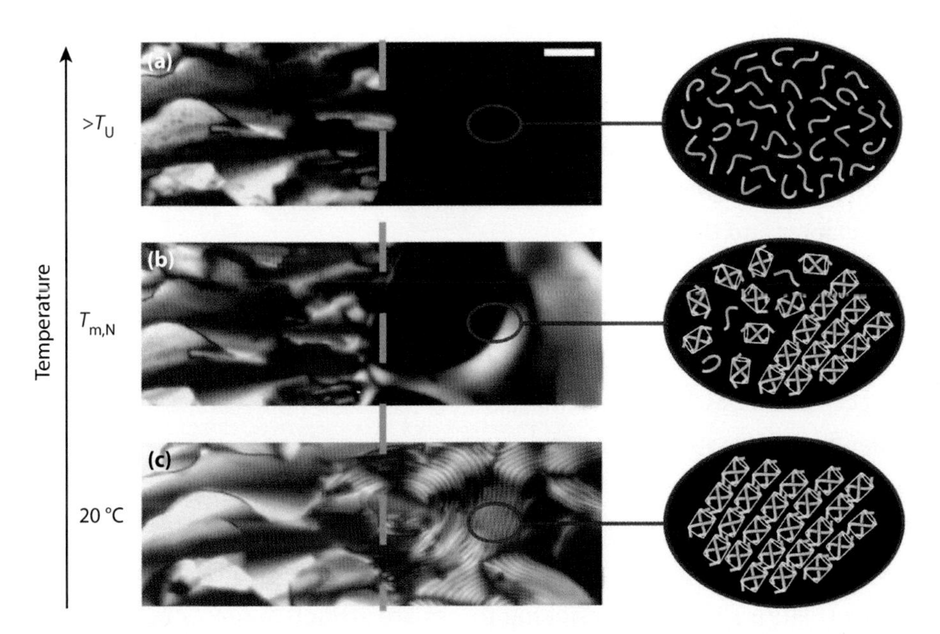

Figure 12.6 Temperature dependence of the columnar-nematic (left-right) phase boundary as observed in depolarized transmission light microscopy in a DNA DD12-CG sample (the melting process of the nematic phase). The orange dashed line indicates the approximate location of the interface between the columnar phase (on the left) and the nematic phase (on the right). The concentration of the sample was chosen so that the nematic and columnar phases could coexist. The scale bar is 20 microns (or 20,000 nm). *(Adapted with permission from G. Zanchetta, M. Nakata, M. Buscaglia, N.A. Clark, and T. Bellini, "Liquid Crystal Ordering of DNA and RNA Oligomers With Partially Overlapping Sequences," J. Phy.: Condens. Matter 20 (2008), 1–6.)*

of the interface. In (b), as the temperature is raised, the nematic melts. It gives way to the isotropic phase (black) so that the nematic is gone once it has exceeded its clearing temperature, $T_{m.N}$. At the clearing temperature, a liquid crystal transforms to an isotropic liquid. Since $T_{m.N}$ is lower than the temperature T_U where isolated duplexes unbind, this melted nematic phase is now an isotropic solution of free duplexes. As the temperature is further increased through T_U these duplexes unbind to give an isotropic solution of dispersed single-stranded DNA. Throughout all of this, the columnar remains an intact phase of duplexes. The researchers gleaned their insights from an interpretation of the micrographs (based on much experience). They also employed fluorescence microscopy and analysis of the optical spectrum of light reflected from the cells in which the solutions were contained to determine several experimental parameters, e.g., DNA (and RNA) concentrations.

Base-Stacking Forces

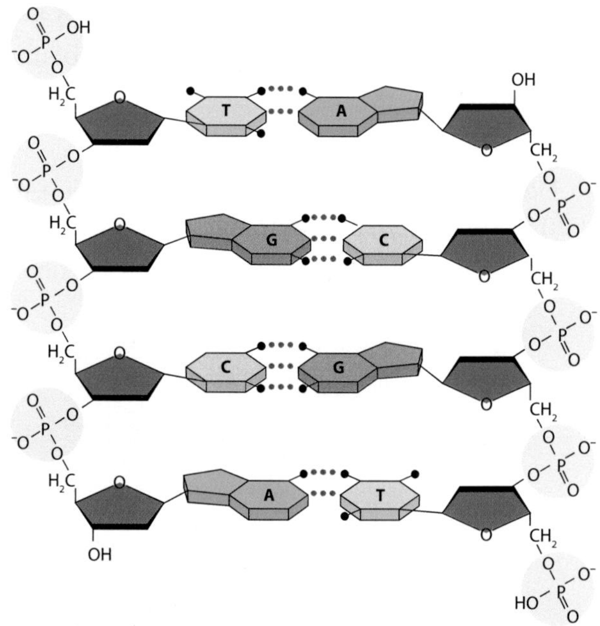

Figure 12.7 Untwisted DNA showing base pairs stacked like flat plates. Forces between the planes of base pairs are vital for the stability of the double helix. *(Reproduced with permission from Neil A. Campbell and Jane B. Reece, Biology, 6th Ed., ©2002, p. 291. Reprinted by permission of Pearson Education, Inc., New York, New York.)*

Before examining one last experiment, we'll say a few words about the forces holding DNA duplexes together. We've spoken frequently of the base-pairing forces mediated by hydrogen bonds that contribute to the stability of the DNA duplex. These are *intra*duplex forces, i.e., within a given DNA duplex. In the previous chapter and the present one, we have talked about forces between the ends of duplexes of nanoDNA that promote elongation and liquid crystal phases. These are *inter*duplex forces. It is appropriate (and perhaps overdue) to mention another force that we've not yet addressed. This is called the base-stacking force and it describes another *intra*duplex interaction that contributes, along with base-pairing forces, to holding the two DNA strands together in a duplex (Figure 12.7). Base-stacking forces are attractive forces between adjacent bases in a given duplex. One can think of each base pair (A-T, G-C) as being shaped like a flat plate as visualized by untwisting the DNA duplex (Figure 12.7). There are both hydrophobic and electrostatic interactions at play between these plates that contribute substantially to the thermal stability of DNA.[17] Later in our tale we will have occasion to return to these base-stacking forces. For now, tuck them away for safekeeping.

The Scope of the Self-Assembly Mechanisms of Nucleic Acids

In the last several experiments we've thrown the spotlight on, Bellini and Clark were exploring the scope of the self-assembly mechanisms of nucleic acids. They weren't done exploring.

When scientists are gathered around the coffee urn and chatting, *parameter space* is a phrase that pops up. The term is often used, albeit non-rigorously, to refer to variables (parameters) that might come into play in a mathematical model or physical experiment. In the work we're examining we could, for example, speak of the form of termination of duplexes as being a parameter. Duplex terminals could be blunt-ended or there might be unpaired bases sticking out from the ends and those bases might or might not be sticky ends for the oligomers in the solution. Bellini and Clark knew that they needed to reconnoiter further in parameter space and did so to great advantage. They boldly ventured into the realm of random-sequence DNA.

Base-pairing forces Attractive forces within a DNA duplex that exist between the base pairs that make up the rungs of the spiral ladder. These forces are in the plane of the bases (visualized as horizontal). They are mediated by hydrogen bonds and contribute to the stability of the double helix.

Base-stacking forces Attractive forces within a DNA duplex that exist between vertically adjacent bases—that is, forces perpendicular to the base planes. These forces are mediated by both hydrophobic and electrostatic interactions and contribute to the stability of the double helix.

Please note: The four sections that follow contain a great many details. *The most important result of these four sections is simply this: random-sequence nanoDNA exhibits the same liquid crystal phases that we found for nanoDNA sequences with specified bases.* Hold on to this result as a guide as you read on.

Random-Sequence NanoDNA

The team recognized that in the natural world, selectivity and reversibility of DNA and RNA association enable crucial biological functions in which oligomers selectively pair to target sequences even within large amounts of nucleic acid chains. Hence, they explored liquid crystal ordering in solutions of DNA oligomers with random sequences where the large body of different competing sequences effectively reduces the selectivity of the interactions.[18] With these results, they showed that the phenomenology of the self-assembly of nucleic acid oligomers was much richer than previously recognized, involving self-selection, linear aggregation, and ordering of fully random chains.

This conclusion strengthened the view that DNA and RNA have unsurpassed capacity for self-structuring. It also suggested self-assembly as the possible key factor for the emergence of nucleic acids from the prebiotic molecular clutter as the coding molecules of life.

In this context Bellini/Clark noted that, "Random-sequence DNA is an important class of systems to understand . . . because pools of heterobase oligomers are likely prebiotic systems emerging from random ligation."[19] By the term *heterobase* they emphasize that any of the four bases A, T, C, G (or A, U, C, G in the case of RNA) can be found at any location in the oligomer.

Random-sequence nanoDNA (ranDNA) DNA oligomers in which base sequences are chosen at random or with varying degrees of randomness. Solutions of oligomers with random sequences are more likely to reflect prebiotic conditions than specifically designed sequences. Remarkably, complementarity still emerges in these solutions and, for a narrow range of oligomer lengths, produces a subtle hierarchical succession of structured self-assembly and organization into liquid crystal phases.

Back in Chapter Two we saw a figure showing DNA sequences synthesized on polymer beads. The researchers did this in constructing random-sequence oligomers (panel (a) of Figure 12.8). N_B denotes the lengths of ranDNA sequences. Bellini and Clark investigated aqueous solutions of DNA oligomers of length $N_B \leq 30$ bp. They introduced various modes of random sequencing such that at selected positions in the sequences the four primary A, C, G, and T bases are found with equal probability. Examples of such ranDNA oligomers are indicated in panel (c) of Figure 12.8, where they denoted a randomly chosen base with the character "*N*" in the sequence.

For example, 8*N* indicates the set of 8-mers having eight randomly chosen bases and C(8*N*)G is the set of 10-mers having complementary C-G terminal base pairs and eight internal randomly chosen bases. In both cases, the family of molecules corresponds to $4^8 \approx 65{,}000$ different sequences mixed together in the same solution. To see this, consider that there are four choices for the first base (at the 5′ end), and four choices for the second base—thus, there are $4 \times 4 = 16 = 4^2$ different ways to select the first two bases. Since the third base also has four choices, there are $4 \times 4 \times 4 = 64 = 4^3$ ways to select the first three bases, and so on.

The Strange World of Random-Sequence NanoDNA

Panels (b) through (f) of Figure 12.8 are full of details. We highlight only a few of the ins and outs but close our once-over with the principal point of the figure.

Panel (b) of Figure 12.8 shows varied duplex motifs in which definite pairing of single strands is achieved by selected sequences. For example, row 1 illustrates blunt-ended duplexes that form liquid crystals. Row 2 demonstrates duplexes with mutually complementary overhangs. Moving to panel (c), the sequence formation becomes bolder—we're now piecing together ranDNA sequences.

Bellini and Clark synthesized these sequences by randomly choosing one of the four bases at given positions along the chains. The resultant solutions are thus mixtures of 4*R* different sequences where *R* is the number of randomly chosen bases. Because of the large number of sequences, ranDNA forms duplexes with a variety of striking pairing

Figure 12.8 (a) Synthesis of random-sequence nanoDNA. **(b)** Duplex motifs of oligomeric DNA: pairing errors, random sequences, and sketches of the behavior of ranDNA oligomers of different lengths in solution. *(**(a)** Courtesy of Tommaso Bellini; **(b)** Reproduced with permission from Tommaso Bellini, Giuliano Zanchetta, Tommaso P. Fraccia, Roberto Cerbino, Ethan Tsai, Gregory P. Smith, Mark J. Moran, David M. Walba, and Noel A. Clark, "Liquid Crystal Self-Assembly of Random-Sequence DNA Oligomers," PNAS 109, No. 4, (24 January 2012), 1110–1115.)*

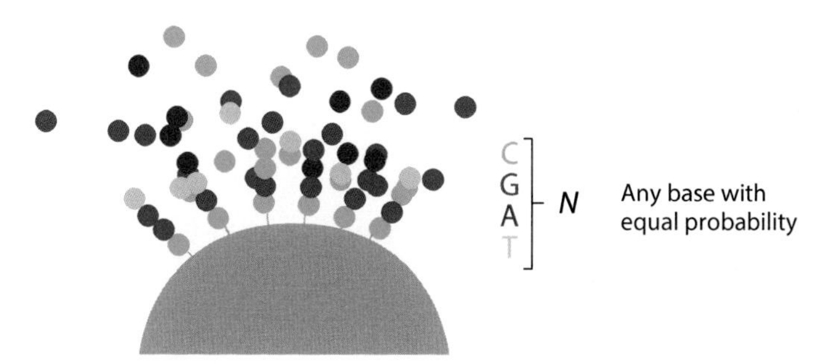
(a)
SOLID PHASE SYNTHESIS
C
G
A
T
N
Any base with
equal probability

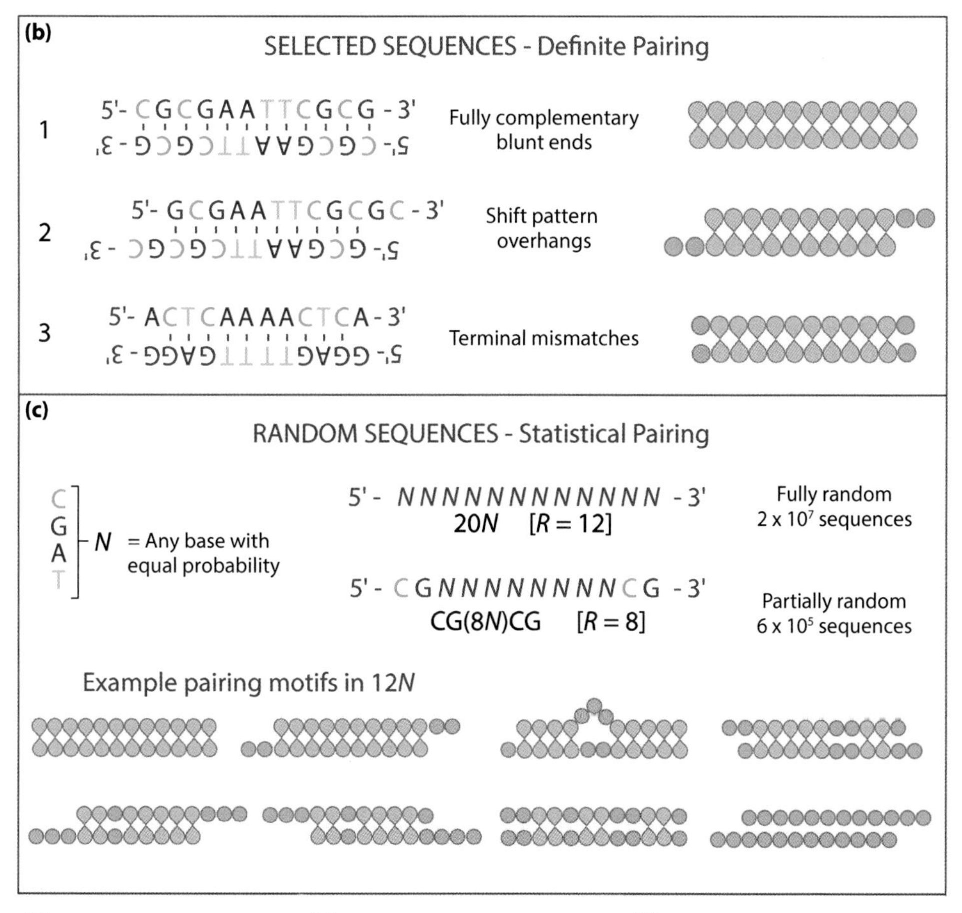
(b)
SELECTED SEQUENCES - Definite Pairing
1
5'- CGCGAATTCGCG - 3'
Fully complementary
blunt ends
2
5'- GCGAATTCGCGC - 3'
Shift pattern
overhangs
3
5'- ACTCAAAACTCA - 3'
Terminal mismatches
(c)
RANDOM SEQUENCES - Statistical Pairing
C
G
A
T
N = Any base with
equal probability
5' - NNNNNNNNNNNN - 3'
20N [R = 12]
Fully random
2 x 10^7 sequences
5' - CGNNNNNNNNCG - 3'
CG(8N)CG [R = 8]
Partially random
6 x 10^5 sequences
Example pairing motifs in 12N

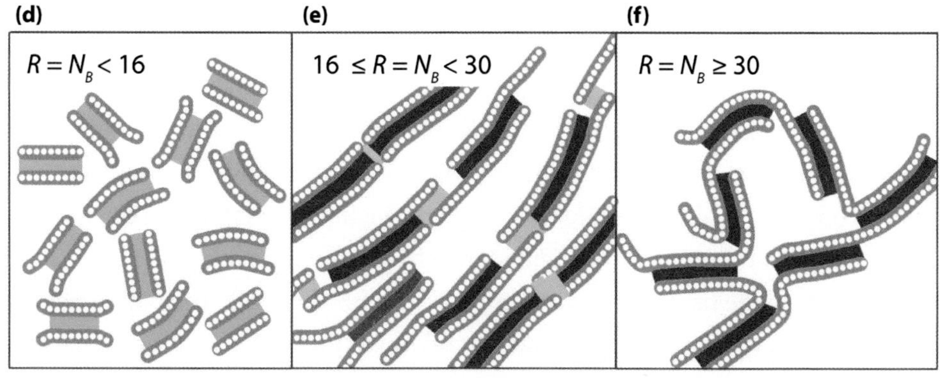
(d)
R = N_B < 16
(e)
16 ≤ R = N_B < 30
(f)
R = N_B ≥ 30

motifs. Panel (c) gives examples of the distribution of pairing motifs for 12*N* ($R = 12$). In both panels (a) and (b), bases in pink are paired to their complementary partners while bases shown in gray are either unpaired (e.g., panel (b), row 2) or mispaired (e.g., panel (b), row 3). Panels (d) through (f) are sketches of the behavior of ranDNA oligomers of different lengths in solution.

The principal point of Figure 12.8 is illustrated in panels (d) through (f). There is an intermediate regime in which liquid crystal ordering exists—even in these mixtures of randomly chosen bases.

This was a big surprise. When the experiments on ranDNA began, it was anything but obvious that this population of liquid crystals would pop up. The collective state of a solution of randomly generated DNA oligomers crucially depends on the role of the polydispersity of interaction. This originates from the wide range of pairing strengths and pairing motifs. To say that these collections of random DNA are polydisperse means that the constituents have a gamut of length, pairing strength, and pairing motifs. This polydispersity of structures results in a polydispersity of interaction.[20]

Because of this non-uniformity, well-matched oligomers may promote liquid crystal phases. In contrast, oligomers may form duplexes with low end-to-end interactions and thus the duplexes won't stack to acquire the correct aspect ratio to form liquid crystal phases. Additionally, there is a significant possibility of chains acting as cross-links (as sketched in panel (f)). In this latter situation, the upshot is a random isotropic gelated state similar to hydrogels formed by long DNA. (A DNA hydrogel is a synthetic biomaterial that has thinly layered porous structures made by enzymatic crosslinking of branched DNA subunits.)[21]

As Bellini and Clark wrote with puckish understatement, "It was thus difficult to anticipate the solution behavior of ranDNA."[22]

Liquid Crystal Ordering of Random-Sequence NanoDNA

What they found was liquid crystal ordering heralded by the appearance of birefringent domains of characteristic color and texture that enabled identification of the phases (seen in Figure 12.9, panels (a) and (b)). As Clark rhapsodized, "The latent complementarity possible if sufficiently well-matched sequences in a random mix find one another is expressed!"[23]

First we touch on a few nuggets in this crowded Figure 12.9 but we conclude with the most important take-away point from the study of random DNA oligomers.

Panels (c) and (d) of Figure 12.9 are maps of the liquid crystal phases observed in ranDNA. The researchers first investigated the phase behavior of oligomers with designed sequence errors in the base pairing (as in row 3 of Figure 12.8 (b)). Their systematic study of the effect of such errors on the liquid crystal phase formation indicated that even a

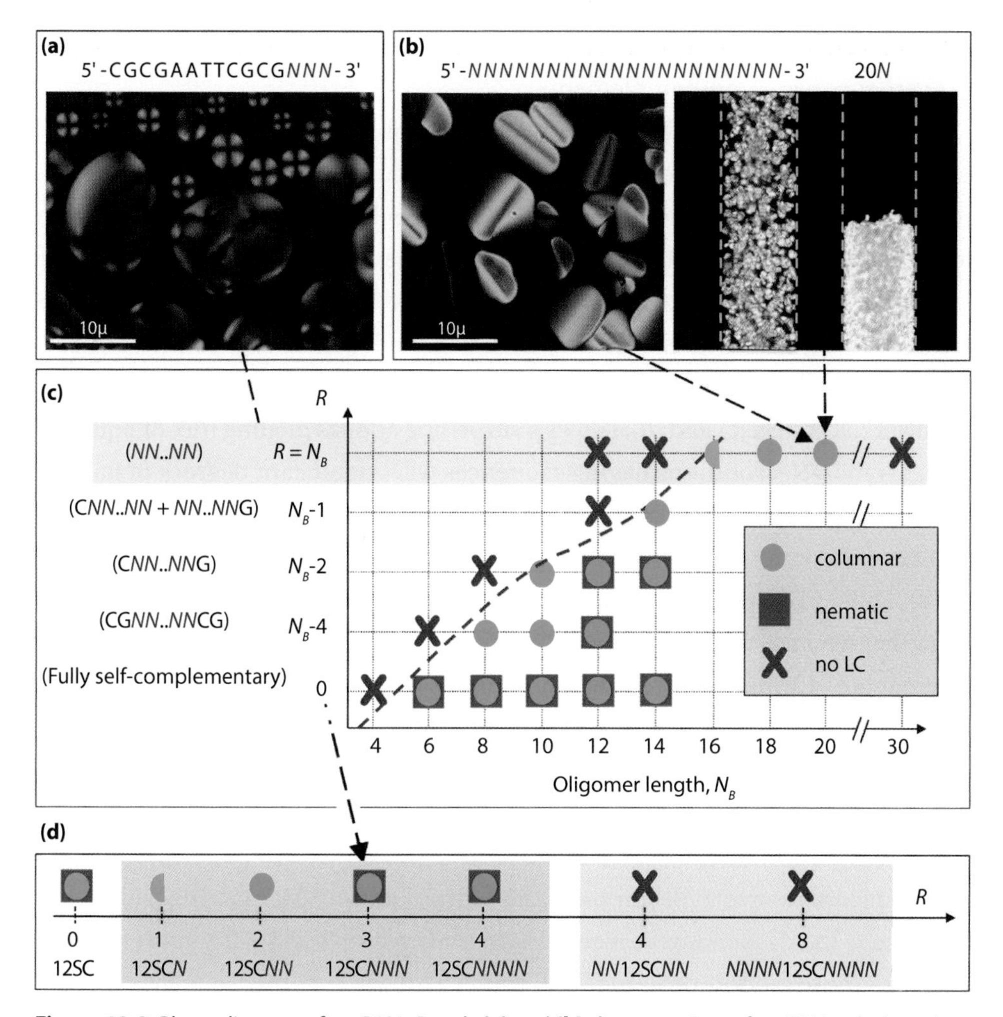

Figure 12.9 Phase diagram of ranDNA. Panels **(a)** and **(b)** show a variety of ranDNA solutions that self-assemble into nematic (a) and columnar (b) liquid crystal phases (see text for far right image in panel (b)). Panels **(c)** and **(d)** are maps of the liquid crystal phases observed in ranDNA of different total number of bases, N_B, and of random bases, R, either internally (c) or at the terminals (d). Symbols indicate the presence of the nematic (blue square) and of the columnar (green circle) phases in some range of concentrations and temperatures of the solution. The half green circle (panel (d), bottom left) indicates that the columnar phase is present in a very narrow range of conditions (low temperature, high DNA concentration). Shaded regions highlight specific phase behavior: light blue—fully random sequences; pink—double helices with a single overhanging random tail per duplex terminal; and orange—two random tails per duplex terminal. *(Reproduced with permission from Tommaso Bellini, Giuliano Zanchetta, Tommaso P. Fraccia, Roberto Cerbino, Ethan Tsai, Gregory P. Smith, Mark J. Moran, David M. Walba, and Noel A. Clark, "Liquid Crystal Self-Assembly of Random-Sequence DNA Oligomers," PNAS 109, No. 4 (January 24, 2012), 1110–1115.)*

single terminal mispairing (Figure 12.8 (b), in row 3 "A" mispaired to "G") reduced the self-association of the duplexes. They found that internal errors (errors not at the ends of the sequences) were less problematic, affecting the phase behavior only when there were more than two in a duplex. In general, the nematic phase was suppressed first, followed by the columnar phase, as the number of pairing errors was increased. In solutions of sequences paired in a shifted mode with overhangs (row 2 in Figure 12.8 (b)), liquid crystal ordering was stabilized if the overhangs were mutually complementary but suppressed if the overhangs were non-complementary.

The images on the far right of Figure 12.9 (b) require clarification. They show that phase coexistence was (further) explored by using centrifugation to force macroscopic isotropic/columnar phase separation of 20*N* in 0.5-mm-diameter capillary tubes. Columnar domains were compacted in the bottom of the capillary, with the isotropic phase floating on top (left—prior to centrifugation, right—after centrifugation). These capillary/centrifugation experiments enabled detailed volume fraction measurements used for other studies that we will not discuss.

Figure 12.9, panels (c) and (d), shows a surprising and fascinating mix of liquid crystal ordering. NanoDNA solutions having sequences with significant degrees of nucleobase randomness still display a rich liquid crystal phase diagram. There is a great deal of information here. We jump to a summary of panel (d) in Figure 12.9 and then return to savor panel (c). Panel (d) summarizes the phase behavior observed when random bases are added to the terminals of a complementary core, in this case the Dickerson dodecamer self-complementary 12-mer that we previously discussed.

Panel (c) of Figure 12.9 summarizes the phase behavior at room temperature of ranDNA as a function of the oligomer length N_B and of the number of random bases, R, when randomness is progressively increased from the center of the sequences. $R = 0$ corresponds to fully self-complementary (SC) sequences. The dashed line in panel (c) indicates the shortest sequence for which Bellini and Clark found liquid crystal ordering (either nematic or columnar). The dashed line reveals that the shortest such sequence increases with R. That is, with more randomness (higher R value) a longer oligomer sequence is required for there to be any liquid crystal ordering. The graph also shows that the nematic phase is more easily suppressed by randomness than the columnar phase.

Most remarkably, columnar ordering can be found in fully random sequences (light blue shading at the top of the graph), but only for $N_B \geq 16$. Thus, the destabilizing effects of randomness are *somehow* mitigated by increase of length. But this phenomenon is limited because liquid crystal ordering is not found for $N_B \geq 30$.

Here is the most telling finding in Figure 12.9, panels (c) and (d) (and anticipated in Figure 12.8) and *the most important take-away point from the study of random DNA oligomers.* The textures and local birefringence observed using depolarized light microscopy indicate that the nematic and columnar liquid crystals of ranDNA are the same phases as those observed in complementary duplexes (of nonrandom DNA) aggregating by end-to-end stacking (examined in Chapter Eleven). We will follow up on the mystery that is entwined with this finding.

And, indeed, initially it was a mystery. In Bellini and Clark's paper they wrote, "The most surprising finding of the lot is the observation of liquid crystal ordering in solutions of 20*N*, the 20-mers having all bases randomly chosen."[24] They were frankly puzzled at first. After considerable thought and effort they figured it out: "we realized that kinetics of pairing plays a crucial role that must be considered in modeling the pairing distribution."[25]

Their comment on the kinetics of pairing carries a lot of weight. The physics involved is challenging. With the reader's indulgence, we will approach the topic gingerly with a human analogy. This will lead us back to terms that we left undefined when we made our broad observations on the strange world of ranDNA oligomers of different lengths. Although the fable that follows is inconclusive, it may give the reader a feel for the perplexing physics that faced Bellini and Clark.

Non-Equilibrium Statistical Mechanics: Kinetic Arrest and Nonergodic Behavior

Jennifer and Michael have been best friends forever, and no wonder: they complement each other perfectly. They decide to meet in Times Square in Midtown Manhattan on New Year's Eve, December 31, 1999, to share a hug to bring in the New Millennium. To be better able to spot one another, Jennifer is wearing a red jacket and Michael a blue one. They know the din from the crowd will make it hard for them to find each other by hollering and it's nighttime so they can't see too far from their own locations. While dressing in this prearranged way is a good step, it's not quite sufficient to enable their get-together. Revelers are shoulder-to-shoulder, spilling over the area bounded north to south by West 53rd Street and West 40th Street and bounded east to west by 6th Avenue and 8th Avenue, an area of over 6,000,000 square feet. More astonishingly, instead of the usual mob of 1,000,000 enthusiastic (but well-behaved) people, crowd estimates for this particular occasion are in the vicinity of 10^{13} folks—that is, 10,000,000,000,000 = 10 trillion individuals. Jennifer is coming from Uptown Manhattan and Michael from Downtown Manhattan. When they each join the perimeter of the throng they are separated by a distance of almost a mile (approximately 1,000 times their own size of about six feet). As the hour nears 12:00 and the big ball is ready to drop, they are still separated. If only they could have availed themselves of a nonergodic process. Huh?

Nonergodic association. That's what Bellini and Clark invoked to help explain the surprising observation of liquid crystal ordering in solutions of 20*N*, where the 20-mers have all bases randomly chosen. A 20*N* sample corresponds to a molecular population of about 10^{12} different sequences. (Each nucleotide in the oligomer can have one of four different bases, so for a 20-mer there are approximately $4^{20} = 10^{12}$ different sequences.) That is, the amount of a particular 20*N* sequence in the entire volume of the solution is about one part in a trillion. In a typical solution used in the experiments a given sequence has a molar concentration of about 0.1 pM = 10^{-13} M (a picomole is 10^{-12} moles) or one ten-trillionth of a mole. Moreover, that particular 20*N* sequence's fully complementary potential partners are at a mean mutual separation of about 10 microns (= 10,000 nm), a separation distance about 1,000 times larger than the oligomer itself. How in the world can 20*N* oligomers order to form liquid crystal phases? This is the Jennifer and Michael quandary played out on the nanoscale. Time to make two additions to our lexicon: kinetic arrest and nonergodic behavior.

Bellini and Clark studied the free energy of ranDNA duplexes with a variety of distinct pairing motifs. In ranDNA solutions, oligomers collide and interact, the actual energy level attained in each interacting pair being determined by the number and location of well-paired bases. Collisions yielding weak binding give rise to short-lived pairs that rapidly separate and proceed to further collisions. Therefore, short sequences ($N_B \leq 16$) whose binding and unbinding take place on time scales shorter than the experimental time, can reach a state of thermodynamic equilibrium (one speaks of an equilibrated ensemble of duplexes). By equilibrium we mean that the free energy of that state is smaller than that of any other state of the system of oligomer sequences (at the same pressure and temperature).

As longer ($N_B \geq 16$) sequences are considered the situation changes. This is because the ranDNA oligomers of these lengths can form duplexes having a lifetime comparable to or larger than the experimental time and are therefore effectively permanently stable. The researchers found that 20*N* sequences were kinetically trapped in defect duplex structures. That is, they reached a state of kinetic arrest. In effect, this prevented these sequences from exploring over time all the possible states potentially available to them. Thus, the 20*N* solution doesn't reach a state of thermodynamic equilibrium. This is an example of nonergodic behavior. The solution is not at a state of minimum free energy. Thus, standard equilibrium statistical mechanics is generally not adequate to describe the association of ranDNA. An explanation requires recourse to the frontier field known as non-equilibrium statistical mechanics that seeks to understand systems far from thermal equilibrium.

Kinetic arrest The sudden slowing down of dynamic systems in which long-lived structures (having a lifetime comparable to or larger than the time of the experiment) form far from equilibrium and thus prevent a system of molecules from reaching its equilibrium state. Kinetic arrest is also referred to by such phrases as dynamical arrest or ergodic-nonergodic transition. In the case of 20-mer ranDNA, kinetic arrest produces duplex pairs that are able to self-assemble into liquid crystal domains.

The word *ergodic* was introduced by nineteenth century Austrian physicist Ludwig Boltzmann and is an amalgamation of the Greek words *ergon* (work) and *odos* (path).[26] A discussion of Boltzmann's ergodic hypothesis would lead us too far from our subject. But to give the reader a taste of the ideas at play, we can say the following: a system is ergodic if its dynamics are such that it will sample in time all possible energy states available to it and will (eventually) reach a state of thermodynamic equilibrium. That is not the case for the random 20-mers because the duplexes they form have a lifetime comparable to or larger than the experimental time and this results in kinetic arrest. (For those who are interested, we offer, without elaboration, the following observation for this behavior: conceptually, the 20-mers are considered as a system existing in a space populated by energy traps that are deep enough to produce nonergodic behavior.)

Thermodynamic equilibrium
A system is in a state of equilibrium if the free energy of that state is smaller than that of any other state of the system (at the same pressure and temperature).

Nonergodic process In the context of this chapter: the kinetic behavior of a system of ranDNA that is unable to explore in time all of its possible states. Thus, its equilibrium distribution is never reached.

Bellini and Clark invoked both kinetic arrest and nonergodic behavior from knowledge of the specialty field of non-equilibrium statistical mechanics. In Chapter Eleven we referred the interested reader to Peter Atkins' very short introduction to (equilibrium) thermodynamics. We note that the author remarks on a domain in which, "I have feared to tread. . . . I have not touched on the still insecure world of non-equilibrium thermodynamics, where attempts are made to derive laws relating to the rate at which a process produces entropy as it takes place."[27]

Time-dependent phenomena are ubiquitous on Earth and in the universe. Technically, one speaks of these phenomena as non-equilibrium. However, both thermodynamics and statistical mechanics were originally devised for static or equilibrium systems that do not change with time.[28] Thermodynamics takes a macroscopic point of view and statistical mechanics looks at the microscopic/molecular level. Discovering how the incorporation of *time* changes the formulation of thermodynamics and statistical mechanics is a challenging subject and remains a work in progress.

Fortified with our primer on kinetic arrest, we conclude by recapping several earlier remarks. We return to what Bellini and Clark called "the most surprising finding of the lot," the formation of liquid crystal ordering in solutions by the 20*N* oligomers with all bases randomly chosen. They sketched in Figure 12.8 (e) the unusual consequences of the combination of factors at play. RanDNA develops liquid crystal long-range ordering only in a limited interval of lengths near $N_B \sim 20$. In the 20*N* case, kinetic arrest produces duplex cores with weakly mutually attractive tails that enable annealing and equilibration into liquid crystal domains. By contrast, shorter ranDNA oligomers ($N_B \leq 16$) form a population of equilibrium duplexes that exhibit ergodic behavior and no kinetic arrest. They lack the interactions necessary for self-assembly and liquid crystal formation (panel (d) of Figure 12.8). And at the opposite extreme, when the sequences are longer ($N_B \geq 30$), the random overhanging tails can form additional kinetically arrested interduplex interactions, leading to oligomer networking and gelation that suppresses liquid crystal formation (panel (f) of Figure 12.8). With these longer sequences, Bellini and Clark

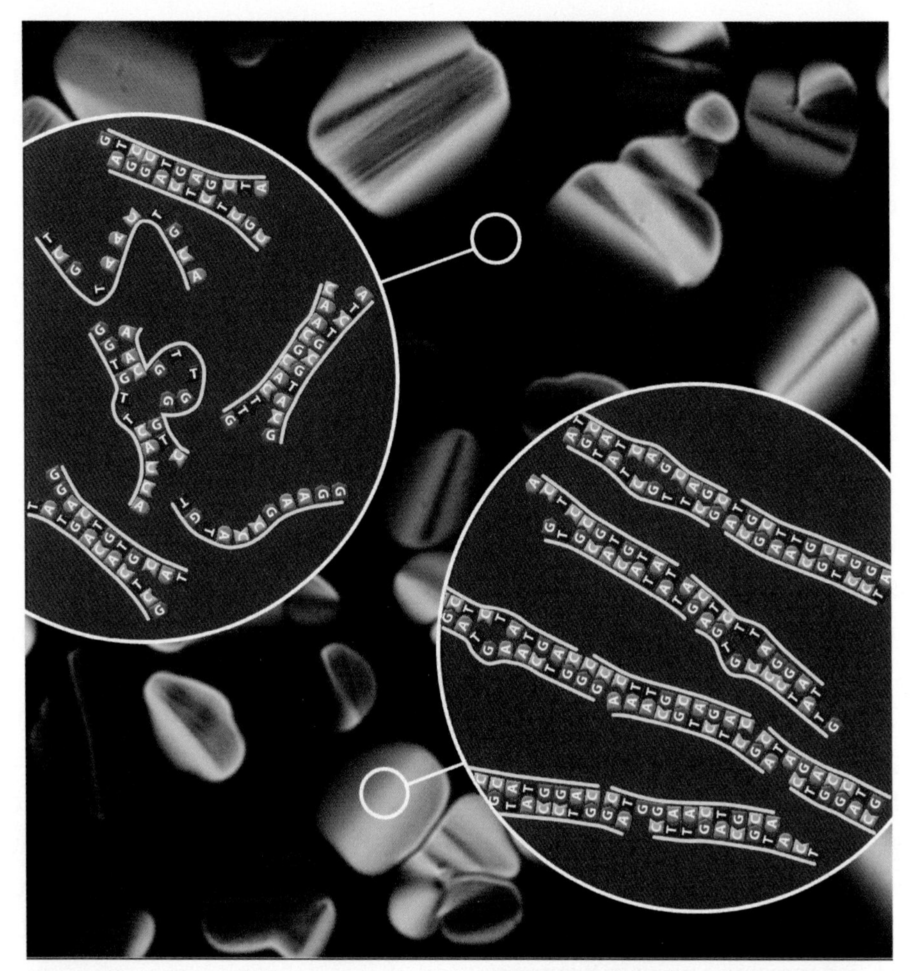

Figure 12.10 The ranDNA sequences in the isotropic phase (black) are those that are unable to order and form liquid crystal phases. But ranDNA sequences in a limited interval of lengths are able to form liquid crystals (gold/orange) because they are in a state of kinetic arrest far from equilibrium—that is, they exhibit nonergodic association. *(Courtesy of Tommaso Bellini; adapted by Kenneth Douglas.)*

performed experiments that showed that ranDNA with $N_B \geq 30$ yields viscous isotropic solutions without liquid crystal domains.

Figure 12.10 speaks volumes.[29] It summarizes the salient insight made by Bellini and Clark in their study of ranDNA. The illustration shows that after spontaneous phase separation, the ranDNA sequences in the isotropic phase (black background) are those that are unable to order and form liquid crystal phases. But ranDNA sequences in a limited interval of lengths are able to form liquid crystals (gold/orange) because they are in a state of kinetic arrest far from equilibrium. That is, they exhibit nonergodic association. This is a remarkable discovery.

We have paraded before the reader a host of Bellini and Clark's investigations that reveal the breadth of the self-assembly mechanisms of nucleic acids. Biochemist, origins researcher, and Nobel laureate Jack W. Szostak also took note of this research and spoke of their "series of elegant experiments."[30] It is a well-deserved tribute.

In 1988, Francis Crick, co-discoverer of the structure of DNA, published his autobiography titled *What Mad Pursuit: A Personal View of Scientific Discovery.* John Cairns, British physician and molecular biologist, wrote a review of Crick's book in which he said, "On certain rare occasions, a group of scientists suddenly find themselves looking, with a wild surmise, at some uncharted sea, and that is a time when their thoughts have to run far ahead of the facts."[31]

Such was the case with Noel Clark, who experienced a moment when his creative and his critical faculties synergized to great advantage. Tommaso Bellini proclaimed it Noel's vision. Whether a vision or a wild surmise, it took many years for Bellini and Clark to gather corroborative evidence, to add facts that would affirm Clark's prescience. Running well ahead of the data, his surmise had the characteristics often found in penetrating scientific conjectures: elegance and deep simplicity. To reshape a remark Francis Crick made about himself, we may say that Clark had found the pass through the interminable mountains of knowledge and could glimpse where he wanted to go.[32]

Exercises for Chapter Twelve: Exercise 12.1

Exercise 12.1

As we've seen, some DNA sequences contain mirror-image sequences called palindromes in which a prime denotes a complementary base pair (Figure 12.11). There are different structures that such molecules might form. A linear structure and a looped-out structure are shown. The sequences are typically 10-30 bases long. Which structure do you think is more stable? Why?

(Exercise reproduced with permission from Physical Biochemistry, 2/e *by David Freifelder, p. 37, Copyright 1982 by W.H. Freeman and Company, New York. Used with permission of the publisher.)*

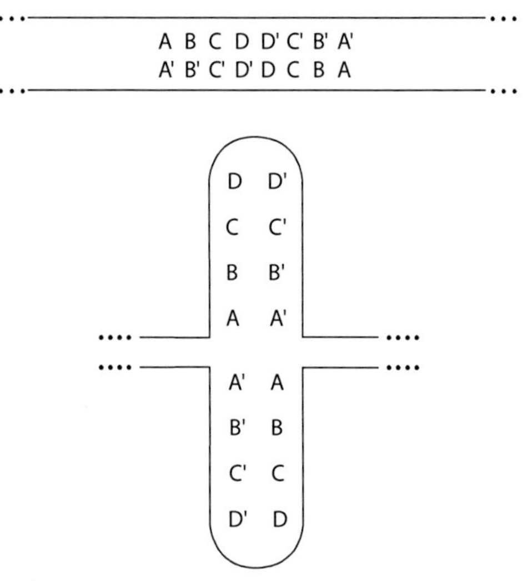

Figure 12.11

(Reproduced with permission from Physical Biochemistry, 2/e by David Freifelder, p. 37, Copyright 1982 by W.H. Freeman and Company, New York.)

Exercises for Chapter Twelve: Exercises 12.2–12.3

Exercise 12.2

The central dogma in molecular biology is: DNA makes RNA makes protein. This means that the information in DNA is somehow transferred into RNA, and that the information in RNA is then used to make protein. In this exercise you will make use of messenger RNA, but we take this opportunity to briefly stretch our knowledge of RNA. There are three main types of RNA (although there are many, many types of RNA and their number is growing): messenger RNA (mRNA), transfer RNA (tRNA), and ribosomal RNA (rRNA). All three of these nucleic acids work together to orchestrate protein synthesis. In cells, the mRNA takes the instructions (from the DNA) in the nucleus to the cytoplasm, where the ribosomes are located. Ribosomes are where the proteins are made. The ribosomes themselves are made out of rRNA molecules and a variety of proteins. The mRNA binds to the ribosome, bringing the instructions to order the amino acids (these are the subunits of proteins) to the site of protein synthesis. Last, the tRNA brings the correct amino acid to the site of protein synthesis.

Having modestly extended our knowledge of RNA, we now present the exercise: what mRNA sequence would the DNA sequence GATCCG code for?

(a) CTAGGC
(b) CTUGGC
(c) CUAGGC

Exercise 12.3

In the DNA Starter Kit in Chapter One and again in Figure 2.4 in Chapter Two, we saw cartoon representations of the DNA double helix. Each depiction showed the chemical bases joining the sugar-phosphate backbones of the two strands. The portrayal of the hydrogen bonds was simplified by using geometrical widgets such as a chevron (V-shape), a small rectangular protrusion, and a small rectangular cavity. These devices were employed only to indicate the specificity of the base complementarity, A bonds with T and C bonds with G. The widgets themselves were chosen arbitrarily. So too were the color choices for the bases. However, readers with a penchant for detail may have noticed that the bases A and G were shown as having longer lengths than the bases T and G. In the present chapter, Figure 12.7 depicted the DNA duplex as unfolded and represented the hydrogen-bonded bases somewhat more explicitly. Can you suggest why the bases A and G were shown as having longer lengths than the bases T and C in the cartoons in Chapters One and Two?

ENDNOTES

[1] Tommaso Bellini, "Exploring Soft Matter With DNA," *Frontiers of Soft Matter 2012 Symposium* (18 May 2012), Boulder, CO, slide 32.

[2] Conversation; Tommaso Bellini to K.D. on 21 May 2013.

[3] Conversation; Noel Clark to K.D. on 12 June 2013.

[4] Unpublished.

[5] Unpublished.

[6] Unpublished: Noel Clark, "Liquid Crystals of Nanonucleic Acids: Hierarchical Self-Assembly as a Route to Prebiotic Selection, Templating, and Autocatalysis," NSF Award Abstract #1207606 (1 September 2012), http://www.nsf.gov/awardsearch/showAward?AWD_ID=1207606&HistoricalAwards=false

[7] Michi Nakata, Giuliano Zanchetta, Brandon D. Chapman, Christopher D. Jones, Julie O. Cross, Ronald Pindak, Tommaso Bellini, and Noel A. Clark, "End-to-End Stacking and Liquid Crystal Condensation of 6– to 20–Base Pair DNA Duplexes," *Science 318* (23 November 2007), 1276–1279.

[8] William Shakespeare, *A Midsummer Night's Dream*, Act V, Scene 1, Lines 1844–1847.

[9] Simon L. Bullock, Inbal Ringel, David Ish-Horowicz, and Peter J. Lukavsky, "A′-form RNA Helices Are Required for Cytoplasmic mRNA Transport in *Drosophila*," *Nature Structural and Molecular Biology 17, No. 6* (2010), 703–709.

[10] Giuliano Zanchetta, Tommaso Bellini, Michi Nakata, and Noel A. Clark, "Physical Polymerization and Liquid Crystallization of RNA Oligomers," *J. Am. Chem. Soc. 130* (2008), 12864–12865.

[11] Ibid., 12864.

[12] Peter J. Collings, *Liquid Crystals: Nature's Delicate Phase of Matter* (Princeton, New Jersey: Princeton University Press, 2002), 9.

[13] Gerald R. Smith, "Meeting Palindromes Head-to-Head," *Genes and Development 22* (2008), 2612–2620.

[14] Giuliano Zanchetta, "Spontaneous Self-Assembly of Nucleic Acids: Liquid Crystal Condensation of Complementary Sequences in Mixtures of DNA and RNA Oligomers," *Liquid Crystals Today 18, No. 2* (December 2009), 40–49.

[15] G. Zanchetta, M. Nakata, M. Buscaglia, N.A. Clark, and T. Bellini, "Liquid Crystal Ordering of DNA and RNA Oligomers With Partially Overlapping Sequences," *J. Phy.: Condens. Matter 20* (2008), 1–6.

[16] Richard Wing, Horace Drew, Tsunehiro Takano, Chris Broka, Shoji Tanaka, Keiichi Itakura, and Richard E. Dickerson, "Crystal Structure Analysis of a Complete Turn of B-DNA," *Nature 287* (23 October 1980), 755–758.

[17] Peter Yakovchuk, Ekaterina Protozanova, and Maxim D. Frank-Kamenetskii, "Base-Stacking and Base-Pairing Contributions Into Thermal Stability of the DNA Double Helix," *Nucleic Acids Research 34, No. 2* (2006), 564–574.

[18] Tommaso Bellini, Giuliano Zanchetta, Tommaso P. Fraccia, Roberto Cerbino, Ethan Tsai, Gregory P. Smith, Mark J. Moran, David M. Walba, and Noel A. Clark, "Liquid Crystal Self-Assembly of Random-Sequence DNA Oligomers," *PNAS 109, No. 4* (24 January 2012), 1110–1115.

[19] Ibid., 1110.

[20] Ibid., 1110 and 1114.

[21] Soong Ho Um, Jong Bum Lee, Nokyoung Park, Sang Yeon Kwon, Christopher C. Umbach, and Dan Luo, "Enzyme-Catalysed Assembly of DNA Hydrogel," *Nature Materials 5* (September 2006), 797–801.

[22] Bellini et al., "Liquid Crystal Self-Assembly," 1111.

[23] Unpublished. Noel Clark, "Liquid Crystals of Nanonucleic Acids: Hierarchical Self-Assembly as a Route to Prebiotic Selection, Templating, and Autocatalysis," NSF Award #1207606 (1 September 2012), 10.

[24] Bellini et al., "Liquid Crystal Self-Assembly," 1112.

[25] Ibid., 1113.

[26] Peter Walters, *An Introduction to Ergodic Theory* (New York: Springer-Verlag, 1982), 2.

[27] Peter Atkins, *The Laws of Thermodynamics: A Short Introduction* (Oxford: Oxford University Press, 2010), 96–97.

[28] Phil Attard, *Non-Equilibrium Thermodynamics and Statistical Mechanics: Foundations and Applications* (Oxford: Oxford University Press, 2012), vi.

[29] Unpublished.

[30] Itay Budin and Jack W. Szostak, "Expanding Roles for Diverse Physical Phenomena During the Origin of Life," *Annu. Rev. Biophys. 39* (2010), 245–263.

[31] John Cairns, "Through a Magic Casement," *Nature 336* (17 November 1988), 268–269.

[32] Francis Crick, *What Mad Pursuit: A Personal View of Scientific Discovery* (New York: Basic Books, 1988), 17.

CHAPTER THIRTEEN

Ligation: Blest Be the Tie That Binds[1]

A tall, slim, young man with short black hair and a reserved demeanor, wearing blue jeans and a long-sleeved shirt buttoned at the wrists, stands at the head of the small seminar room. He does not seem completely at ease. Tommaso ("Tommy") Fraccia, a graduate student of Tommaso Bellini, is in Boulder, Colorado, on May 15, 2013, to present his results on ligation of stacked nanoDNA duplexes. He speaks with a heavy Italian accent, his delivery something of a lulling monotone punctuated by brief pauses to search, not unerringly, for the correct word. Ten minutes into his talk he says "Sorry for my English." His humility helps to break the ice. The audience is sympathetic and eager to help. Tommy becomes more relaxed. In fact, he acquits himself quite well.

About fifteen minutes into the talk, after the prefatory material is over, an audience member in the first row interrupts to clarify a point. And again several minutes later. He's not the only one to do so. But he's the most (gently) persistent. Yet again he questions; he probes. Often his interruptions begin with a quick "Sorry" before he plunges into his question. If Tommy isn't on track with his answer, the questioner speedily redirects him—he's after the details. All the details. The questioner is Noel Clark.

Because of so much back and forth with the audience, Tommy's seminar runs way past its allotted time. Moderator Matt Glaser rises and suggests that, in view of the time, the formal talk should be considered at an end. The audience claps and leaves. Clark and a few others remain. Clark is tall and rangy, bald with a longish fringe of

[1] Expression taken from: *Blest Be the Tie That Binds,* a Christian hymn text written by John Fawcett in 1782.

black hair, dark-complexioned with a salt-and-pepper beard and bespectacled. He huddles with Tommy for 45 minutes, relentlessly digging to unearth more information, to get to the core. His focus is incandescent, like the burning point of a lens. And no wonder. The material Tommy is disclosing is crucial to the origins of life proposal of Bellini and Clark.

NanoDNA Stacking: Weak Physical Attractive Forces Versus Chemical Ligation

To begin, let's back up the bus. Before diving into the ligation experiment, the lynchpin in the origins proposal of Bellini and Clark, we need to be clear on the importance of chemical ligation. We know that nanoDNA duplexes spontaneously self-assemble in solution by the process of stacking as the temperature of the solution they're in is lowered. This stacking produces an aspect ratio of the stacked duplexes—the ratio of their length to their width, *L/D*—that enables liquid crystal formation in accord with Onsager's criterion. However, as we've said before, weak physical forces mediate this stacking. The duplexes are *not* strongly chemically bound. There are no covalent bonds between duplexes in the stack.

So what? We answer this with an analogy. When explaining DNA origami, Paul Rothemund used cylinders to represent double-helix DNA (Figure 8.1 of Chapter Eight). We borrow from Rothemund and idealize a nanoDNA duplex as a can (Figure 13.1). To whet your appetite for our discussion, imagine it's a can of Primordial Soup. Microbiologist James W. Brown of North Carolina State University has an anecdotic label for a can of primordial soup on his website.[1]

Figure 13.1 An idealized nanoDNA duplex shown as a can. *(Shutterstock Image ID: 72840952.)*

If we simply stack several cans on top of one another we expect them to remain as a stack, but not a terribly stable stack. It wouldn't be hard to disrupt this arrangement and then, with a din and a clatter, the cans would no longer form an elongated column. But if we decided to solder or weld the top of one can to the bottom of another we could fashion a more robust elongated array. Think of chemical ligation of stacked nanoDNA duplexes as cans soldered together. Think of stacked nanoDNA duplexes held in physical proximity only by weak attractive forces as cans just placed on top of one another.

Abiotic Ligation Experiments With EDC

Bellini and Clark took on the challenge of demonstrating chemical ligation of nanoDNA duplexes sequestered in liquid crystals. They were determined to do this under plausible prebiotic conditions. If they could show this was experimentally possible in the laboratory it would give their proposal for the origin of life a huge boost. Such experiments would provide compelling evidence that liquid crystal formation served as a template to guide the prebiotic appearance of systems of linear molecules with the capacity for self-replication. We've already backed up the bus. We have a sense of the significance of chemical ligation. We now put the bus in forward gear. We start to learn how ligation might be achieved without recourse to enzymes or any other biomolecules that were not available many billions of years ago on a barren Earth, an Earth devoid of life.

Carbodiimides are a popular class of ligation agents. The one called 1-ethyl-3-(3-dimethylaminopropyl)carbodiimide—commonly known as EDC—is an attractive choice. Part of its appeal is that it's a ligation agent of zero length. A zero length ligation agent is a chemical compound that mediates the binding of two molecules by forming a bond between them that contains *no* additional atoms. Thus, one atom of a molecule is covalently attached to an atom of a second molecule with no intervening linker or spacer. There is direct linkage of the two molecules.[2]

Another crucial attribute of EDC is that it is an abiotic ligation agent, meaning that it mediates ligation without the use of biomolecules. As we will see, Bellini and Clark used EDC to couple the terminals of their nanoDNA molecules to one another. In so doing they employed a means of connection that is consistent with a prebiotic environment, an environment in which there were no biotic ways to attain ligation.

Abiotic Describes a process that occurs in the absence of biomolecules.

Zero length ligation agent
A chemical compound that mediates the joining of two molecules by the formation of a bond that contains no additional atoms. There is direct linkage of the two molecules.

For those interested in the detailed mechanism of EDC-mediated ligation, take a look at the cartoon at the bottom left of Figure 13.2, beneath other cartoons that depict the molecular species used in the experiments. The details are not essential for an appreciation of the ligation experiments, but for those who would like to know, the cartoon shows that EDC *activates* the phosphate terminals of the dodecamers so that they can react with hydroxyl terminals to produce a covalent phosphodiester bond. This bond links the 3′ carbon atom of one sugar molecule and the 5′ carbon atom of another. This is the same bond, the native bond, that we've seen before in the backbone of DNA (Figure 6.8 in Chapter Six). The EDC reaction produces a compound called isourea as a by-product.

The Boulder/Milan collaborators first investigated ligation in solutions of EDC using the self-complementary Dickerson dodecamer that we encountered in Chapter Twelve, for example, in Figure 12.5 of that chapter. The Dickerson dodecamer has the base sequence CGCGAATTCGCG with polarity: 5′ end on the left and 3′ end on the right, written as 5′- CGCGAATTCGCG - 3′. For technical reasons, Bellini and Clark added a so-called 3′- phosphate termination so that they formed 5′- CGCGAATTCGCGp - 3′.[3] They referred to this molecule simply as D1p—and we will, too.

In work we saw in the previous chapter, the Boulder/ Milan team found that the Dickerson dodecamer formed duplexes that self-organized into nematic or columnar liquid crystal phases in aqueous solutions depending on the concentration of the dodecamer and the temperature. So this molecule was a suitable candidate for the ligation experiments. However, in their initial ligation experiments they found that the solubility of EDC in the liquid crystal phases was limited and this presented an obstacle. Before we flesh out the problem, recall a term from high school chemistry. A mole is a scientific unit for measuring large quantities of very small entities such as atoms or molecules.[4] Thus, a molar ratio, R, is a ratio between the amounts, in moles, of any two compounds involved in a chemical reaction. For example, we can speak of the molar ratio of the amount of EDC and the amount of D1p (Dickerson dodecamers) in a chemical reaction as $R = [EDC] / [D1p]$.

D1p Symbol representing the Dickerson dodecamer, CGCGAATTCGCG, with the addition of a phosphate group (a phosphorus atom surrounded by four oxygen atoms) that is bound to the 3′- end of the dodecamer.

Bellini and Clark found that the solubility of EDC in the liquid crystal phases was limited to molar ratios $R = [EDC] / [D1p] < 3$. This condition prevented a large enough presence of EDC to generate polymerization worthy of attention. An expression that comes into play here is the polymerization yield denoted by p where $0 < p < 1$. Think back to our discussion of yield in Chapter Eight. The context in Chapter Eight was different but the principles are the same. When we speak of polymerization yield we are still talking about a reaction yield, that is, the fraction of ligated Dickerson dodecamers out of the total number of Dickerson dodecamers present. The team found that when $R < 3$ the value of the polymerization yield, $p \leq 0.13$. This was nothing to write home about.

Bellini and Clark needed to find an effective scheme to escape the solubility limitation on the polymerization yield. And they *were* able to dodge the deterrent created by the low R value.

They capitalized on a phenomenon that we've often seen—phase separation. We now turn to their plan of action.

Mole A scientific unit for measuring large quantities of very small entities such as atoms or molecules.

Molar ratio, R A ratio between the amounts, in moles, of any two compounds involved in a chemical reaction.

Polymerization yield The fraction of polymerized molecules out of the total number of molecules available for polymerization (after a specified time). In the present case, the fraction of ligated Dickerson dodecamers out of the total number of Dickerson dodecamers present.

Please note: The six sections that follow may appear challenging. They contain many niceties that can be best taken in by carefully chewing on them rather than trying to wolf them down. *The most important result of these six sections is just this: liquid crystal ordering markedly enhances ligation efficiency that leads to long chains of DNA.* Keep this most important conclusion in mind as you read on.

The Scheme: Polyethylene glycol (PEG)-Induced Phase Separation

Bellini, Clark, and colleagues wrote of their scheme, “This limitation was effectively bypassed by exploiting the self-selection properties of phase separation. We thus diluted DNA by a solution of polyethylene glycol (PEG), a chemically inert water-soluble [random coil] polymer often used as a depletant to control osmotic pressure *Π*.”[5]

We’ll break down this statement into pieces. The King of Hearts instructed the White Rabbit to “Begin at the beginning and go on till you come to the end: then stop.”[6] We, however, begin at the end: what is osmotic pressure?

Jacobus H. van’t Hoff was a Dutch chemist and the first winner of the Nobel Prize in Chemistry in 1901. Early in his Nobel Lecture, he too asked the (rhetorical) question, “What is osmotic pressure? When a solution, e.g., of sugar in water, is separated from the pure solvent—in this case water—by a membrane which allows water but not sugar to pass through it, then water forces its way through the membrane into the solution. This process naturally results in greater pressure on that side of the membrane to which the water is penetrating, i.e., to the solution side. This pressure is osmotic pressure.”[7]

Osmotic pressure Suppose a membrane separates two fluid volumes: one volume is a solution (i.e., a solute dissolved in a solvent) and the other is the pure solvent. Osmotic pressure is the pressure on the solution side of the membrane that is necessary to prevent the flow of pure solvent through the membrane and into the solution.

Expanding on van’t Hoff: the movement of the solvent (water) across the membrane is the phenomenon of osmosis. The pressure that builds up on the solution side of the membrane eventually stops the osmosis, and this value of the pressure of the solution is called the osmotic pressure.

Using PEG to control osmotic pressure is conceptually similar to the depletion interaction that we discussed in Chapter Eleven. This is why Bellini and Clark referred to PEG as a *depletant* to control the osmotic pressure, denoted by the Greek symbol *Π* (pi). When DNA is mixed with PEG, an entropy-driven phase separation occurs (just as we described the entropy-driven separation of small and large particles in Chapter Eleven). Their experiments showed that the osmotic pressure provided by the PEG phase separates the solution into DNA-rich columnar (COL) liquid crystal domains surrounded by a PEG-rich isotropic (ISO) fluid.[8]

This is illustrated in the micrographs of panel (f) of Figure 13.2. In addition to the instructional cartoons, notice the colorful birefringence in the left micrograph. This is the now familiar signature of liquid crystals transitioning into an ordered phase. The left side image is taken through a depolarized light microscope. The image on the right side is a micrograph using fluorescence microscopy where the DNA duplexes are selectively marked with the fluorescent stain ethidium bromide.

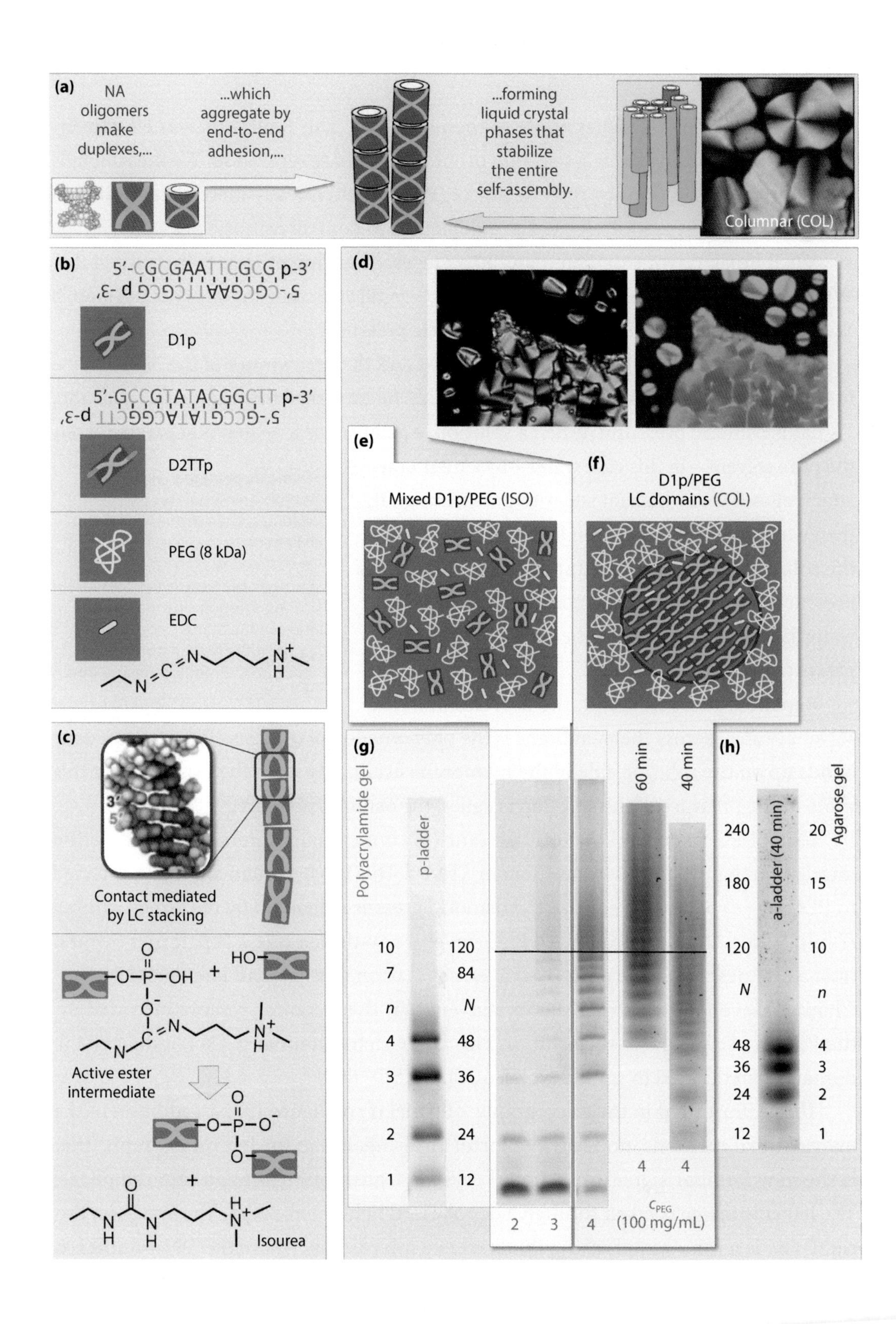

(a)
NA oligomers make duplexes,...
...which aggregate by end-to-end adhesion,...
...forming liquid crystal phases that stabilize the entire self-assembly.
Columnar (COL)
(b)
5'-CGCGAATTCGCG p-3'
D1p
5'-GCCGTATACGGCTT p-3'
D2TTp
PEG (8 kDa)
EDC
(c)
Contact mediated by LC stacking
Active ester intermediate
Isourea
(d)
(e)
Mixed D1p/PEG (ISO)
(f)
D1p/PEG
LC domains (COL)
(g)
Polyacrylamide gel
p-ladder
60 min
40 min
c_{PEG} (100 mg/mL)
(h)
a-ladder (40 min)
Agarose gel

The phenomenon of phase separation is of vital importance. We use the symbol c_{DNALC} to denote the concentration of DNA necessary to induce liquid crystal ordering in a PEG-free water-D1p solution.

Consider DNA solutions with a concentration, c_{DNA}, much smaller than c_{DNALC}. Suppose we add a small amount of PEG to the solution. If the PEG concentration, c_{PEG}, is low we find that the DNA duplexes are homogeneously mixed with PEG. This is shown in cartoon panel (e) of Figure 13.2 and is labeled as mixed D1p/PEG (ISO). However, for sufficiently large PEG concentrations, D1p columnar liquid crystal domains will appear, and they are surrounded by an isotropic phase that contains EDC at values of R that can be orders of magnitude larger than what we saw earlier, typically in the range $10 < R < 50$. This is shown in cartoon panel (f) of Figure 13.2, sketching a COL domain of D1p/PEG liquid crystal domains using the symbols from panel (b).

Reflect on what the scheme of Bellini and Clark has thus far accomplished. They have obtained a system in which liquid crystal domains—constituting only a small fraction of the entire sample volume—act as microreactors. The term *(micro)reactor* is used in a variety of ways in the chemical literature. In general, reactors are large vessels used in chemical and refinery plants.[9] By contrast, microreactors are typically devices etched in metal or glass and they can fit in the palm of your hand. But Bellini and Clark are using the term *microreactor* to denote the self-assembled liquid crystal domains in their experiments. These phase-separated domains have an internally fluid structure bounded by an aqueous/aqueous interface that enables easy transport of material. They are maintained at constant osmotic pressure by the PEG depletant. Such domains are an ideal arrangement to carry out the ligation reactions that the team is looking for.

Moreover, think about this: in the previous two chapters we've seen that phase separation is inherently selective. As an example, phase

c_{PEG} The concentration of PEG in a solution.

c_{DNALC} The concentration of DNA necessary to induce liquid crystal ordering in a pure water-D1p solution.

Microreactors Small vessels for carrying out chemical reactions. In the context of the work of Bellini and Clark, self-assembled liquid crystal domains serve as naturally occurring microreactors.

Figure 13.2 Enhanced abiotic ligation afforded by the condensation of DNA into liquid crystal domains. **(a)** Stacked DNA oligomer duplexes order into liquid crystal phases that stabilize the stacking; NA = nucleic acid. **(b)** Molecular species used in the experiments. **(c)** Mechanisms of autocatalysis and ligation. **(d)** Phase behavior and ligation in D1p/PEG/EDC/water mixtures. **(e)** Sketch at low PEG concentration, where the mixture is a uniform isotropic solution. **(f)** For $c_{PEG} > 400$ mg/mL, the mixture phase separates into columnar droplets of stacked duplex DNA surrounded by an isotropic solution of PEG; micrographs above show depolarized (left) and fluorescence microscopy (right). **(g)** Polyacrylamide gel stained by ethidium bromide. **(h)** Agarose gel runs of the $c_{PEG} = 400$ mg/mL sample at two different run times. The picture of the 60-min run gel has been compressed and shifted to match the band position and spacing of the 40-min run gel (details in Supplementary Methods online). Numbers running vertically along the lanes show DNA ladders with oligomer lengths in number of base pairs, synthesized by repetition of the D1p sequence; polyacrylamide (left), agarose (right). The straight line is a guide for the eyes marking the condition of $N_b = 120$ ($n = 10$). *(Reproduced with permission from Tommaso P. Fraccia, Gregory P. Smith, Giuliano Zanchetta, Elvezia Paraboschi, Youngwoo Yi, David M. Walba, Giorgio Dieci, Noel A. Clark, and Tommaso Bellini, "Abiotic Ligation of DNA Oligomers Templated by Their Liquid Crystal Ordering," Nature Communications 6, Article No. 6424 (10 March 2015), 1–7.)*

separation rejects the entry of single DNA strands into liquid crystal domains while admitting duplexes with sufficiently adhesive tails. We saw this in the previous chapter in solutions of complementary duplexes with overhangs. And we saw this with 20-mers of random sequence DNA oligomers admitted into liquid crystal domains.

In the current ligation experiments, the selection based on these molecular features we've spoken of also controls entry into the microreactors, adding an important self-sorting mechanism to the liquid crystal ordering.

Gel Electrophoresis of D1p Oligomers With Polyacrylamide and Agarose Gels

Back now to using PEG to facilitate nanoDNA ligation. We focus on the results of gel electrophoresis of the samples. Bellini and Clark dissolved small amounts of D1p in a solution of water and EDC to obtain a homogeneous solution with $c_{PEG} \sim 0.05\, c_{DNALC}$ and molar ratio, $R = 50$. Next they added PEG to the solution giving various values of final concentration up to $c_{PEG} = 400$ mg/mL. They performed these experiments in parallel in both small plastic tubes and in flat cells—the latter to enable visual confirmation of liquid crystal formation (the birefringence we pointed out earlier) using depolarized optical microscopy.

Typical polyacrylamide gel scans of DNA extracted from such solutions are shown in panel (g) of Figure 13.2 where the fluorescence of the gel bands is provided by the same ethidium bromide staining used in the fluorescence micrograph we saw earlier.[10] Ethidium bromide is what is called a DNA intercalator. That means it inserts itself into the spaces between the base pairs of the double helix. Ethidium bromide absorbs ultraviolet light and re-emits it as visible light, so it's a convenient stain for DNA.

Apropos of panel (g), notice the numbers running vertically along the lanes. They show DNA ladders where the oligomer lengths are expressed in number of base pairs. These ladders serve as standards to enable rapid approximation of the oligomer length of the experimental samples that are found as bands in the various lanes. The different bands we're seeing in the various gel lanes are accumulations of oligomers of different lengths and hence different molecular weights.

Bellini and Clark examined the gel scans and found that ligation in the low c_{PEG} homogeneous mixtures is limited even at a relatively high concentration of EDC (c_{EDC})—i.e., molar ratio, $R = 50$.

DNA ladder In the case of Bellini and Clark, synthesized sequences of multiple repetitions of the Dickerson dodecamer. In gel electrophoresis, they serve as standards to enable rapid approximation of the oligomer length of the experimental samples that are found as bands in the various lanes.

However, the appearance of condensed droplets with liquid crystal ordering that we saw in panel (f) produces a qualitatively new behavior that is revealed in panel (g): the growth of a large peak in the polymer probability distribution with a mean degree of polymerization, $\langle n \rangle > 10$. We return to these terms in the next section. But briefly, this means that if we consider each Dickerson dodecamer, D1p, as a monomer (a building block), the EDC successfully ligated monomers together in the liquid crystal droplets to form polymers and the mean (i.e., average) length of the polymers formed was $\langle n \rangle \sim 10$ monomers, that is ~ 10 dodecamers. This is a tenfold enhancement over what they found in the isotropic outside the liquid crystal droplets. But $\langle n \rangle \sim 10$ is a preliminary result based on polyacrylamide gels.

To better visualize the oligomer length distribution, they ran the same samples in agarose gels, a technique we first mentioned in Peng Yin's work on DNA bricks in Chapter Eight. The reason Bellini and Clark switched to agarose is that the bands associated with longer DNA oligomers are more easily distinguished in this type of gel, as seen in panel (h) of Figure 13.2. Agarose and polyacrylamide have different attributes as gel material. Agarose gels have a large range of separation but relatively low resolving power. They're a suitable choice for DNA with a large number of base pairs. Polyacrylamide gels, by contrast, have a comparatively small range of separation but very high resolving power. They're a good choice for separating fragments of shorter DNA oligomers.[11] The lesson: not all gels are created equal.

We'll soon see the true extent of the polymerization that was revealed by the agarose gels. But first, we need just a little more preparation.

Another Stellar Contribution by Chemist Paul J. Flory

In Chapter Eleven we presented the work of the remarkable chemist Paul J. Flory in explaining the nanoDNA phase separation observed by Bellini and Clark. There we used the subheading Flory's Model and this was slightly misleading. Flory was all over the map in tackling most of the major problems in the physical chemistry of polymeric substances and he came up with *many* models. His Stanford University chemistry colleague Henry Taube wrote, "He had an extraordinary capacity to penetrate to the heart of a scientific problem and to isolate the essential features of even complex systems. . . . I . . . remember . . . thinking to myself, 'Flory can make scientific sense even out of glue.' "[12] We'll now convey the gist of another of Flory's models.

Flory wanted a general way of predicting the molecular size distribution of polymers formed by the ligation of monomers (i.e., polymer subunits).[13] We'll confine ourselves to the situation that reflects the experiments we're interested in, namely, when a polymer is formed from *identical monomeric subunits.* Flory was able to use a simple and intuitive mathematical argument to determine the probability for the formation of a polymer composed of an arbitrary number, n, of monomeric subunits (panel (a) of Figure 13.3). He did a swell job with the mathematics and made it so accessible that we can readily follow his flow.

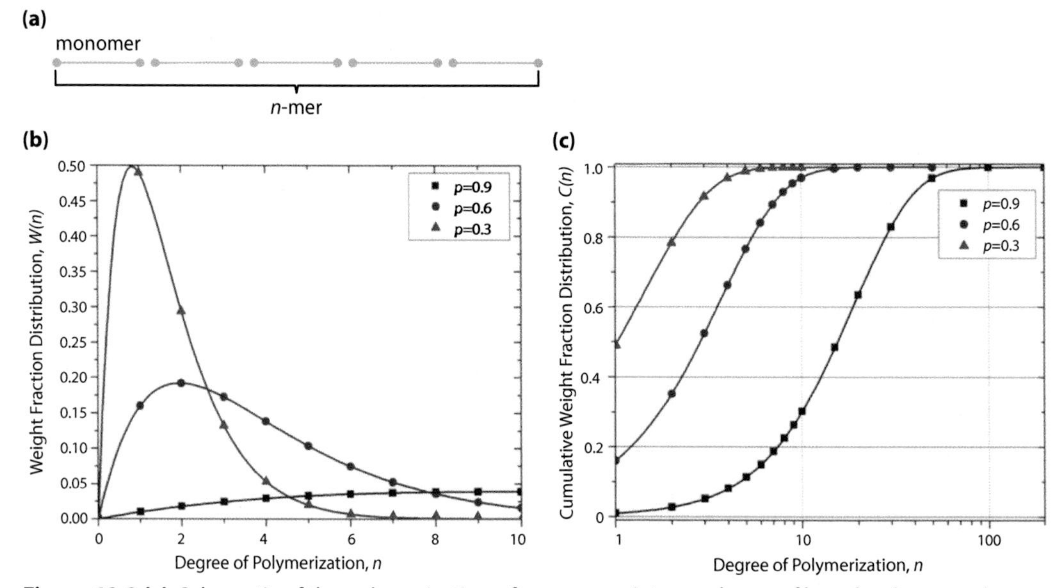

Figure 13.3 (a) Schematic of the polymerization of monomers into a polymer of length *n* (an *n*-mer). **(b)** Flory weight fraction distribution for three different polymerization yields, *p*. **(c)** Flory cumulative weight fraction distribution for the same three values of *p*. *(Bottom portion reproduced with permission from Online Supplementary Information for Tommaso P. Fraccia, Gregory P. Smith, Giuliano Zanchetta, Elvezia Paraboschi, Youngwoo Yi, David M. Walba, Giorgio Dieci, Noel A. Clark, and Tommaso Bellini, "Abiotic Ligation of DNA Oligomers Templated by Their Liquid Crystal Ordering," Nature Communications 6, Article No. 6424 (10 March 2015), 1–7.)*

Flory determined that the probability that an n-length oligomer (an n-mer, where $n = 1, 2, 3\ldots$) exists, is mathematically dependent on the polymerization yield, p, which we introduced earlier. For a given value of polymerization yield there is a probability distribution of the products of polymerization. This is known as the weight fraction distribution, $W(n)$. It can be displayed as a curve plotted against the degree of polymerization, n. From the $W(n)$ curves we can determine the relative quantities (by weight) for each value of n. This is panel (b) of Figure 13.3. That's the précis. Next we expand.

Starting with an initial weight of monomer, $W(n)$ is the fraction of this weight found as n-mers—as measured by the p-value achieved—after the polymerization has proceeded for a certain time. $W(n)$ is shown for three p values in Figure 13.3 (b). At the smaller values of p the weight fraction curve has a sharp maximum that is very close to the mean degree of polymerization, $\langle n \rangle$. But as p approaches 1.0 (which would mean that all the monomers have ligated to form polymers), this maximum becomes lower and shifts to larger values of n.[14]

We also note that the sum of the weight fractions of all constituents is, of necessity, always equal to unity. Its value is approximately equal to the area under one of the p-curves in panel (b).[15]

Panel (c) of Figure 13.3 shows the cumulative weight fraction distribution, $C(n)$. This is the fraction of the total weight of monomers found in oligomers of length n and smaller. $C(n)$ is useful because $1 - C(n)$ directly gives the fraction of the initial weight ending up in oligomers longer than n. The plots of $C(n)$ have been displayed for convenience on a semi-logarithmic graph with the horizontal axis using intervals that correspond to factors of ten. We saw this used in Figure 11.3 in Chapter Eleven. The idea is to compress the horizontal axis so we can more easily see the behavior of the *C(n)* curve for small values of *n* as well as large values of *n*. One point to notice is that as we consider higher values of *p*, the curve will reach unity (its maximum value) at higher values of *n*. This reflects the fact that as a greater amount of the monomer ligates to form polymers, the occurrence of longer polymers increases.[16]

Weight fraction distribution, *W(n)* A probability distribution curve that shows the ratio of the total weight of polymers that exist as *n*-mers (polymers composed of a particular number, *n*, of monomers) divided by the total weight of all the molecules.

Cumulative weight fraction distribution, *C(n)* An S-shaped distribution curve that shows the sum of the weight fractions of all polymeric constituents (all values of *n*). *C(n)* includes monomers ($n = 1$) that have not been polymerized.

Analysis of Gel Profiles: The Experimental Data Are Well Described by the Flory Model

It's taken some doing for us to get the savvy to absorb the analysis of the gel profiles that we present in Figure 13.4. But now the pieces are in place. For the moment, we'll confine ourselves to panels (a) and (b) of Figure 13.4 that show polymer length distributions extracted from the gel profiles. We delay a look at panels (c) and (d) until the next section.

What we see are profiles of the fluorescent emissions of the gel runs. The variable i_F is the fluorescence intensity, shown as a continuous line. This is plotted as a function of *n*, the position along the gel converted to the degree of polymerization. The variable C_F is the cumulative weight fraction distribution. C_F is derived by integrating i_F (i.e., summing under the i_F peaks) and is also plotted as a function of *n*. These profiles are analyzed by exploiting the approximate proportionality between electrophoretic mobility and the logarithm of the DNA chain length.[17] We won't discuss the details but take them as a given.

The black dotted lines in panels (a) and (b) and the black dashed line in panel (b) are fits of the experimental data to the Flory model. By using these fits, Bellini and Clark were able to determine the polymerization yield *p*. In this context, *p* is referred to as a fitting parameter. A small point: in Figure 13.3, we used a logarithmic axis only for the display of the cumulative weight fraction distribution. In Figure 13.4, both the weight fraction distribution and the cumulative weight fraction distribution are plotted against a logarithmic axis for *n*.

The plots show that the chain length distributions are well described by Flory's model that predicts the molecular sizes of polymers formed by the ligation of monomer subunits (in this case, the D1p Dickerson dodecamer). The qualitatively new behavior that we called out earlier is displayed by the growth of a large peak in the oligomer population. The data confirm that due to the phase separation and liquid crystal condensation, within the liquid crystal domains there is a sharp increase in ligation efficiency. This is evident by eyeballing

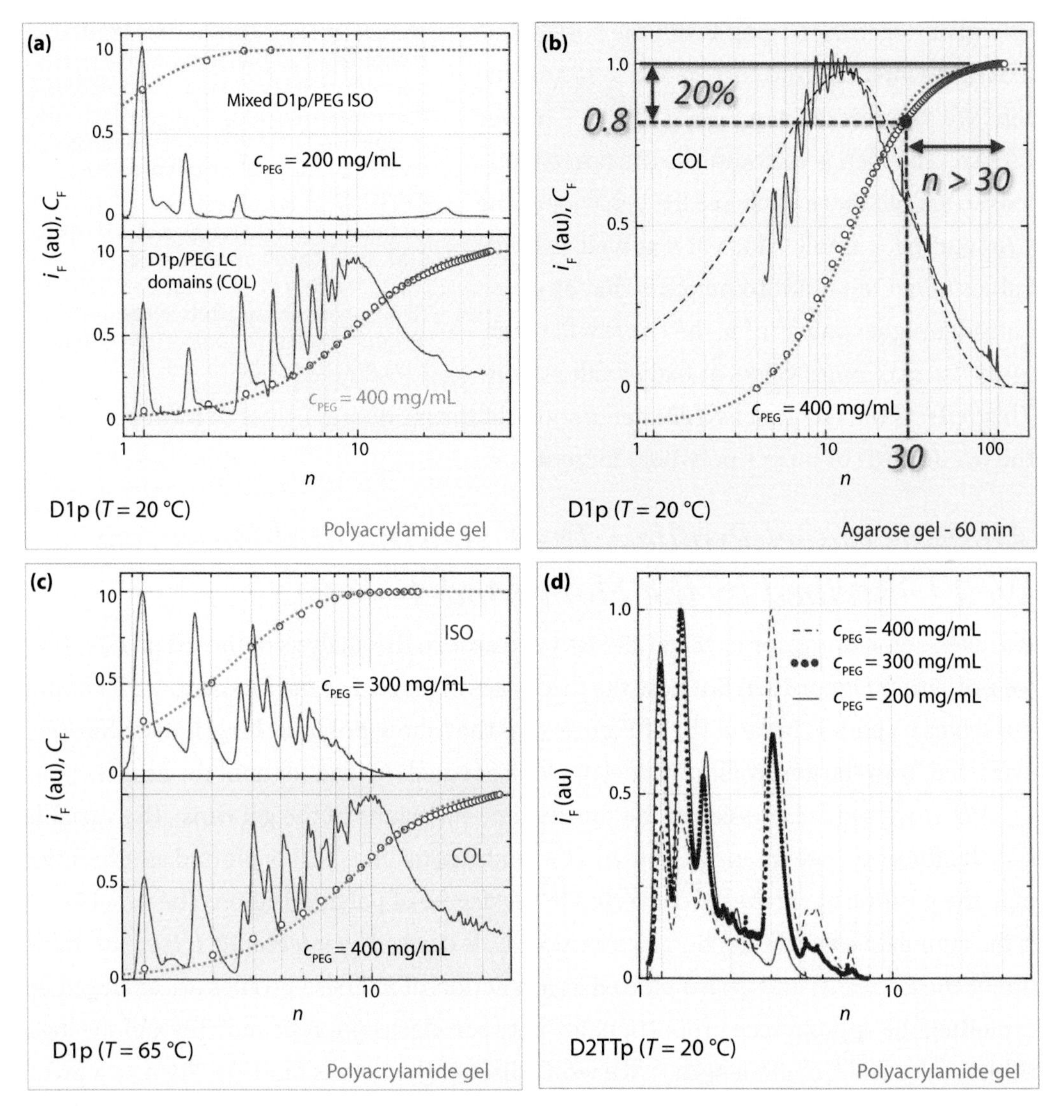

Figure 13.4 Fluorescence intensity profiles of the gel electrophoresis runs performed on DNA/PEG/EDC mixtures. Plots are shown as a function of *n*, the position along the gel converted to the degree of polymerization. i_F (continuous lines in arbitrary units, au) is the fluorescence intensity. C_F (open dots) is the measured cumulative weight fraction distribution. Black dotted lines in panels (a), (b), and (c) and the black dashed line in panel (b) show the fit to the experimental data of the Flory model, from which the polymerization yield, *p*, has been determined. **(a)** D1p/PEG/EDC mixtures measured in a 15% polyacrylamide gel for increasing PEG concentration. **(b)** D1p/PEG/EDC mixtures measured in 3.5% agarose gels for 60 min running time. **(c)** D1p/PEG/EDC mixtures measured at T = 65 °C. **(d)** D2pTT/PEG/EDC mixtures at various PEG concentrations. *(Reproduced with permission from Tommaso P. Fraccia, Gregory P. Smith, Giuliano Zanchetta, Elvezia Paraboschi, Youngwoo Yi, David M. Walba, Giorgio Dieci, Noel A. Clark, and Tommaso Bellini, "Abiotic Ligation of DNA Oligomers Templated by Their Liquid Crystal Ordering," Nature Communications 6, Article No. 6424 (10 March 2015), 1–7.)*

the lower plot of panel (a). There is a mean degree of polymerization $\langle n \rangle > 10$. Quantitative analysis of panel (a) reveals that the polymerization yield, *p*, is enhanced from $p \sim 0.13$ to $p \sim 0.84$ in going from the mixed D1p/PEG (ISO) isotropic phase (the upper plot) to the D1p (COL) liquid crystal phase (the lower plot). Quantitative evaluation also shows the average degree of polymerization increases from $\langle n \rangle < 2$ (upper plot) to $\langle n \rangle \sim 11$ (lower plot).

The agarose gel in panel (b) enables better detection of longer ligation products. The S-shaped curve in panel (b) is approximately equal to the area under the weight fraction distribution profile (which is the dashed curve to its left). We made this point in discussing Flory's model. The fraction of the weight polymerized into chains longer than n can be read from Figure 13.4 (b). This is illustrated by the mark-up and text that we've added in red to the graph to highlight the $C(n{=}30)$ point, as an example. The double-headed red arrow indicates that at $n = 30$, $1\text{-}C(n) = 0.2$. This means that in the columnar liquid crystal domains, fully 0.2 or 20% of the nanoDNA weight is in chains with a degree of polymerization, $n > 30$. Moving up the $C(n)$ curve similarly shows that about 10% of the DNA weight is part of chains with $n > 50$.

Finally, additional analysis of the agarose gel data in panel (b) indicates that in the liquid crystal droplets the polymerization yield is boosted to $p \sim 0.90$. Phrased differently, the arched profile of the weight fraction distribution curve (dashed line) fits the Flory model when $p \sim 0.90$. And the weight fraction curve maximum—which is very close to the average degree of polymerization, $\langle n \rangle$—gives the value $\langle n \rangle \sim 19$. Since our monomer has 12 base pairs, this results in polymers of $12n = 228$ base pairs. If the reader is wondering about the earlier statement that $\langle n \rangle > 10$ and the current claim of $\langle n \rangle \sim 19$, the explanation is in the gel medium, as we alluded to already. The $\langle n \rangle > 10$ observation came from polyacrylamide gels, but when the same samples were run on agarose gels it was possible to obtain more reliable information on the longer polymer chains.

Using the fluorescence profiles of the gels (plotted on the vertical axis) as a quantitative measure of the degree of polymerization does introduce some intrinsic uncertainty. The reason for this is that the amount of ethidium bromide stain that is intercalated in each strand is not controllable and so the fluorescence signal is not an absolute measurement of the number of strands. Nonetheless, assuming that ethidium bromide intercalates in the same way for each DNA strand of the same length, it's possible to obtain a relative measurement of DNA weight in each gel band.[18] Regardless of the intrinsic uncertainty in the quantitative information carried by the fluorescence profiles of the gels, the product length distribution extracted from the profiles is described well by the Flory model.

The Lowdown on Ligation Efficiency

The results we've discussed show that the combination of DNA condensation (i.e., the compacting of duplexes driven by osmotic pressure) and DNA liquid crystal formation are decisive in templating the chemical ligation of DNA oligomers. However, it remains uncertain to what extent such ligation efficiency may be due to the local increase in DNA concentration, c_{DNA}, or to the liquid crystal ordering. We must reflect on the critical implications of this last sentence.

The Bellini/Clark proposal for the origins of life—specifically, for the formation of long-chain polymers of DNA—is at stake here. To reiterate their envisioned process of the stages of hierarchical self-assembly: DNA duplexing, end-to-end stacking of duplexes, phase segregation of stacked duplexes, and enhanced ligation of stacked duplexes within liquid crystal domains. They proposed this as the route by which the liquid crystal phase of nanonucleic acids could autocatalytically select, template, and replicate its constituent molecules on the primordial Earth.

Template (verb)
To act as a template.

But essential to Clark's vision is that the *liquid crystal ordering of DNA* act as the gatekeeper that selects the appropriate molecules (DNA duplexes) and enhances their ligation. In the Supporting Material for the team's initial publication in 2007, they showed experimentally that increasing complementary oligomer length enhances liquid crystal phase stability (Figure 13.5).[19] They took this as a clear indication that introducing abiotic ligation in the liquid crystal phase would produce elongated molecules that would also act to stabilize liquid crystal ordering.

Let the thrust of this figure sink in. Consider the isotropic-liquid crystal phase transition temperatures, notably, isotropic-nematic and isotropic-columnar in Figure 13.5. (These plots are for a fixed nanoDNA concentration.) Read from right to left on the *N*-axis. You see that as the length in base pairs (bp) of the nanoDNA oligomers becomes larger, the liquid crystal phase exists for a higher temperature before melting into the isotropic. This is true for both nematic and columnar. This increasing temperature range signals the increased thermal stability of the liquid crystal phase. Since the liquid crystal ordering is more stable for these longer oligomers, ligation promotes a positive feedback loop between the formation of liquid crystals and the elongation of their constituent molecules.

In sum, Bellini and Clark propose that on the prebiotic Earth, subject to conditions of cycled temperature or hydration, liquid crystalline ordering both selects and templates molecules that, in turn, enhance the liquid crystal order and thermal stability.

In the following section, *The Liquid Crystal Phase as Gatekeeper*, we carefully review two experiments that the teams from Milan and Boulder performed to cinch that *liquid crystal ordering (as opposed to the*

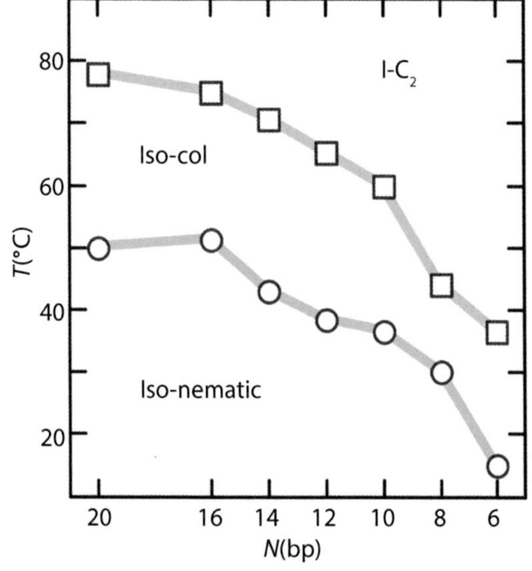

Figure 13.5 Liquid crystal thermal melting plotted against length of nanoDNA oligomers. Horizontal axis, *N*, is the length of the nanoDNA oligomers in base pairs. *(Reproduced with permission, Online Supporting Material, Michi Nakata, Giuliano Zanchetta, Brandon D. Chapman, Christopher D. Jones, Julie O. Cross, Ronald Pindak, Tommaso Bellini, and Noel A. Clark, "End-to-End Stacking and Liquid Crystal Condensation of 6– to 20–Base Pair DNA Duplexes," Science 318 (23 November 2007), 1276–1279.)*

local DNA concentration) was responsible for the enhanced ligation efficiency within the liquid crystal domains. The experiments are intricate and the reader may wish to accept their conclusion at face value and move on to the next section, *Cascaded Phase Separation.* Alternatively, learning the details behind the discoveries—although this requires close scrutiny—may bring additional satisfaction.

The Liquid Crystal Phase as Gatekeeper

We now plunge ahead and see how Bellini and Clark were able to discriminate between the local increase in DNA concentration, c_{DNA}, and the presence of DNA liquid crystal ordering. To tease these influences apart, they needed to compare ligation efficiencies under conditions where the only significant difference is the liquid crystal ordering. They implemented this in two distinct experiments. There are numerous details in these experiments and our commentary will present specifics for those who wish to have them.

Each experiment exploited a new finding that is illustrated in Figure 13.6. Bellini and Clark discovered that depending on temperature, c_{PEG}, and the structure of the DNA oligomers (the use of the word *structure* will be explained), PEG can phase separate droplets of concentrated duplexed DNA oligomers in either the isotropic (ISO) or the liquid crystal columnar (COL) phase. The depolarized light micrographs of panels (b) and (c) show this. There is no colorful birefringence in (b) when the samples are viewed with crossed polarizers—cyan and magenta arrows at right angles. But in (c), the birefringent signature of liquid crystals is present when viewed with crossed polarizers.

Now check out the cyan and magenta arrows in the insets to panels (b) and (c). They are parallel to one another. Here the polarizers aren't crossed. This reveals that the circular domains—the isotropic and columnar liquid crystal domains—are about the same physical size.

In the first of the two experiments, they used a temperature of 65 °C. This is just about where the liquid crystal columnar phase melts. But they chose this elevated temperature because they found that at this value, condensation of D1p into an isotropic phase domain is possible. This behavior contrasts with that at 20 °C where, when increasing c_{PEG}, D1p condenses directly into the liquid crystal columnar phase.

By maintaining a temperature of 65 °C they were able to select and study two cases in parallel. The first is the ligation of D1p at a concentration of c_{PEG} = 300 mg/mL. Here the DNA phase separates from PEG into isotropic droplets (panel (b) of Figure 13.6). This is rather different than what we've seen previously, where duplexed DNA organized into an ordered liquid crystal phase. The cartoon in panel (b) shows duplexed DNA but it remains isotropic. *Note*: 3, 4, etc., below gel lanes denote c_{PEG} of 300 mg/mL, 400 mg/mL, etc.

The parallel study was done at c_{PEG} = 400 mg/mL, where columnar liquid crystal domains are found (panel (c) of Figure 13.6).

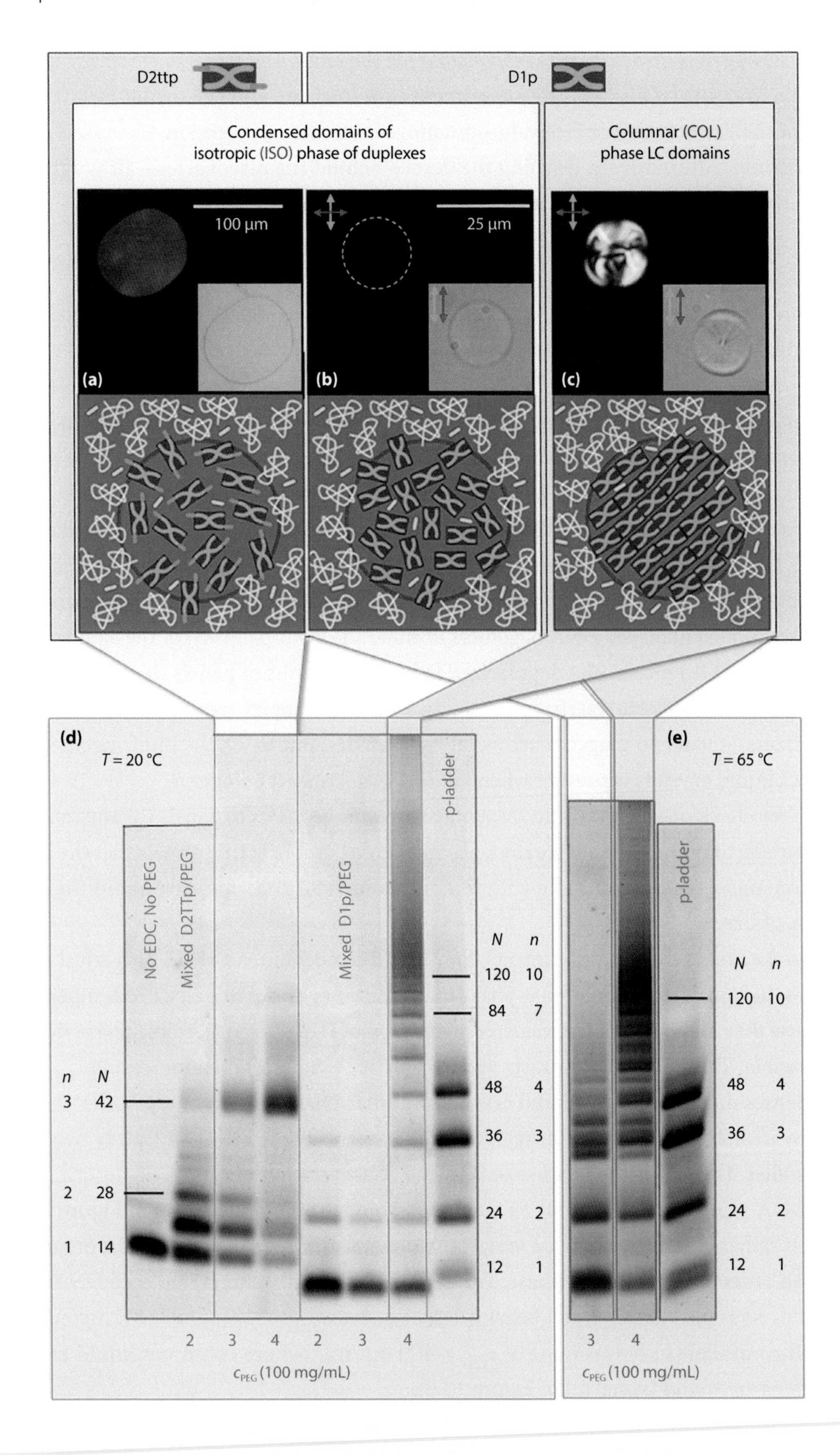

D2ttp
D1p
Condensed domains of isotropic (ISO) phase of duplexes
Columnar (COL) phase LC domains
100 μm
25 μm
(a)
(b)
(c)
(d)
T = 20 °C
No EDC, No PEG
Mixed D2TTp/PEG
Mixed D1p/PEG
p-ladder
n N
3 42
2 28
1 14
N n
120 10
84 7
48 4
36 3
24 2
12 1
2 3 4 2 3 4
c_{PEG} (100 mg/mL)
(e)
T = 65 °C
p-ladder
N n
120 10
48 4
36 3
24 2
12 1
3 4
c_{PEG} (100 mg/mL)

Because this first experiment is indeed sophisticated and even a bit tricky, we elaborate further. Consider panels (b) and (c) of Figure 13.6. Panel (c) shows ordered, stacked, nanoDNA duplexes that form a liquid crystal domain surrounded by a PEG-rich isotropic fluid. But in panel (b), while there is DNA condensation—the nanoDNA is shown as compacted into a domain of comparable size to that in (c)—there is no liquid crystal ordering. Panel (b) shows that PEG can phase separate droplets of concentrated DNA oligomers that, although sequestered in a domain, remain in an isotropic phase within that domain.

So we have the makings for the first experiment in which ligation efficiencies can be compared between environments where the only significant difference is the liquid crystal ordering. The previous proclamation may not be writ large simply from the cartoons. But know this: both optical observation and a detailed computer simulation revealed that c_{DNA}, the concentration of DNA in the domains of panels (b) and (c), are similar in these two condensed phases, being no more than 10% larger in the columnar liquid crystal domain of panel (c).[20]

The data from polyacrylamide gel runs on the samples characterized in panels (b) and (c) are on display in panel (e) of Figure 13.6. The intensity profiles of the data are plotted in panel (c) of Figure 13.4, the graphical analysis of the gel profiles—which we finally return to. And it was worth the long wait. Because the graphical analysis shows that the difference in the product length distribution of polymers that is directly attributable to liquid crystal ordering is quite marked. Analyses of the curves give the mean degree of polymerization, $\langle n \rangle \approx 3.5$ in the ISO phase and $\langle n \rangle \approx 10$ in the COL phase.

At the outset of this section we said that the Milan/Boulder collaborators used two distinct experiments to compare ligation efficiencies between conditions where the only significant difference is the liquid crystal ordering. We now present the second experiment. This was performed at 20° C and at equal osmotic pressures, Π, in PEG-induced condensed phases. However, the subtlety in this experiment was that two different DNA dodecamers were used. The first was the Dickerson dodecamer DNA sequence D1p. The second was the DNA sequence 5′- <u>GCCGTATACGGC</u>TTp - 3′ that was abbreviated as D2TTp. D2TTp was synthesized to be a self-complementary dodecamer sequence (<u>underlined section</u>) with two additional T bases on the 3′ end. What does this buy the experimenter?

Figure 13.6 Ligation in condensed liquid crystal and isotropic DNA droplets. **(a)** Sketch and bright field microscope image of condensed D2TTp (ISO) droplets. Duplexes are selectively marked by a green dye. **(b)** Sketch and depolarized microscope images of D1p (ISO) droplets. **(c)** Sketch and depolarized microscope images of D1p (COL) droplets at $T = 65$ °C. **(d)** Polyacrylamide gels comparing ligation products in D1p and D2TTp in identical conditions. **(e)** Same as (d) but at $T = 65$ °C. Depending on c_{PEG}, the system is either uniformly mixed, or phase separated into two coexisting ISO phases, (b), or phase separated into coexisting COL and ISO phases, (c). Numbers along the lanes are oligomer lengths expressed in number of bases (N_b) and polymerization number $\langle n \rangle$. The straight lines are used as a visual guide to help to identify bands corresponding to selected N_b. *(Reproduced with permission from Tommaso P. Fraccia, Gregory P. Smith, Giuliano Zanchetta, Elvezia Paraboschi, Youngwoo Yi, David M. Walba, Giorgio Dieci, Noel A. Clark, and Tommaso Bellini, "Abiotic Ligation of DNA Oligomers Templated by Their Liquid Crystal Ordering," Nature Communications 6, Article No. 6424 (10 March 2015), 1–7.)*

D2TTp forms duplexes terminating in non-pairing TT overhangs that suppress end-to-end duplex adhesion and thus suppress liquid crystal ordering. Bellini and Clark explored PEG/D2TTp/EDC mixtures with molar ratio $R = 50$ and variable c_{PEG} using similar protocols to those described in the first experiment. They characterized the results with polyacrylamide gel electrophoresis. They were rewarded for their initiative with phase separation of isotropic droplets composed of D2TTp molecules for $c_{PEG} > 300$ mg/mL as shown in the cartoon of panel (a), Figure 13.6.

The light brown tails on the D2TTp molecules (seen in isolation at the top left of Figure 13.6) symbolize the TT overhangs that suppress liquid crystal formation. The condensation of DNA-rich isotropic droplets of D2TTp brings about a much smaller effect in the ligation efficiency than the condensation of the D1p liquid crystal droplets. This is visible in the gel runs of panel (d) in Figure 13.6 and in their intensity profiles in panel (d) of Figure 13.4. The authors noted that the appearance of bands in the gel not corresponding to integral multiples of D2TTp probably arise from the formation of circular products and this made it impossible to use the semi-quantitative analysis with the Flory model as had been done with other samples. Thus, there are no Flory fits in panel (d) of Figure 13.4. Nonetheless, they were able to use the intensity profiles to extract a value of the degree of polymerization, $\langle n \rangle < 3$ even at the largest values of $c_{PEG} = 400$ mg/mL.

The two experimental strategies attest to liquid crystal formation rather than the local increase in DNA concentration, c_{DNA}, as being decisive in templating the chemical ligation of DNA oligomers.

D2TTp Symbol representing a dodecamer, with the DNA sequence GCCGTATACGGCTT. This is a self-complementary sequence (underlined section) with the exception of two additional thymine bases, TT, and a phosphate group (a phosphorus atom surrounded by four oxygen atoms) that is bound to the 3′- end of the dodecamer.

In the lowdown on ligation we've laid out, the experiments showcase phase separation and liquid crystal ordering. These are the essential factors in the self-assembly-induced enhancement of abiotic EDC-based ligation that has been the subject of this chapter. Liquid crystal ordering provides continuous close contact between the reacting duplex terminals. Without that, no pronounced elongation is observed. However, phase separation is crucial as well. It selectively confines the DNA liquid crystals into domains whose internal fluid structure is bounded by an aqueous/aqueous interface that enables easy transport of material. (By the way, the team has extended the experiments we've described here that used blunt-ended duplexes (Dickerson dodecamers). They found the same results with sticky-ended duplexes with two-base-long mutually interacting overhangs.[21] Again they demonstrated liquid crystal–enhanced chain lengthening and showed that the elongation greatly enhanced the thermal stability of the liquid crystal phase by significantly raising the melting temperature of the phase.)

Cascaded Phase Separation

As a topper, Bellini and Clark performed an experiment that used a three-component mixed solution of D1p and D2TTp and PEG. It led to a result that is best summed up in Figure 13.7. The depolarized light images are shown with crossed polarizers, seen in panel (a), and parallel polarizers, panel (b). In panel (a), the dramatic birefringence boasted by the columnar liquid crystal domains leaps from the page without a shred of modesty.

But panel (b) tells a story of its own. The colors disappear when the polarizer and analyzer are parallel to one another, but we are afforded a lucid look at a phase separation phenomenon that becomes a type of staged or cascaded selection mechanism. This is manifest in a tumble of thermodynamic instabilities.

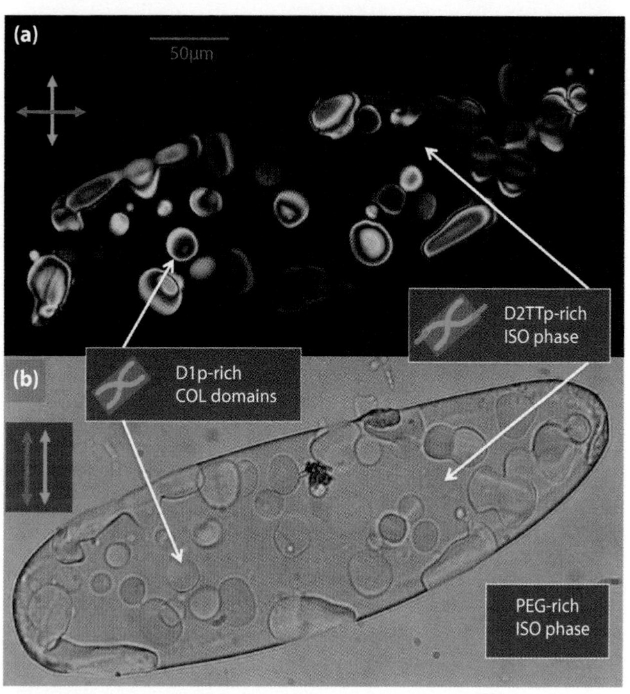

Figure 13.7 Depolarized optical microscope images showing phase separation in a D1p/D2TTp/PEG mixture at a molar ratio ≈ 1 and c_{PEG} = 300 mg/mL. The same cell is imaged through crossed polarizers in **(a)** and parallel polarizers in **(b)**. Colorful birefringent COL liquid crystal domains are visible in panel (a). The colors depend mainly on the domain thickness. Panel (b) shows the coexistence of three phases: the D1p liquid crystal COL domains seen as circular or oval structures, the D2TTp ISO phase—black in panel (a), pale yellow in panel (b)—and the PEG-rich ISO phase background. *(Reproduced with permission from Tommaso P. Fraccia, Gregory P. Smith, Giuliano Zanchetta, Elvezia Paraboschi, Youngwoo Yi, David M. Walba, Giorgio Dieci, Noel A. Clark, and Tommaso Bellini, "Abiotic Ligation of DNA Oligomers Templated by Their Liquid Crystal Ordering," Nature Communications 6, Article No. 6424 (10 March 2015), 1–7.)*

The bottom panel displays the phase coexistence observed in a three-component mixed solution of D1p, D2TTp, and PEG. At low PEG concentration these solutions are single phase, consisting of random coil PEG and duplexed DNA oligomers. But for $c_{PEG} > 300$ mg/mL both species of DNA duplexes phase separate, a result of their compact rigid structure, originating in their hybridized portions. This mode of selection is based on sequence since unpaired strands retain their flexibility, favoring their separation from the rigid duplexes. As shown in panel (b) of Figure 13.7, the duplexed DNA that has been condensed at high concentration can undergo a second phase separation between the isotropic (ISO) and columnar (COL) liquid crystal phases. In the mixture considered here, the COL phase is principally composed of blunt-end D1p while the ISO is mainly composed of D2TTp, which cannot form liquid crystals owing to the TT overhangs that inhibit the stacking of duplexes.

This second phase separation is carried out in a context that is created by the first and acts as a cascaded selection mechanism. Such staged partitioning is a hierarchical mechanism that condenses together the duplexes that enable liquid crystal ordering—*precisely those duplexes in which contacting terminals promote enhanced ligation.*

When we saw Tommy Fraccia giving his ligation seminar in Boulder he was still a graduate student in Tommaso Bellini's laboratory at the University of Milan. Although the results he presented that day were noteworthy, they were not what Bellini and Clark were hoping for. Nor were they only a skosh off the mark; certainly nothing that could be made right by a slew of probing questions from Clark. That was in the spring of 2013. Bellini and Clark were still left wondering if their origins proposal would unfold or unravel.

On that lovely spring afternoon,
with scientists filling the room,
Tommy's answer was not one of rapture or gloom,
'twas still far off as the moon

But by the end of the year Tommy had written his Ph.D. thesis and was awarded his degree. As often happens, once a student has passed that hurdle he has a resurgence of zeal in his quest to get the very best outcome from his experiments, now as a newly minted postdoc. Tommy got on a roll.

In early 2014 the Milan lab produced the conclusive data we've seen in this chapter. They showed amplification of ligation efficiency within the liquid crystals (as contrasted with the isotropic). They also devised clever new experiments showing that amplification was attributable to the liquid crystal environment, evidence crucial to the Bellini/Clark origins proposal.

In the next chapter we will learn about other origins of life proposals in which self-assembly is often invoked as the only available mechanism to bridge the "enormous kinetic obstacle to nucleic acid chain elongation."[22] This means that molecular selection, self-assembly, and polymerization require a lot of (free) energy. The kinetic obstacle to be overcome on the early Earth was that which connects basic carbon-based molecules and their development into the simplest structures capable of enzymatic activity. For example, if RNA is presumed to be the first molecule with the capability to support life, then the kinetic obstacle might have been overcome by *ribozymes*. This is a word made from *ribo*nucleic acid en*zyme* and denotes RNA molecules that perform catalytic activities.

What Bellini and Clark have experimentally established with DNA (but which is likely extendable to RNA or other nucleic acid precursor molecules) is how the path to the simplest molecules with the capability to support life could have actually taken place. What we've seen in these last three chapters is that the interplay of fluid ordering, distinctive modes of hierarchical and sequence-directed self-assembly, and phase separation, can direct chemical reactions to produce long-chain nucleic acids. Bellini and Clark envision their findings as a paradigm of what could have happened on the prebiotic Earth based on the fundamental and illuminating assumption that *the origin of nucleic acids is written in their structure.*

As Bellini quipped to Clark, they've found "a definitive link in a missing chain."[23] Future research will focus on the remainder of the chain. What could have preceded the processes they have delineated? For instance, clarifying the minimal ingredients for condensation of DNA liquid crystal domains in complex molecular milieus. What circumstances on the early Earth were present to enable the necessary concentration of appropriate molecular feedstock that is inherent in the Bellini and Clark studies? Also, what could have followed the processes they have uncovered? For example, determining the first crucial enzymatic activities that enhanced and perhaps guided the liquid crystal-assisted ligation.

Though Bellini and Clark tackled a question that had flummoxed scientists for decades, there is no doubt that unsolved problems remain.

That said, however, discovering a real example of what could have happened on the prebiotic Earth is as good as it gets in a scientific endeavor. Finding a definitive link in a missing chain.

Exercises for Chapter Thirteen: Exercises 13.1–13.2

Exercise 13.1

When we introduced the molar ratio, *R*, we were in the realm of *stoichiometry*. Stoichiometry (derived from the Greek words for element and measure) is concerned with the proportions in which elements or compounds react with one another.

Consider the reaction between the solid copper (II) oxide and the gas ammonia at high temperatures. This reaction produces nitrogen gas with the additional reaction products being solid copper and water vapor. The balanced equation is

$$2NH_3 + 3CuO \rightarrow N_2 + 3Cu + 3H_2O.$$

What is the molar ratio between CuO and NH_3?

(Exercise reproduced with permission from S. Zumdahl and S. Zumdahl. Chemistry, 7E. © 2007 Brooks/Cole, a part of Cengage Learning, Inc. Reproduced by permission. www.cengage.com/permissions)

Exercise 13.2

We've seen many gel runs in this chapter employing either polyacrylamide or agarose as the gel medium. Will two molecules having the same molecular weight and charge have the same mobility when run in a gel? Please explain your answer.

(Exercise reproduced with permission from Physical Biochemistry, 2/e *by David Freifelder, Copyright 1982 by W.H. Freeman and Company, New York. Used with permission of the publisher.)*

ENDNOTES

[1] http://www.mbio.ncsu.edu/jwb/soup.html

[2] Greg T. Hermanson, *Bioconjugate Techniques* (Oxford: Academic Press, 2013), 259.

[3] Tommaso Pietro Fraccia, "Exploring the Role of Liquid Crystal Ordering of DNA Oligomers in the Prebiotic Synthesis of Nucleic Acids," Ph.D. dissertation, Università Degli Studi Di Milano (2013), 96.

[4] http://www.britannica.com/science/mole-chemistry

[5] Tommaso Fraccia, Greg Smith, Giuliano Zanchetta, Elvezia Paraboschi, Youngwoo Yi, David M. Walba, Giorgio Dieci, Noel Clark, and Tommaso Bellini, "Abiotic Ligation of DNA Oligomers Templated by Their Liquid Crystal Ordering," *Nature Communications 6, Article No. 6424* (10 March 2015), 1–7.

[6] Lewis Carroll, *Alice's Adventures in Wonderland* (New York: Dover Publications, 1993), 81.

[7] Jacobus H. van't Hoff, "Osmotic Pressure and Chemical Equilibrium," Nobel Lecture, 13 December 1901, and http://www.nobelprize.org/nobel_prizes/chemistry/laureates/1901/hoff-lecture.pdf

[8] Fraccia, "Exploring the Role," 101–105.

[9] http://faculty.washington.edu/finlayso/che475/microreactors/Group_A/whatmr1.htm

[10] http://www.lifetechnologies.com/us/en/home/life-science/dna-rna-purification-analysis/nucleic-acid-gel-electrophoresis/dna-stains/etbr.html

[11] http://idtdna.com/pages/decoded/decoded-articles/pipet-tips/decoded/2011/06/17/running-agarose-and-polyacrylamide-gels

[12] William S. Johnson, Walter H. Stockmayer, and Henry Taube, "Paul John Flory," *Biographical Memoirs, National Academy Press 82* (2002), 22.

[13] Paul J. Flory, "Molecular Size Distribution in Linear Condensation Polymers," *J. Am. Chem. Soc. 58, No. 10* (October 1936), 1877–1885.

[14] Ibid., 1879.

[15] Ibid., 1880.

[16] Ibid., 1879.

[17] Tom Maniatis, Andrea Jeffrey, and Hans van deSande, "Chain Length Determination of Small Double- and Single-Stranded DNA Molecules by Polyacrylamide Gel Electrophoresis," *Biochemistry 14, No. 17* (1975), 3787–3794.

[18] Fraccia, "Exploring the Role," 28–29.

[19] Michi Nakata, Giuliano Zanchetta, Brandon D. Chapman, Christopher D. Jones, Julie O. Cross, Ronald Pindak, Tommaso Bellini, and Noel A. Clark, "End-to-End Stacking and Liquid Crystal Condensation of 6– to 20–Base Pair DNA Duplexes," *Science 318* (23 November 2007), Supporting Online Material, Figure S2, 17 (linked to the online version of the paper as a PDF download).

[20] Tatiana Kuriabova, M.D. Betterton, and Matthew A. Glaser, "Linear Aggregation and Liquid-Crystalline Order: Comparison of Monte Carlo Simulation and Analytic Theory," *J. Mater. Chem. 20* (2010), 10366–10383.

[21] Tommaso P. Fraccia, Giuliano Zanchetta, Valeria Rimoldi, Noel A. Clark, and Tommaso Bellini, "Evidence of Liquid Crystal-Assisted Abiotic Ligation of Nucleic Acids," *Orig Life Evol Biosph 45* (2015), 51–68.

[22] Christian de Duve, *Singularities: Landmarks on the Pathways of Life* (New York: Cambridge University Press, 2005), 79.

[23] Conversation; Noel Clark to K.D. on 9 July 2014.

BRIEF INTERLUDE

The Handedness of Life

Nineteenth-century French chemist and microbiologist Louis Pasteur was a legendary character during his lifetime and remains a scientific icon. Pasteurization of beverages and food, the germ theory of disease, a vaccine against rabies—how much can one person contribute? Well . . . before any of these achievements, Pasteur—in 1848, at the tender age of 25—laid the foundations for a new field of chemistry known as stereochemistry.[1] This is the study of molecules that are *stereoisomers* of one another—they have the same chemical composition and the atoms of the molecules have the same connections to one another but the atoms differ in their position in three-dimensional space. Pasteur was able to prove that organic molecules having identical chemical content and connectivity can exist in uniquely different forms. Pasteur referred to his findings as molecular asymmetry. Stereochemistry has huge practical implications for the manufacture of safe drugs, and, as we will see, it is inseparably linked with life.

In Pasteur's era, there was only a rudimentary grasp of the idea of chemical bonds between the atoms that form a molecule. Pasteur worked with empirical evidence but little else.[2] French chemist and physicist Joseph Louis Gay-Lussac had discovered that two chemical compounds, tartaric acid and racemic acid, had the same composition (but different properties, e.g., different melting points). Swedish chemist Jöns Jakob Berzelius coined the term *isomer* (from the Greek *isos* meaning equal and *meros* meaning parts) to denote molecules composed of equal parts. Berzelius described tartaric acid and racemic acid as isomers of one another. French physicist Jean Baptiste Biot showed that a solution of tartaric acid derived from plants rotated the plane of polarization of light that was passed through it. Such a solution is said to show *optical activity.* The solution possesses a type of birefringence called circular birefringence. However, solutions of racemic acid did not display optical activity. Like other scientists of the day, Pasteur was downright perplexed: "How could one conceive of two substances that resembled each other so much without being identical?"[3]

Isomers Two molecules are called isomers of one another if they have the numbers of the same kinds of atoms, i.e., the same chemical formula, but differ in their chemical and physical properties.

Stereoisomers Two molecules are called stereoisomers of one another if they have the same chemical formula and the atoms of the molecules have the same connections to one another but the atoms differ in their position in three-dimensional space.

Optical activity A medium shows optical activity if it causes the plane of polarization of a plane-polarized light wave to rotate as it travels through the medium.

Pasteur was working in the laboratory of Professor of Chemistry Antoine Jérôme Balard at the École Normale Supérieure in Paris, an elite graduate school and research center. Consumed by the conundrum, he examined crystals of each chemical compound and noticed an unusual anomaly that no one else had spotted. In racemic acid he found two discernibly different crystals. Working with a microscope and tweezers, he meticulously separated the two crystal types, made solutions of each, and passed light through them. He found that one solution rotated the plane of polarization of the light to the left (counterclockwise to its direction of travel) and one solution rotated the plane of polarization to the right (clockwise).

This would explain why there would be no optical activity in a solution of racemic acid that was presumed to be of uniform composition but in which equal amounts of two different isomers of tartaric acid were present. In effect, in such a solution the optical activity of the two chemical compounds canceled each other out and this accounted for the lack of optical activity. For the record, there is some disagreement as to the exactitude of Pasteur's experimental results and how he arrived at his interpretation of them.[4]

Following his experiment, Pasteur, despite his usual reserve, could not contain his excitement and enthusiasm. He rushed out of his laboratory, ran into a physics instructor in the hallway, embraced him, and dragged him into the nearby Luxembourg Gardens public park to explain his discovery to him.[5]

Pasteur also observed that the two stereoisomers he had discovered were mirror images of one another. (This is not true of all stereoisomers.) He was the first person to postulate a molecular explanation for optical activity derived from this observation. It took additional decades before other chemists found a rigorous theory for the relationship between molecular structure and optical activity.

With this background we are now ready to learn of the chirality—that is, the handedness—of molecules. We will examine the initial endeavors of Bellini and Clark to study the handedness of nanoDNA. Their results to date are striking but await a decisive interpretation. Thus, we have placed this discussion in a Brief Interlude by itself, separated from the other Bellini/Clark studies. Regardless, this is the first inquiry into chirality on such short lengths of DNA and contributes to the ongoing homochirality discourse in science.

Chirality

A discussion of a proposal for the origin of life would not be complete without mention of the chirality of molecules—and by extension, the chirality of life (first mentioned in Brief Interlude I). The word *chirality* derives from the Greek *kheir* meaning "hand." It describes the property characterizing three-dimensional forms that are not superposable on their mirror images.[6] Such as . . . your hands. Your hands are chiral objects because they are mirror images of one another; they are said to lack mirror symmetry. (If an object is achiral, then it is an exact copy of its mirror image.)

Chirality A property describing three-dimensional forms that are not superposable on their mirror images, such as your hands.

Your hands have the same number of fingers and the fingers of each hand are connected in the same way, but if you try to superpose your right hand on top of your left, your right hand's thumb overlays your left hand's pinky finger. Another example that is more evocative of the DNA molecule we've been so keen on is shown in Figure I2.1. A mirror plane separates two spiral staircases of a department building at Chalmers University in Gothenburg, Sweden.[7] The rightmost staircase is suggestive of the right-handed helices of typical long DNA.

Most biologically relevant molecules are chiral (i.e., there is a right-handed version and a left-handed version) and biointeractions are typically highly sensitive to the molecular handedness. One important effect of chiral interactions is the transmission of chirality from the molecular structure to supramolecular assemblies—such as liquid crystal phases. However, the propagation of chirality from molecules to supramolecular structures doesn't mean that the same handedness will be expressed at each level of organization.

Life is Homochiral

We need to grow our vocabulary a little more. An enantiomer is one of a pair of stereoisomer molecules that are mirror images of each other. It is from the Greek *enantios* meaning "opposite." Enantiomers are chiral molecules: an enantiomer is not superposable on its mirror image. You cannot place one of the enantiomeric pair on the other so that all the atoms of one molecule coincide with the identical atoms of the other.

Figure I2.1 Two spiral staircases at Chalmers University in Gothenburg, Sweden, are mirror images of one another, an example of chirality. *(Reproduced with permission from* Ingo Dierking, *Textures of Liquid Crystals (Weinheim, GR: Wiley-VCH Verlag, 2003), 55.)*

The basic macromolecules of life are DNA, RNA, and proteins. The components of these macromolecules, specifically the sugars (in the case of DNA and RNA) and the amino acids (in the case of proteins), can exist in two enantiomeric forms—that is, mirror images of one another. However, in living organisms, only one enantiomer is selected: right-handed sugars and left-handed amino acids (with a few exceptions).[8] Because the constituents of DNA, RNA, and proteins are of only one enantiomeric form, it is said that life is homochiral. This means that the biomolecules of life possess only a single handedness. For example, the right-handedness of the classic B-form DNA is determined by the right-handed configuration of the deoxyribose sugars.

Again, we must acknowledge an exception. In Chapter Seven we became acquainted with Nadrian Seeman's nanomechanical device that employed the Z-form of DNA. Z-DNA is left-handed. That this is so must come under the heading: "As often in Nature nothing is absolute."[9]

To summarize: exceptions aside, it is commonplace to say that RNA and DNA are made up of only right-handed molecules (and proteins are made up of only left-handed molecules). Why should this be so? Why aren't amino acids right-handed and sugars left-handed? There is no consensus on the answer to this question. Chemist Jay S. Siegel has written: "Rationalizations of the origins of (homo)chirality in nature have reached extraordinary levels."[10]

Origins researchers express wide-ranging opinions. Some hold that homochirality in nature is deterministic—due to a physical law, not happenstance. Others see it differently. Some believe that biomolecular homochirality might have arisen from chiral self-selection. Swiss chemist Albert Eschenmoser worked with variants of RNA molecules (variations in their sugar-phosphate backbones). He found that some of these molecular chains would self-assemble with greater than 90% enantiopurity—i.e., molecules of only one chirality. He speculated that this capacity of selecting only one enantiomeric form, referred to as chiroselectivity, might arise whenever the complexity of a system of oligonucleotides exceeds a critical level.[11,12] (We'll revisit the idea of a critical level in Chapter Fifteen.)

Enantiomer An enantiomer is one of a pair of molecules that are mirror images of each other. Thus, enantiomers are chiral molecules. Enantiomers are a subset of stereoisomers—i.e., all enantiomers are stereoisomers but not all stereoisomers are enantiomers.

Homochiral Molecules that occur in only one enantiomeric form (i.e., molecules of the same chirality). The biomolecules of life—DNA, RNA, and proteins—possess only a single-handedness because they are made of building blocks that have only one-handedness. For example, the right-handedness of the classic B-form DNA is determined by the right-handed configuration of its deoxyribose sugars.

Enantiopure A sample in which all molecules have the same chiral sense. Enantiopure is not entirely synonymous with homochiral because it applies to an experimental attribute, not to a concept.

Chiroselective A chiroselective process is one that enables the proliferation of molecules with a specific handedness.

Macroscopic Chiral Helical Precession of Molecular Orientation

Before finally turning to Bellini and Clark's preliminary findings on nanoDNA chirality, we will concretize our earlier mention of the propagation of chirality from molecules to supramolecular assemblies of molecules. By now we are quite familiar with concentrated solutions of duplex-forming DNA oligomers that organize into various liquid crystalline phases, one of which is the nematic phase. If the individual molecules that make up the nematic phase are chiral, then this molecular chirality affects their collective structure on larger length scales, e.g., ordered phases.

In the nematic phase, the molecules have an orientational order. In writing about Figure 5.2 of Chapter Five we identified the molecular director as the descriptor we give to the average local alignment of the molecules. Also recall that in Figure 4.8 of Chapter Four we saw how this director orientation in a nematic can be forced to execute a helical twist. This was accomplished by rubbing the two glass slides confining the nematic to define a preferred direction to the molecules near the slides and rotating one slide by 90° with respect to the other. We speak of this as imposing mutually orthogonal orientations on the two bounding surfaces.

However, a common feature of chiral nematic phases is that they exhibit such a helical twist even in the complete absence of surfaces. That is, the chiral nematic phase spontaneously forms a helix as a macroscopic consequence of the molecular-scale chirality. Because of the chirality of the DNA molecule, its nematic phase exhibits a macroscopic chiral helical precession of molecular orientation. Although this phrase is a mouthful, its meaning is not complicated.

The explanation is that when the individual molecules are chiral, they impart a small twist to their local alignment and this averages to the macroscopic helical precession. The specific handedness of this precession is determined by the chirality of the individual molecules. So just as the chirality of the sugar deoxyribose imparts chirality to the DNA molecule of which it is a part, so too does the chirality of the DNA molecule impart a chiral character to the nematic liquid crystal phase formed by DNA duplexes. (This subject comes up again in the Appendix in our perusal of liquid crystal texture.)

Bellini and Clark Examine NanoDNA Chirality

A large number of experiments indicate that there is no simple rule connecting the handedness of individual helices to that of their phase. In all their previous studies of nanoDNA, Bellini and Clark "did not focus on the handedness of the phase, taking for granted that it matched the one of long DNA solutions. Quite surprisingly, this is not at all the case."[13]

Sequence	Nickname	lpl range (μm)	*H*
CGAATTCG	8sc1	1–2	L
CGCATGCG	8sc2	0.45–0.8	L
CGCAATTGCG	10sc	0.4–∞	R
ATAAATTTAT	10allAT	1–3	R
CGCGAATTCGCG	DD	0.7–3	R
pCGCGAATTCGCG	pDD	0.35–1	R
CGCGAATTCGCGp	DDp	0.3–1	R
GCGCTTAAGCGC	antiDD	1–2	R
GGAGTTTTGAGG + CCTCAAAACTCC	12mc	0.7–2	R
ACCGAATTCGGT	ACC	1–3	R
AACGAATTCGTT	AAC	∞	A
AATGAATTCATT	AAT	1–2	L
AATAAATTTATT	allAT	0.5–1	L
CCGGCGCGCCGG	allCG1	1–3	L
CGCGCCGGCGCG	allCG2	0.3–∞	L,R
GCGCGAATTCGC	sDD	0.3–1	L
ACGCGAATTCGCGT	14Aterm	1–3	L
CGCGAAATTTCGCG	14sc	1–3	L
pCGCGAAATTTCGCG	p14sc	1–3	L
GCCGCGAATTCGCG	GC-DD	0.35–1	L
CGCGAATTCGCGGC	DD-GC	1–3	L
CGCGCGAATTCGCG	CG-DD	1–3	L
ATCGCGAATTCGCG	AT-DD	1–3	L
ACGCAGAATTCTGCGT	16Aterm	1–4	L
CGCGAAAATTTTCGCG	16sc	1–4	L
AACGCAAAGATCTTTGCGTT	20sc	1–4	L
CGCGAAUUCGCG	DD-RNA	0.3–1	R
–	long DNA	2–4	L

Figure I2.2 Chiral nematic helix handedness (*H*) and pitch (|*p*|) for a selection of nanoDNA oligomers. In the column showing handedness, A indicates achiral, while L and R indicate left- and right-handedness. Long DNA (bottom row) has a nematic phase with a left-handed helix and pitch ~ 2–4 microns. The table reveals that nanoDNA forms nematic liquid crystal phases exhibiting varying chirality—A, L, or R handedness—and a wide range of pitch. *(Adapted with permission from Giuliano Zanchetta, Fabio Giavazzi, Michi Nakata, Marco Buscaglia, Roberto Cerbino, Noel A. Clark, and Tommaso Bellini, "Right-Handed Double-Helix Ultrashort DNA Yields Chiral Nematic Phases With Both Right- and Left-Handed Director Twist," PNAS 107, No. 41 (12 October 2010), 17497–17502.)*

The Boulder/Milan team used a quantitative analysis of the transmission spectra from depolarized light microscopy to determine the handedness of the chiral nematic helix for a large number of DNA base sequences (and one RNA sequence) ranging from 8 to 20 bases (Figure I2.2).[14] They were able to assign a handedness to each sequence by placing the sample between a crossed polarizer and analyzer and then rotating the analyzer by 10° to each side (where 0° is defined as the position of crossed polarizer and analyzer). The transmission spectra of chiral nematic phases of opposite handedness were modified in opposite ways upon rotating the analyzer by 10° in opposite directions. This technique also enabled them to measure the helical pitch.

The helical pitch is the distance measured along the helix axis to complete one full helix turn. Remember that in the present context, the helix we're talking about is the one described by macroscopic chiral helical precession of molecular orientation of the molecules that collectively comprise the liquid crystal phase.

Helical pitch The pitch of a helix is the distance measured along the helix axis to complete on full helix turn.

"q" A quantity inversely related to the helical pitch.

Conventional, long B-DNA exhibits a right-handed molecular double-helix structure that always yields nematic liquid crystal phases with left-handed pitch in the micron range. But the Boulder/Milan team found that nanoDNA has an extremely diverse behavior, with both left- and right-handed nematic helices and pitches ranging from the macroscopic (i.e., several microns = several thousands of nanometers) down to 300 nanometers. The behavior is much richer than the nematic phase of long DNA. For nanoDNA, minor changes in oligomer length or sequence, or modifications in the mode of end-to-end aggregation (i.e., sticky ended or blunt ended) strongly affect the handedness and the pitch of the phase formed by the helices.

In Figure I2.2, look at the column labeled "*H*" and you'll see the marked variations in handedness; R is right-handed, L is left-handed, and A is achiral. Examination of the column labeled "$|p|$ range (μm)" shows that the pitch varies widely. Overall, the results of this study are quite complex and not well understood as yet (Figure I2.3).

The Boulder/Milan collaboration is continuing to pursue their study of the origin of handedness by using the self-assembly of nanoDNA to tune the chirality of the columnar aggregates of nanoDNA duplexes and study its effect on the collective ordering. They've used mixtures of natural right-handed DNA oligomers forming right-handed nanoDNA double helices. They've also used mirror symmetric oligomers forming left-handed nano-DNA helices.

Nematic ordering of nanoDNA duplexes is mediated by their end-to-end aggregation into linear columns, and the stacking interaction is not chirally selective.

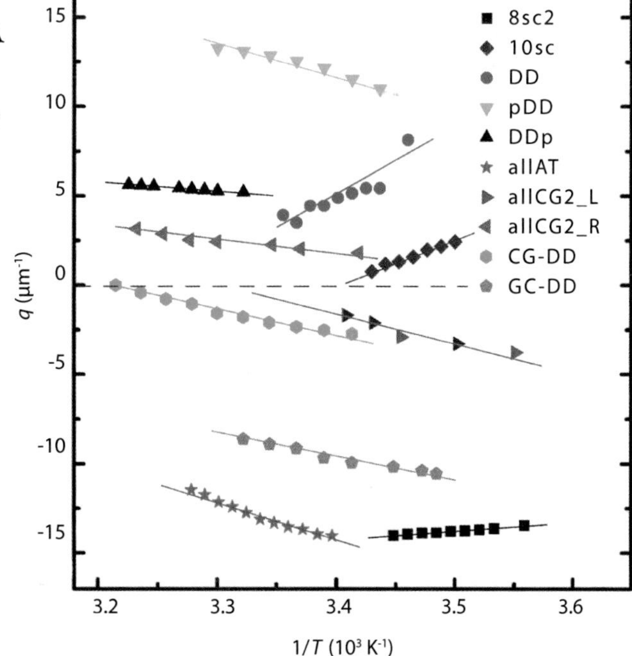

Figure I2.3 A quantitative analysis of the temperature dependence of the inverse pitch (q) in chiral nematic nanoDNA samples. The vertical axis plots "q," a quantity inversely related to pitch, p. Liquid crystals formed by sequences shown in the upper half (i.e., $q > 0$) have right-handed chirality and those in the lower half ($q < 0$) have left-handed chirality. *(Adapted with permission from Giuliano Zanchetta, Fabio Giavazzi, Michi Nakata, Marco Buscaglia, Roberto Cerbino, Noel A. Clark, and Tommaso Bellini, "Right-Handed Double-Helix Ultrashort DNA Yields Chiral Nematic Phases With Both Right- and Left-Handed Director Twist," PNAS 107, No. 41 (12 October 2010), 17497–17502.)*

Thus, by controlling the terminals of both enantiomers they are able to study the propagation of chirality in solutions where the right-handed and left-handed species form mixtures of homochiral columns, and in solutions in which heterochiral columns are formed (Figure I2.4).

Bellini and Clark have found that the two systems behave in a markedly different fashion.[15] For example, the chirality of the nematic liquid crystal phase formed by homochiral columns is left-handed, whereas the chirality of the nematic formed by heterochiral columns is right-handed. A complete understanding is still to come.[16]

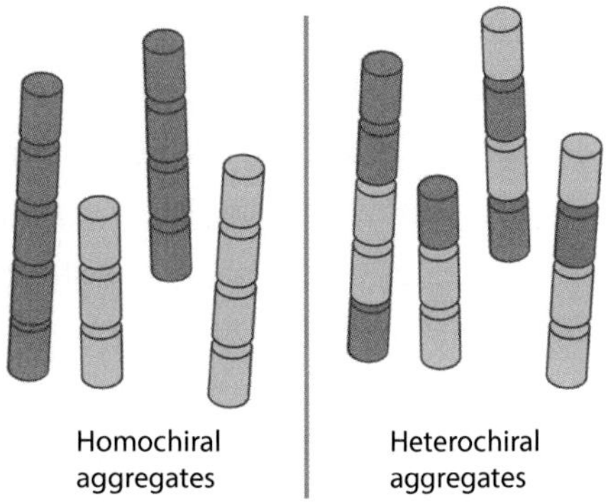

Homochiral aggregates

Heterochiral aggregates

Figure I2.4 NanoDNA duplexes of different color have different chirality. By suitable choice of terminal groups both homochiral and heterochiral stacks of duplexes can be formed and the chirality of their liquid crystal phases can be studied. *(Reproduced with permission from Marina Rossi, Giuliano Zanchetta, Sven Klussmann, Noel A. Clark, and Tommaso Bellini, "Propagation of Chirality in Mixtures of Natural and Enantiomeric DNA Oligomers," Physical Review Letters 110 (2013), 107801-1–107801-5.)*

A Lighter Take on Chirality

We began this Interlude with a look at Louis Pasteur's discovery of enantiomers of tartaric acid. His work showed that molecules could exist as asymmetric pairs. Because the direction of optical rotation was correlated with a particular asymmetry and optical rotation had been observed in many biological materials, it became clear that the occurrence of molecules of a particular asymmetry was a fundamental property of life. Hence, the

Figure I2.5 *Peanuts* weighs in on chirality. *(Reproduced with permission from PEANUTS © 1996 Peanuts Worldwide LLC. Dist. by UNIVERSAL UCLICK. Reprinted with permission. All rights reserved.)*

question of the origin of homochirality: how did life choose between the two enantiomeric forms with which it was almost always presented?[17]

As Robert Hazen has written, "In spite of a century and a half of study, the origin of . . . biochemical 'homochirality' remains a central mystery of life's emergence."[18] In order to discuss the preliminary contributions of Bellini and Clark to this difficult subject, we have made demands on the reader by introducing their work-in-progress with few definitive conclusions but requiring many new words and concepts. It seems only fair to conclude on a lighter note (Figure I2.5).[19]

Exercises for Brief Interlude: Exercise I2.1

Exercise I2.1

Figure I2.6 shows two molecules and their mirror images. The top part of the figure shows the molecule CHBrClF (bromochlorofluoromethane). The bottom part of the figure shows the molecule CH_3Cl (chloromethane). Your job is to decide whether each molecule is chiral or achiral.

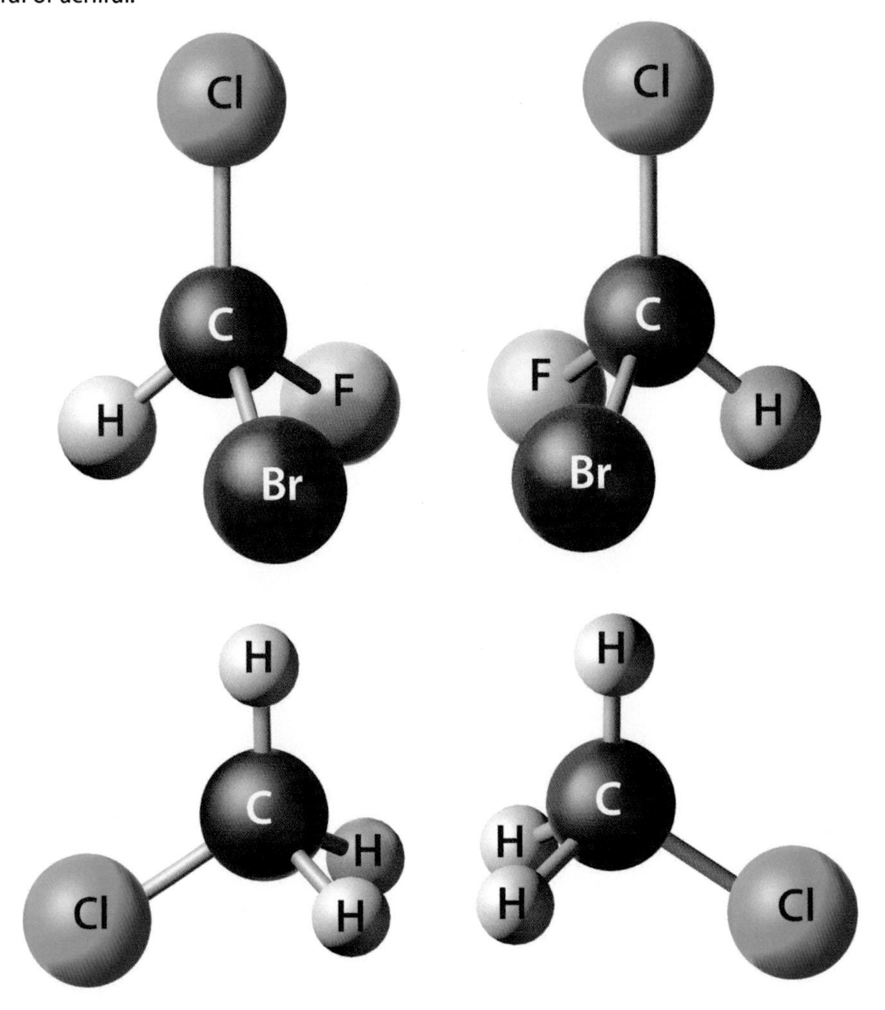

Figure I2.6 Two molecules and their mirror images. *(Courtesy of David M. Walba)*

ENDNOTES

[1] George B. Kauffman and Robin D. Myers, "Pasteur's Resolution of Racemic Acid: A Sesquicentennial Retrospective and a New Translation," *The Chemical Educator 3, No. 6* (1998), 1–18.

[2] https://classesv2.yale.edu/wiki/site/chem124_f08/isomers_of_tartaric_acid.html

[3] Patrice Debré, *Louis Pasteur,* translated by Elborg Forster (Baltimore: The Johns Hopkins University Press, 1998), 45.

[4] William C. Summers, reply to M.F. Perutz, "Pasteur's 'Private Science'," *The New York Review of Books* (6 February 1997).

[5] Debré, *Louis Pasteur,* 47.

[6] L.L. Whyte, "Chirality," *Nature 182* (19 July 1958), 198.

[7] Ingo Dierking, *Textures of Liquid Crystals* (Weinheim, GR: Wiley-VCH Verlag, 2003), 55.

[8] Louise N. Johnson, "Asymmetry at the Molecular Level in Biology," *European Review 13, Supp. No. 2* (2005), 77–95.

[9] Ibid., 84.

[10] Jay S. Siegel, "Homochiral Imperative of Molecular Evolution," *Chirality 10* (1998), 24–27.

[11] Martin Bolli, Ronald Micura, and Albert Eschenmoser, "Pyranosyl-RNA: Chiroselective Self-Assembly of Base Sequences by Ligative Oligomerization of Tetranucleotide-2′, 3′-Cyclophosphates (With a Commentary Concerning the Origin of Biomolecular Homochirality)," *Chemistry and Biology 4* (April 1997), 309–320.

[12] Robert M. Hazen, *genesis: The Scientific Quest for Life's Origins* (Washington, DC: Joseph Henry Press, 2005) 171.

[13] Giuliano Zanchetta, Fabio Giavazzi, Michi Nakata, Marco Buscaglia, Roberto Cerbino, Noel A. Clark, and Tommaso Bellini, "Right-Handed Double-Helix Ultrashort DNA Yields Chiral Nematic Phases With Both Right- and Left-Handed Director Twist," *PNAS 107, No. 41* (12 October 2010), 17497–17502.

[14] Ibid., 17499 and Figure S5, page 7 in Supporting Information online at www.pnas.org/lookup/suppl/doi:10.1073/pnas.1011199107/-/DCSupplemental

[15] Marina Rossi, Giuliano Zanchetta, Sven Klussmann, Noel A. Clark, and Tommaso Bellini, "Propagation of Chirality in Mixtures of Natural and Enantiomeric DNA Oligomers," *Physical Review Letters 110* (2013), 107801-1–107801-5.

[16] Tommaso Bellini, Roberto Cerbino, and Giuliano Zanchetta, "DNA-Based Soft Phases," *Top. Curr. Chem. 318* (2012), 225–280.

[17] John Cronin and Jacques Reisse, "Chirality and the Origin of Homochirality," in *Lectures in Astrobiology I,* ed. Muriel Gargaud, Bernard Barbier, Hervé Martin, and Jacques Reisse (Berlin, Heidelberg: Springer-Verlag, 2005), 482–483.

[18] Hazen, *genesis,* 167.

[19] PEANUTS © 1996 Peanuts Worldwide LLC. Dist. by UNIVERSAL UCLICK. Reprinted with permission. All rights reserved.

CHAPTER FOURTEEN

All the World's a Stage and Life's a Play— Did It Arise From Clay?

Montpellier in southern France is a lovely city only 10 kilometers from the Mediterranean Sea. It is early in July 2011 and Montpellier is host to a big scientific meeting with a big name: The ISSOL (International Society for the Study of the Origin of Life) and Bioastronomy Joint International Conference. The six-day meeting is packed with *invited* lectures, *contributed* oral presentations, and *accepted* posters. Invited lectures are given by some of the most eminent names in the field. Those scientists giving contributed (rather than invited) talks are no less respected, as are those who are standing stalwartly by their poster exhibition boards, keen to chat with any and all who express the desire to learn more than can be fit on a 119 x 84 cm surface. Tommaso Bellini submitted a request to give a contributed oral presentation of his research. His petition was denied.[1]

Instead, along with a total of 324 similarly situated scientists, he hovers next to his poster, titled *"Liquid-crystalline self-assembly of nucleic acids: A new pathway for the prebiotic elongation of RNA?"* The posters are grouped according to a list of nine different topics. His is in the category: *P6. Prebiotic Chemistry.*[2] Other poster categories include *Origins: From Stars to Life, Biosignatures and Clues of Life, Early Earth Processes, Exoplanets and Habitability,* and *Early and Minimal Life.* It is not surprising that his request to give an oral presentation was declined. His approach is about as far out as the *Stars* or the *Exoplanets.* In 2011, he is a nonentity in the field of origins research. That is going to change.

Before we consider some of the origins theories that were presented at Montpellier—and many other venues—we note that the very definition of life is highly debated. There are many candidates, and we won't provide a list but we will present an often-cited example. A 1994 NASA panel chaired by Gerald Joyce of the Scripps Research Institute came up with the following: "Life is a self-sustaining chemical system capable of undergoing Darwinian evolution."[3] The salient points of this description of life are (1) It must be a chemical system; (2) It sustains itself via chemical reactions, energy transforming processes that comprise metabolism; and (3) It is able to make competitive and adaptive changes leading to variation.

Joyce has since revised his own take on the definition to read, "Life is a self-sustaining chemical system capable of *incorporating novelty* and undergoing Darwinian evolution."[4] The reason for the amendment is that not long ago the Joyce laboratory provided an experimental demonstration of the original description. They did this by using two RNA enzymes that catalyze each other's synthesis.[5] This resulted in a chemical system that can undergo Darwinian evolution in a self-sustained manner. Self-sustained in this context means that all the heritable bits are part of the system that is evolving. But the bits are not susceptible to mutation and selection.[6] Joyce himself has said of the system he created, "I hardly think that it is alive."[7]

Emergence and Complexity

O.K. Despite our reservations, suppose we move beyond our working description of life and embark on a survey of the intriguing, lively, and oftentimes fractious field of origins research. We start by calling out a theme that is embedded in every one of the disparate origins hypotheses that we mention: emergence.

As pervasive and important as this concept is, it "will not submit meekly to a concise definition," in the words of University of Michigan computer scientist John H. Holland.[8] Nonetheless, an informal impression might be that emergence describes processes in which novel, complex phenomena arise spontaneously and often unpredictably from simpler systems. As Holland put it, "The hallmark of emergence is this sense of much coming from little."[9]

The relevance to origins research is: life on Earth arose through a sequence of many emergent phenomena. The Earth was once a place of water, organic (carbon-based) molecules, rocks (and their constituent minerals), and energy flows (e.g., solar radiation, hydrothermal vents). Meteorites brought organic molecules to Earth, but there is no evidence that they brought nucleic acids.[10] But from this primordial, geochemical early Earth came biochemical complexity. Even the simplest forms of life are immensely more complex than any nonliving components from which they have arisen.[11] All origins research tries to explain how such a transition might have occurred.

You're correct if you complain that in the previous paragraphs we've used the words *complex* and *complexity* a bunch of times without an explanation. We again offer a colloquial conjecture. Complexity refers to intricate, often patterned structures. Chemist Pier Luigi Luisi of the Roma Tre University demurs from an attempt to define complexity, but rather stays within the following framework. "A complex system is seen as a hierarchic system, i.e., a system composed of subsystems, which in turn have their own subsystems, and so on."[12] The concept of a complex emergent system—life, for example—is recondite. Plopping it down here in a few hundred words is an expedient. We will return to emergence in Chapter Fifteen. For the present, keep the idea in the back of your head. But we would be remiss if we had failed to broach the subject now.

Emergence Processes by which more complex systems arise from simpler systems sometimes spontaneously and in an unpredictable fashion. Emergent phenomena exhibit self-organization. Life is sometimes cited as the quintessential emergent phenomenon.

Complex systems Systems that display novel collective behaviors that arise from the interactions of many simple components. Such systems are hierarchical. Complex emergent systems arise when energy flows through a collection of many interacting particles.

To appreciate the diverse investigations of life's origins presented at Montpellier we must look back. Not quite so far back as the primordial Earth, but rather to Chicago in the mid-twentieth century.

Miller-Urey Experiment

The inquiry into life's origins was transformed from speculation to experimental science in the early 1950s. Beginning in 1953, Stanley Miller, a graduate student in the Chemistry Department of the University of Chicago, published a series of papers on the synthesis of organic compounds from inorganic precursors under possible primitive Earth conditions.[13] Although Miller is the sole author of the initial papers recounting his experiments, they are generally referred to as the Miller-Urey experiments to acknowledge the vital role of Miller's Ph.D. advisor, chemist and Nobel laureate Harold Urey, in the conception and design of the study.[14]

Miller built an apparatus to circulate the gases methane (CH_4), ammonia (NH_3), water vapor (H_2O), and hydrogen (H_2) past an electrical discharge. He chose these gases because others had suggested they were the constituents of the atmosphere of the early Earth. Miller boiled water in a five-liter flask and the resulting water vapor mixed with the other gases, circulated past the electrodes, condensed, and emptied back into the boiling flask (Figure 14.1). Electrical discharge (in the form of lightning) was believed to play a significant role in the formation of compounds in the primitive atmosphere. The water in the flask was boiled and the electrical discharge was run continuously for one week. At the end of that time a residue had accumulated and it was extracted and analyzed (by paper chromatography). Analysis revealed several amino acids (the subunits of proteins) and other organic compounds—that is, compounds containing both carbon and hydrogen. Miller took steps to prevent contaminants in the air from surviving in the extract. He noted, "The amino acids are not due to living organisms because their growth would be prevented by the boiling water during the run. . . ."[13]

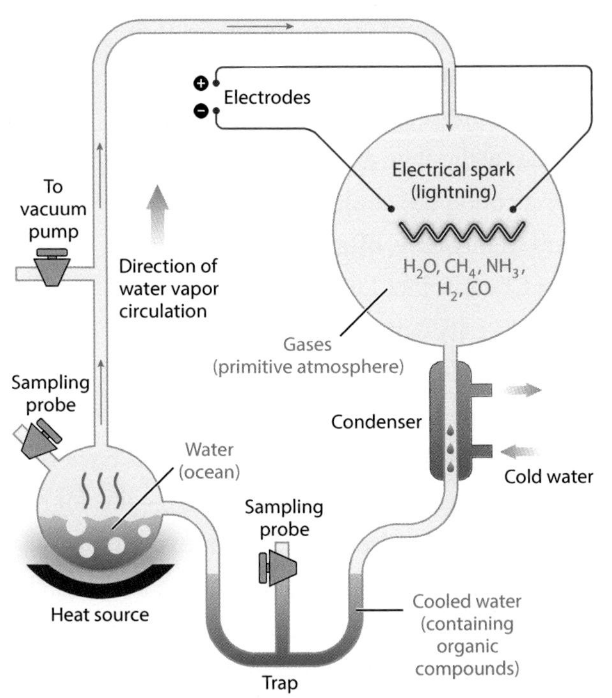

Figure 14.1 Miller-Urey experiment. Under conditions that were thought to have existed on the prebiotic Earth, Miller demonstrated the synthesis of organic compounds from inorganic precursors. *(Reproduced with permission from Thomas M. McCollum, "Miller-Urey and Beyond: What Have We Learned About Prebiotic Organic Synthesis Reactions in the Past 60 Years?" Annu. Rev. Earth Planet. Sci. 41 (2013), 207–229.)*

Chemist Jeffrey L. Bada, a former student of Miller's, said of the extraction residue, "Gerald Wasserburg and Harmon Craig were also graduate students in Urey's laboratory (both would also become members of the National Academy of Sciences) when Stanley did the first experiment. When Stanley showed them that the solution in the spark discharge experiment had turned brown after a couple of days of sparking the gas mixture, they started snickering. When Stanley asked them what they were laughing at, they told him he had a lot of fly excrement (they did not use that polite word) in his apparatus and that he should have cleaned it out better."[14]

Fly excrement notwithstanding, the experiment has become a landmark in the modern study of prebiotic chemistry. For the present, we will not discuss the many experiments that followed or the issues that have been raised about the supposed content of the atmosphere of the primitive Earth. However, we will point out that Miller published his work within months of other publications that described the first amino acid sequence of a protein (leading to Frederick Sanger's first Nobel prize) and the double-helix structure of DNA (Watson and Crick). Thus, the Miller-Urey experiments came at a time of steeply increasing knowledge of the structure and function of proteins and nucleic acids. The production of amino acids in the Miller-Urey experiment provided a possible link between the environment from which life emerged and the types of organic compounds that could conceivably represent the first steps toward life. In the scientific community there was an upsurge in the belief that theories of the origins of life were not just matters of speculation but that they could be tested by scientific investigation.[15]

RNA World Hypothesis

The Miller-Urey experiment is one element that provides flooring for our discussion of origins research. Another piece of the parquetry is the RNA World hypothesis. Miller's experiment and others like it were able to produce nucleobases, notably adenine (A), with remarkable ease.[16] But the availability of biomolecular precursors does not in itself indicate a path for the emergence of life. Note that these experiments did not produce nucleotides—and so they certainly did not produce nucleic acids. (Recall that a nucleotide consists of a nucleobase bound to a five-carbon sugar and one or more phosphate groups.)

However, subsequent to Miller-Urey, if one takes as a starting point the presence of such complex organic molecules as nucleobases, then—*somehow*—RNA (or something like RNA) might come into being. *Somehow* is the subject of the Bellini/Clark research as well as alternative proposals we'll soon speak about. The RNA World hypothesis is a concept of the early Earth that has many proponents but that is also disputed for various reasons.

Physicist and molecular biologist Walter Gilbert, a Nobel prizewinner in Chemistry, first coined the term RNA World. He suggested that prior to the existence of both DNA and proteins, RNA molecules performed the catalytic activities necessary to assemble themselves from a nucleotide soup.[17] However, the idea that RNA might have been the primordial molecule goes back to Francis Crick, Leslie Orgel, and Carl Woese.[18–20] The suggestion was that catalysts made entirely of RNA were likely to have been important at the beginning of life on Earth. Why were catalysts of such interest? Recall that a catalyst is any substance that increases the rate of a reaction without itself being altered in the process. Enzymes are naturally occurring catalysts responsible for many essential biochemical reactions.

The molecules that interact in cellular processes are generally quite stable, so without help they are ponderously slow to react. Enzymes speed up these interactions and bring them into a time scale compatible with life. Each protein enzyme typically accelerates a biochemical reaction by a factor ranging from a million to a trillion.[21]

It had long been believed that a protein enzyme catalyzed every cellular reaction. But in 1981 and 1982, chemist Thomas Cech of the University of Colorado and his colleagues first demonstrated that RNA—rather than a protein enzyme—could act as a catalyst.[22,23] (For this contribution he shared the Nobel Prize in Chemistry with Yale chemist Sidney Altman.) In contemporary biology, nucleic acids and proteins are interdependent. The nucleic acids DNA and RNA orchestrate the synthesis of proteins but they do so with the assistance of enzymes which themselves are proteins. Because of this interdependence, it had been argued that nucleic acids and proteins must have evolved together. Cech's finding that RNA can be a catalyst as well as an informational molecule suggested that when life originated, RNA might have functioned without DNA or proteins.[24] This work reignited earlier speculation on the possible role of RNA in the origin of life.[25] In fact, it produced a seismic upheaval in origins research. The proposed sequence of events is shown in Figure 14.2.

The RNA world hypothesis conjectures that a hypothetical period existed when information storage (that is, genetics) and chemical catalysis (that is, enzymatic activity) was once the sole responsibility of RNA, before the advent of DNA and proteins. But do keep in mind the *somehow* we called attention to earlier—the lack of a plausible prebiotic pathway for the spontaneous formation of the RNA polymers, the kingpins of the RNA world.

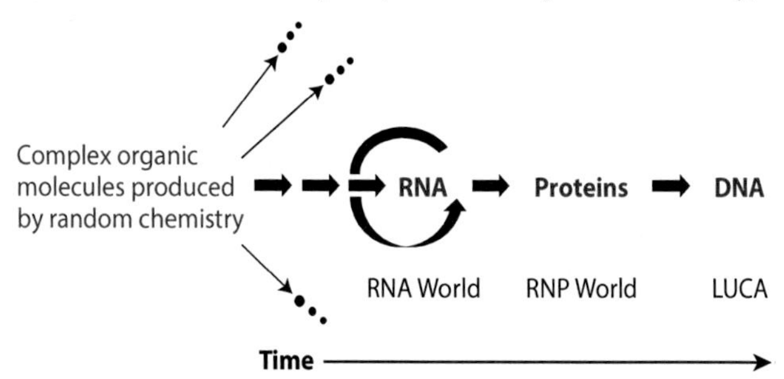

Figure 14.2 An RNA World model for the successive appearance of RNA, proteins, and DNA on the Earth. Many isolated mixtures of complex organic molecules failed to achieve self-replication, and therefore died out, as shown by the arrows leading to the diminishing dots that denote extinction. The three arrows to the left of RNA (RNA self-replication is suggested by the circular arrow closing upon itself) represent the likely self-replicating systems that preceded RNA. Ribonucleoproteins (RNPs) are complexes of RNA and proteins. DNA later took over the role of information carrier (genome). LUCA (Last Universal Common Ancestor) had a DNA genome and catalyzed reactions using both protein enzymes as well as RNP enzymes. *(Reproduced with permission from Thomas R. Cech, "The RNA Worlds in Context," Cold Spring Harbor Perspectives in Biology 4 (2012), 1–5.)*

Maybe instead of "RNA first" there was "PNA first"? This alternative hypothesis is based on a PNA, a peptide nucleic acid, which is a molecule with nucleobases attached to a protein-like backbone. Because PNA is simpler and chemically more stable than RNA, some researchers believe it could have been a self-replicating polymer that preceded RNA, and was thus the centerpiece of a Pre-RNA World.[26]

From the RNA World hypothesis we have the suggestion that RNA (or an RNA-like precursor) might have been the first molecule with the capability to support life based on RNA genomes that are copied and maintained through the catalytic function of RNA itself, later to be replaced by DNA and proteins. However, enigmas exist. Various investigations have been carried out to identify ribozymes—a hybrid word made from *ribo*nucleic acid en*zyme*—RNA molecules that perform catalytic activities. In modern times ribozymes are certainly known. But the Holy Grail of ribozymes is an RNA replicase ribozyme, an ancestral RNA molecule that replicated everything, including itself.[27,28] This molecule would have been crucial in the RNA world. To date it has not been found via in vitro and molecular engineering techniques. Perhaps it will be discovered.

Genome The total genetic information of an organism as encoded in DNA, or, for many viruses, encoded in RNA.

Ribozyme An RNA molecule that can act as an enzyme.

LUCA (Last Universal Common Ancestor) The most recent organism from which all contemporary organisms are believed to have descended.

Other Plausible Venues

In a recent review, Tommaso Bellini wrote, "Past the season of enthusiasm for Miller's discovery [the Miller-Urey experiment] of the abiotic synthesis of simple organic compounds, the growing awareness that random chemistry couldn't have assembled functional biomolecules . . . stimulated creative thinking in a wide community of chemists, biologists, physicists and geologists."[29] Abiotic means non-living and refers to Miller's synthesis of organic compounds from inorganic precursors. In this chapter we'll have a non-exhaustive overview of the fruits of such creative thinking as well as how some of these propositions can be categorized. All origins models share a common aim: to reduce the "fantastic luck" implied by the fortunate assembly of functional molecules by introducing mechanisms implanting a stronger degree of necessity. [29] The very possibility that such molecular properties exist is at least amazing, if not fantastic.

By way of easing into our topic, Figure 14.3 shows a picture of the Earth's history painted in very broad strokes.[29] The most recent portion (green) shows the range of time in which we have paleontological evidence of life on Earth. At the other extreme, we peer back more than four billion years. This earliest interval (purple) is a time when no life was present because of the planet's condition. We're interested in the stretch in the middle of these two periods, the time frame indicated in orange.

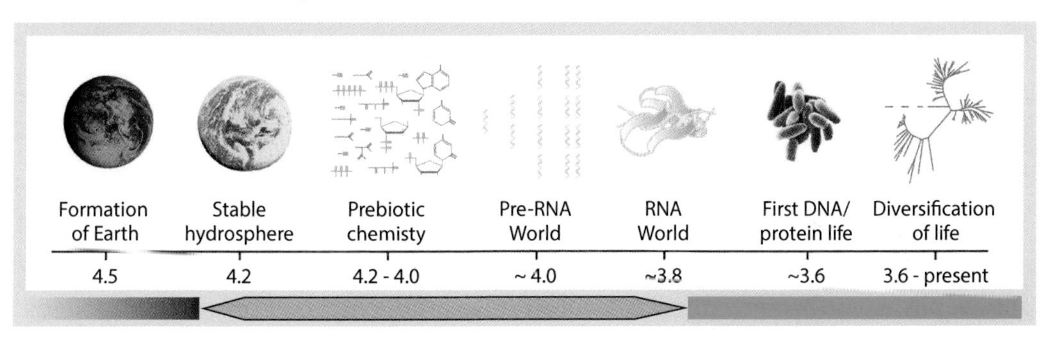

Figure 14.3 Timeline, expressed in units of billions of years, of early life events. Colored bars indicate the era when the Earth could not have hosted life (purple), the era when life was certainly present (green), and the ~ 500 million year interval (orange) when the origin of life took place. *(Reproduced with permission from Gerald F. Joyce, "The Antiquity of RNA-Based Evolution," Nature 418 (11 July 2002), 214–221; adaptation courtesy of Tommaso Bellini.)*

This is a gap of about 500 million years where *somehow* the inanimate became animated. It's here, ca. four billion years ago, that the study of the origins of life focuses its efforts. A cadre of researchers prod this gap with a horde of strategies, like so many seals poking at blowholes.

The Miller-Urey experiment jump-started experimental origins research. Later, another major driver was the discovery in the mid-1970s of oceanic hydrothermal systems and the suggestion that they could have been an alternative venue for life's beginnings.[30,31] Submarine hydrothermal systems are compelling for origins hypotheses because as hydrothermal fluids are ejected into seawater they produce large physical and chemical gradients—e.g., there are huge temperature changes within a small distance. Vents provide sources of chemical energy to drive reactions, many essential ingredients for those reactions, and also mineral surfaces that can serve as potential catalysts.[15] While the Miller-Urey picture is consonant with an RNA World view, such hydrothermal vents are suggestive of alternative possibilities (Figure 14.4).

Figure 14.4 Undersea hydrothermal vents where hot water jets upward into the cool surrounding ocean have been proposed as sites for the origin of life. *(Corbis Images, stock image WH001289.)*

In 1988 the German chemist Günter Wächtershäuser published a detailed theory in which he pounced on the iron-sulfide mineral surfaces abundant at hydrothermal vents as the birthplace for life.[32] His thesis is often called the Iron-Sulfur World (in contrast to the RNA World). He rejected the premise of a prebiotic soup and proposed that early life was autotrophic not heterotrophic, as most other theories maintain and that chemical energy from mineral sources, not sunlight or lightning, was the driver.[33] We need to explain the big words.

According to Wächtershäuser, the prebiotic soup was so dilute that early life made its own organic molecules from inorganic sources. Its metabolism was thus autotrophic from the Greek for "self," *auto,* and "feed," *trophos.* Modern examples of autotrophs are plants that use carbon dioxide, water, and minerals from the soil and derive their energy source from sunlight in the familiar process of photosynthesis. Plants are thus *photo*-autotrophs. Some bacteria are *chemo*autotrophs, producing their organic compounds by oxidizing inorganic substances such as sulfur and ammonia. The first life forms in Wächtershäuser's theory are chemoautotrophic.

This is in distinct contrast to the view that early life obtained organic material by scavenging from its surroundings. Its metabolism would then have been heterotrophic from the Greek for "other," *hetero,* and "feed," *trophos.* In this scenario, early life relied on molecules such as amino acids present in the prebiotic soup. In our present world the most obvious form of heterotrophy occurs when animals eat plants or other animals, but many fungi and bacteria consume organic litter, e.g., fallen leaves or feces.

Metabolism preceding replication is a prominent feature of Wächtershäuser's theory. Many in the origins field believe

Autotroph An organism that is itself capable of making energy-containing molecules from inorganic (non-carbon-containing) raw materials by using basic energy sources (e.g., sunlight).

Heterotroph An organism that cannot make its own energy-containing molecules and must make use of energy that originates in other organisms.

life began with a self-replicating molecule similar to RNA. According to Wächtershäuser, life began with metabolism and it was a cycle of chemical reactions that replicated itself. Only later did informational molecules appear. Note that other researchers who also favor a metabolism-first hypothesis differ in the nature of the chemical reactions they envision. But they concur with Wächtershäuser in the belief that life began as the chemical reactions evolved in sophistication.

The particulars of Wächtershäuser's ideas have undergone a considerable revision over the years, although he and his coworkers have provided documentation for his theory of evolving chemical reactions.[34] Recent evidence points to molecular interaction between the organic products of carbon monoxide fixation (the conversion of carbon monoxide to organic compounds—specifically, certain amino acids) and the process of further fixation under aqueous conditions at high temperatures and pressures.[35,36] However, the feasibility of Wächtershäuser's proposed carbon-fixation cycle as the correct theory for the origin of life remains uncertain at best.[37] Questions persist, for example, on the amounts of products produced and also the kinetics of the cycle (each of the steps involved must occur rapidly enough for the cycle to be useful).

In contrast to Wächtershäuser's views, supporters of an RNA World have long contended that metabolism could not develop independently of a genetic material.[38]

Replicator-First Versus Metabolism-First

As alluded to in the previous paragraph the dispute over metabolism-first versus replicator-first has been a major schism in origins research. We'll now give a flavor of that debate. For many years researchers had been unable to show that nucleotides (the essential components of both RNA and DNA) could be formed under plausible prebiotic conditions. And they faced the additional problem of how those nucleotides—even supposing that they had been available—could have polymerized into long chains. These were huge obstacles to the picture of the RNA World.

It is only recently that noteworthy investigations led by chemists Matthew Powner, John Sutherland, and Jack Szostak have shown a possible prebiotic route for the synthesis of the RNA nucleotides cytosine and uracil and a reaction path to precursors of the remaining two ribonucleotides, those incorporating the bases guanine and adenine.[39–42] Powner, Sutherland, and a team of colleagues have also announced a possible prebiotic route for chemically driven ribonucleotide ligation.[43] (The nucleic acid polymerization/ligation question is one that has been addressed by Bellini/Clark via physical mechanisms. We will return to this in the next chapter.) We will also return to the work of Powner, Sutherland, Szostak, and their colleagues in Chapter Fifteen.

Figure 14.5 presents a cartoon with one take on the replicator versus metabolism models (but there are many variations on the two themes portrayed). Both start from

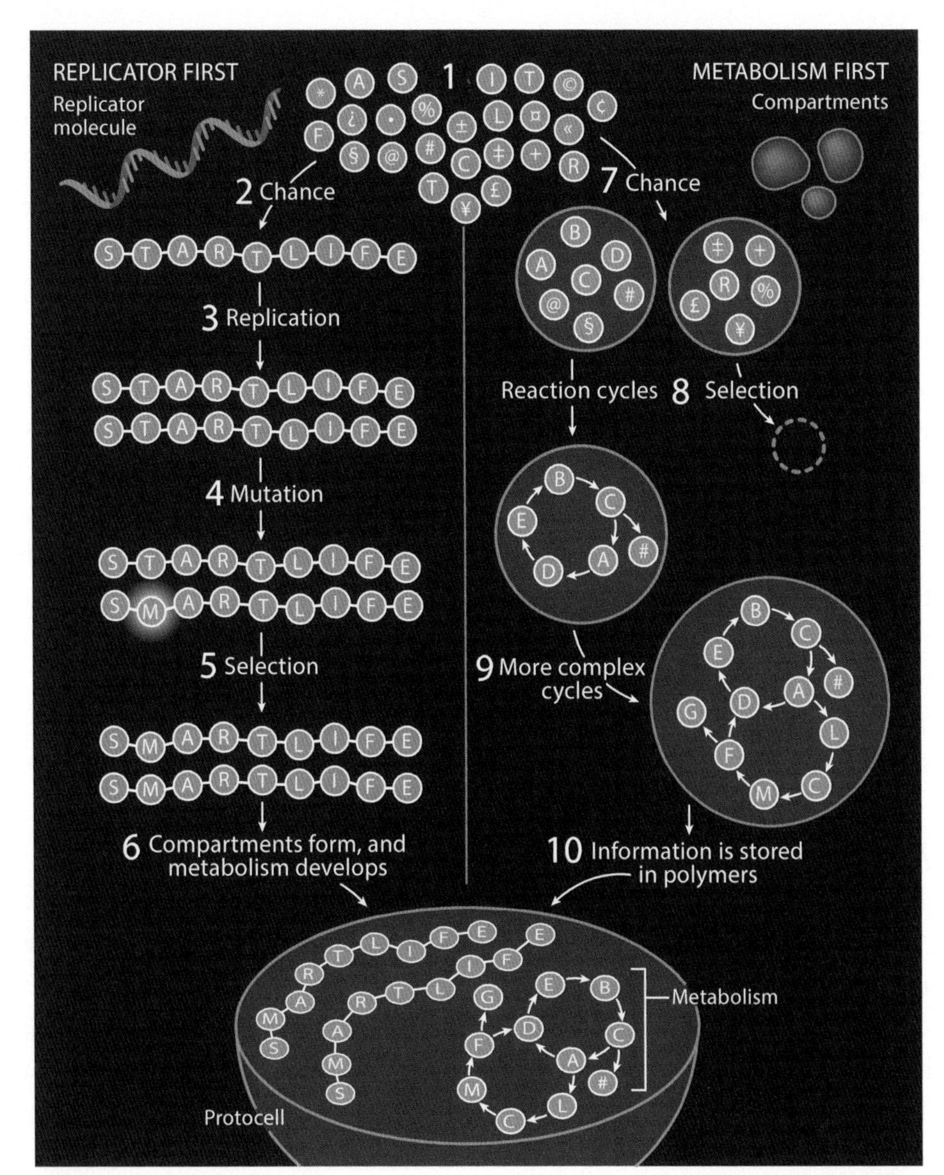

Figure 14.5 Replicator-first theories of how life originated posit that a large molecule capable of replication (e.g., RNA) formed first. Metabolism-first theories suggest that life began when small molecules formed an evolving network of chemical reactions driven by an energy source. *(Reproduced with permission from Robert Shapiro, "A Simpler Origin for Life," Scientific American (June 2007), 46–53 and artist Andrew Swift.)*

molecular compounds represented by balls labeled with symbols, shown in "1" at the top of Figure 14.5.[44] In the replicator-first model, some of these compounds join together in a chain by random chemistry (labeled "2" in the figure) and form a larger molecule (maybe a precursor to RNA?) that is capable of self-replication ("3" in the figure). The molecule makes many copies of itself. Some of these are mutant versions that can also self-replicate as shown in "4". Molecules that are better adapted to the ambient conditions are favored, step "5". As the evolutionary process continues, cell-like compartments called protocells form and metabolism develops—smaller molecules use energy to perform useful processes, shown in "6".

By contrast, the metabolism-first model starts with the spontaneous formation of compartments, "7". Some of the compartments contain mixtures of the starting compounds (from "1") that undergo chemical reaction cycles, "8," and the cycles become more complex over time, shown in "9". In "10" the system becomes sophisticated enough that it can store information in polymers. (These are the self-replicating molecules in the metabolism-first view.)

Replicator-first theories of life's origin Such scenarios suggest that a large molecule (e.g., RNA) capable of self-replication formed first. Then metabolism developed as smaller molecules used energy to perform useful processes.

Metabolism-first theories of life's origin Such scenarios suggest that preceding a large, self-replicating molecule, small molecules formed an evolving network of chemical reactions driven by an energy source.

The replicator-first and the metabolism-first lines of approach have been part of the mix in origins research for many decades. However, it should be appreciated that the majority of origins experts do not support a purely metabolic life form; they favor replicator-first.[45] Reproduction involves the passage of an enormous amount of information from one generation to the next; the only known way to both store and copy so much data is with a molecule similar to DNA or RNA.

Feats of Clay

In our sampling of the origins field we now turn to clay. Several researchers have invoked the role of clay and we will mention two examples. The first is the work of chemist James Ferris of Rensselaer Polytechnic Institute (RPI). Ferris's most well-known contribution is the finding that the clay mineral called montmorillonite can catalyze the formation of RNA polymers by serving as a surface template.[46] Using techniques such as PAGE for analysis, he found that the mineral surface enabled the formation of oligomers considerably longer than those that were formed in solution. The minerals' ability to bind nucleotides brought reactive molecules close together and this facilitated the formation of bonds (Figure 14.6).[47] His result is supportive of the replicator-first school of thought and also of the suggestion that life may have started on mineral surfaces—not necessarily near hydrothermal vents, but perhaps in clay-rich muds at the bottom of pools of water. Several concerns persist with this work. One issue is that to accomplish the

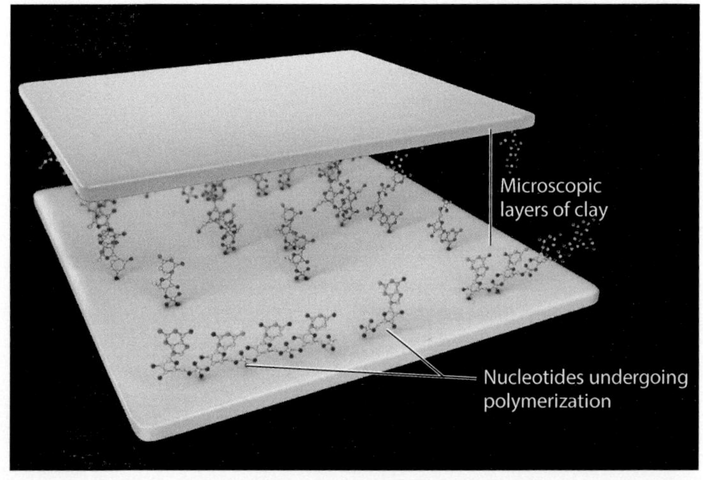

Figure 14.6 Microscopic layers of clay minerals such as montmorillonite can bind nucleotides. By bringing nucleotides close together, clay surfaces may have facilitated the formation of bonds between nucleotides and linked them up into single-stranded polymers similar to modern RNA. *(Reproduced with permission from Alonso Ricardo and Jack W. Szostak, "Origin of Life on Earth," Scientific American 301, No. 3 (Sept. 2009), 54–61 and artist Andrew Swift.)*

polymerization, Ferris had to *activate* the nucleotides. This means that he made the nucleotide chemically reactive—and thus more likely to form a polymer—by the addition of a reactive molecule (specifically, imidazole) to the phosphate of the nucleotide.[48] However, this activation step doesn't fit in with the plausible prebiotic conditions understanding that we've spoken of before.

Alternatively, chemist A.G. Cairns-Smith of the University of Glasgow provided a novel clay scheme. Cairns-Smith was the first person to emphasize the complexity of RNA and how improbable it is that RNA could have formed de novo on the primitive Earth.[49] He suggested that the first system capable of evolving by natural selection was a self-reproducing clay.[50] We've already seen minerals invoked as catalytic and organizing surfaces for biomolecules. Cairns-Smith went further in arguing that fine-grained clay crystals themselves may have been the first life on Earth. Clay crystals grow and their layered structure can carry and replicate what can be thought of as genetic information. Just as DNA and RNA code information using four nucleotides, clay minerals such as kaolinite have thin layers that stack on one another in three possible relative orientations—0°, 120°, or 240°—providing three coding elements.[51] In addition, there can be aperiodic stacking (layers of different thickness and/or chemical composition) that potentially provides further information coding (Figure 14.7).[52] Cairns-Smith proposed that reproduction could occur when clay crystals grow, copy a given layer or stacking sequence, and then cleave—the cleavage being the act of self-replication. If a sequence of layers or a particular chemical composition is particularly stable it will have an evolutionary advantage and will win out and proliferate. And what is the tie-in to biomolecules?

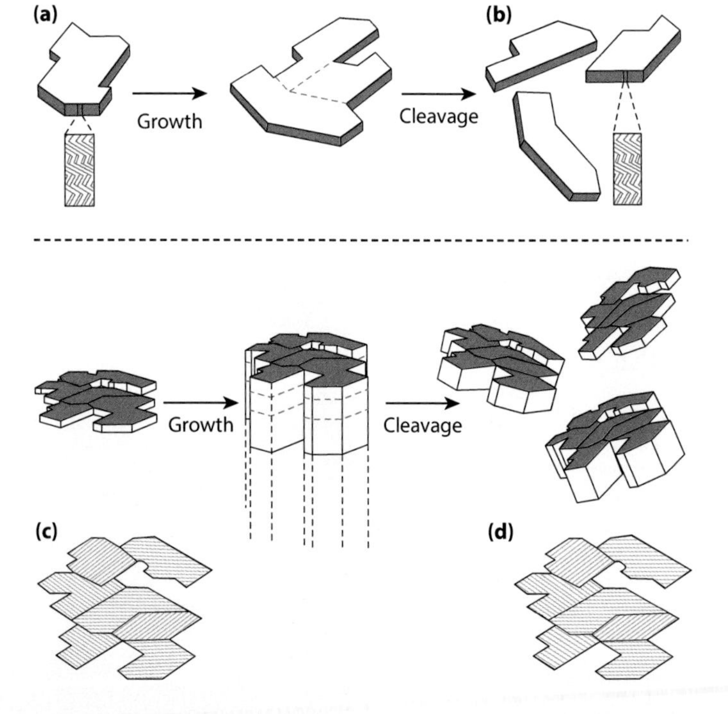

Figure 14.7 Crystal genes would require the right combination of structural, growth, and cleavage attributes. Information might be stored in one or two dimensions. In a one-dimensional gene (top) information would be in the detailed structure of a sequence of stacking layers (red). Details remain constant, **(a)** and **(b)**, as the gene replicates via cleavage. The layers could vary physically or in chemical composition. In two-dimensional genes (bottom) information is in a physical or chemical pattern on one crystal face (red) that remains constant, **(c)** and **(d)**, as the gene replicates. Three relative orientations (see text) are the striations in (c) and (d). *(Reproduced with permission, A.G. Cairns-Smith, "The First Organisms," Scientific American 252, No. 6 (1985), 90–100, artist George V. Kelvin.)*

In Cairns-Smith's model, specific organic molecules are attracted to different exposed layer edges and/or chemical moieties (in effect, particular parts) of the clay crystal. One interesting case is the intercalation (i.e., insertion) of small biomolecules between the layers of clay minerals. For example, the ionized form of the nucleobase adenine will enter the interlayer regions of montmorillonite.[53] More generally, Cairns-Smith believes that genetic takeover occurs—the gradual replacement of one genetic material by another quite different material.[54] The first genetic material might be microcrystalline clay holding information in various features, e.g., electric charge distribution, a pattern of surface grooves, a particular stacking sequence of layers, a defect structure such as a screw dislocation. These replicate through crystal growth and cleavage such as shown in Figure 14.7. The features in the clay crystals act as a scaffold of sorts for the biomolecules that eventually take over the role of genetic material. Thus, an organic biochemical genetic material gradually takes over from what was initially a geochemical genetic material. Hence, a transition from the inanimate to the animate.

Cairn-Smith's unusual crystals-as-genes concepts have long resisted experimental testing; clay crystals are difficult to grow reproducibly, sizably, and with well-defined characteristics. Chemist Bart Kahr (then at the University of Washington) and co-workers did evaluate one proposed mechanism whereby crystals act as a primitive source of transferable information. They used a combination of light microscopy, atomic force microscopy, and luminescence labeling. They examined growth hillocks not in a clay mineral but rather in a model crystal system called KAP or often KHP (the compound potassium hydrogen phthalate—although the acronym derives from potassium *acid* phthalate) better suited for testing.[55] They found that there was more mutation than there was inheritance in successive generations of KAP crystals. However, they concluded that alternatives to KAP and to hillock dynamics might be better candidates to test the idea of genetic takeover.

The Lipid World

Creative thinking about the origins question has generated many ways in which molecules other than nucleic acids might have had important prebiotic roles, including the ability to catalyze their own replication, and the ability to store and propagate information. Several researchers have drawn attention to lipids, notably biochemists Pier Luigi Luisi, David Deamer, and Jack Szostak. Luisi, of the faculty of the Roma Tre University, has focused for decades on several means of lipid encapsulation including micelles and bilayer vesicles. He has shown that micelles can

Examples of scenarios for the origin of life:

RNA World RNA (in the form of ribozymes) serves as both catalyst and gene.

Iron-sulfur World Minerals composed of iron, nickel, and sulfur catalyze metabolic reactions.

Clay surfaces Clay surfaces catalyze synthesis of polymers.

Lipid World Organized lipid structures have catalytic properties and contain information in their composition.

catalyze their own replication. He has described a system in which autocatalytic micelles are formed from amphiphiles in solution.[56] His self-replicating micelles have geometrically defined boundaries and achieve population growth within the boundaries of the parent micellar structures. A structure that is self-bounded and is able to self-generate due to reactions that take place within the boundary meets the criteria of *autopoiesis.* The term *autopoiesis* is from the Greek *auto,* "self," and *poiesis,* "formation." Luisi's micellar self-replicating systems are attempts to model with simple synthetic systems the basic chemical mechanisms of life. This is done without DNA or proteins; the autopoietic system can be considered as a cell that metabolizes low molecular weight components.

Micelle A spherical formation caused by an amphiphilic substance in a solution. The hydrophilic end of the molecule tends to orient itself toward the outside of the sphere while the hydrophobic end tends to orient itself toward the inside of the sphere.

David Deamer, of the University of California Santa Cruz, also has a stake in the Lipid World. Part of his effort has been to find conditions under which mononucleotides will self-assemble (non-enzymatically) to form RNA-like polymers in a lipid environment.[57–59] He did not chemically activate the mononucleotides (as Ferris had done) to synthesize polymers, but rather he devised a prebiotic environment in the laboratory where wet and dry periods were cycled and heat provided activation energy during the dry cycles. He found polymer formation when lipids were present in the medium under investigation, but when no lipids were present he did not observe polymerization. Deamer refers to this as "guided polymerization in an anhydrous lipid environment."[59] Among the tools that he used to detect the presence of RNA-like polymers was the nanopore translocation technique employing alpha-hemolysin that we discussed at length in Chapter Ten.

We will return to Deamer shortly to compare and contrast his ideas with those of Bellini and Clark. But we now turn to some of the lipid-related work of Harvard biochemist Jack Szostak. Among Szostak's contributions is the demonstration of the permeability of early cells (protocells).[60] As we know from Chapter Four, phospholipids form the membranes of modern cells. In order to permit the exchange of molecules with the external environment, the membranes are equipped with channels and pumps. The question is: what did primitive cells look like? Without complex transport machinery how could simple cells take in complex nutrients? If one considers lipid vesicles—small cavities enclosed by a lipid bilayer—as candidates for the earliest life forms, how did nutrients cross their membranes without help from transmembrane proteins?

Until the experiments of Szostak and his colleagues, some researchers believed that these challenges precluded a heterotrophic lifestyle for primitive cellular life forms. Szostak demonstrated the permeability of prebiotically plausible lipid membranes (Figure 14.8). This suggests that primitive protocells made from simple amphiphiles that form bilayer vesicles could have obtained complex nutrients from their environment without macro-

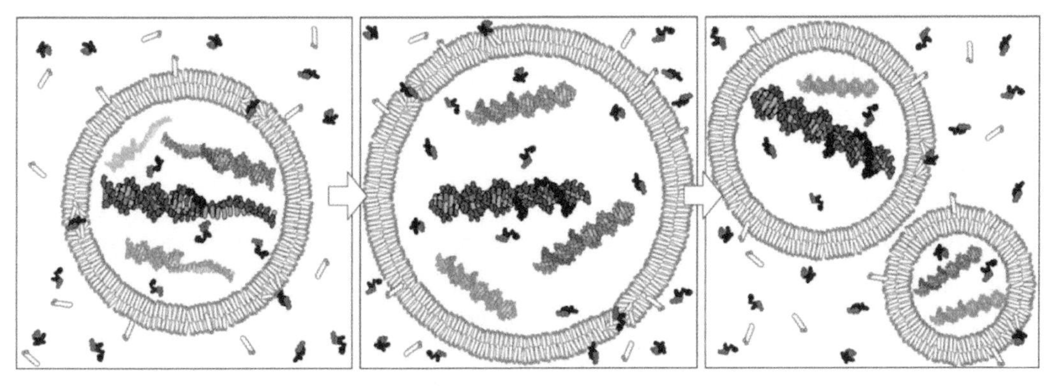

Figure 14.8 Conceptual model of a heterotrophic protocell. Growth of the protocell membrane results from incorporation of externally supplied amphiphiles. Externally supplied nucleotides cross the protocell membrane via local membrane deformations and act as substrates for the non-enzymatic copying of internal templates. *(Reproduced with permission from Sheref S. Mansy, Jason P. Schrum, Mathangi Krishnamurthy, Sylvia Tobé, Douglas A. Treco, and Jack W. Szostak, "Template-Directed Synthesis of a Genetic Polymer in a Model Protocell," Nature 454 (July 2008), 122–126.)*

molecular transport machinery. Thus, they could well have been heterotrophs, taking in the nutrient energy they needed by transport through local membrane deformations. He also showed that (activated) nucleotides added to the outside of a model protocell cross the membrane and take part in template copying in the protocell interior. Demonstrating a possible heterotrophic origin for primitive cellular life may be the key point of the study.[61]

David Deamer wrote a commentary on Szostak's paper in which he said, "there is increasing reason to think that the first form of life was a primitive version of a cell, rather than a replicating molecule supported by a metabolic network."[61]

Liquid Crystals in the Work of Deamer and the Work of Bellini/Clark

And that brings us back to Deamer. Since, as we will see, he invokes liquid crystals in his vision of a Lipid World, it's important to contrast his perspective with that of Bellini and Clark. As we saw in Chapter Eleven, Bellini and Clark call out liquid crystals as the structural gatekeeper that is at the heart of hierarchical self-assembly (a theme to be revisited in Chapter Fifteen).

Deamer specifies liquid crystalline nanostructures as organizing matrices for non-enzymatic nucleic acid polymerization.[58] He invokes liquid crystallinity in the context of what he calls "organized fluid lipid matrices (liquid crystals)."[57] He described his hypothesis in a paper in which he reports the non-enzymatic synthesis of single-stranded RNA-like polymers from mononucleotides in lipid environments. "Fluid lipid microenvironments [liquid crystals] impose order on mononucleotides in such a way that they are able to form extensive phosphodiester bonds and thereby produce RNA-like polymers sufficiently long to have catalytic activity."[62] Phosphodiester bonds are the bonds connecting adjacent sugars in nucleic acids; more precisely, they are the bonds formed between the 3′ hydroxyl group of one sugar and the 5′ phosphate group of the adjacent sugar. They occur on the left and right sides of Figure 12.7 in Chapter Twelve.

Deamer goes on to say that "The reaction conditions in which RNA-like polymers form are relatively complex, and further research will be required before a mechanism can be put forward."[63] He makes several suggestions as to how the RNA-like polymers may have formed.

Additionally, in a paper on DNA templating, Deamer showed that a single-stranded DNA oligomer acts as a template for synthesis of complementary products under conditions simulating a plausible prebiotic environment.[64] He combined (1) single-stranded DNA molecules (64 bases in length), (2) a mixture of nucleoside monophosphates (i.e., subunits of nucleic acids—in this case, subunits of DNA—consisting of a deoxyribose sugar linked to a nitrogen-containing organic ring compound, either adenine, guanine, cytosine, or thymine) in equal amounts of all four DNA bases, and (3) lipids. (A nucleoside lacks the phosphate group that would make it a nucleotide. Hence, a nucleotide is also referred to as a nucleoside monophosphate, such as adenosine monophosphate.) He found that single-stranded DNA acted as a template for the formation of complementary strands that could then pair specifically with the original strands. Here too he invokes an important property of liquid crystals when he contrasts his study with that of Ferris (in which montmorillonite clay surfaces were used as matrices for the synthesis of nucleic acids). Deamer observes, "However, potential reactants bound to solid clay surfaces have limited diffusional mobility [Ferris' experiment], while lipid matrices are liquid crystalline structures that allow diffusion even in the dry state."[65]

Nucleoside A nucleoside is a structural subunit of a nucleic acid. It consists of a molecule of sugar linked to a nitrogen-containing ring compound called a nitrogenous base, also called a nucleobase or just a base. In DNA, the sugar is deoxyribose and the base is adenine, cytosine, guanine, or thymine. In RNA, the sugar is ribose and the base is adenine, cytosine, guanine, or uracil.

Nucleotide A nucleotide consists of one or more phosphate groups bonded to a nucleoside. When only one phosphate group is bonded to the nucleoside, the nucleotide thus formed is referred to as a nucleoside monophosphate. An example is adenosine monophosphate (AMP).

Phosphate group A phosphorous atom bonded to four oxygen atoms at least one of which is bonded to another atom.

Phosphodiester bonds These connect adjacent sugars in nucleic acids.

Figure 14.9 is a cartoon that embraces Deamer's RNA and DNA research.[66] The small inset in the upper right shows a particular nucleotide monomer, namely adenosine monophosphate, in both a top view and a side view (for illustrative purposes). These nucleotides appear, in side view, in the middle of the figure. In the DNA templating work, Deamer proposes that mononucleotides are organized and concentrated when they are trapped during dehydration of lipid vesicles. Then, as the vesicles are dried, the lipids fuse to form multilamellar structures (multiple, concentric lipid bilayers), and any solutes that are present—specifically, the mononucleotides—can be captured between alternating bilayers of lipid. Figure 14.9 shows a cross-sectional view of such a multilamellar structure with lipid bilayers on either side of the captured nucleotides. Since the lipid is a liquid crystal, the mononucleotides are free to diffuse within the two-dimensional plane of the multilayer. Thus, Deamer proposes, if single-stranded DNA templates are also present, the concentrated mononucleotides undergo complementary base pairing on the single-stranded template molecule to produce double-stranded DNA.[67]

To summarize Deamer's results: in the case of his RNA studies he uses a mixture of ribonucleotides and lipids to show that in the microenvironment of a liquid crystalline matrix, the ribonucleotides will form phosphodiester bonds and thus produce RNA-like polymers. In his DNA studies he employs a mixture of deoxynucleotides, lipids, and single-stranded DNA oligomer templates. He shows that in the liquid crystalline microenvironment, conditions are amenable to hybridization to create double-stranded DNA.

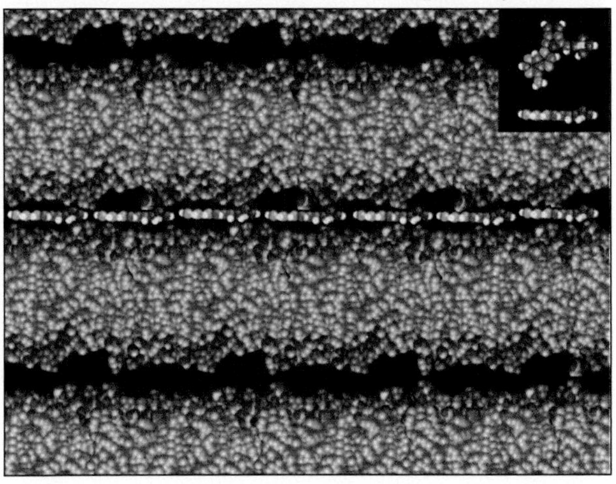

Figure 14.9 Nucleotides such as adenosine monophosphate (inset) can become organized within the two-dimensional plane between lipid bilayers in a multilamellar lipid matrix. Wet and dry cycling can promote polymerizaton of a suite of mononucleotides. *(Reproduced with permission from Felix Olasagasti, Hyunsung John Kim, Nader Pourmand, and David W. Deamer, "Non-Enzymatic Transfer of Sequence Information Under Plausible Prebiotic Conditions," Biochimie 93 (2011), 556–561.)*

Bellini and Clark see the role of liquid crystals quite differently than does Deamer. They suggest that starting out with very short oligomers of duplex nucleotides (in the case of both nanoRNA and nanoDNA), the polymerization of those chains occurs by the mechanisms of hierarchical self-assembly that we have previously discussed. It is the liquid crystal phase that selects, templates, and replicates its constituent molecules and thus promotes a cyclic process. And, as we saw in the previous chapter, it *is* the liquid phase that does this and not merely the concentration of nanoDNA. The process then proceeds to the appearance and evolution of molecular species that in turn enhance the stability of the liquid crystal phase.

Deamer's experiments suggest lipid-assisted formation of polymeric nucleic acids and lipid-assisted templating to form duplexed nucleic acids. Bellini and Clark have shown that liquid crystal droplets provide the microenvironment in which, via chemical ligation, nanoDNA (or nanoRNA) oligomers can elongate and ultimately become polymers. Furthermore, they have shown that liquid crystals themselves are the templating agent. That is, the liquid crystalline matrix provides a flexible template favoring the linear aggregation and chemical ligation of nanonucleic acids. This results in liquid crystal

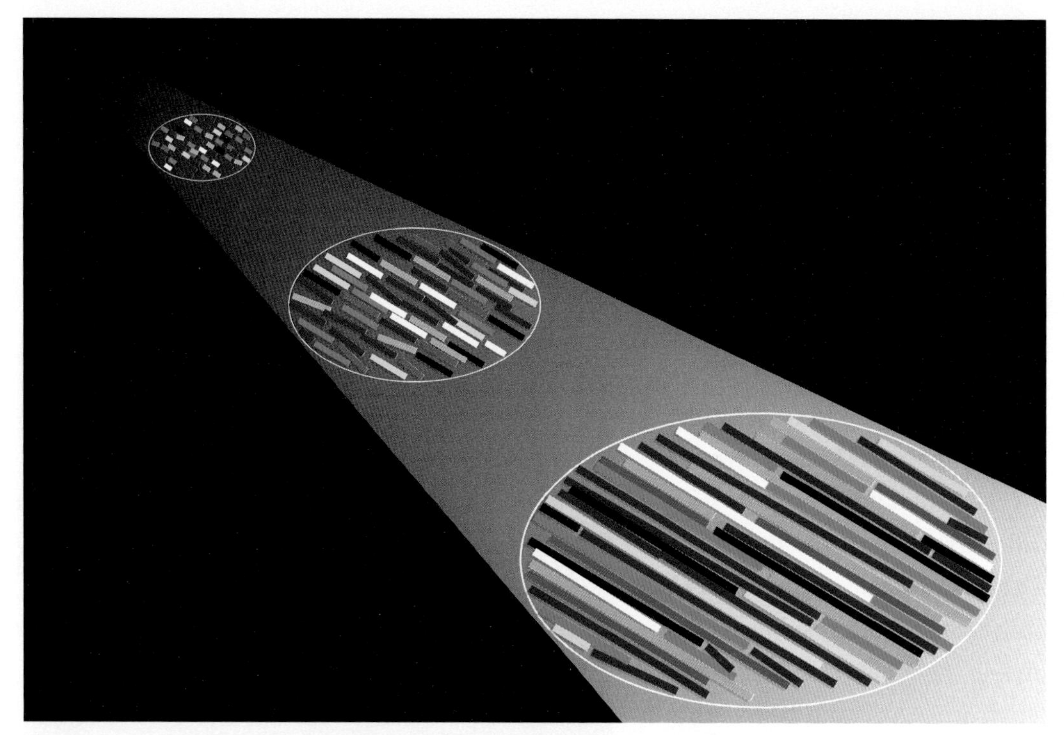

Figure 14.10 Liquid crystal autocatalysis → Nucleic acid polymerization → Origin of life. Liquid crystals, shown as ellipses, provide supramolecular ordering and promote an autocatalytic cycle favoring the growth of elongated nucleic acid chains from few-base long oligomers. After many iterations over a long span of time, the polymerization of nucleic acids will reach the lengths required for catalytic activity. These long chains become feedstock for the RNA world. *(© Kenneth Douglas)*

autocatalysis in which longer and longer complementary nucleic acid sequences are favored because they stabilize the liquid crystal ordering.

Thus, Bellini and Clark have experimentally demonstrated a mechanism in a presumed prebiotic environment for the elongation of nanoRNA and nanoDNA into long, polymeric chains such as are found in our contemporary world (Figure 14.10).

Since we have frequently employed the abbreviated phrase "Bellini and Clark," before closing this chapter it's important to elaborate on that phrasing by calling out their talented graduate students and postdocs whose skill and effort were essential for the success of the work. Among these contributors were Tommaso Fraccia, Christopher Jones, Mark Moran, Michi Nakata, Elvezia Paraboschi, Gregory Smith, Ethan Tsai, Youngwoo Yi, Dong Ki Yoon, and Giuliano Zanchetta. Any mention of contributors to the Bellini and Clark origins work would be incomplete without the name of Professor David M. Walba, Department of Chemistry and Biochemistry at the University of Colorado.

Manfred Eigen and Stuart Kauffman

We conclude our sketch of origins theories with a brief mention of two scientists whose theoretical efforts have contributed to the scientific dialogue and whose ideas are much discussed. Manfred Eigen is a German physical chemist and Nobelist in chemistry. Eigen and co-author Peter Schuster developed a theory (along with experimental work) of the self-organization of biological macromolecules.[68] They coined the term *hypercycle* to describe a self-reproducing system in which there is cooperation between short RNA sequences and enzymes. Hypercycles are seen as predecessors of the first cells. The proposed hypercycles are autocatalytic chemical reaction cycles that are postulated to have evolved so as to contain RNA strands that themselves then functioned as enzymes.[69] For interested readers, Eigen has written a brief book that contains what he calls "Vignettes from Molecular Biology" integrated into a general understanding of nature.[70]

Stuart Kauffman is a physician and theoretical biologist who has proposed an origin of life theory in the larger context of the behavior of complex systems displaying emergent properties. His analysis is built upon mathematics and computer simulations. The focus is on the emergence of chemical complexity by means of the competition of autocatalytic sets in which chemicals catalyze their own formation.[71] Kauffman believes that life is a natural property of complex chemical systems. According to Kauffman, when the number of different kinds of molecules in a chemical soup passes a certain threshold, a self-sustaining network of reactions—an autocatalytic metabolism—will emerge.[72] An important requirement in his theoretical framework is that catalytic closure is achieved. This means that every member of the autocatalytic set has at least one of the possible last steps in its formation catalyzed by some member of the set and Kauffman describes the conditions required for catalytic closure.[73] For the interested reader, his most accessible book is that cited in endnote 72.

In November 2010, eight months prior to the large origins meeting in Montpellier, France, Noel Clark flew to Boston to attend the Materials Research Society's annual fall meeting. While there, he gave an invited talk (on the liquid crystal mechanism for the self-assembly of nucleic acids) to Jack Szostak's research group at Massachusetts General Hospital. Bit by bit, both Bellini and Clark have been getting the word out about their unusual take on the origins question. In Cairo, Egypt, at a liquid crystals workshop; in Santa Fe, New Mexico, at a conference on complexity; at Tokyo University and the University of Wisconsin; at scientific meetings in Jeju, Korea, in El Paso, Texas, in New London, New Hampshire. In Chapter Twelve we mentioned a review paper co-authored by Szostak in which he highlighted a "series of elegant experiments" by Bellini and Clark. Since the time of that paper, Bellini and Clark have strengthened their case impressively. In stages, like climbers using carabiners to pull themselves up a slope of resistance on belay, they have progressed beyond obstacles of increasing complexity and subtlety. In the previous chapter we focused on the realization of the long-sought jewel in the crown, ligation. With this achievement in hand, you can expect Clark or Bellini to give an invited lecture at an upcoming international origins meeting. Their way of thinking is no longer far out. It may just be spot on.

Exercises for Chapter Fourteen: Exercises 14.1–14.2

Exercise 14.1

Back in Chapter Four we first discussed the amphiphilic molecules that form spherical micelles. Micelles have come up again in this chapter so it's time to do another exercise on these common structures. Let "v" denote the volume of the hydrocarbon tail and let "a" denote the area per polar head group of an archetypal amphiphilic molecule. (You may ignore the volume of the polar head group.) Calculate the radius, r, of the spherical micelle formed by these amphiphiles assuming that you have an aggregation number of N; that is, the total number of amphiphiles that make up the micelle = N.

Note: In performing this exercise you will be replicating a (minor) calculation—on page 1534 (endnote 74) for those interested!—in a classic paper that helped to lay the theoretical foundations for the self-assembly of hydrocarbon amphiphiles into micelles and bilayers.[74]

Exercise 14.2

Here's an exercise that brings together a potpourri of terms and concepts that are now part of your toolkit—e.g., phosphodiester bonds, transcription of DNA by messenger RNA, the enzyme RNA polymerase—and perhaps stretches your ingenuity.

(a) During transcription, which of the following events is earliest to occur?

A. Local disassembly of a DNA's double-helical structure

B. Binding of an enzyme to a DNA molecule

C. Location of ribonucleoside triphosphates at the start site

D. Phosphodiester bond formation

(b) Going further, try to place the four events in the order in which they occur.

(Exercise adapted with permission from Cracking the MCAT, *2013-2014 Edition, The Princeton Review.)*

ENDNOTES

[1] Conversation; Tommaso Bellini to K.D. on 21 May 2013.

[2] http://www.origins2011.univ-montp2.fr/PublSchedAbstr/P6.html

[3] Gerald F. Joyce, "Foreword," in *Origins of Life: The Central Concepts,* ed. David W. Deamer and Gail R. Fleischaker (Boston: Jones and Bartlett, January 1994), xi–xii.

[4] Robert M. Hazen, *The Story of Earth* (New York: Viking, 2012), 130.

[5] Tracey A. Lincoln and Gerald F. Joyce, "Self-Sustained Replication of an RNA Enzyme," *Science 323* (27 Februrary 2009), 1229–1232.

[6] Gerald F. Joyce, "Bit by Bit: The Darwinian Basis of Life," *PLoS Biology 10, No. 5* (May 2012), http://www.plosbiology.org/article/info:doi/10.1371/journal.pbio.1001323

[7] Personal communication; email from Gerald Joyce to K.D. on 5 July 2013.

[8] John H. Holland, *Emergence: From Chaos to Order* (New York: Basic Books, 1999), 3.

[9] Ibid., 2.

[10] Sandra Pizzarello and Everett Shock, "The Organic Composition of Carbonaceous Meteorites: The Evolutionary Story Ahead of Biochemistry," *Cold Spring Harbor Perspectives in Biology 2, No. 3* (March 2010), 1–19.

[11] Robert M. Hazen, "Emergence and the Experimental Pursuit of the Origin of Life," in *Exploring the Origin, Extent, and Future of Life: Philosophical, Ethical and Theological Perspectives,* ed. Constance M. Bertka (New York: Cambridge University Press, 2009), 23.

[12] Pier Luigi Luisi, *The Emergence of Life: From Chemical Origins to Synthetic Biology* (Cambridge, UK: Cambridge University Press, 2006), 113.

[13] Stanley L. Miller, "A Production of Amino Acids Under Possible Primitive Earth Conditions," *Science 117* (15 May 1953), 528–529.

[14] Jeffrey L. Bada and Antonio Lazcano, "Stanley L. Miller," Biographical Memoirs, *National Academy of Sciences (2012),* 1–40, http://www.nasonline.org/publications/biographical-memoirs/memoir-pdfs/miller-stanley.pdf

[15] Thomas M. McCollum, "Miller-Urey and Beyond: What Have We Learned About Prebiotic Organic Synthesis Reactions in the Past 60 Years?" *Annu. Rev. Earth Planet. Sci. 41* (2013), 207–229.

[16] Michael P. Robertson and Gerald F. Joyce, "The Origins of the RNA World," *Cold Spring Harbor Perspectives in Biology 4 (2012),* 1–22, http://cshperspectives.cshlp.org/content/4/5/a003608.full.pdf+html?sid=680fdf0f-bdd0-442f-b5b7-b8be8b6e8976

[17] Walter Gilbert, "Origin of Life: The RNA World," *Nature 319* (20 February 1986), 618.

[18] Carl Woese, *The Genetic Code: The Molecular Basis for Genetic Expression* (New York: Harper & Row, 1967), 179–195.

[19] F.H.C. Crick, "The Origin of the Genetic Code," *J. Mol. Biol. 38* (1968), 367–379.

[20] L.E. Orgel, "Evolution of the Genetic Apparatus," *J. Mol. Biol. 38* (1968), 381–393.

[21] Thomas R. Cech, "RNA as an Enzyme," *Scientific American 255* (November 1986), 64–75.

[22] Thomas R. Cech, Arthur J. Zaug, and Paula J. Grabowski, "In Vitro Splicing of the Ribosomal RNA Precursor of Tetrahymena: Involvement of a Guanosine Nucleotide in the Excision of the Intervening Sequence," *Cell 27* (December 1981, Part 2), 487–496.

[23] Kelly Kruger, Paula J. Grabowski, Arthur J. Zaug, Julie Sands, Daniel E. Gottschling, and Thomas R. Cech, "Self-Splicing RNA: Autoexcision and Autocyclization of the Ribosomal RNA Intervening Sequence of Tetrahymena," *Cell 31* (November 1982), 147–157.

[24] Thomas R. Cech, "The RNA Worlds in Context," *Cold Spring Harbor Perspectives in Biology 4* (2012), 1–5, http://cshperspectives.cshlp.org/content/4/5/a003608.full.pdf+html?sid=680fdf0f-bdd0-442f-b5b7-b8be8b6e8976

[25] Thomas R. Cech, "Self-Splicing and Enzymatic Activity of an Intervening Sequence RNA From *Tetrahymena*," *Nobel Lecture* (8 December 1989), 669.

[26] Peter Egil Nielsen, "Peptide Nucleic Acid (PNA): A Model Structure for the Primordial Genetic Material?" *Origins of Life and Evolution of the Biosphere 23* (1993), 323–327.

[27] Aniela Wochner, James Attwater, Alan Coulson, and Philipp Holliger, "Ribozyme-Catalyzed Transcription of an Active Ribozyme," *Science 332* (8 April 2011), 209–212.

[28] Michael Yarus, "Climbing in 190 Dimensions," *Science 332* (8 April 2011), 181–182.

[29] Tommaso Bellini, Marco Buscaglia, Andrea Soranno, and Giuliano Zanchetta, "Origin of Life Scenarios: Between Fantastic Luck and Marvelous Fine-Tuning," *Euresis Journal 2* (Winter 2012), 113–139.

[30] J.B. Corliss, J.A. Baross, and S.E. Hoffman, "An Hypothesis Concerning the Relationship Between Submarine Hot Springs and the Origin of Life on Earth," *Oceanologica Acta 4* (Supplement) 1981, 59–69.

[31] J.A.Baross and S.E. Hoffman, "Submarine Hydrothermal Vents and Associated Gradient Environments as Sites for the Origin and Evolution of Life," *Orig. Life Evol. Biosph. 30* (1985), 327–345.

[32] Günter Wächtershäuser, "Before Enzymes and Templates: Theory of Surface Metabolism," *Microbiological Reviews 52, No. 4* (December 1988), 452–484.

[33] Ibid., 469.

[34] Leslie E. Orgel, "The Implausibility of Metabolic Cycles on the Prebiotic Earth," *PloS Biology 6, No. 1* (January 2008), 0005–0013.

[35] Juli Peretó, "Out of Fuzzy Chemistry: From Prebiotic Chemistry to Metabolic Networks," *Chem. Soc. Rev. 41* (2012), 5394–5403.

[36] Claudia Huber, Florian Kraus, Marianne Hanzlik, Wolfgang Eisenreich, and Günter Wächtershäuser, "Elements of Metabolic Evolution," *Chem. Eur. J. 18* (2012), 2063–2080.

[37] H. James Cleaves II, Andrea Michalkova Scott, Frances C. Hill, Jerzy Leszczynski, Nita Sahai, and Robert Hazen, "Mineral-Organic Interfacial Processes: Potential Roles in the Origins of Life," *Chem. Soc. Rev. 41* (2012), 5502–5525.

[38] Leslie E. Orgel, "The Origin of Life—A Review of Facts and Speculations," *Trends in Biochem. Sci. 23* (December 1998), 491–495.

[39] Matthew W. Powner, Béatrice Gerland, and John D. Sutherland, "Synthesis of Activated Pyrimidine Ribonucleotides in Prebiotically Plausible Conditions," *Nature 459* (14 May 2009), 239–242.

[40] Matthew W. Powner, John D. Sutherland, and Jack W. Szostak, "The Origins of Nucleotides," *SYNLETT 14* (2011), 1956–1964.

[41] Matthew W. Powner, John D. Sutherland, and Jack W. Szostak, "Chemoselective Multicomponent One-Pot Assembly of Purine Precursors in Water," *J. Am. Chem. Soc. 132* (2010), 16677–16688.

[42] Erratum: Matthew W. Powner, John D. Sutherland, and Jack W. Szostak, "Chemoselective Multicomponent One-Pot Assembly of Purine Precursors in Water," *J. Am. Chem. Soc.133* (2011), 4149–4150.

[43] Frank R. Bowler, Christopher K.W. Chan, Colm D. Duffy, Béatrice Gerland, Saidul Islam, Matthew W. Powner, John D. Sutherland, and Jianfeng Xu, "Prebiotically Plausible Oligoribonucleotide Ligation Facilitated by Chemoselective Acetylation," *Nature Chemistry 5* (May 2013), 383–389.

[44] Robert Shapiro, "A Simpler Origin for Life," *Scientific American 296, No. 6* (June 2007), 46–53.

[45] Robert M. Hazen, *genesis: The Scientific Quest For Life's Origins* (Washington, DC: Joseph Henry Press, 2005), 215.

[46] James P. Ferris, Aubrey R. Hill Jr., Rihe Liu, and Leslie E. Orgel, "Synthesis of Long Prebiotic Oligomers on Mineral Surfaces," *Nature 381* (2 May 1996), 59–61.

[47] Alonso Ricardo and Jack W. Szostak, "Origin of Life on Earth," *Scientific American 301, No. 3* (September 2009), 54–61.

[48] James P. Ferris, "Mineral Catalysis and Prebiotic Synthesis: Montmorillonite-Catalyzed Formation of RNA," *Elements 1* (June 2005), 145–149.

[49] Leslie E. Orgel, "Prebiotic Chemistry and the Origin of the RNA World," *Critical Reviews in Biochemistry and Molecular Biology 39* (2004), 99–123.

[50] A.G. Cairns-Smith, "The Origin of Life: Clays," in *Frontiers of Life, Volume 1,* ed. D. Baltimore, R. Dulbecco, F. Jacob, and R. Levi-Montalcini (New York: Academic Press, 2001), 169–192.

[51] S.W. Baily, "Layer Silicate Structures," in *Clay Minerals and the Origin of Life,* ed. A.G. Cairns-Smith and H. Hartman (Cambridge, UK: Cambridge University Press, 1986), 33–34.

[52] A.G. Cairns-Smith, "The First Organisms," *Scientific American 252, No. 6* (June 1985), 90–100.

[53] Cairns-Smith, "The Origin of Life: Clays," 173.

[54] Ibid., 182.

[55] Theresa Bullard, John Freudenthal, Serine Avagyan, and Bart Kahr, "Test of Cairns-Smith's 'Crystals-as-Genes' Hypothesis," *Faraday Discuss. 136 (*2007), 231–245.

[56] Pascale Angelica Bachmann, Pier Luigi Luisi, and Jacques Lang, "Autocatalytic Self-Replicating Micelles as Models for Prebiotic Structures," *Nature 357* (7 May 1992), 57–59.

[57] Sudha Rajamani, Alexander Vlassov, Seico Benner, Amy Coombs, Felix Olasagasti, and David Deamer, "Lipid-Assisted Synthesis of RNA-like Polymers From Mononucleotides," *Orig. Life Evol. Biosph. 38* (2008), 57–74.

[58] David Deamer, "Liquid Crystalline Nanostructures: Organizing Matrices for Non-Enzymatic Nucleic Acid Polymerization," *Chem. Soc. Rev. 41* (2012), 5375–5379.

[59] Laura Toppozini, Hannah Dies, David W. Deamer, and Maikel C. Rheinstädter, "Adenosine Monophosphate Forms Ordered Arrays in Multilamellar Lipid Matrices: Insights Into Assembly of Nucleic Acid for Primitive Life," *PLOS ONE 8, No. 5* (2013), 1–8.

[60] Sheref S. Mansy, Jason P. Schrum, Mathangi Krishnamurthy, Sylvia Tobé, Douglas A. Treco, and Jack W. Szostak, "Template-Directed Synthesis of a Genetic Polymer in a Model Protocell," *Nature 454* (July 2008), 122–126.

[61] David W. Deamer, "How Leaky Were Primitive Cells?" *Nature 454* (July 2008), 37–38.

[62] Rajamani et al., "Lipid-Assisted Synthesis," 58–59.

[63] Ibid., 71.

[64] Felix Olasagasti, Hyunsung John Kim, Nader Pourmand, and David W. Deamer, "Non-Enzymatic Transfer of Sequence Information Under Plausible Prebiotic Conditions," *Biochimie 93 (2011), 556–561.*

[65] Ibid., 556.

[66] Ibid., 557.

[67] Ibid., 557.

[68] M. Eigen and P. Schuster, *The Hypercycle—A Principle of Natural Self-Organization* (Heidelberg: Springer, 1979).

[69] Iris Fry, *The Emergence of Life on Earth* (New Jersey: Rutgers University Press, 2000), 109.

[70] Manfred Eigen with Ruthild Winkler-Oswatitsch, *Steps Towards Life: A Perspective on Evolution* (New York: Oxford University Press, 1996).

[71] Stuart A. Kauffman, "Autocatalytic Sets of Proteins," *J. Theor. Biol. 119* (1986), 1–24.

[72] Stuart Kauffman, *At Home in the Universe* (New York: Oxford University Press, 1995), 47.

[73] Stuart A. Kauffman, *The Origins of Order* (New York: Oxford University Press, 1993), 299.

[74] Jacob N. Israelachvili, D. John Mitchell, and Barry W. Ninham, "Theory of Self-Assembly of Hydrocarbon Amphiphiles Into Micelles and Bilayers," *J. Chem. Soc., Faraday Trans. 2 Vol.72* (1976), 1525–1568.

CHAPTER FIFTEEN

The Passover Question: Why Is This Origins Proposal Different From All Other Proposals?

Like smashing a champagne bottle against the bow of an ocean liner, the 1950s christened the two storylines in our tale of DNA nanoscience. In 1953, Watson, Crick, Wilkins, and Franklin brought the molecular structure of DNA to center stage. That gave Nadrian Seeman the foundation on which to build his structural DNA nanotechnology and gave Bellini and Clark the custom-made nucleic acid sequences with which to indulge their curiosity. Coincidently, 1953 also saw Stanley Miller publish his experimental findings that launched the scientific study of the origins of life. Let's go back to the early fifties once more, but this time to look at the commonplace, not the extraordinary.

You're a young child. School is out. It's raining. What to do? (Remember now, this is the mid-twentieth century. It may be hard to fathom, but *there are no video games.* The available options might be said to be a bit simpler—not necessarily better because of a wistful nostalgia for the past. But the choices available to while away the time are different.) So, again, the question: what to do? Mr. Potato Head. Hmm. Finger Painting. Hmm. Silly Putty. Nah. Pick-up-sticks. Uh-uh. How about blowing bubbles? Mother helps. A small bowl of water. Add a little dishwashing liquid. Not quite enough. Add a little more. And again a little more. Pick up the bubble wand, the stick with the ring at the end. Dip the ring into the soap solution. Blow. Wondrous.

If we didn't know by observation that soap bubbles can self-assemble, how could we anticipate that they would suddenly appear? We know we need to get to a certain concentration of soap molecules in solution. Then we need to create a flow of energy—we blow. Voila. We have produced an *emergent* phenomenon, a phenomenon that exhibits self-organization.[1]

Emergence and Broken Symmetry

Consider several points in this simple example. There is a critical concentration of molecules (soap, in this case) and there is a flow of energy (breath). Implicit too is some sort of interaction between the soap molecules. And to complete our description of *emergence,* we add one more idea: the cycling of the energy flow. This is a subtle but profound concept. The distinguished Yale University biochemist Harold Morowitz contributed this idea, his cycling theorem as it is called, a concept that some scientists refer to as the fourth law of thermodynamics.[2] We note in passing that Morowitz's seminal paper in the *Journal of Theoretical Biology* had a single reference: work done by none other than Lars Onsager (prior to his liquid crystals papers). Examples of the cycling of the flow of energy abound in the natural world and—most germane to our interest—they were present on the primordial Earth. Cycles of temperature variation that resulted in sequences of freezing and thawing; cycles of wetting and drying that led to concentrations of molecules from initially dilute solutions; diurnal/nocturnal cycles; the cycles of the tides. Stash these descriptors of emergent phenomena in an accessible place in your thoughts.

The discussion of emergence has been around for a long time. Emergence was implicit in Darwin's theory of biological evolution.[3] More recently, physicist and Nobel laureate Philip W. Anderson published a landmark paper on the subject titled "More is Different."[4] Anderson described the makings of a general theory of how entirely new properties appear as systems become more complex. He wrote about a theory of broken symmetry that addresses those changes that require new laws or concepts when we shift from quantitative to qualitative differentiation. The theory of broken symmetry is not a topic suitable for this book. However, in the context of the subject of liquid crystals we will briefly touch on symmetry and symmetry breaking.

One of the important characteristics of liquid crystal phase transitions is the symmetry of the phases on each side of the phase transition.[5] The symmetry of the phase is determined by how that phase can be translated, rotated, and so on, and still look exactly the same. For instance, in the isotropic phase there is continuous rotational symmetry about every possible axis. If you think about a liquid crystal in its isotropic phase as an ordinary fluid, it's evident that you can, in principle, rotate a portion of the isotropic liquid about any axis you choose and the phase will look exactly as it did before the rotation.

Now consider the nematic phase where there is a degree of orientational order, a preferred direction for the molecules. If you consider the axis that represents that preferred direction, then yes, you can rotate the nematic phase through any angle about that particular axis and not change the phase at all. But rotation about any other axis usually results in a nematic sample with its preferred orientation pointing in a different direction. Because the phase is no longer identical after the rotation, continuous rotational symmetry does not exist about that other axis—that is, about any other axis other than the one that defines the preferred direction of the molecules. So when a phase transition occurs from isotropic to nematic, some of the symmetry that was initially present is lost. In fact, this is a general feature of most phase transitions; a moment's reflection will convince you that a loss of symmetry also occurs, in a transition from the nematic phase to the columnar phase. This general feature of phase transitions is referred to as spontaneously broken symmetry. Cornell University biophysicist James P. Sethna put it nicely: "In nematic liquid crystals, the broken rotational symmetry introduces an orientational elastic stiffness (it pours, but resists bending!)."[6]

About-Face

In the context of the Bellini/Clark discoveries, the liquid crystal state is an environment of broken symmetry and the fluidity of this setting enables templating to emerge. The molecules that are selected in such a process necessarily reflect the ordering of the liquid crystal phase. Since liquid crystal ordering must break the orientational symmetry (in the nematic phase) and, in some cases, the translational symmetry (in the columnar phase) of the isotropic liquid, we find that chemistry within the liquid crystal phase is templated by the broken symmetry of the phase. In essence, the liquid crystal phase condensation selects the appropriate molecular components and ligation in the condensed phase then produces elongated molecules that stabilize the liquid crystal. This is liquid crystal autocatalysis—*unknown prior to Bellini/Clark*—whereby a liquid crystal templates the structure of its own molecules.

Their conclusion is that the linear polymer shape of DNA (and RNA) is itself a vestige of templating by liquid crystal order—*a total about-face of twentieth century wisdom*. The entrenched belief is: because of their linear, chain-like shape, our molecular carriers of genetic information can form liquid crystal phases.

Emergence Processes by which more complex systems arise from simpler systems sometimes spontaneously and in an unpredictable fashion. Emergent phenomena exhibit self-organization. Life is sometimes cited as the quintessential emergent phenomenon.

Symmetry breaking Symmetry means uniformity or invariance, the existence of different viewpoints from which a system appears the same. Symmetry breaking is the process by which such uniformity is broken.

Spontaneously broken symmetry A loss of symmetry that occurs without external intervention.

Template (verb) To act as a template.

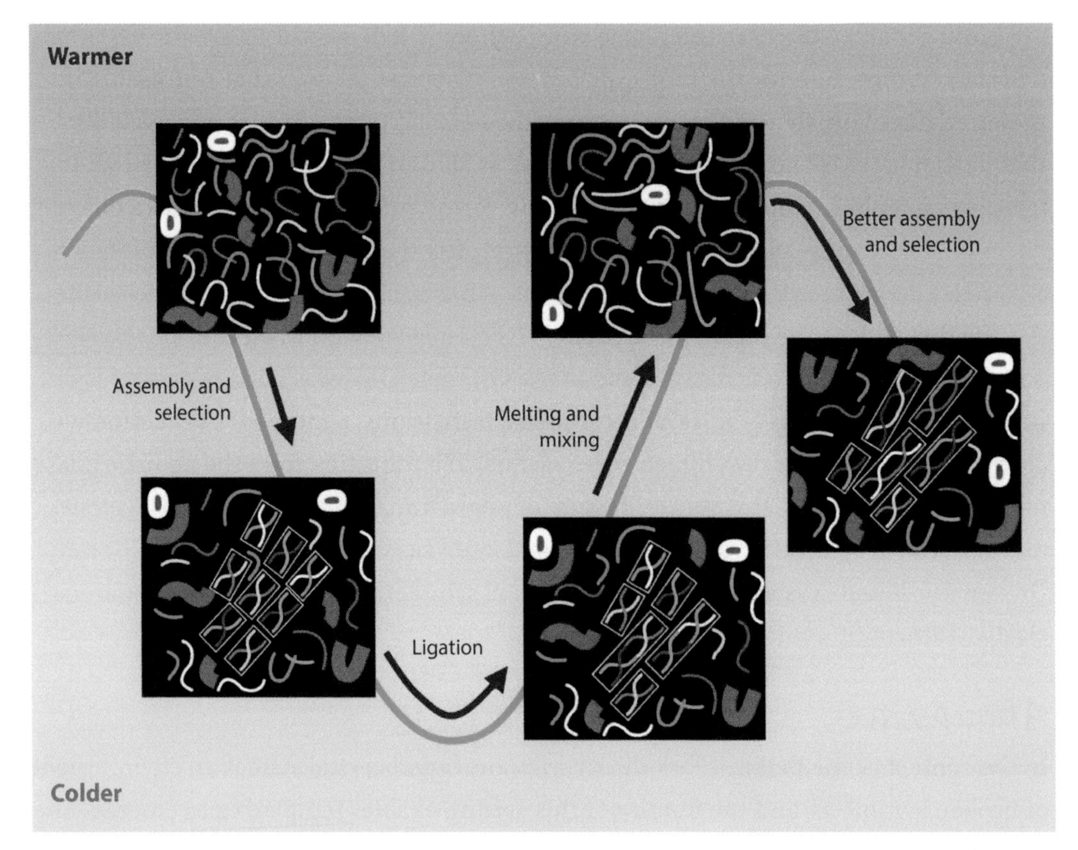

Figure 15.1 Prebiotic scenario. Bellini and Clark envision that from the prebiotic clutter, single-stranded nucleic acids with complementary base pairs will hybridize and segregate into liquid crystal phases as temperatures lower—a part of the cyclical flow of energy through the system. When the physically bound but chemically segmented chains are thus concentrated, liquid crystalline self-assembly promotes covalent ligation of duplexes, shown as elongated duplex helices. The liquid crystal assembly itself serves as a catalytic element promoting ligation (liquid crystal autocatalysis). As energy flow cycling continues, the process is iterated many times. Ligation within the liquid crystal phase eventuates in long nucleic acid polymers that further stabilize that phase. *(Courtesy of Tommaso Bellini.)*

The discoveries of Bellini and Clark are quintessential examples of emergent phenomena. How many emergences comprise their process of hierarchical self-assembly? We have previously called out the stacking of nanoDNA duplexes, the phase separation/condensation of duplexes into liquid crystal droplets, and so on. In fact, as Harold Morowitz observed, "There is . . . a certain arbitrariness in counting emergences. . . . The unfolding of life involves many, many emergences. . . ."[7] Figure 15.1 illustrates our discussion and also incorporates Morowitz's cycling of energy flow that is characteristic of emergence.[8]

Occam's Razor

How is the proposal of Bellini and Clark different from other proposals? One reflection is this: there is no other plausible prebiotic hypothesis for the polymerization of mononucleotides to form nucleic acids that is so devoid of baggage, so free of accoutrements for its accomplishment. Occam's razor never gave a smoother shave than with the theory of Bellini and Clark. Occam's razor, also spelled Ockham's razor, is a principle named for William of Ockham, an English philosopher and theologian of the fourteenth century. The principle gives precedence to simplicity; trim off the unnecessary. When two or more theories make the same predictions, the simplest one is the best.[9] Since we're speaking of nucleic acids, Francis Crick's notable caveat to the principle comes to mind. He cautioned that while Occam's razor is a useful tool in the physical sciences, it can be a very dangerous implement in biology.[10] In biology, the simplest possible solution may not be attainable through natural selection/biological evolution. However, Bellini and Clark have proposed a materials science solution to the enigma of nucleic acid formation—and that is part of the uniqueness of their approach.

One more observation is in order before moving on. We might best summarize the point with a famous line penned by Stuart Kauffman, the theoretical biologist whose work we briefly mentioned in Chapter Fourteen. "Anyone who tells you that he or she knows how life started on the sere earth some 3.45 billion years ago is a fool or a knave. Nobody knows."[11] Kauffman says that we may never recover the actual historical sequence of molecular events that led to the first self-reproducing, evolving molecular systems. However, we can still develop theories and experiments to show how life might realistically—i.e., in a prebiotically plausible way—have established itself. It is in this spirit that we value the work of Bellini and Clark.

The RNA World Revisited

Taking the RNA World as the primary exemplar of the replicator-first viewpoint, we consider the Bellini/Clark hypothesis in the context of the RNA World. For 40 years chemists attempting to reenact the prebiotic synthesis of the nucleotide constituents of RNA did so based on the assumption that these ribonucleotides must have been assembled from their three molecular components: a nucleobase (adenine, guanine, cytosine, or uracil), a ribose sugar, and phosphate.[12] As we mentioned in Chapter Fourteen, in 2009 chemists Matthew Powner, Béatrice Gerland, and John Sutherland eschewed such an approach in favor of a different pathway.[13]

They considered a synthesis of two of the four nucleotides (the pyrimidine ribonucleotides) in which both the sugar and nucleobase emerge from a common precursor (Figure 15.2).[14] In this pathway, the structure forms without using free sugar and nucleobase molecules as intermediates. Equally satisfying, they found that ultraviolet radiation destroyed the majority of undesired by-products of the reaction sequence but the desired ribonucleotides were not destroyed by the ultraviolet radiation. This happens because of a protective

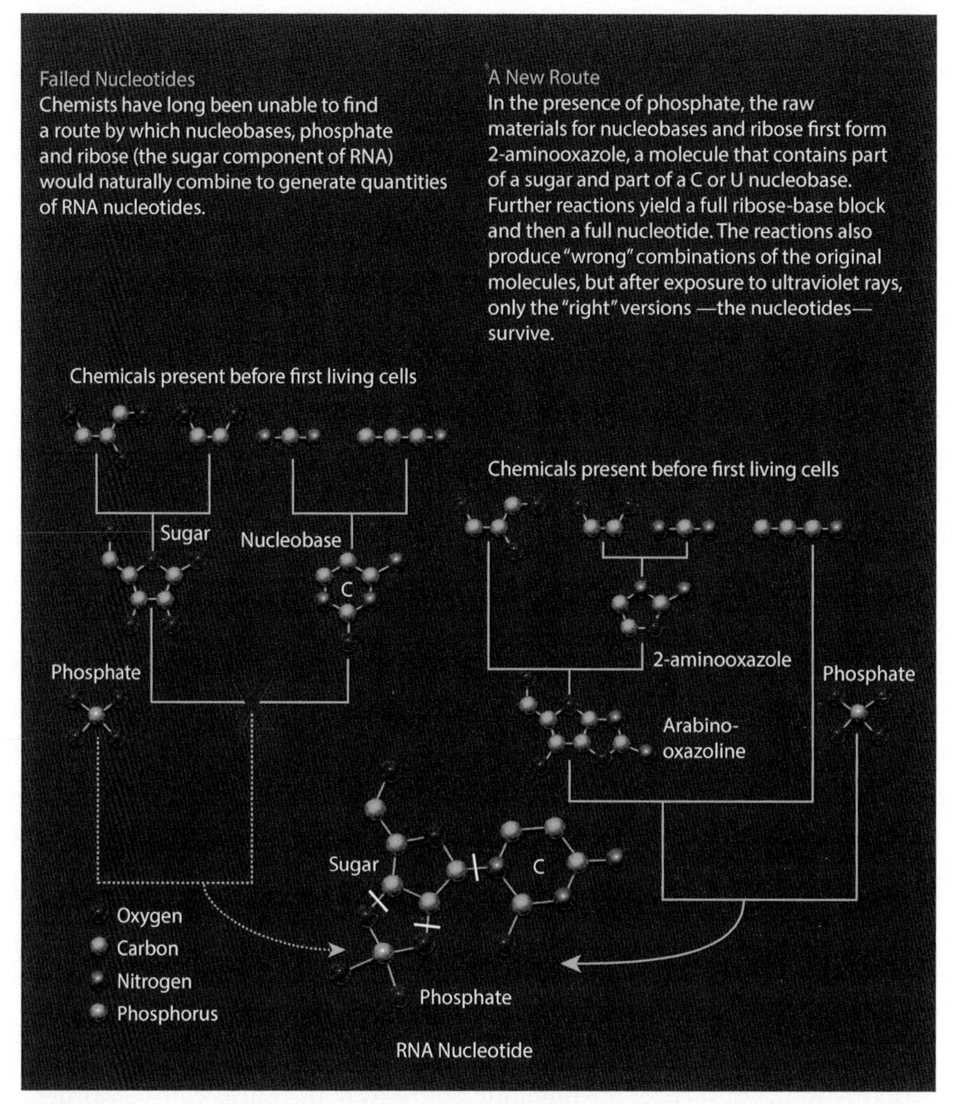

Figure 15.2 A team of researchers have recently found a means to form two of RNA's four nucleotides (those incorporating cytosine and uracil) under prebiotically plausible conditions, a synthesis that had eluded chemists for decades. *(Reproduced with permission from Alonso Ricardo and Jack W. Szostak, "Origin of Life on Earth," Scientific American 301, No. 3 (September 2009), 54–61 and artist Andrew Swift.)*

photochemical mechanism that serendipitously operates in the case of the desired end products but does not operate with the undesired by-products. Thus, the two activated pyrimidine ribonucleotides necessary for RNA synthesis can be enriched relative to the other end products of the process. The researchers have proposed a potential mechanism to account for this selective protection.[15]

This breakthrough resulted in a solution to the vexing problem of prebiotic ribonucleotide synthesis in the case of the pyrimidine nucleotides. Nobelist Jack Szostak said of this research, "it will stand for years as one of the great advances in prebiotic chemistry."[12] Such outside-the-box thinking has also led Powner, Sutherland, and Szostak working together to find a prebiotically plausible pathway to precursors of the purine ribonucleotides, those incorporating the bases guanine and adenine.[16–18] Moreover, Sutherland, Powner, and colleagues have recently demonstrated prebiotically plausible oligoribonucleotide ligation.[19] These are all impressive results. However, it's time to step back and examine a phrase that has appeared numerous times in this book.

Sticky Business, Part I: What Constitutes Plausible Prebiotic Conditions?

Albert Eschenmoser of the Swiss university ETH Zürich is widely regarded as a master of organic chemistry.[20] Speaking in general terms of the current state of prebiotic chemistry, he recently observed, "As to whether such experiments are relevant to our conceptions of the origin of life—in distinct contrast to the early days of prebiotic chemistry—there no longer is any general sense of agreement."[21] Specifically, Eschenmoser praised the work of Sutherland and colleagues but expressed reservations as to its etiological relevance by which he meant the applicability of the results to the origin of life.[22] Others have raised similar questions.

Plausible prebiotic conditions Chemical or environmental precursors to the origin of life are spoken of as prebiotic conditions. In the laboratory, scientists try to simulate what they believe to be such conditions and to determine which are the most conducive to the self-assembly processes that led up to the origin of life.

Noted Georgia Institute of Technology chemist Nicholas Hud considers the experiments of Sutherland and colleagues to be "elegant as a synthetic organic achievement." But he points to the fact that the synthesis involves the sequential addition of two reactive carbohydrates—glycolaldehyde and glyceraldehyde—and the synthesis fails if these precursor molecules are added at the same time or in a different order. Hud points out that quantities of each molecule are required, and because both would be present through known reaction schemes, "it is challenging to rationalize how high-yielding separation and ordered timely addition could be realized (i.e., through geophysical means)."[23] Other origins researchers have also wondered aloud about the prebiotic legitimacy of the work.[24,25] It should be noted that the Sutherland team acknowledged concern in their original work: "the issue of temporally separated supplies of glycolaldehyde and glyceraldehyde remains a problem."[26]

The point is that, as Eschenmoser has said, there is not a consensus on the thorny issue of what is prebiotically plausible. And don't think for a minute that this subject only comes up in the work of Powner, Sutherland, and colleagues. For instance, in the work of Bellini and Clark, some have questioned the high concentration of nucleic acids that are required for liquid crystal formation.[27] However, as we said in Chapter Thirteen, Bellini and Clark view their contribution as a definitive link in a missing chain. The question of the necessary concentration of appropriate molecular feedstock inherent in their proposal is part of the missing chain.

It seems that any investigation of the origins topic, however adroit, will have both supporters and critics. It certainly adds to the challenge for both laymen and scientists to evaluate the merits of competing experimentally based theories. And that leads to a second point: the differing perspective of scientists from different disciplines.

Sticky Business, Part II: The Origins Question—Whose Home Turf Is It?

Of course, the notion of home turf is fatuous. The origins puzzle is open to all comers and the quest benefits from the diversity of the backgrounds—geology, biology, chemistry, physics, astronomy—of those in pursuit. However, more often than not chemistry is regarded as being at the heart of the matter. Consider Philip Ball, a former editor at the journal *Nature* and a prolific science writer who took his Ph.D. in theoretical physics (after taking a B.A. in chemistry).

The United Nations dubbed 2011 as the International Year of Chemistry. Ball wrote an article for *Scientific American* that addressed the ten most profound questions in chemistry.[28] Number one on his list of unsolved mysteries was: "How Did Life Begin?" Ball wrote, "How did relatively simple molecules in the primordial broth give rise to more and more complex compounds? And how did some of those compounds begin to process energy and replicate (two of the defining characteristics of life)? At the molecular level, all of those steps are, of course, chemical reactions, which makes the question of how life began one of chemistry."[29]

One dissenter from this view is a respected expert on the origin of life, David Deamer. We discussed aspects of Deamer's research in Chapter Fourteen in which we went to some length to delineate the distinctions between his work and that of Bellini and Clark. We can now look at the convergence of Deamer's viewpoint and that of the Milan/Boulder collaboration.

Deamer, who is himself a chemist, has said that because life as we know it is so much a phenomenon of chemistry, it has been mostly chemists who are attracted to the question of how life began. He observed that chemists perceive the origin of life as a chemical process. As the first microscopic organisms started to grow and reproduce on the early Earth, chemical reactions associated with growth, metabolism, and replication were paramount to much of what we perceive as being alive. Deamer asked the rhetorical question, "But how did that chemistry begin?" His reply: "I believe the answer will be found in the realm of physics, and more specifically biophysics, defined as the physical processes that we now associate with the living state. *The chemistry of life only became possible after certain physical processes permitted specific chemical reactions to occur.* Life can emerge where physics and chemistry intersect"[30] [italics added].

Discovering the Physical Processes That Enabled the Chemistry of Life

When writing about different versions of the RNA World hypothesis, accomplished origins researcher Robert Hazen said, "Sometimes you have to place your bets and put your cards on the table."[31] We will borrow Hazen's line and place our bet on Deamer's aforementioned insight. Specifically, we will place our bet that Bellini and Clark *have discovered the physical processes that permitted specific chemical reactions to occur.* For well over 40 years chemists have attempted to achieve polymerization of nucleotides to form the nucleic acids RNA or DNA under plausible prebiotic conditions. Success has eluded them.

In 1992, Stanley Miller (of the Miller-Urey experiment) said this: "The first step, making the monomers, that's easy. We understand it pretty well. But then you have to make the first self-replicating polymers. That's *very* easy. Just like it's easy to make money in the stock market—all you have to do is buy low and sell high. [He laughs.] Nobody knows how it's done."[32]

The strategy of Bellini and Clark in realizing the polymerization of nucleic acids is as outside the box as that of Powner and his colleagues (in synthesizing the pyrimidine ribonucleotides) and equally successful in accomplishing its objective. In their facile synthesis of nucleic acid polymers, Bellini and Clark do not use activated nucleotides, nor do they rely on reactive mineral surfaces, nor are any of the mechanisms that lead to polymerization left unexplained. Following on the heels of Powner et al., if (all four) monomeric nucleotides existed in sufficient concentrations on the prebiotic Earth, then the Bellini and Clark scenario maintains that these nucleotides will hybridize to form double-stranded nanoDNA. (The same step applies in the case of nanoRNA but with the nucleobase uracil replacing thymine.) These duplexes stack on one another and phase separate to form liquid crystals, leaving single-stranded oligomers in solution. The liquid crystal condensation brings the duplex stacks into close physical proximity, and subsequent chemical ligation results in elongated duplexes. The autocatalytic process is relentlessly reiterated to the rhythm of cyclic energy flow—and there we are, back at the prebiotic scenario of Figure 15.1.

Metabolism-First Revisited

We next examine how the Bellini and Clark work might fit in the metabolism-first school of thought. To enrich our inventory of this point of view, we provide a brief summation of a metabolism-first thesis (with a caveat in endnote 33) that we did not cover in Chapter Fourteen. It is the Thioester World proposed by Belgian chemist and Nobelist Christian de Duve.[33] His hypothesis centers on the bond between sulfur and a carbon-containing

Another scenario for the origin of life: Thioester World
Sulfur bonds contain chemical energy to drive metabolic reactions.

entity called an acyl group, which yields a compound called a thioester.[34] Thioesters play an important role in present-day metabolic pathways and were thought to be common in the primordial soup of ancient Earth. De Duve contends that they could have played an important role in the development of a primordial metabolism, which he speaks of as a protometabolism. The thioester bond is what biochemists call a high-energy bond, and can be thought of as similar to the phosphate bonds in adenosine triphosphate (ATP), the main source of energy in all living organisms. Thus, the Thioester World represents a posited early stage in the development of life that could have provided the energetic (and catalytic) framework of a protometabolic set of primitive chemical reactions.

But regardless of the belief that metabolism came first, one must still account for the development of a replicator—and the emergent steps that led to its development. That is, whether one advocates replicator-first or metabolism-first, the critical step for the advent of RNA as the information carrier lies in its elongation from single nucleotides, or oligomers, to the long polymer we know nowadays.[35] De Duve has been eloquent on this subject.

"How RNA could possibly have emerged from the clutter without a 'guiding hand' would baffle any chemist; it seems explainable only by selection, a process that presupposes replication. . . . The need seems inescapable for some autocatalytic process such that each lengthening step favors subsequent lengthening. Only in this way could the enormous kinetic obstacle to chain elongation be surmounted. . . . Any invoked catalytic mechanism must accommodate the participation of a template, for there can have been no emergence of true RNA molecules without replication."[36]

Bellini and Clark have responded to De Duve's reasoning with the phenomena we've previously described for nanoDNA and nanoRNA. These phenomena are base pairing, duplex stacking, phase segregation of single-strands and duplexed strands, liquid crystal ordering of duplexes, and enhanced chemical ligation of stacked duplexes within the liquid crystalline environment. They provide at a single shot (1) a mechanism of selection, because only paired strands segregate to form liquid crystals; (2) a locally enhanced concentration of complementary strands and proximity of terminals within the confines of the liquid crystal; (3) a liquid crystalline matrix that provides a flexible template, favoring linear aggregation and chemical ligation; and (4) an autocatalytic process by which longer and longer complementary sequences are favored, because they stabilize liquid crystal ordering.[37] Thus, the Milan/Boulder collaboration argues that modern nucleic acids emerged as the most capable of self-structuring via a liquid crystal–mediated cascade of pairing, stacking, and elongation that led to self-replication and information storage.

Computer Simulations and Mathematical Modeling

Computer and mathematical modeling lends support to Bellini and Clark's body of work. When speaking of emergence earlier in this chapter we introduced theoretical physicist Philip W. Anderson. He wrote a brief opinion piece on computers and emergence in which he said that the basic laws of physics, though not incorrect, are inadequate.[38] The reason he gave was emergence: the constituents of a sufficiently large and complex system self-organize into arrangements that one could never deduce a priori even though the laws of physics are obeyed. (The obvious example is life.) And in an attempt to cope with such problems, Anderson continued, some scientists try to follow all the atoms or electrons as they interact using massive computer simulations. He also made the provocative remark that, in addition, they use various assumptions and tricks that tend to predetermine the outcome. He concluded his critique by counseling that a simulation, even if correct, can't really prove anything.

Anderson's remarks serve as something of a cautionary introduction before we present computer simulation results that have been done in the wake of the Bellini and Clark results. Anderson's caveats notwithstanding, carefully performed simulations can be of great value, complementing and informing experimental studies. In fact, because computer chips are silicon based, it has become fashionable to employ the term *in silico* to refer to a study performed on computer or via computer simulation, by analogy with the terms *in vivo, in vitro,* and *in situ*. Several different computational publications have appeared that examine the Bellini/Clark experimental discoveries.[39–42] We will sample the work of chemist Aleksei Aksimentiev whose group is in the physics department of the University of Illinois at Urbana-Champaign.[40]

Aksimentiev and his colleagues used a molecular dynamics simulation to characterize the end-to-end interactions of duplex nanoDNA. They quantitatively described the forces, free energy, and kinetics of the end-to-end association process for both blunt-ended and sticky-ended nanoDNA duplexes. Confirming Bellini and Clark's experiments, they found that nanoDNA duplexes spontaneously aggregate end-to-end. Initially, Aksimentiev and colleagues described the interaction of two DNA fragments in isolation. They then simulated multifragment aggregation. In Figure 15.3 we see the result of a simulation that begins with 458 duplex nanoDNA fragments (each 10 base pairs in length) as shown in panel (a). These spontaneously aggregate into long rod-like structures, shown in panels (b) and (c), within a salt solution that is shown as a semi-transparent cube 23.8 nm on each side. For those interested, there are movies showing the molecular dynamics simulations and the last of these Supplementary Data mpeg files shows a movie of the

Molecular dynamics simulation
A computational technique that allows one to predict the time evolution of a system of interacting particles (atoms, molecules, etc.). The method can also be used to explore macroscopic properties of the system by averaging over all the atoms or molecules in the system.

simulation trajectory that is captured in still images in Figure 15.3.[43]

In Figure 12.7 of Chapter Twelve we saw base pairs stacked like flat plates and we tackled base-stacking forces in a given duplex. Aksimentiev's simulation provides the first direct estimate of the base-stacking free energy in the end-to-end association of nanoDNA. His results are in agreement with the estimate of Bellini and Clark obtained by other means.[44,45] The base-stacking free energy is an important parameter because the end-to-end adhesion of nanoDNA duplexes is promoted by a delicate combination of pairing and stacking forces. Moreover, simulations such as Aksimentiev's corroborate a key element of the self-assembly processes that Bellini and Clark have discovered—the end-to-end aggregation of duplex nanoDNA into rodlike structures.

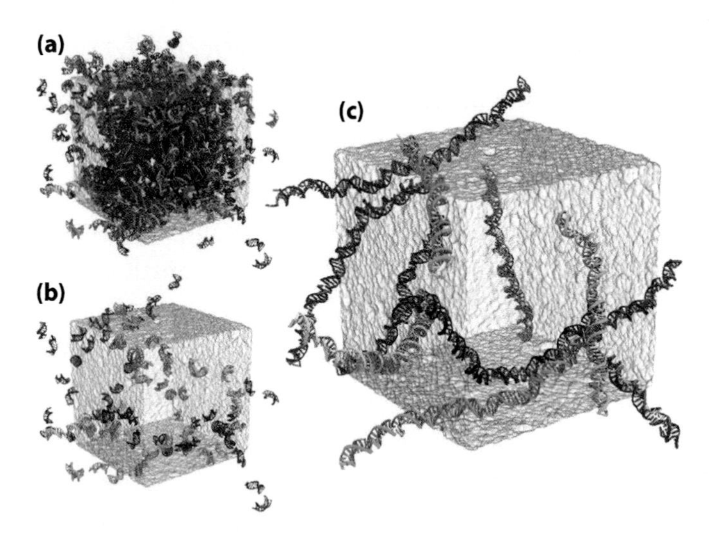

Figure 15.3 Molecular dynamics computer simulation of the spontaneous aggregation of duplex nanoDNA into long rod-like structures. The simulation partially mimics the experimental assay of Bellini and Clark. *(Reproduced with permission from Christopher Maffeo, Binquan Luan, and Aleksei Aksimentiev, "End-to-End Attraction of Duplex DNA," Nucleic Acids Research 40, No. 9 (2012), 3812–3821.)*

An Ancient "Liquid Crystal World"

In 1989, Thomas R. Cech and Sidney Altman shared the Nobel Prize for Chemistry for their discovery of the catalytic properties of RNA. In making their announcement, The Royal Swedish Academy of Sciences said, "This discovery, which came as a complete surprise to scientists, concerns fundamental aspects of the molecular basis of life. Many chapters in our textbooks have to be revised."[46]

In that same year, M. Mitchell Waldrop, a physicist working as a journalist, wrote an article entitled "Did Life Really Start Out in an RNA World?"[47] In that piece he spoke about unease within the community of origin-of-life researchers regarding the conventional wisdom of the RNA World view. "The fact is that the RNA-World scenario fails a crucial plausibility test: No one has yet figured out where the RNA itself came from."[48] More than twenty-five years later there is still no consensus on where the RNA came from. Bellini and Clark believe it may have come from an ancient "Liquid Crystal World."[49]

Earlier in this chapter, in *The RNA World Revisited,* we saw how the basic constituents of nucleotides may have been available in the prebiotic environment.[15] They may have even joined to form oligomers under appropriate conditions.[50,51] But recall the ribozymes we spoke of in previous chapters, the RNA molecules that perform catalytic activity. The appearance of chains having the length and chemical homogeneity required to make functional ribozymes—at sufficient concentration and quantity to act as an effective feedstock for the first self-sustaining cycles—remains without convincing explanation. According to current knowledge, the required length is more than 30 bases.[52] Bellini and Clark have confronted this conundrum.

They contend that the onset of the RNA World had to be anticipated by a more ancient series of developments. Such developments were necessary to enable the formation of nucleotide-like chains long enough to adopt a secondary structure and become catalytically active.[53] Some researchers describe conditions that were favorable for local high concentrations of nucleic acid precursors.[50,54,55] Even so, Bellini and Clark maintain that random ligation would not have produced molecules with sufficient homogeneity.[56] The awareness of this need prompted many origins scientists to search for substrates acting as templates to guide ligation of different nucleotides.[50,57–59] This is analogous to modern enzymes whose efficiency comes from selective recognition of molecular species as well as constraint into physical contact, both acting to enhance reaction rates. However, the Milan/Boulder team has forged a different path (Figure 15.4).

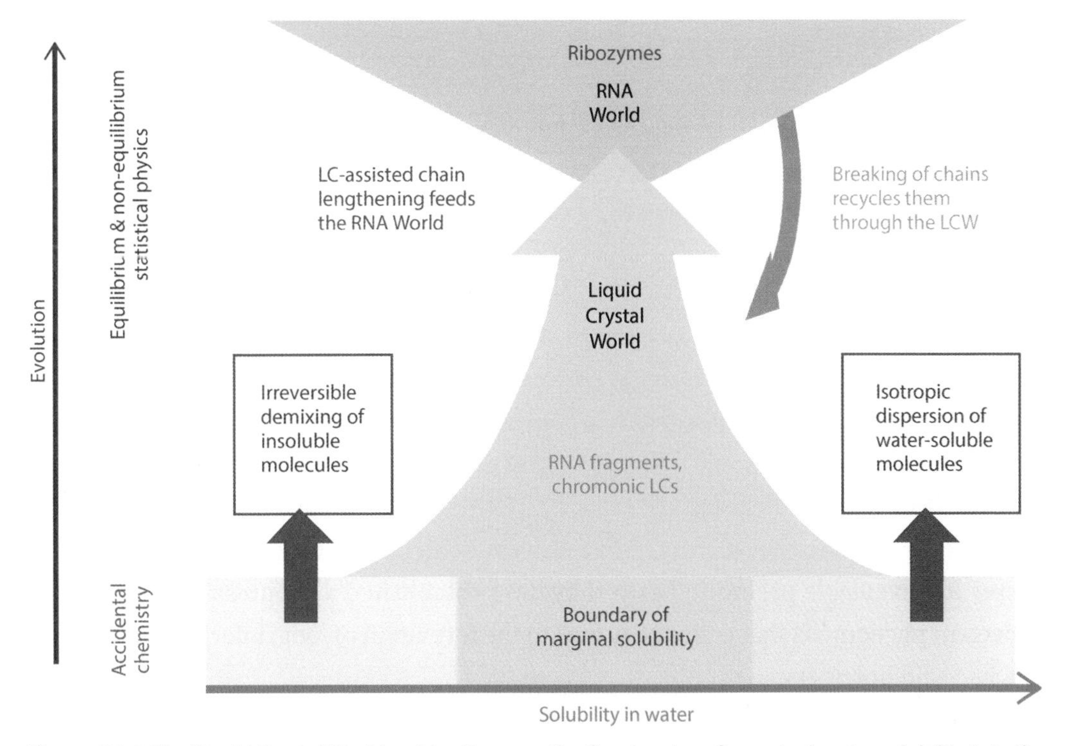

Figure 15.4 The Liquid Crystal World guides the growth of molecules of marginal water solubility into the fluid-structured environment of a liquid crystal phase. This holds the molecules in relative positions that promote ligation into linear chains. These chains provide a continuous production of molecular candidates for the RNA World. *(Courtesy of Noel A. Clark.)*

They have put forward the Liquid Crystal World. This preceded and gave rise to the RNA World, providing its basic molecular design and feedstock. In this way, they've introduced autocatalysis at an earlier prebiotic stage (Figure 15.4). A term that appears in the figure is "chromonic" liquid crystals. Chromonic phases are lyotropic liquid crystals (Chapter Four) formed by soluble molecules such as drugs, dyes, and nucleic acids.[60] In early publications they were pictured as molecules aggregated like piles of pennies or stacks of cards.[61] The molecular structure of nanonucleic acids makes them chromonic. This puts them at the boundary of marginal solubility (they are easy to melt and to efficiently mix).[62] They must then form collective assemblies or be bonded to a water-soluble chain in order to stay in solution.

Liquid crystal ordering condenses such molecules, selecting those compatible with an overall structural theme. Such ordering incorporates the molecules into a soft but highly structured low entropy environment where ligation can occur. An autocatalytic feedback loop forms in which longer chains stabilize the liquid crystal phase and vice versa. Bellini and Clark propose the Liquid Crystal World as providing the feedstock for the RNA World. Perhaps textbook chapters will have to be revised once again.

In one of his last papers, published posthumously in 2008, British chemist and esteemed origins researcher Leslie Orgel of the Salk Institute critiqued origins theories based on metabolic cycles.[63] His appraisal was particularly concerned with the metabolism-first theories of Wächtershäuser and of theoretical biologist Stuart Kauffman, both of whom we met in Chapter Fourteen. He felt that neither scheme was likely to describe any system relevant to the origin of life. Although Orgel was of the replicator-first school of thought, he concluded his article with the even-handed comment that "solutions offered by supporters of geneticist [replicator-first] or metabolist [metabolism-first] scenarios that are dependent on 'if pigs could fly' hypothetical chemistry are unlikely to help."

The Bellini and Clark premise is the very antithesis of an "if pigs could fly" proposition. It is entirely evidence based, rooted in a succession of relevant, systematic, measurable steps. In a plausible prebiotic Earth they have established a progressive hierarchy of emergent phenomena that collectively lead to the formation of long, information-bearing nucleic acid polymers.

Figure 15.5 is a famous painting by artist Peter Sawyer. This depiction of the Archean World makes its home in the Smithsonian Institution National Museum of Natural History. (*Photosynthetic life is depicted in the mural, so the scene is not prebiotic.*) It gives one pause to look at the mural and realize that Bellini and Clark can travel back in time

Figure 15.5 The Archean World, a mural by artist Peter Sawyer and housed in the Smithsonian Institution National Museum of Natural History. The scene depicts the Earth about 3.5 billion years ago. In this representation, early photosynthetic life exists and is represented in the form of stromatolites. These are the rock-like structures formed by photosynthetic microbial mats in which layers of either sand or precipitated minerals are present. Other algal life forms lend a green tint to the waters of hot springs. Thus, the painting is of an era that followed the prebiotic period. *(Reproduced with permission from the Smithsonian Institution, National Museum of Natural History.)*

some hundreds of millions of years prior to that world via their thoughtful examination of tiny glass cells filled with liquid crystals made up of nanoDNA. And their journey began with simple curiosity-driven research.

In 2006, three future Nobelists, molecular biologists Elizabeth Blackburn of the University of California, San Francisco, Carol Greider of Johns Hopkins University, and biochemist/molecular biologist Jack W. Szostak (whose name has come up before in our tale) co-authored a paper. A lengthy quote from that publication is pertinent to the work of Bellini and Clark. "Our understanding of the way the world works is fragmentary and incomplete, which means that progress does not occur in a simple, direct and linear manner. It is important to connect the unconnected, to make leaps and to take risks, and to have fun talking and playing with ideas that might at first seem outlandish. *Fundamental advances are often stimulated by unexplained observations made in the course of applied work,* and the resulting growth in understanding directs our attention to new avenues of research. This cycle of interaction between fundamental and applied research is autocatalytic and contributes greatly to the explosive increase in scientific knowledge. If the cycle is broken, through a failure to recognize the importance of basic science, the continuation of progress in more applied domains of science, medicine and engineering will surely be limited"[64] [italics added].

Recall the title of Chapter Eleven. *Chance Findings.* Like children at play with their bubble wands, Bellini and Clark have blown a string of emergent bubbles. These emergences are arresting. And they have striking implications. They connect the unconnected; the origin of life on Earth and the hierarchical self-organization of tiny threads of matter like those found in our every cell. They whisper of being at one with the universe.

ENDNOTES

[1] David Deamer, Foreword, in Robert M. Hazen, *genesis: The Scientific Quest for Life's Origins* (Washington, DC: Joseph Henry Press, 2005), ix–x.

[2] Harold J. Morowitz, "Physical Background of Cycles in Biological Systems," *J. Theoret. Biol. 13* (1966), 60–62.

[3] Andrew Assad and Norman H. Packard, "Emergence," in *Emergence: Contemporary Readings in Philosophy and Science,* ed. Mark A. Bedau and Paul Humphreys (Cambridge: Massachusetts Institute of Technology, 2008), 231.

[4] P.W. Anderson, "More Is Different: Broken Symmetry and the Nature of the Hierarchical Structure of Science," *Science 177* (4 August 1972), 393–396.

[5] Peter J. Collings, *Liquid Crystals: Nature's Delicate State of Matter* (Princeton, New Jersey: Princeton University Press, Second edition, 2002), 174–175.

[6] James P. Sethna, "Order Parameters, Broken Symmetry, and Topology," in *Lectures in Complex Systems, Santa Fe Institute Studies in the Sciences of Complexity,* Proc. Vol. XV, ed. L. Nagel and D. Stein (Boston: Addison-Wesley, 1992), 5.

[7] Harold Morowitz, *The Emergence of Everything: How the World Became Complex* (New York: Oxford University Press, 2004), 84.

[8] Tommaso Bellini, "Exploring Soft Matter With DNA," *Frontiers of Soft Matter 2012 Symposium* (18 May 2012), Boulder, CO, slide 33.

[9] http://www.britannica.com/topic/Occams-razor

[10] Francis Crick, *What Mad Pursuit: A Personal View of Scientific Discovery* (New York: Basic Books, 1990), 138.

[11] Stuart Kauffman, *At Home in the Universe* (New York: Oxford University Press, 1995), 31.

[12] Jack W. Szostak, "Systems Chemistry on Early Earth," *Nature 459* (14 May 2009), 171–172.

[13] Matthew W. Powner, Béatrice Gerland, and John D. Sutherland, "Synthesis of Activated Pyrimidine Ribonucleotides in Prebiotically Plausible Conditions," *Nature 459* (14 May 2009), 239–242.

[14] Alonso Ricardo and Jack W. Szostak, "Origin of Life on Earth," *Scientific American 301, No. 3* (September 2009), 54–61.

[15] Matthew W. Powner, Béatrice Gerland, and John D. Sutherland, "Synthesis of Activated Pyrimidine Ribonucleotides in Prebiotically Plausible Conditions," *Nature 459* (14 May 2009), Supplementary Information, 28–33 (linked to the online version of the paper as a PDF download).

[16] Matthew W. Powner, John D. Sutherland, and Jack W. Szostak, "Chemoselective Multicomponent One-Pot Assembly of Purine Precursors in Water," *J. Am. Chem. Soc. 132* (2010), 16677–16688.

[17] Erratum: Matthew W. Powner, John D. Sutherland, and Jack W. Szostak, "Chemoselective Multicomponent One-Pot Assembly of Purine Precursors in Water," *J. Am. Chem. Soc. 133* (2011), 4149–4150.

[18] Matthew W. Powner, John D. Sutherland, and Jack W. Szostak, "The Origins of Nucleotides," *SYNLETT 14* (2011), 1956–1964.

[19] Frank R. Bowler, Chistopher K.W. Chan, Colm D. Duffy, Béatrice Gerland, Saidul Islam, Matthew W. Powner, John D. Sutherland, and Jianfeng Xu, "Prebiotically Plausible Oligoribonucleotide Ligation Facilitated by Chemoselective Acetylation," *Nature Chemistry 5* (May 2013), 383–389.

[20] Bernhard Kräutler, "Congratulations to Professor Albert Eschenmoser on His 85th Birthday," *Heterocycles 82, No. 1* (2010), 1–4.

[21] Albert Eschenmoser, "Etiology of Potentially Primordial Biomolecular Structures: From Vitamin B12 to the Nucleic Acids and an Inquiry Into the Chemistry of Life's Origin: A Retrospective," *Angew. Chem. Int. Ed. 50* (2011), 12412–12472.

[22] Ibid., 12453.

[23] Nicholas V. Hud, Brian J. Cafferty, Ramanarayanan Krishnamurthy, and Loren Dean Williams, "The Origin of RNA and 'My Grandfather's Axe'," *Chemistry and Biology 20* (18 April 2013), 466–474.

[24] Steven A. Benner, Hyo-Joong Kim, and Matthew A. Carrigan, "Asphalt, Water, and the Prebiotic Synthesis of Ribose, Ribonucleosides, and RNA," *Accounts of Chemical Research 45, No. 12* (2012), 2025–2034.

[25] Thomas M. McCollum, "Miller-Urey and Beyond: What Have We Learned About Prebiotic Organic Synthesis Reactions in the Past 60 Years?" *Annu. Rev. Earth Planet. Sci. 41* (2013), 207–229.

[26] Powner, Gerland, and Sutherland, "Synthesis of Activated Pyrimidine," 242.

[27] Eric D. Horowitz, Aaron E. Engelhart, Michael C. Chen, Kaycee A. Quarles, Michael W. Smith, David G. Lynn, and Nicholas V. Hud, "Intercalation as a Means to Suppress Cyclization and Promote Polymerization of Base-Pairing Oligonucleotides in a Prebiotic World," *PNAS 107, No. 12* (23 March 2010), 5288–5293.

[28] Philip Ball, "10 Unsolved Mysteries," *Scientific American 305* (2011), 48–53.

[29] Ibid., 48.

[30] David Deamer, *First Life: Discovering the Connections Between Stars, Cells, and How Life Began* (Berkeley and Los Angeles: University of California Press: 2011), 192.

[31] Robert M. Hazen, *genesis: The Scientific Quest for Life's Origins* (Washington, DC: Joseph Henry Press, 2005), 219.

[32] Peter Radetsky, "How Did Life Start?" *Discover 13, No. 11* (November 1992), 74–82.

[33] It has been said that de Duve's position is not metabolism-first but rather "preparatory metabolism" because metabolism-first ignores his position that a genetic polymer is necessary for the onset of natural selection; Iris Fry, "The Role of Natural Selection in the Origin of Life," *Orig. Life Evol. Biosph. 41* (2011), 3–16.

[34] Christian de Duve, "The Beginnings of Life on Earth," *American Scientist 83* (September–October 1995), 428-437.

[35] Giuliano Zanchetta, "Liquid Crystalline Phases in Oligonucleotide Solutions," Ph.D. dissertation, Università Degli Studi Di Milano (2007), 123.

[36] Christian de Duve, *Singularities: Landmarks on the Pathways of Life* (New York: Cambridge University Press, 2005), 78–80.

[37] Giuliano Zanchetta, "Spontaneous Self-assembly of Nucleic Acids: Liquid Crystal Condensation of Complementary Sequences in Mixtures of DNA and RNA Oligomers," *Liquid Crystals Today 18, No. 2* (2009), 48.

[38] Philip Anderson, "On Computers and Emergence," *Seed Magazine* (9 July 2008), http://seedmagazine.com/content/article/philip_anderson_on_computers_and_emergence/

[39] Tatiana Kuriabova, M.D. Betterton, and Matthew A. Glaser, "Linear Aggregation and Liquid-Crystalline Order: Comparison of Monte Carlo Simulation and Analytic Theory," *J. Mater. Chem. 20* (2010), 10366–10383.

[40] Christopher Maffeo, Binquan Luan, and Aleksei Aksimentiev, "End-to-End Attraction of Duplex DNA," *Nucleic Acids Research 40, No. 9* (2012), 3812–3821.

[41] Cristiano De Michele, Tommaso Bellini, and Francesco Sciortino, "Self-Assembly of Bifunctional Patchy Particles With Anisotropic Shape Into Polymer Chains: Theory, Simulations, and Experiments," *Macromolecules 45* (2012), 1090–1106.

[42] Cristiano De Michele, Lorenzo Rovigatti, Tommaso Bellini, and Francesco Sciortino, "Self-Assembly of Short DNA Duplexes: From a Coarse-Grained Model to Experiments Through a Theoretical Link," *Soft Matter 8* (2012), 8388–8398.

[43] http://nar.oxfordjournals.org/content/40/9/3812/suppl/DC1

[44] Maffeo, Luan, and Aksimentiev, "End-to-End," 3819–3820.

[45] Michi Nakata, Giuliano Zanchetta, Brandon D. Chapman, Christopher D. Jones, Julie O. Cross, Ronald Pindak, Tommaso Bellini, and Noel A. Clark, "End-to-End Stacking and Liquid Crystal Condensation of 6– to 20–Base Pair DNA Duplexes," *Science 318* (23 November 2007), Supporting Online Material, Estimates of the Short DNA-Short DNA Stacking Energy, 11 (linked to the online version of the paper as a PDF download).

[46] http://www.nobelprize.org/nobel_prizes/chemistry/laureates/1989/press.html

[47] M. Mitchell Waldrop, "Did Life Really Start Out in an RNA World?" *Science 246, No. 4935* (8 December 1989), 1248–1249.

[48] Ibid., 1248.

[49] Tommaso P. Fraccia, Giuliano Zanchetta, Valeria Rimoldi, Noel A. Clark, and Tommaso Bellini, "Evidence of Liquid Crystal-Assisted Abiotic Ligation of Nucleic Acids," *Orig. Life Evol. Biosph. 45* (2015), 51–68.

[50] Sudha Rajamani, Alexander Vlassov, Seico Benner, Amy Coombs, Felix Olasagasti, and David Deamer, "Lipid-Assisted Synthesis of RNA-like Polymers From Mononucleotides," *Orig. Life Evol. Biosph. 38* (2008), 57–74.

[51] Giovanna Costanzo, Samanta Pino, Fabiana Ciciriello, and Ernesto Di Mauro, "Generation of Long RNA Chains in Water," *J. Biol. Chem. 284* (2009), 33206–33216.

[52] Quentin Vicens and Thomas R. Cech, "A Natural Ribozyme with 3′,5′ RNA Ligase Activity," *Nature Chemical Biology 5, No. 2* (February 2009), 97–99.

[53] Fraccia et al., "Evidence of Liquid Crystal-Assisted," 66.

[54] Christof B. Mast, Severin Schink, Ulrich Gerland, and Dieter Braun, "Escalation of Polymerization in a Thermal Gradient," *PNAS 110, No. 20* (2013), 8030–8035.

[55] Sheref S. Mansy, Jason P. Schrum, Mathangi Krishnamurthy, Sylvia Tobé, Douglas A. Treco, and Jack W. Szostak, "Template-Directed Synthesis of a Genetic Polymer in a Model Protocell," *Nature 454* (July 2008), 122–126.

[56] Noel A. Clark, Renewal Proposal to the National Science Foundation for Award DMR-1207606, 9 November 2015, D8.

[57] Itay Budin and Jack W. Szostak, "Expanding Roles for Diverse Physical Phenomena During the Origin of Life," *Annu. Rev. Biophys. 39* (2010), 245–263.

[58] Olga Taran, Oliver Thoennessen, Karin Achilles, and Günter von Kiedrowski, "Synthesis of Information-Carrying Polymers of Mixed Sequences From Double Stranded Short Deoxynucleotides," *Journal of Systems Chemistry 1, No.9* (18 August 2010), 1–16.

[59] J.P. Ferris and G. Ertem, "Oligomerization of Ribonucleotides on Montmorillonite: Reaction of the 5′-Phosphorimidazolide of Adenosine," *Science 257, No. 5075* (September 1992), 1387–1389.

[60] J.E. Lydon, "The DNA Double Helix—The Untold Story," *Liquid Crystals Today 12, No. 2* (2003), 1–9.

[61] John Lydon, "Chromonic Liquid Crystalline Phases," *Liquid Crystals 38, Nos. 11–12* (November–December 2011), 1663–1681.

[62] Fraccia et al., "Evidence of Liquid Crystal-Assisted," 58.

[63] Leslie E. Orgel, "The Implausibility of Metabolic Cycles on the Prebiotic Earth," *PLoS Biology 6, No. 1* (January 2008), 0005–0013.

[64] Elizabeth H. Blackburn, Carol W. Greider, and Jack W. Szostak, "Telomeres and Telomerase: The Path From Maize, Tetrahymena, and Yeast to Human Cancer and Aging," *Nature Medicine 12, No. 10* (October 2006), 1133–1138.

Epilogue

The principals in this book have already contributed science of Homeric proportions. What might lie ahead in the work of Nadrian Seeman (Figure E.1) and in the collaboration of Tommaso Bellini and Noel Clark?

As Seeman said in Chapter Nine in speaking of his successful construction of three-dimensional crystals of DNA, that is a milestone and not an end point. He is determined to place periodically arrayed macromolecules as guests in DNA lattices and then use X-ray diffraction to find the detailed structure of the guest molecules.

Figure E.1 Nadrian Seeman. *(Courtesy of Nadrian C. Seeman.)*

Seeman's self-assembly process (and most self-assembly techniques) relies on weak, reversible interactions. If a building block attaches to an incorrect site, it can detach again because of the reversibility of the process. It follows that for such a self-correcting mechanism to work effectively, the inter-unit interactions must be sufficiently weak under assembly conditions. However, stability of the final structures requires the same interactions to be strong. These are two conflicting requirements for the inter-unit interactions. This poses something of a dilemma.

The Seeman and Mao labs are working to develop a strategy for the stabilization of self-assembled DNA crystals by strengthening the inter-unit interactions after crystal assembly.[1] To date, the significance of their success is that a solution environment similar to physiological conditions can now be used for DNA crystals. In such an environment, potential guests behave more as they would in their native conditions. For instance, enzymes will have higher activities, and inorganic nanoparticles will be less likely to experience irreversible aggregation.

Another ambition is to develop *self-replicating* artificial materials. Along with physicists Paul Chaikin, David Pine, and other colleagues, Seeman has made a first development in this direction in the self-replication of seeds made from DNA tile motifs.[2] The DNA tile motifs can recognize and bind complementary tiles in a pre-programmed fashion. After design of the initial seed tiles, Seeman used the seeds to instruct the formation of a first generation of complementary daughter sequences. Then the daughter sequences instructed the formation of granddaughter tiles that are identical to the initial seeds. Since DNA can organize inorganic matter, the technique has the potential for multiplicative copying of nanoscale structures and devices.

Though similar to self-replication of DNA in the cell, this process is distinct because no biological components, particularly enzymes, are used in its execution—even the DNA is synthetic. Chaikin said, "This is the first step in the process of creating artificial self-replicating materials of an arbitrary composition. The next challenge is to create a process in which self-replication occurs not only for a few generations, but long enough to show exponential growth."[3]

Recently, the authors of a review paper on structural DNA nanotechnology took a forward-looking view of the frontier endeavor of self-replicating DNA nanostructures.[4] They described self-replication as an astounding process by which a molecule in a dynamic system makes an identical copy of itself. We might wonder about the word *astounding*. After all, biological cells reproduce by cell division in a suitable environment. In the process, linear DNA autonomously replicates itself by an enzyme-mediated activity and this DNA is transmitted to offspring. However, as the authors of the review observed, "It is a considerable challenge to design and construct autonomous structures that mimic the action of nucleic acid polymerases [such as we've seen as early as Chapter Two] and are capable of replicating entire synthetic DNA systems nonenzymatically."[5] Figure E.2 cartoons the desired procedure using DNA nanostructures that we saw in Chapter Eight.

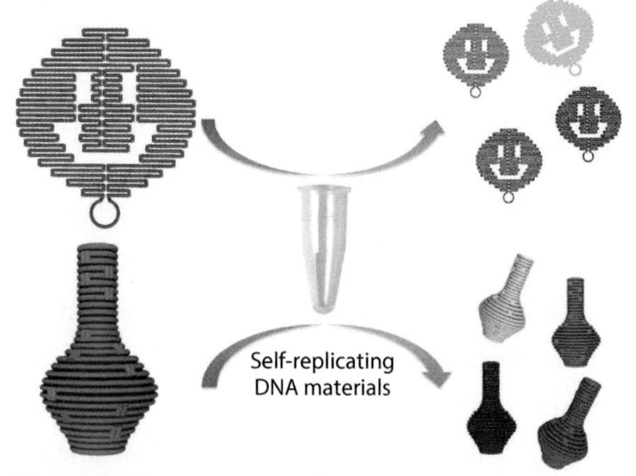

Figure E.2 Autonomous self-replicating DNA materials. *(Reproduced with permission from Fei Zhang, Jeanette Nangreave, Yan Liu, and Hao Yan, "Structural DNA Nanotechnology: State of the Art and Future Perspective," J. Am. Chem. Soc. 136 (2014), 11198–11211; http://pubs.acs.org/doi/pdf/10.1021/ja505101a)*

We turn now to Bellini and Clark (Figure E.3). The demonstration of liquid crystal autocatalysis that we presented in Chapter Thirteen starts with self-complementary 12bp oligomers and generates oligomers that are much longer (up to about 1,000bp) but always complementary. The principal challenge of this line of work will be to diversify the liquid crystal autocatalysis process at both the input and output stages. They must explore how to start it from more primitive and heterogeneous pools of molecules, and must find ways to produce extended chains with varying degrees of mutual- and self-complementarity, some with a capability of self-folding into more complex three-dimensional structures.

Figure E.3 Noel Clark (right) and Tommaso Bellini (left). *(Courtesy of Noel A. Clark.)*

At the end of Chapter Fifteen we saw Bellini and Clark's intuition for the Liquid Crystal World feeding the RNA World. Along these lines, they plan to investigate the potential enzymatic activity of the chains resulting from liquid crystal autocatalysis polymerization. Initially, they will use liquid crystal phases with sequences that, when chemically combined, yield known ribozymes whose action on target substrates can be detected (for example through fluorescent signals). If this is successful, they intend to progress to random RNA ligation. They will study the evolution of sets of oligomers by testing the effect of ligation during repeated temperature and/or concentration cycles on the resulting length and sequence distributions. This will be a search for effects such as the emergence of specific sequence patterns, the dependence on different initial seeds, and the role of emerging secondary structures such as hairpins, relevant in the function of ribozymes. Indeed, they aim to understand and control the liquid crystal–mediated emergence of ribozymes, the RNA enzymes nowadays considered to be the simplest form of life. If successful, this will demonstrate the formation of long chemically active nucleic acid chains from pools of short fragments and monomers. This would arguably become the best available explanation for the appearance of life on Earth.[6]

Bellini and Clark have recently succeeded in forming liquid crystals from many nano-DNA tetramer solutions—duplexes consisting of only four base pairs (Figure E.4).[7] In their initial discovery a decade ago the smallest lengths were six base pairs. They wish to start with single base nucleotides and demonstrate liquid crystal formation under plausible prebiotic conditions. A deeper question concerns the origin, that is, the emergence, of complementarity itself. The RNA World model of self-replicating, information-carrying molecules requires precursor families of highly selected and templated molecules, the origin of which requires self-replication.[8] The mode of information transmittal by the hydrogen bonding of side group sequences of linear polymers has obvious advantages and function in biotic systems and has been shown to enable the templating of companion chains on, for example, mineral surfaces. However, the compelling requirement for hydrogen bonding of side group sequences to appear at all before the RNA World is not clear.

Figure E.4 (a) Cartoons showing duplexing and stacking of GTAC oligomers. The leftmost cartoon shows the 4-mers in solution, the middle shows the hybridization of the 4-mers, and the rightmost cartoon is a more abstract version of the middle one. **(b)** Depolarized transmission light microscopy image of the nanoDNA 4-base oligomer GTAC. *(Courtesy of Noel A. Clark.)*

Thus an important theme in their future work is the emergence of complementarity in the context of hierarchical self-assembly and templating. Specifically, Bellini and Clark are exploring chromonic liquid crystal systems. These are lyotropic systems of polycyclic aromatic hydrocarbons (PAHs). In an aqueous environment such molecules spontaneously stack into columns to reduce contact with water by their aromatic parts. This leads to a hierarchical self-assembly sequence of nematic or columnar phases markedly similar to that of nanoDNA. Exploring chromonic liquid crystal systems (e.g., the dyes sunset yellow and chromolyn) may prove to be a fruitful line of inquiry, especially because PAHs are known components of the pool of organic molecules present on the early Earth.[9]

It has been a privilege and a pleasure to write about the imaginative research of such superlative scientists. Oxford University theoretical physicist Steven H. Simon worked at Bell Laboratories in New Jersey for 11 years and was the director of theoretical physics research there from 2000 to 2009. Simon's last day at Bell Labs was Thursday, December 18, 2008, and he wrote about it in a blog.[10] He said, "Someone once described my job as being like the sport of curling—sweeping the ice so the stone can go in the right direction. . . . What they needed from me most, was just the removal of various impediments. . . ." In a more modest capacity, the chronicler of this book hopes to have performed such a service for the reader.

ENDNOTES

[1] Jiemin Zhao, Arun Richard Chandrasekaran, Qian Li, Xiang Li, Ruojie Sha, Nadrian C. Seeman, and Chengde Mao, "Post-Assembly Stabilization of Rationally Designed DNA Crystals," *Angew. Chem. Int. Ed. 54* (2015), 9936–9939.

[2] Tong Wang, Ruojie Sha, Rémi Dreyfus, Mirjam E. Leunissen, Corinna Maass, David J. Pine, Paul M. Chaikin, and Nadrian C. Seeman, "Self-Replication of Information-Bearing Nanoscale Patterns," *Nature 478* (13 October 2011), 225–229.

[3] New York University Press Release. "NYU Scientists' Creation of Artificial Self-Replication Process Holds Promise for Novel Production of New Materials," (12 October 2011), http://www.nyu.edu/about/news-publications/news/2011/10/12/nyu-scientists-creation-of-artificial-self-replication-process-holds-promise-for-novel-production-of-new-materials-.html

[4] Fei Zhang, Jeanette Nangreave, Yan Liu, and Hao Yan, "Structural DNA Nanotechnology: State of the Art and Future Perspective, *J. Am. Chem. Soc. 136* (2014), 11198–11211.

[5] Ibid., 11206.

[6] Tommaso Bellini, Noel Clark and Hans-Achim Wagenknecht, "How Did Molecules Become Alive?" a proposal to the Volkswagen Foundation (VolkswagenStiftung), 2016.

[7] Tommaso P. Fraccia, Gregory P. Smith, Lucas Bethge, Giuliano Zanchetta, Sven Klussmann, Noel A. Clark, and Tommaso Bellini, "Liquid Crystal Ordering and Isotropic Gelation in Solutions of 4-Base-Long DNA Oligomers," submitted to *ACS NANO.*

[8] Christian de Duve, *Singularities: Landmarks on the Pathways of Life* (New York: Cambridge University Press, 2005), 78–80.

[9] Unpublished: Noel Clark, "Liquid Crystals of Nanonucleic Acids: Hierarchical Self-Assembly as a Route to Prebiotic Selection, Templating, and Autocatalysis," NSF Award Number 1207606, 11.

[10] http://complexmatters.blogspot.com/2008/12/my-last-day-at-bell-labs.html

APPENDIX

Texture of Liquid Crystal Optical Images

Near the end of Chapter Eleven we referred the reader to a concise primer on equilibrium thermodynamics. In discussing the second law of thermodynamics the author writes, "There are various little wrinkles in the foregoing about which we now need to own up."[1] So too must we.

To maintain narrative momentum, we have refrained several times from greater elaboration on the subject of the texture of liquid crystal optical images. Now it's time to do a little explaining. In this Appendix we will do just that—a little explaining. We will try to impart a sense of the connection between the liquid crystal images and the molecular arrangement of the molecules making up those images. There are a large number of distinct liquid crystal phases. We will confine ourselves to a select few that have made their appearance in the previous chapters. We will have something to say about (one subtype of) a smectic phase, a nematic phase, and also a columnar phase.

Smectic Phase Liquid Crystal Texture

It seems reasonable to begin with the very first liquid crystal photograph that caught our eye—the striking image that appears as Figure 1.4 in Chapter One. As it happens, this is probably the most abstruse liquid crystal image in the entire book. Moreover, this is not an image of a DNA liquid crystal. However, it is dazzling and cries out for some commentary on its beautiful colors and shapes.

The caption says that we're seeing liquid crystal domains nucleating from the compound P10PIMB. That's true as far as it goes. P10PIMB is shown in a smectic phase.

We've talked about this liquid crystal phase. For example, in Figure 4.1 of Chapter Four we saw a multi-panel view of the phase transition (of a different chemical compound) from the smectic A phase to the cholesteric phase. (The cholesteric phase is an alternative name for the chiral nematic phase.) In that same chapter we also mentioned that cell membranes are liquid crystals in the smectic phase. One last prefatory remark is that in Chapter Four we touched on the existence of bent-core (also called banana-shaped) molecules and they form smectic liquid crystal phases. Now we're ready for some commentary.

The smectic phases of a liquid crystal are layered structures. Figure A.1 shows the difference between molecular ordering in a smectic phase and in a nematic phase. Lines represent the individual molecules. In the nematic phase the molecules tend to orientationally align on average. In the smectic phase there is, in addition, a degree of translational order because the molecules tend to form layers. The layers should be understood to be coming out of the page and we are looking at the layers from the side. The layer spacing is comparable to the molecular length, i.e., a few nanometers. The molecular layers in a smectic are nearly incompressible. Thus, unlike nematics, there is a strong constraint on the structures the liquid crystal can adopt, viz., structures that have constant layer spacing.[2]

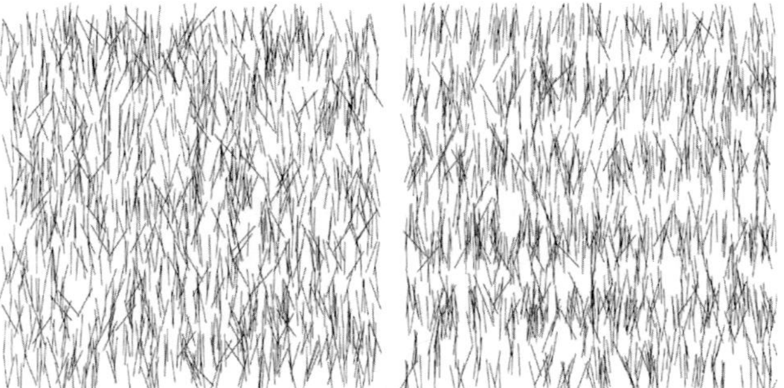

Figure A.1 Molecular alignment in nematic (L) and smectic (R) phases. *(Courtesy of Timothy J. Atherton.)*

In the absence of any constraints the layers would be flat and the molecules tend to lie parallel to one another because their side-to-side attractions are relatively strong. However, in general when a smectic is placed between microscope slides or cooled from the isotropic liquid, it does not assume the simple form (with flat layers), but rather the layers bend in order to conform to the boundary conditions. From a study of the optical properties of the smectic texture, it's known that the surfaces of the layers form a variety of shapes.[3] Figure A.2 shows some of the structures that smectic liquid crystal layers can adopt.[2]

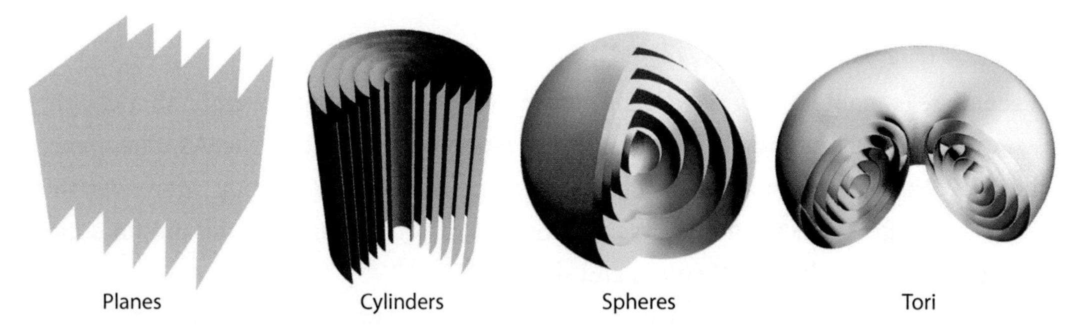

Figure A.2 Examples of smectic layer structures with constant layer spacing. *(Courtesy of Timothy J. Atherton.)*

In Figure 1.4 of Chapter One, we have smectic domains nucleating from the isotropic as the compound is cooled in its temperature-controlled cell while being observed in the depolarizing microscope. In this example, the liquid crystal domains develop as nested cylinders (as in Figure A.2) and we are seeing the nested cylinders from one end. That is, the nested cylinders are at right angles to the page and so we see them as circles. The concentric cylindrical layers are only separated by a few nanometers and so the optical microscope can't resolve all the cylinders. Instead, in general we see a single circle—the outermost of the cylinders as viewed on end.

Bent-Core Molecules

The variety of birefringent colors in Figure 1.4 of Chapter One comes about because the compound is composed of molecules with bent-cores. Bent-core compounds self-organize in a number of unique ways with the result that the tilt angle of the molecule can change with respect to the layer normal. Let's explore this a little. Figure A.3 shows a generic bent-core molecule using chemical symbols (left) and its simplified geometrical representation as a bow and arrow (middle). This analogy proves useful since one defines the molecular director, *n*, in phases of these molecules as being along the bowstring.[4] (More generally, in liquid crystals the molecular director defines the average direction of the long axis of the molecule.) The molecular bow plane is the plane containing the bowstring and the arrow and is shown parallel to the page (in the center of Figure A.3). However, it is useful to discuss the construct when viewed from a point away from the molecule but in the bow plane. Thus, if the bow and arrow are rotated 90° such that the arrow is pointing directly away from the viewer, a cross inside a circle is used to represent the tail of the arrow. In this view the projection of the bow is simply a straight line. A 90° rotation in the opposite direction has the arrow pointing directly at the viewer, as indicated by a circle with a bullet inside.

The variation in the colors of the circles (that is, varying degrees of birefringence) in Figure 1.4 of Chapter One comes about because

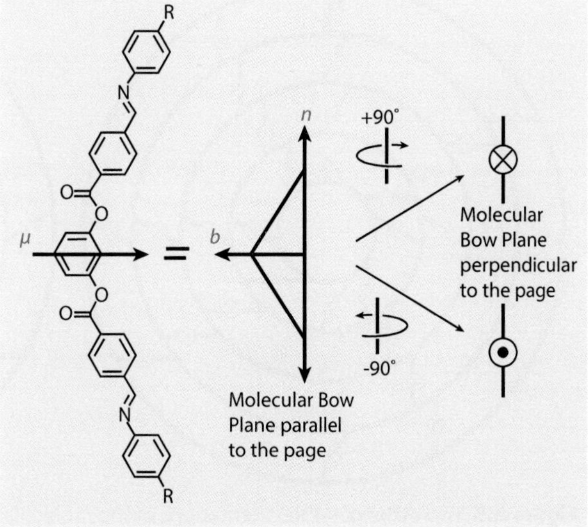

Figure A.3 Example of a bent-core molecule. The molecule can be thought of as a bow (as in "bow and arrow"). The middle drawing shows the molecular bow plane parallel to the page. The rightmost drawing shows the two possible orientations of the bow plane when it is perpendicular to the page. *(Reproduced with permission from David M. Walba, "Ferroelectric Liquid Crystal Conglomerates" in Materials-Chirality: Volume 24 of Topics in Stereochemistry, ed. Mark M. Green, R.J.M. Nolte, and E.W. Meijer (Hoboken, New Jersey: John Wiley & Sons, 2003), 457–518.)*

of different smectic states that are present in the same sample. These arise from different molecular orientation within the layers. As we noted earlier, the bent-core molecules have a tilt relative to the layer normal; they don't simply point radially outward. Moreover, in some places they have the same tilt and in other places they have alternating tilts. Figure A.4 gives an example of the case of alternating tilts. In other cases, the tilt of the molecules with respect to the layer normal (the direction that is perpendicular to the circles we see as we look at the nested cylinders end on) does *not* alternate. Two examples of this are shown in Figure A.5.

Back in Figure 1.4 of Chapter One we also find cases in which the color changes abruptly as you proceed from the center of the circle to its perimeter. For example, in the lower left quadrant of the photograph we have a domain where the color is largely green and then changes to a band of red all around the perimeter. This is a result of the smectic state spontaneously changing from one tilt arrangement to another and thus changing the birefringence.

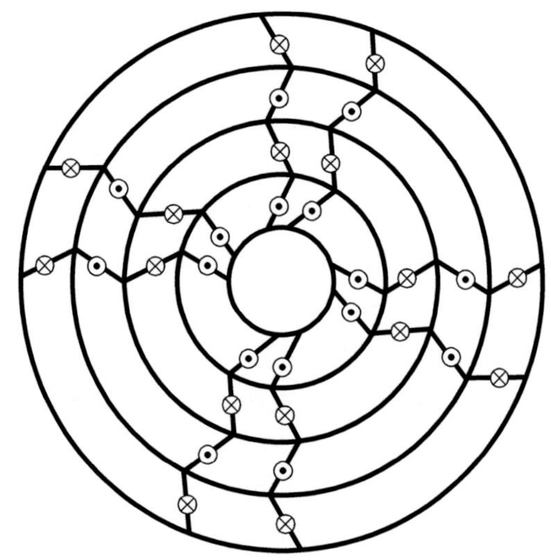

Figure A.4 An illustration of how the bent-core molecules can change their tilt angle (with respect to the layer normal—that is, a line that would be radially outward from the center and normal to the layers).

Extinction Brushes

And what about the black crosses that appear within the circles, dividing them up into four quadrants? These are called extinction brushes. Let's consider how they arise. Remember that the photograph

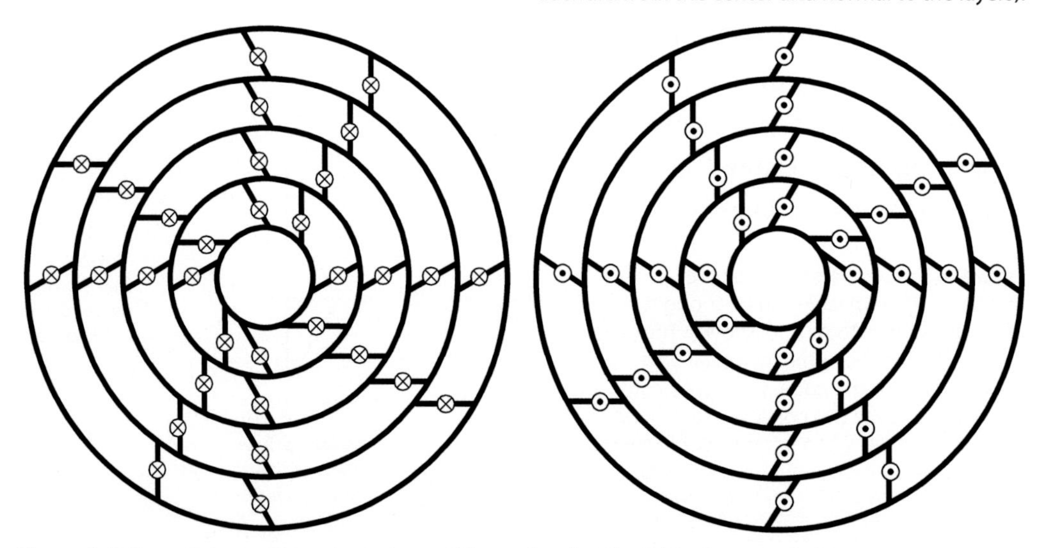

Figure A.5 Two other possible orientations of the molecular tilt within the smectic layers.

we're studying was taken with the sample placed between a polarizer and an analyzer—the crossed polarizers that we've discussed in several chapters. And in Chapter Four in the section on liquid crystal displays, we addressed birefringence. We noted that if the incident light is at an angle of 0° or 90° with respect to the molecular direction of a material between crossed polarizers, then the birefringence of the material (the existence of two different indices of refraction) does not cause any rotation of the electric field of the light. That means that where there are molecules that are aligned with the direction of either the polarizer or the analyzer no light exits the liquid crystal cell—the light is extinguished. In these regions the light passing through the sample only experiences one refractive index and so behaves as if it were passing through an isotropic liquid. This is what produces the extinction brushes.

To concretize this, consider Figure A.6. We've taken one of the liquid crystal domains that appears in Figure 1.4 of Chapter One and shown how those molecules in the smectic layers that happen to be aligned with either the polarizer or the analyzer determine the angular orientation of the extinction brushes. By the way, the fact that the brushes become wider the further we go out from the center is because the radius is getting larger while the angular tilt of the molecules with respect to the layer normal remains constant. This broadening of the width as they progress radially outward, gives the extinction brushes the nickname Maltese Crosses.

One final remark: in Figure A.4 we showed the case of alternating tilts of the molecules in consecutive smectic layers. In this case, the average long axis of the molecules *is* a radial line (that is, it is normal to the layer surfaces). So the extinction brushes they generate would be vertical and horizontal assuming those were the directions of the polarizer and the analyzer. In Figure 1.4 of Chapter One this is very nearly the case for the red circle at the far right of the photograph.

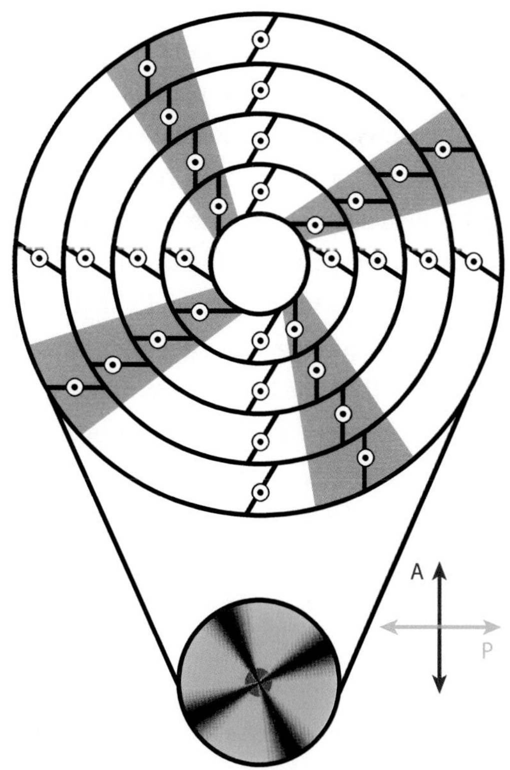

Figure A.6 An example of the origin of extinction brushes.

Chiral Nematic Texture of NanoDNA Liquid Crystals

We now move on to the texture of the nematic phase of DNA. As we've learned, in the case of DNA this is a chiral nematic (although not all nematic liquid crystal phases of molecules are chiral). The chirality turns out to have a big influence on the structure, producing an intrinsic twisted (i.e., helical) director field. The texture of such a helical director depends on the orientation of its axis, *c*, relative to the viewing direction and the magnitude of its pitch, *p*, relative to the wavelength of the illuminating light.

There are many different chiral nematic textures of DNA. We will confine ourselves to just one of these. It's called the fingerprint texture and there are several examples of it in this book. For instance, the bottom right panel of Figure 12.3 in Chapter Twelve shows the fingerprint texture in nanoRNA liquid crystals made up of oligomers that are 22 base pairs in length. Also in Chapter Twelve, panel (c) of Figure 12.6 shows nanoDNA 14-mers (14 base-pair strands) that form the nematic fingerprint texture. Both images are small and the fingerprint pattern may be hard to see. So we'll use a portion of a figure that appears in the Bellini and Clark paper on chirality that we dipped into in Brief Interlude II.[5]

We'll illustrate the means by which the fingerprint texture comes about in Figure A.7 using two papers that dealt with conventional, long DNA rather than nanoDNA, although the essential ideas are the same for nanoDNA.[6,7] At the top of the figure, panels (a) and (b), we indicate the observation plane—i.e., the viewer is looking down at the helix, and we see a schematic representation of the helix from that perspective. The axis of the helix, denoted by C, is lying parallel to the planes of the coverslip, c, and the glass slide, s (to simplify, we will refer to this as the cell plane or the viewing plane), that bound the sample of DNA. Let's first comment on the way the helix is drawn. When the helix axis, C, is lying in the preparation plane, the DNA chains present different orientations relative to the viewing plane, alternatively parallel, oblique, and perpendicular. In the helix we represent these orientations by lines (molecules parallel to the viewing plane), nail-shaped marks (molecules oblique to the viewing plane), and points (molecules perpendicular to the viewing plane). The tip of each nail indicates the extremity of the molecule that points toward the observer. The length of each nail corresponds to the projection of a unit segment onto the viewing plane. Note that this is a left-handed helix. In Brief Interlude II on the handedness of life we discussed the propagation of chirality in supramolecular assemblies. In conventional, long DNA, even though the duplex double helix is right-handed, the supramolecular helical structure formed by the liquid crystal phases of DNA is left-handed.

In panel (c), we show the continuous precession of the DNA chain orientation in a three-dimensional perspective along the helix, representing the DNA chains as if they were on a series of imaginary planes. The defining length scale of this chiral nematic phase is the pitch of the helix, *p*, the distance for which the director rotates through a complete 360° turn. As shown in the inset to panel (c), each of the lines represents the DNA duplex chains.

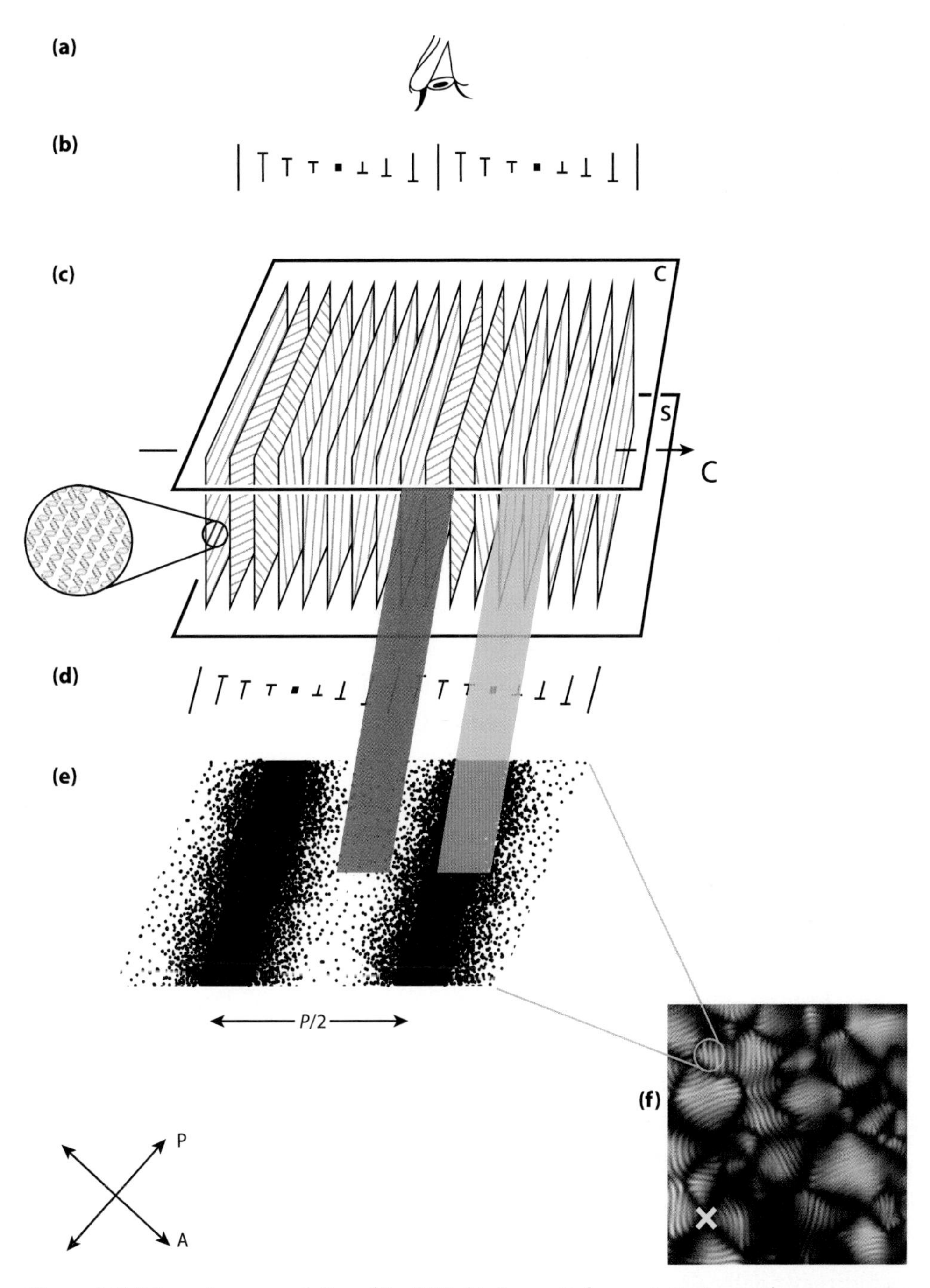

Figure A.7 Schematic representation of the DNA chiral nematic fingerprint texture with accompanying photograph. Looking down at the cell containing nanoDNA, the observer views a helix. The helix is continuous and the colored planes shown in panel (c) are imaginary (a visual aid). The pitch, p, of the helix is the distance for which the director rotates 360°. For pitch >1 μm, panel (e) shows lines of birefringence (bright) and extinction (dark) with a periodicity *p*/2. This fingerprint texture is apparent in the photograph (gray circle), and so too are extinction brushes (yellow cross). *Note*: photograph is about 70 μm on a side. *(Panels **(b)–(e)** adapted with permission from Françoise Livolant and Amélie LeForestier, "Condensed Phases of DNA: Structures and Phase Transitions," Prog. Polym. Sci. 21 (1996), 1115–1164; Panel **(f)** adapted with permission from Giuliano Zanchetta, Fabio Giavazzi, Michi Nakata, Marco Buscaglia, Roberto Cerbino, Noel A. Clark, and Tommaso Bellini, "Right-Handed Double-Helix Ultrashort DNA Yields Chiral Nematic Phases With Both Right- and Left-Handed Director Twist," PNAS 107, No. 41 (October 12, 2010), 17497–17502.)*

Between crossed polarizers (P and A), the transmitted light is a maximum for molecules lying in the preparation plane and a minimum for molecules normal to this plane. If the pitch is less than 1 μm (= 1000 nm), one cannot microscopically resolve the resulting optical modulation.[8] However, in chiral nematics with $p > 1$ μm, one observes a regular alternation of bright lines where n is parallel to the viewing plane. The birefringence produces transmission for the orientations of P and A indicated, and dark extinction lines where the director, n, is normal to the viewing plane and there is no effect of birefringence. The periodicity of these equidistant lines corresponds to a rotation of 180° of the molecular orientations; that is, $p/2$ where p is the helix pitch. Overlaying bars of purple and green emphasize those portions of the helix that result in birefringence and those that result in extinction, as seen in panel (e).

In panel (f), these alternating, equidistant stripes are spotlighted in an image of chiral nematic liquid crystals of 12bp nanoDNA from the Bellini and Clark paper on chirality that we mentioned. Note that the orientation of the helix axis can change in the viewing plane because of the presence of defects. Thus, we see a multiplicity of domains in the photograph where the stripes (which are perpendicular to the duplex chain direction axis) abruptly change their direction. Note too the extinction brushes in the photo (as shown by overlaying yellow lines). These are the locations at which the light is polarized along the helix axis, i.e., in directions parallel to the polarizer and analyzer. Here the polarization/analyzer is along an optic axis and does not couple to the birefringence.

Columnar Texture of NanoDNA Liquid Crystals

So far we have given some physical insight into how a (non-nanoDNA) smectic texture comes about and also how a chiral nematic nanoDNA textures arises. To conclude this brief look at the molecular origins of the appearance of liquid crystal films in depolarized light microscopy, we now consider a texture of the columnar phase of nanoDNA. We saw our first example of this in Figure 1.5 of Chapter One, and subsequently in Figures 5.2 and 5.9 of Chapter Five and elsewhere in the book. In the case of the columnar phase as opposed to the (chiral) nematic phase, the solution is more concentrated. Molecules are unidirectionally aligned with a lateral hexagonal order, as shown in Figure A.8.[9] (Note that unlike the chiral nematic phase, twist is prevented—there is no long helix present. There can be local twist owing to screw dislocation defect lines, but we will not discuss this.) However, this is a liquid crystal and not a true crystal: molecules present a degree of disorder around their position in the hexagonal array and the columns of molecules generally show a parallel bend and columns are able to slide with respect to each other. The various numbers and letters in Figure A.8 pertain to symmetries of the columnar hexagonal phase that we will not discuss.

Textures of the columnar phase often have many defects. A frequent type of defect is called a disclination line. These disclination lines are perpendicular to the viewing plane (i.e., coming straight out of the page). We won't venture into the subject of defect structures except as they are apropos of our examination of textures. Figure A.9 shows a photograph using depolarized light microscopy. It shows a columnar hexagonal liquid crystal phase of long DNA.[10] Although the birefringence has a marked blue color, the appearance is similar to that of the columnar hexagonal phase of nanoDNA that we saw, for example, in Figure 1.5 of Chapter One.

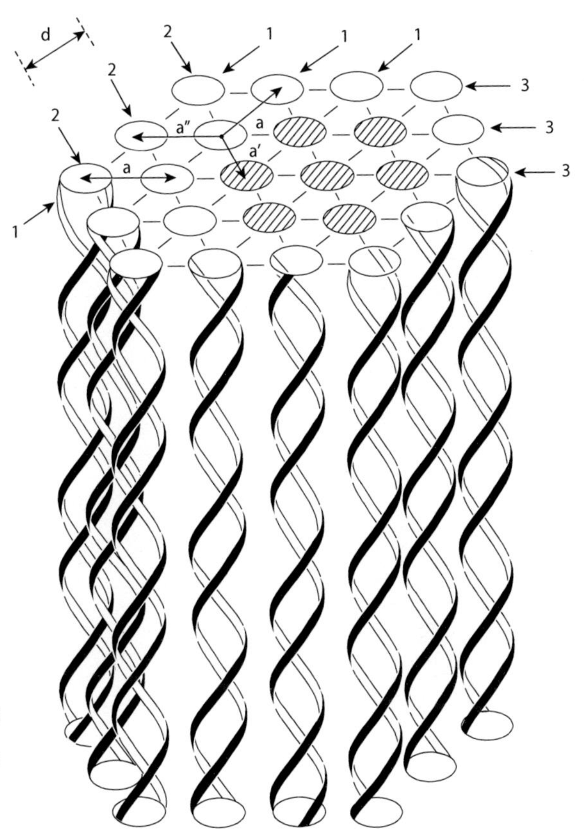

Figure A.8 Molecular arrangement of DNA helices in the columnar hexagonal phase. *(Reproduced with permission from Françoise Livolant and Amélie LeForestier, "Condensed Phases of DNA: Structures and Phase Transitions," Prog. Polym. Sci. 21 (1996), 1115–1164.)*

In Figure A.9 the author has marked the photo to indicate "+ π" disclination lines. (These are locations in which the molecular direction, n, rotates by about 180°; recall that π = 180°.) The rotation often occurs with the formation of a so-called wall, marked by "w" in the photograph. Other walls frequently join to form what are known as "- π" disclination lines. Figure A.10 is a schematic representation of the molecular organization of the director complete with defect lines and walls. Notice how the director, n, showing the local orientation of the columns of DNA duplexes, winds around defect lines (shown as black vertical lines). This is what we first saw in the case of nanoDNA in Figure 5.2 of Chapter Five on one of those occasions in which we referred the ambitious reader to the Appendix.

As was true in the nematic phase, nanoDNA in the columnar hexagonal phase *is* chiral. However, the columnar phase appears when the DNA concentration (whether long DNA or nanoDNA) is high. This leads to a competition between different molecular interactions that are involved in the packing of helical polymers. In this case there is an antagonism between chirality and hexagonal packing.[11] As the interhelical distance between

Figure A.9 Long DNA in the columnar hexagonal phase. The scale bar is 10 μm (= 10,000 nm). *(Reproduced with permission from Françoise Livolant and Amélie LeForestier, "Condensed Phases of DNA: Structures and Phase Transitions," Prog. Polym. Sci. 21 (1996), 1115–1164.)*

DNA duplex chains progressively decreases owing to the higher concentrations, the hexagonal packing prevents the formation of chiral structures such as the helix we saw in the nematic phase. As Bellini and Clark nicely phrased it in the context of nanoDNA, "fluctuations of local columnar order tend to expel the twist, and parallel packing prevails over chiral torque."[12]

Finally, we've thrown out the term *defect* a bunch of times and the reader might be a little bothered. Now we hope to provide a little balm. Defects come in different flavors; there are point defects, line defects, screw dislocation defects, and so on. We will stick to line defects. It happens that there are lines along which the molecular director is forced to change quickly and eventually is

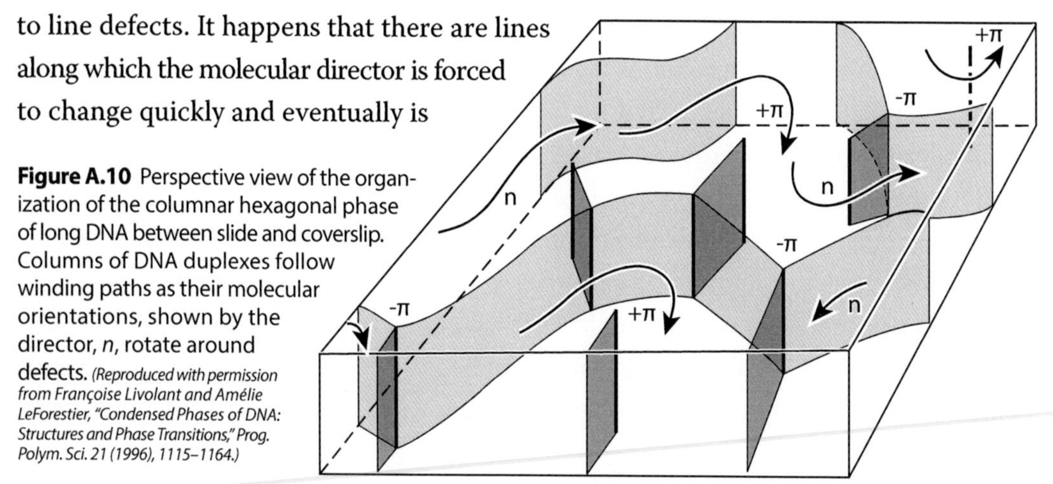

Figure A.10 Perspective view of the organization of the columnar hexagonal phase of long DNA between slide and coverslip. Columns of DNA duplexes follow winding paths as their molecular orientations, shown by the director, *n*, rotate around defects. *(Reproduced with permission from Françoise Livolant and Amélie LeForestier, "Condensed Phases of DNA: Structures and Phase Transitions," Prog. Polym. Sci. 21 (1996), 1115–1164.)*

simply not defined. An analogy might be water running down a plug, where at the center of the vortex the flow of water is turbulent.[13] Line defects were given a special name early in the history of liquid crystals research: disclinations. The name comes about from the fact that the line represents a *dis*continuity in the in*clination* of the director.[14] That is about as far as we will go on the subject.

We became involved in topological issues in Nadrian Seeman's work. Here we only mention that there is a robust topological literature of defects in liquid crystals. However, that subject would lead us too far afield—even in an appendix endeavoring to provide a wee bit extra.

ENDNOTES

[1] Peter Atkins, *The Laws of Thermodynamics: A Very Short Introduction* (New York: Oxford University Press, 2010), 55.

[2] Timothy Atherton, "Pattern Formation in Liquid Crystals," *Tufts Soft Matter Theory Group* (15 November 2011), http://sites.tufts.edu/softmattertheory/2011/11/15/research-highlight -pattern-formation-in-liquid-crystals/

[3] Michael J. Stephen and Joseph P. Straley, "Physics of Liquid Crystals," *Reviews of Modern Physics 46, No. 4* (October 1974), 617–704.

[4] David M. Walba, "Ferroelectric Liquid Crystal Conglomerates" in *Materials-Chirality: Volume 24 of Topics in Stereochemistry,* ed. Mark M. Green, R.J.M. Nolte, and E.W. Meijer (Hoboken, NJ: John Wiley & Sons, 2003), 457–518.

[5] Giuliano Zanchetta, Fabio Giavazzi, Michi Nakata, Marco Buscaglia, Roberto Cerbino, Noel A. Clark, and Tommaso Bellini, "Right-Handed Double-Helix Ultrashort DNA Yields Chiral Nematic Phases With Both Right- and Left-Handed Director Twist," *PNAS 107, No. 41* (12 October 2010), 17497–17502.

[6] Amélie Leforestier and Françoise Livolant, "Supramolecular Ordering of DNA in the Cholesteric Liquid Crystal Phase: An Ultrastructural Study," *Biophysical Journal 65* (July 1993), 56–72.

[7] Françoise Livolant and Amélie LeForestier, "Condensed Phases of DNA: Structures and Phase Transitions," *Prog. Polym. Sci. 21* (1996), 1115–1164.

[8] Giuliano Zanchetta, "Liquid Crystalline Phases in Oligonucleotide Solutions," Ph.D. dissertation, Università Degli Studi Di Milano (2007), 70.

[9] Livolant and LeForestier, "Condensed Phases of DNA," 1115–1164.

[10] Ibid., 1126.

[11] Ibid., 1161.

[12] Giuliano Zanchetta, Fabio Giavazzi, Michi Nakata, Marco Buscaglia, Roberto Cerbino, Noel A. Clark, and Tommaso Bellini, "Right-Handed Double-Helix Ultrashort DNA Yields Chiral Nematic Phases With Both Right- and Left-Handed Director Twist," *PNAS 107, No. 41* (12 October 2010), Supporting Information 1.10, 3 (linked to the online version of the paper as a PDF download).

[13] David Dunmur and Tim Sluckin, *Soap, Science, and Flat-Screen TVs: A History of Liquid Crystals* (New York: Oxford University Press, 2011), 82.

[14] Peter J. Collings, *Liquid Crystals: Nature's Delicate Phase of Matter* (Princeton, New Jersey: Princeton University Press, 2002), 178.

Glossary

Abiotic Describes a process that occurs without the participation of biomolecules, for example, abiotic ligation.

Adatom An atom adsorbed on a surface.

Addressable DNA structures DNA nanostructures that can be labeled. Such labeling provides uniquely identifiable positions so that, for example, the initial structure can act as a substrate and additional patterns can be formed on top of it.

Amphiphile A molecule with a hydrophilic head and a hydrophobic tail. Thus, it's a molecule with one end that attracts water and one end that repels water.

Anaglyph A type of stereoscopic illustration—i.e., an illustration that appears to have three-dimensional depth and solidity. This is achieved with a pair of images of the same subject that are superimposed. The images are printed in different colors (usually red and green) and viewed with appropriate filters over each eye.

AND gate Thinking in terms of locks and keys, both aptamer locks must be opened simultaneously to activate the nanorobot described in Chapter Ten. The robot remains inactive when only one of the two aptamer locks is opened. The keys to these locks are the cell surface antigens. When the locks open in response to the keys, the nanorobot can deliver its antibody (or gold nanoparticle) cargo.

Anisotropic An adjective describing properties that vary depending on the direction of measurement. In liquid crystals, this is due to the alignment and the shape of the molecules. Anisotropic is the opposite of isotropic.

Annealing of DNA Using temperature changes to cause complementary sequences of single-stranded DNA to pair by the formation of hydrogen bonds.

Antibody Proteins produced by the body's immune system in response to the presence of a foreign substance called an antigen.

Antigen Any substance that causes the body's immune system to produce antibodies against it. Antigens may have entered the body from the outside or may have been generated within the body.

Apoptosis A genetically determined process of cell self-destruction (sometimes called programmed cell death). It is a process that eliminates damaged or unneeded cells. If apoptosis is halted, e.g., by genetic mutation, the result may be uncontrolled cell growth and tumor formation.

Aptamers Short single-stranded nucleic acids (or amino acid polymers) that recognize and bind to molecular targets with high affinity and selectivity.

Aspect ratio The ratio of two dimensions of an object. For example, it is common to refer to the ratio of the length, L, to the diameter, D, of a rigid rod (L/D) as its aspect ratio. The aspect ratio is a measure of the object's shape anisotropy.

Atomic force microscope (AFM) An instrument for imaging surfaces in which a fine tip attached to a cantilever moves back and forth over the surface of the sample. The AFM measures the forces between the atoms at the end of the tip and the atoms on the surface of the sample. The forces influence the up-and-down motion of the cantilever and this information is converted to an image of the surface topography. In contrast to the scanning tunneling microscope (STM), the sample need not be an electrical conductor; it can be an insulator or a semiconductor.

Autocatalysis A process in which the reaction product is itself the catalyst for that reaction. For example, in liquid crystal autocatalysis, the liquid crystal phase is able to catalyze its own formation.

Autotroph An organism that is itself capable of making energy-containing molecules from inorganic (non-carbon-containing) raw materials by using basic energy sources, e.g., sunlight.

Base complementarity A (adenine) bonds to T (thymine) and G (guanine) bonds to C (cytosine) in a pair of strands of DNA that form a double helix, also called a duplex. The base pairs A-T and C-G form hydrogen bonds and hold the two strands of the duplex together.

Base-pairing forces Attractive forces within a DNA duplex that exist between the base pairs that make up the rungs of the spiral double-helical ladder. These forces are in the plane of the bases (and can be thought of as oriented horizontally in a vertical DNA ladder). They are mediated by hydrogen bonds and contribute to the stability of the double helix.

Base ratios Erwin Chargaff found that in DNA the amount of adenine (A) is equal to the amount of thymine (T) and the amount of guanine (G) is equal to the amount of cytosine (C). James Watson and Francis Crick later discovered that this indicated the specificity of base bonding in DNA. This became known as base complementarity: A bonds to T and G bonds to C in a pair of strands of DNA that form a double helix (a duplex).

Base-stacking forces Attractive forces within a DNA duplex that exist between vertically adjacent bases, that is, forces perpendicular to the base planes. These forces are mediated by both hydrophobic and electrostatic interactions and contribute to the stability of the double helix.

Bend persistence length A quantity that characterizes the bending rigidity, or alternatively, the flexibility, of DNA. Choose a segment of the DNA. If the segment is shorter than the bending persistence length, then along that segment the DNA can be thought of as a rigid rod with ~ a single direction. If the chosen segment of DNA is longer than the bend persistence length, then that segment can be thought of as a flexible rod with a direction that varies randomly from one location (within the chosen segment) to the next.

B-form DNA also called B-DNA This is the most common form of DNA. It occurs as a right-handed double helix (the double helix winds to the right—clockwise—when viewed end on).

Birefringent A material that has two indices of refraction is called birefringent. Liquid crystals are birefringent because of their anisotropic nature.

Blunt-end DNA A fragment of duplex DNA in which there are no unpaired bases (overhangs) at either terminus.

Bragg's law The relation between the spacing of atomic planes in crystals and the angles of incidence at which these planes produce the most intense reflections (i.e., constructive interference) of electromagnetic radiation, specifically, X-rays. The atomic planes of the crystal act on the X-rays in the same way that a ruled grating acts on a beam of light to produce optical diffraction.

Branch point migration See DNA branch point migration.

B-Z transition DNA of an appropriate nucleotide sequence is capable of forming both the right-handed double-helical form (B-form) and the left-handed double-helical form (Z-form). The two forms can be converted from one to the other by several means including changing the solution conditions, specifically, by the addition or removal of the salt known as hexaamminecobalt(III) chloride.

Cassette See DNA cassette

Catenane A group of organic compounds in which two or more ring structures are interlocked with each other (like a chain).

Cell cycle The growth, replication, and division of cells consists of a series of processes collectively known as the cell cycle.

Cell membrane A semi-permeable structure that surrounds the interior of a cell.

Cell membrane channels Pores composed of proteins that form passageways to enable ions to pass through lipid bilayer membranes.

Chain terminator A modified nucleotide that terminates DNA strand replication by DNA polymerase.

Chirality A three-dimensional form has chirality if it is not superposable on its mirror image. For example, your hands have chirality. They are spoken of as being chiral. More generally, objects can be chiral or achiral. If it is achiral, the object is superposable on its mirror image.

Chiroselective A chiroselective process is one that enables the proliferation of molecules with a specific handedness.

Clay surfaces Clay surfaces catalyze the synthesis of oligonucleotides.

Columnar phase A liquid crystal phase consisting of molecules arranged in parallel columns. The columns are free to bend and to shift with respect to one another along the columnar direction.

Complementary See Base complementarity

Complex systems Systems that display novel collective behaviors that arise from the interactions of many simple components, e.g., molecules formed out of atoms. *Complex emergent systems* arise when energy flows through a collection of many interacting particles.

Constructive and destructive interference of light For simplicity, consider a pair of light waves from the same source and with the same wavelength and the same amplitude. If the electric field vibrations associated with each wave occur in the same plane, then the light waves may interfere. Constructive interference occurs when the crests of the electric field vibrations of the two waves coincide. This interference doubles the amplitude of the electric field of the resultant wave. Destructive interference occurs when the crests of one wave coincide with the troughs of the other. In this case the two waves will eliminate each other; the electric field vibration will have an amplitude of zero.

Covalent bond A strong chemical bond in which one or more pairs of electrons are shared by two atoms.

Crossed polarizers also referred to as Crossed polarizer and analyzer In depolarized light microscopy, one polarizer is placed before the sample in the optical path and one after the sample. These two polarizers are typically oriented at 90° with respect to one another. These are referred to as crossed polarizers. Often, the second polarizer is referred to as the analyzer.

Cumulative weight fraction distribution, *C*(*n*) An S-shaped distribution curve that shows the sum of the weight fractions of all polymeric constituents (polymers consisting of all possible numbers, n, of monomers where $n = 1, 2, 3 \ldots$). C(n) includes monomers ($n = 1$) that have not been polymerized.

Current Blockade A transient reduction of the ionic current through a nanopore that is produced when large molecules such as DNA impede the flow of small ions.

de Broglie wavelength French physicist Louis de Broglie proposed that matter displays wavelike behavior. The wavelength of a particle is derived from an equation postulated by de Broglie.

Denaturing gel electrophoresis Electrophoretic gels that are run under conditions that denature the sample (that is, disrupt its molecular conformation).

Depletion interaction In a solution that contains solutes of two distinct sizes—large and small—the small particles will exert pressure on the large particles from all sides. But when two large particles come close together, the small ones are excluded from the volume between the two. This produces a force on opposite sides of the two large particles that can be thought of as an effective attraction (the depletion interaction) between the two.

Depolarized light microscopy also called Polarized light microscopy A method of illumination used to create contrast in birefringent materials such as liquid crystals. Samples are placed in a temperature-controlled stage between two polarizers that are typically oriented at 90° with respect to one another (crossed polarizers). Polarized light interacts strongly with the birefringent sample and enhances visualization.

Dickerson dodecamer A 12-base oligonucleotide with the sequence CGCGAATTCGCG. This dodecamer was used to build B-DNA duplexes that provided the first detailed structural description of a right-handed DNA double helix.

Dielectric property of a liquid crystal The response of liquid crystal molecules to an applied electric field.

DNA A long, unbranched polymer composed of four types of subunits called nucleotides each consisting of the sugar deoxyribose, a phosphate group, and one of the four chemical bases adenine (A), cytosine (C), guanine (G), or thymine (T). The DNA polymer forms two chains (the sugar-phosphate backbones) that twist into a double helix. The base parts of the four nucleotides connect the chains and match up in pairs, called complementary base sequences. Each base pair is held together by hydrogen bonds, that is, shared hydrogen atoms. The base shapes and chemical makeup are such that A only fits with T and G only fits with C.

DNA branch point migration As an example, consider four oligonucleotide strands with complementary base sequences that can form a branched DNA structure. The place where the four strands converge is known as a branch point or a junction. Branch point migration occurs if this junction can move (by the stepwise breakage and reformation of hydrogen bonds) and hence the DNA structure can change its shape.

DNA bricks A generalization of the concept of DNA single-stranded tiles (SSTs) to DNA single-stranded bricks that takes the modular-assembly method into three dimensions. One can use DNA bricks to construct complex three-dimensional nanostructures by using only short, synthetic DNA strands as modular components. Fabrication with DNA bricks is mediated solely by local binding interactions (sticky ends).

DNA cassette A structure built from duplex DNA that enables the insertion of a DNA nanomechanical device at a specific site in a two-dimensional crystalline DNA array.

DNA hairpin It's not unusual for a single strand of DNA to curl back on itself to become self-complementary. This results in a partial double helix with a bend in it; several bases remain unpaired in the bend before the strand loops back on itself. In the classic case, the structure resembles a real hairpin and hence the name. These structures occur naturally and can also be synthesized to perform various functions in DNA nanotechnology. A dumbbell hairpin is a variation of the simpler structure.

DNA ladder A set of known DNA fragments with different sizes in base pairs. For example, in the case of Bellini and Clark, synthesized sequences of multiple repetitions of the Dickerson dodecamer. In gel electrophoresis, ladders serve as standards to enable rapid approximation of the oligomer length of the experimental samples that are found as bands in the various lanes.

DNA ligase An enzyme that catalyzes the formation of a covalent bond between two strands of DNA by the formation of a phosphodiester bond.

DNA machine A machine is spoken of as a particular type of device in which component parts change their relative positions as a result of some external stimulus. By using strand replacement to convert from one form to another, the paranemic motif (see paranemic molecule) provides a device that constitutes a DNA machine. Such DNA machines facilitate the donation of cargo to DNA walkers in a DNA assembly line.

DNA nanomechanical device A nucleic acid-based nanoconstruction that is capable of controlled mechanical movement.

DNA origami See Scaffolded DNA origami.

DNA origami with complex curvature Several methods can exploit DNA base pairing to create complex, closed shapes. The method of Yan et al. first decomposes the shape into circular contour lines. Then DNA double helices are designed that can bend and follow these contours while crossovers join DNA along adjacent contours. The method succeeds by a delicate balance of slightly overtwisted and slightly undertwisted helices. This produces different bending angles resulting in a combination of structural flexibility and stability.

DNA polymerase An enzyme that uses a DNA template to replicate that molecule by assembling its complementary nucleotides.

DNA sequence symmetry minimization This technique for the design of synthetic nucleic acid aims to minimize sequence similarities between segments of DNA strands. The goal is to decrease the chances of forming undesired structures instead of the desired target nanostructure.

DNA sequencing with nanopores A developing technique using naturally occurring protein nanopores to read DNA base sequences. This approach may offer advantages —in speed, cost, and length of strands read—compared to present methods. Synthetic nanopores constructed by DNA nanotechnology might possibly be used for this purpose.

DNA strand polarity The nucleotide subunits of a DNA strand are lined up in the same direction giving the strand a chemical polarity. Moreover, the two ends of a strand are chemically distinguishable. The end having a phosphate group at its terminus is called the 5′ end and the end having a hydroxyl group at its terminus is called the 3′ end. One speaks of the direction of the chain as running from the 5′end to the 3′ end, or simply 5′→ 3′.

DNA tensegrity triangle Three self-assembling double helices of DNA that point in three independent directions so as to define a structure having threefold rotational symmetry. Each of the ends of the helices has two unpaired bases (sticky ends) that bind to complementary pairs of bases on other triangles. This enables the construction of macroscopic three-dimensional crystals. These crystal lattices contain cavities that might be used to host biomolecules in a three-dimensional periodic arrangement. In that event, X-ray diffraction could be used to determine the structure of the guest molecules.

DNA tiles also called Single-Stranded Tiles (SSTs) A tile in the present context is a single-stranded DNA molecule with four different sticky-ended binding domains that specify which four other tiles can bind to it as neighbors. SSTs are used as a modular style of nanostructure design in two dimensions.

DNA walkers Molecules composed of DNA that can walk along tracks also composed of DNA. In the work presented in this book we only consider non-autonomous walkers. Single-stranded DNA feet move along a DNA track in response to the sequential addition to the aqueous solution of single-stranded DNA called fuel strands. Energy for motion is derived from DNA hybridization (the binding together of strands having complementary base sequences).

Dormancy During the cell cycle, a cell may enter a state of dormancy (called the G_0 phase). When a cell is dormant it is in a non-dividing state. In cancer cells, dormancy is a state in which the cells may survive but growth is arrested.

Double-crossover (DX) molecule The DX molecule consists of two double helices aligned side by side with strands crossing between the helices yoking them together. There are several types of double crossover molecules differentiated by the relative orientation of their helix axes, parallel or antiparallel, and by the number of double helical half-turns (even or odd) between the two crossovers. While the parallel versions are usually not well behaved, the antiparallel versions are used extensively in structural DNA nanotechnology. They provide excellent rigidity for fabrication of extended nanostructures and nano-mechanical devices.

Duplex A double-stranded DNA molecule, i.e., double-helical DNA.

Effective diameter A parameter that characterizes the excluded volume of DNA by accounting for the conformational properties of actual electrically charged DNA, rather than idealized hard rods. The effective diameter is a measure of the electrostatic repulsion between segments of DNA molecules

Effective length, L_{eff} A length introduced to make the phase behavior of nanoDNA obey the Onsager criterion. The simplest way to do this is to assume that the nanoDNA duplexes stack into units of total length L_{eff}.

Eigenvalues and eigenfunctions For many situations in quantum mechanics the system of interest has a fixed, unchanging total energy. In these cases the general Schrödinger equation reduces to a simpler equation called the time-independent Schrödinger equation. Standing waves are the solutions (the eigenfunctions) of the time-independent Schrödinger equation and the corresponding allowed energies are the energy eigenvalues.

Electron density map A three-dimensional description of the electron density in a crystal structure determined from X-ray diffraction.

Emergence Processes by which more complex systems arise from simpler systems often spontaneously and in an unpredictable fashion. Emergent phenomena exhibit self-organization. Life is sometimes cited as the quintessential emergent phenomenon.

Enantiomer An enantiomer is one of a pair of molecules that are non-superposable mirror images of each other. Thus, enantiomers are chiral molecules. Enantiomers are a subset of stereoisomers—i.e., all enantiomers are stereoisomers but not all stereoisomers are enantiomers.

Enantiopure A sample in which all molecules have the same chiral sense. Enantiopure is not entirely synonymous with homochiral because it applies to an experimental attribute, not to a concept.

Endonuclease An enzyme that cuts the sugar-phosphate backbone of DNA within the length of the chain (rather than at the end of the chain).

Entropy A quantitative way to measure the dispersal of energy in a process.

Enzymatic ligation of nucleic acids The covalent bonding of two DNA fragments (or two RNA fragments) catalyzed by the use of enzymes.

Excluded volume The volume that is impenetrable to other molecules in the solution because of the presence of the first molecule.

Exonuclease An enzyme that cuts nucleotides from the end of a DNA chain.

Ferguson plot A mathematical technique applicable to gel electrophoresis that was devised by veterinary scientist Kenneth Ferguson. This graphical analysis measures the electrophoretic mobility (essentially, the rate of migration) of a molecule at several different gel concentrations. From this one can learn about the molecule's size, its net charge density, and its shape.

Flow cytometry Flow cytometry is a technology used to measure and analyze multiple physical characteristics of cells as they flow in a fluid stream through a laser beam. The light that is scattered in several directions is captured and analyzed by software to uncover cellular properties such as size and internal complexity.

Fluid mosaic model Cell membranes are viewed as two-dimensional fluids in which proteins are inserted into lipid bilayers.

Forward- versus side-scatter dot plot In a scatter dot plot, each cell is represented by a dot that is positioned on the horizontal and vertical axes according to the intensities of scattered light detected for that cell. Light scattering information is collected in the forward direction (light reflected by cells at angles less than 90° from the direction of the incident light) and in the side direction (light reflected at 90°). The forward-versus side-scatter dot plot provides information about both the size of the cells and the granularity, that is, the complexity of the cells. Scatter dot plots reveal quantitative percentages of cells with various properties and show discrete subpopulations of cells with different intensities.

Fourier Transform Any waveform can be decomposed into a set of pure sine waves of shorter and shorter wavelength and thus of higher and higher frequency that sum to the original waveform. This mathematical tool is known as the Fourier Transform.

Free energy Energy available to do work rather than just be dispersed as heat.

FRET (Fluorescence resonance energy transfer) A distance-dependent transfer of energy from a donor molecule to an acceptor molecule. In our context, FRET is a technique that is used to measure the distance between fluorescent dye molecules (applicable only when they are separated by 10 nm or less).

Fuel strand A single strand of DNA that acts as a chemical fuel for a DNA device, e.g., a DNA walker. The fuel strand hybridizes with another DNA single strand thus lowering the free energy of the system. This provides energy to do work, e.g., powering structural change or motion.

Functionalize (verb) The addition of functional groups, that is, an atom (or groups of atoms) that has similar chemical properties whenever it occurs in different compounds.

Genome The total genetic information of an organism as encoded in DNA, or, for many viruses, encoded in RNA.

Hairpin See DNA hairpin.

Helical pitch The pitch of a helix is is the distance measured along the helix axis to complete one full helix turn.

Heterotroph An organism that cannot make its own energy-containing molecules and must make use of energy that originates in other organisms.

Hierarchical self-assembly Stages of self-assembled, supramolecular organization in which one level of self-assembly guides the next.

Homochiral Molecules that occur in only one enantiomeric form (i.e., molecules of the same chirality). The biomolecules of life—DNA, RNA, and proteins—possess only a single handedness because they are made of building blocks that have only one handedness. For example, the right-handedness of the classic B-form DNA is determined by the right-handed configuration of its deoxyribose sugars.

Hybridization The process of joining two complementary strands of nucleic acids, e.g., two single strands of DNA, to form a double-stranded molecule (a duplex). In DNA, the process utilizes the base complementarity of A with T and G with C (while in RNA, duplexes are formed by the complementarity of A with U and G with C).

Hydrogen bonds Hydrogen atoms mediate an attractive force between two bases in a DNA duplex. The base pair G-C is stabilized by three hydrogen bonds while the base pair A-T is stabilized by two hydrogen bonds. Hydrogen bonds form links between an atom like oxygen or nitrogen present in one base to an oxygen or nitrogen present in another base.

Hydrophilic, Hydrophobic "Water loving"—describes a molecule that is attracted to water; "Water fearing"—describes a molecule that is repelled by water.

Immobile DNA branched junction A convergence of strands of synthetic DNA at a junction such that the intersection point does not change. In principle, such immobile junctions can be combined using sticky-ended cohesion into rigid lattices.

Ion Channel Current Gating Gating describes conformational changes (i.e., changes in shape or structure) in an ion channel. For example, a voltage can induce these changes. Gating currents result from the movement of small electrically charged molecules in the ion channel. The current will change as the channel conformation changes.

Iron-Sulfur World Minerals composed of iron, nickel, and sulfur catalyze metabolic reactions.

Isomers Two molecules are called isomers of one another if they have the same number of the same kinds of atoms, i.e., the same chemical formula, but differ in their chemical and physical properties.

Isotropic An adjective describing properties that do *not* vary depending on the direction of measurement.

Isotropic-Nematic phase transition (I-N) A liquid crystal transition from a phase in which the properties of the molecules do not change with direction (isotropic phase) to a phase in which there is a preferred direction of orientation—a degree of orientational order—of the molecules (nematic phase).

Kinetic arrest The sudden slowing down of dynamic systems in which long-lived structures (having a lifetime comparable to or larger than the time of the experiment) form far from equilibrium and thus prevent a system of molecules from reaching its equilibrium state. Kinetic arrest is also referred to by such phrases as dynamical arrest or ergodic-nonergodic transition. In the case of 20-mer ranDNA, kinetic arrest produces duplex pairs that are able to self-assemble into liquid crystal domains.

Langmuir trough An apparatus used to compress a monolayer of molecules on the surface of a subphase (e.g., water). The apparatus can measure surface phenomena such as surface tension—i.e., the attractive force between molecules at the water surface that resists extension of the surface.

Ligation In the context of nucleic acids, ligation is the joining together of DNA fragments by strong covalent (electron-sharing) bonds.

Ligation-closure experiments DNA strands are ligated and their ligation products are assayed (typically via denaturing gel electrophoresis) to determine if cyclization has resulted in a unique outcome or in many different products. The presence of many cyclic products suggests that the angles at the junction points of the tested molecules are not rigid. A three-dimensional molecule made from such ligated strands is said to be floppy.

Lipid bilayer A two-layered structure composed of amphiphilic molecules in which the hydrophobic portions of each layer point toward each other.

Lipid World Organized lipid structures have catalytic properties and contain information in their composition.

Liquid crystals (LCs) A state of matter whose order is intermediate between that of a liquid and that of a crystalline solid. LCs straddle the fence between fluidity and rigidity. The molecules are often rod-shaped and how they arrange themselves—their ordering—is a function of temperature. These different types of ordering are called phases and LCs will abruptly change phase (they will go through a phase transition) at certain temperatures.

Liquid crystal phase separation also referred to as Liquid crystal phase condensation Upon cooling solutions of a mixture of complementary and non-complementary nanoDNA, the nanoDNA capable of forming duplexes separates into liquid crystal droplets, leaving the unpaired single strands in isotropic solution.

Long DNA (lDNA) This denotes conventional long DNA, as opposed to nanoDNA.

LUCA (Last Universal Common Ancestor) The most recent organism from which all contemporary organisms are believed to have descended.

Lyotropic liquid crystals Materials in which liquid crystalline phases appear because of the presence of a solvent. The concentration of the solvent as well as the temperature will determine the type of phase (e.g., nematic).

Macrocyclic products A macrocycle is a cyclic (ring-like) macromolecule. Macrocyclic products refer to a multiplicity of different size molecules having cyclic/ring structures.

Messenger RNA (mRNA) Information from DNA is transcribed (copied) by RNA polymerase onto a molecule of RNA. The molecule that is produced is called messenger RNA.

Metabolism-first theories of life's origin Such scenarios suggest that preceding a large, self-replicating molecule, small molecules formed an evolving network of chemical reactions driven by an energy source.

Micelle A spherical formation caused by an amphiphilic substance in a solution. The hydrophilic end of the molecule tends to orient itself toward the outside of the sphere while the hydrophobic end tends to orient itself toward the inside of the sphere.

Microbeam X-ray diffraction Synchrotron X-ray diffraction using X-ray beams focused down to beam diameters smaller than one micron (1 micron = 1000 nm).

Micron A micron is a metric unit of length measurement that is more formally called a micrometer, that is, one-millionth of a meter. One micron equals 1000 nanometers. (One inch equals 25,400 microns.)

Microreactors Small vessels for carrying out chemical reactions. In the context of the work of Bellini and Clark, self-assembled liquid crystal domains serve as naturally occurring microreactors.

Molar ratio, *R* A ratio between the amounts, in moles, of any two compounds involved in a chemical reaction.

Mole A scientific unit for measuring large quantities of very small entities such as atoms or molecules.

Molecular canvas (in two and three dimensions) In two dimensions, hundreds of different SSTs act as a master library of tiles. When mixed together they self-assemble into a rectangular molecular canvas. Each SST serves as a molecular pixel. One designs a shape by selecting only its constituent pixels on the canvas and annealing them to form the target shape. Similarly, starting with a thousand-brick block, this master brick collection defines a cuboid molecular canvas in three dimensions. (A cuboid is a box-like structure composed of six rectangular faces. A cube is a special case in which all six faces are squares.) One selects subsets of bricks and then anneals them to form the target.

Molecular dynamics simulation A computational technique that allows one to predict the time evolution of a system of interacting particles (atoms, molecules, etc.). The method can also be used to explore macroscopic properties of the system by averaging over all the atoms or molecules in the system.

Motif In structural DNA nanotechnology, a motif is a distinctive design that specifies how DNA strands will link together. This pattern serves as the elemental building block for desired structures or devices.

NanoDNA Short duplexes of B-form DNA of lengths less than 20 base pairs (bp).

Nanometer A nanometer is a metric unit of length measurement that equals one-billionth of a meter. (One inch equals 25.4 million nanometers.)

Nanotechnology A broad term encompassing the creation, exploration, and manipulation of materials measured in nanometers (billionths of a meter).

Nematic-Columnar (N-C_u) phase transition A liquid crystal phase transition from a phase in which the molecules exhibit a degree of orientational order (nematic phase) to a phase in which the molecules exhibit a degree of translational order as well as a degree of orientational order (columnar phase).

Nematic phase A liquid crystal phase in which the molecules show a degree of orientational order—the molecules tend to align themselves with respect to one another. This phase does not have any positional order; the molecules are not constrained to occupy only specific positions.

Nonergodic process In our context, the kinetic behavior of a system of ranDNA that is unable to explore in time all of its possible states. Thus, its equilibrium distribution is never reached.

Nucleic acid A biopolymer, that is, a long biological molecule composed of monomers (single chemical units). The monomers are called nucleotides. The major types of nucleic acids are DNA and RNA.

Nucleoside A nucleoside is a structural subunit of a nucleic acid. It consists of a molecule of sugar linked to a nitrogen-containing ring compound called a nitrogenous base, also called a nucleobase or just a base. In the case of DNA, the sugar is deoxyribose and the base is adenine, guanine, cytosine, or thymine. In the case of RNA, the sugar is ribose and the base is adenine, thymine, cytosine, or uracil.

Nucleotide A nucleotide consists of one or more phosphate groups bonded to a nucleoside. When only one phosphate group is bonded to the nucleoside, the nucleotide thus formed is often referred to as a nucleoside monophosphate. An example is adenosine monophosphate, written as AMP.

Oligomers A molecule that consists of a small number of monomers.

Oligonucleotides Short, single-stranded DNA or RNA molecules, i.e., short nucleic acid polymers.

Onsager's Criterion For molecules considered as hard rods, the shape anisotropy, L/D, must be such that $L/D > \sim 5$ for the molecules to exhibit liquid crystal phases.

Optical activity A medium shows optical activity if it causes the plane of polarization of a plane-polarized light wave to rotate as it travels through the medium.

Osmotic pressure Suppose a membrane separates two fluid volumes: one volume is a solution (i.e., a solute dissolved in a solvent) and the other is the pure solvent. Osmotic pressure is the pressure on the solution side of the membrane that is necessary to prevent the flow of pure solvent through the membrane and into the solution.

PAGE (Polyacrylamide gel electrophoresis) A method to separate and/or identify biological macromolecules, typically proteins or nucleic acids. Molecules are separated on the basis of their electrophoretic mobility, that is, their rate of migration through a porous material under the influence of an applied electric field.

Paranemic molecule A type of DNA crossover molecule that occurs in two different forms, PX and JX_2. Conversion from one form to the other takes place in a two-step process mediated by short DNA fuel strands of two classes, set and unset. Such conversion produces measurable work.

Patterson function Arthur L. Patterson derived the first successful solution to the phase problem (in 1934) and colleagues dubbed it the Patterson function.

(The) Phase problem This refers to an issue in X-ray diffraction. The diffraction pattern only shows the intensity of the diffracted X-rays. To work back to the molecular structure of the molecules that make up the crystal, it's also necessary to have the phase data, that is, the position of the wave crests and troughs of the diffracted X-rays relative to one another.

Phase transition A phase transition is a change from one state of matter to another without a change in chemical composition. Familiar examples include the melting of a solid to a liquid or the vaporization of a liquid to a gas. In liquid crystals, an example of a phase transitions is the change from a state in which the properties of the liquid crystal molecules do not vary with direction (isotropic phase) to a state in which there is a preferred direction of orientation—a degree of orientational order—of the molecules (nematic phase).

Phosphate group A phosphorous atom bonded to four oxygen atoms at least one of which is bonded to another atom.

Phosphodiester bonds The bonds that connect adjacent sugars in nucleic acids.

Pixel A small rectangular dot on a display screen. The word is derived from picture element. Pixels arranged in a grid formation create a digital representation of an image.

Plausible prebiotic conditions Chemical or environmental precursors to the origin of life are spoken of as prebiotic conditions. In the laboratory, scientists try to simulate what they believe to be such conditions and to determine which are the most conducive to the self-assembly processes that led up to the origin of life.

Polarity Refers to a separation of electrical charge and can be used to describe an entire molecule (a polar molecule). This also applies to solvents. For example, a polar solvent like water is composed of molecules whose electric charges are unequally distributed, leaving one end of each molecule more positive than the other. If such charge separation is absent the molecule or solvent is said to be non-polar.

Polarization of light Describes the direction of the electric field vibrations in a light wave.

Polarizer (linear) A device that transmits electromagnetic radiation (e.g., light) and confines the vibration of the electric field to one plane.

Polymer A molecule consisting of many small units known as monomers.

Polymerase See DNA polymerase.

Polymerization of nucleic acids The covalent bonding of nucleic acid monomers to form a polymer. Noel Clark and Tommaso Bellini contend that DNA and RNA became long, chain-like molecules because ultrashort lengths of DNA and RNA were able to self-assemble into liquid crystals.

Polymerization yield The fraction of polymerized molecules out of the total number of molecules available for polymerization (after a specified time).

Primer A segment of single-stranded DNA consisting of a small number of nucleotides. A primer is designed so that its base sequence is complementary to the first portion of DNA that one wishes to sequence.

Programmable nucleic acid strands By building nucleic acid strand sequences with appropriate combinations of complementary bases, the strands will self-assemble into a predesigned architecture.

"*q*" A quantity inversely related to the helical pitch.

Quantum Corrals An STM can be used to slide atoms across a surface and position them to form ring-like structures—quantum corrals—that enable the wave-like properties of electrons to be seen.

Radius of gyration A way to measure the dimension of a flexible polymer chain in which its effective size dynamically changes owing to its flexibility. The radius of gyration gives a better sense of the size of a flexible polymer than does the end-to-end distance.

Random-sequence nanoDNA (ranDNA) DNA oligomers in which base sequences are chosen at random or with varying degrees of randomness. Solutions of oligomers with random sequences are more likely to reflect prebiotic conditions than specifically designed sequences. Complementarity still emerges in these solutions and, for a narrow range of oligomer lengths, produces a subtle hierarchical succession of structured self-assembly and organization into liquid crystal phases.

Raster fill pattern A type of digital image that uses small rectangular pixels (a word derived from picture elements) arranged in a grid formation to represent an image. The raster, or scanning pattern, typically adds the pixels in a series of straight parallel lines progressing from left to right and then line-by-line from top to bottom.

Refraction The bending of light when it passes from one medium (e.g., air) to a second medium (e.g., water) in which the velocity of the light is different.

Refractive index For a given medium, the refractive index is the ratio of the speed of light in a vacuum divided by the speed of light in the medium.

Replicator-first theories of life's origin Such scenarios suggest that a large molecule (e.g., RNA) capable of self-replication formed first. Then metabolism developed as smaller molecules used energy to perform useful processes.

Resolution See Spatial resolution

Restriction enzymes Enzymes that recognize and bind to specific DNA sequences. Once they bind to their recognition site/sequence, restriction enzymes cut the sugar-phosphate backbones of the DNA strands.

Restriction site A sequence of DNA that is recognized by a restriction enzyme.

Ribozyme An RNA molecule that can act as an enzyme.

RNA A long polymer composed of four types of subunits called nucleotides each consisting of the sugar ribose, a phosphate group, and one of the four chemical bases adenine (A), cytosine (C), guanine (G), or uracil (U). Unlike DNA, RNA is often found single stranded (but it can be double stranded).

RNA polymerase A molecule that copies a DNA sequence into an RNA sequence.

RNA World RNA (in the form of ribozymes) serves as both catalyst and gene.

Robust DNA nanomechanical device A nanomechanical device that behaves like a macroscopic device in that it has well-defined endpoints and does not undergo component-changing reactions, e.g., dissociation.

Scaffolded DNA origami A method of directed molecular self-assembly to build custom-shaped nanoscale objects composed of DNA. Naturally occurring DNA (e.g., from a harmless virus) is typically used for the single-stranded scaffold DNA molecule. The scaffold can be many thousands of bases in length. Hundreds of short staple strands of DNA (tens of bases in length) are designed so they can connect (via hybridization) distant locations on the scaffold strand to fold it into the desired shape. Staples are synthesized and mixed in solution with the scaffold. The strands collectively self-assemble when annealed —i.e., the temperature is lowered to allow hydrogen bonds to form between complementary base sequences. (*Note*: crossover molecules find extensive use in DNA origami.)

Scanning Tunneling Microscope (STM) An instrument for imaging surfaces at the atomic level based on the concept of quantum tunneling. A fine probe scans across the surface and the variation in the electron tunneling current between the sample and probe produces three-dimensional images of atomic topography and structure.

Scatter dot plot See Forward- versus side-scatter dot plot.

Schrödinger equation Based on the research of French physicist Louis de Broglie, Austrian physicist Erwin Schrödinger formulated an equation that describes the motion of particles in terms of wavelike properties. (In its full generality it is properly called the time-dependent Schrödinger equation.) This is the quantum mechanical analog of Newton's equation of motion in classical mechanics (Newton's Second Law) that describes the behavior of a particle in terms of its motion as a function of time.

Self-assembly of two-dimensional crystals Double-crossover (DX) molecules can be designed to self-assemble into two-dimensional crystals. Each corner of each DX unit has a single-stranded sticky end with a unique sequence. Choosing sticky ends according to the familiar base complementarity rules controls the bonding of DX units.

Shape anisotropy also called Aspect ratio The ratio of a molecule's length to its diameter, L/D.

Solute, Solvent A solute is the component in a solution that is present in the lesser amount while a solvent is the component in a solution that is present in the largest amount.

Spatial resolution The ability of a detection system (e.g., a microscope) to record details of the objects under study. It is often defined in terms of how close two features can be within an image and still be recorded as distinct.

Spherical Nucleic Acids (SNAs) Three-dimensional conjugates (i.e., coupled objects) consisting of densely functionalized and highly oriented nucleic acids covalently attached to the surfaces of nanoparticles.

Spontaneously broken symmetry A loss of symmetry that occurs without external intervention.

Standing waves Standing wave patterns are the result of the repeated interference of two waves of the same frequency that are moving in opposite directions. In a quantum mechanical context, standing wave patterns can result from the interference between incident and reflected surface-state electrons hitting atoms adsorbed on the surface.

Stereoisomers Two molecules are called stereoisomers of one another if they have the same chemical formula *and* the atoms of the molecules have the same connections to one another *but* the atoms differ in their position in three-dimensional space.

Sticky end DNA A fragment of duplex DNA in which a terminal portion of one strand has a stretch of unpaired bases—that is, an overhang—that extend beyond the end of the other strand.

Strand invasion also referred to as Strand displacement In DNA nanotechnology, strand displacement refers to the hybridization of two partially or fully complementary DNA strands by means of the displacement of one or more pre-hybridized strands. Strand displacement can be initiated at toeholds (complementary single-stranded domains) and then progress through a branch migration process.

Supramolecular A complex of molecules held together by noncovalent bonds.

Surface tension The attractive force between molecules at the surface of a fluid that resists extension of the surface.

Symmetry breaking Symmetry means uniformity or invariance, the existence of different viewpoints from which a system appears the same. Symmetry breaking is the process by which such uniformity is broken.

Synchrotron X-ray diffraction Particle accelerators called synchrotrons provide tunable and high-intensity X-rays—features that are lacking in standard laboratory X-ray apparatus.

Template (verb) To act as a template.

Tensegrity triangle See DNA tensegrity triangle.

Texture A term used to describe the principal optical properties of a liquid crystal phase when viewed through crossed polarizers in a microscope. Texture is a qualitative term.

Thermodynamic equilibrium A system is in a state of equilibrium if the free energy of that state is smaller than that of any other state of the system (at the same pressure and temperature).

Thermotropic liquid crystals Liquid crystal molecules that exhibit temperature-dependent liquid crystalline behavior.

Thioester world Sulfur bonds contain chemical energy to drive metabolic reactions.

Topological equivalence When two objects can be continuously deformed from one to the other, they are said to be topologically equivalent. The classic illustration of two such objects is a donut and a coffee cup.

Transcription The process in which the information (the base sequence) stored in a molecule of DNA is copied (transcribed) into a new molecule called messenger RNA (mRNA). This transcription is done by a molecule called RNA polymerase.

Transmission electron microscope (TEM) A TEM operates on the same basic principles as a light microscope but uses an electron beam instead of light. The much lower wavelength of electrons compared to the wavelength of light enables a much higher spatial resolution. Objects less than a nanometer can be visualized in a TEM. Electromagnetic lenses (an example of electron optics) are used instead of glass lenses.

Twisted nematic effect The controllable, reversible alignment of liquid crystal molecules from one ordered molecular configuration to another in response to an applied electric field.

Two-dimensional nanoparticle arrays Metallic and semiconductor nanoparticles have optical and electronic properties that might be used in future nanoelectronic devices. Double-crossover molecules have been used to precisely organize nanoparticles into designed structural arrangements in two dimensions.

Voxel A volumetric pixel element.

Wave-particle duality A principle in quantum mechanics which maintains that matter and light exhibit both wave-like and particle-like behavior.

Weight fraction distribution, *W*(*n*) A probability distribution curve that shows the ratio of the total weight of polymers that exist as n-mers (polymers composed of a particular number, n, of monomers) divided by the total weight of all the molecules.

Work The energy transfer that occurs when an object is moved over a distance by a force at least part of which is applied in the direction of the displacement.

X-ray diffraction A technique in which a beam of X-rays interacts with the atoms of a crystal. Recall the phenomenon of wave-particle duality. The uniform spacing of the atoms produces an interference pattern owing to the wave-like nature of the X-rays. Incident X-rays exit the crystal as a complex group of beams. The intensities, angles of exit, and phases of these beams can be used to determine the detailed structure of the repeating unit of the crystal.

Yield The ratio of the number of desired target structures produced to the total number of structures produced.

Zero-length ligation agent A chemical compound that mediates the binding of two molecules by forming a bond that contains no additional atoms, i.e., there is direct linkage of the two molecules.

Z-form DNA also called Z-DNA A naturally occurring, left-handed, double-helical form of DNA in which the double helix winds to the left (counterclockwise when viewed end on).

Index

E

M

N